Practical UML Statecharts
in C/C++

Practical UML Statecharts in C/C++

Event-Driven Programming for Embedded Systems

2nd Edition

Miro Samek

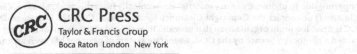

CRC Press
Taylor & Francis Group
Boca Raton London New York

CRC Press is an imprint of the
Taylor & Francis Group, an **informa** business

CRC Press
Taylor & Francis Group
6000 Broken Sound Parkway NW, Suite 300
Boca Raton, FL 33487-2742

© 2009 by Taylor & Francis Group, LLC
CRC Press is an imprint of Taylor & Francis Group, an Informa business

No claim to original U.S. Government works

ISBN-13: 978-0-7506-8706-5 (pbk)

**Visit the Taylor & Francis Web site at
http://www.taylorandfrancis.com**

**and the CRC Press Web site at
http://www.crcpress.com**

Table of Contents

Part I: Uml State Machines.. *1*

**Chapter 1: Getting Started with UML State Machines and
Event-Driven Programming**... *3*
1.1 Installing the Accompanying Code 4
1.2 Let's Play...5
    1.2.1 Running the DOS Version 7
    1.2.2 Running the Stellaris Version 8
1.3 The main() Function...11
1.4 The Design of the "Fly 'n' Shoot" Game..16
1.5 Active Objects in the "Fly 'n' Shoot" Game..20
    1.5.1 The Missile Active Object 21
    1.5.2 The Ship Active Object 24
    1.5.3 The Tunnel Active Object..27
    1.5.4 The Mine Components...29
1.6 Events in the "Fly 'n' Shoot" Game 32
    1.6.1 Generating, Posting, and Publishing Events.........................36
1.7 Coding Hierarchical State Machines 39
    1.7.1 Step 1: Defining the Ship Structure......................................39
    1.7.2 Step 2: Initializing the State Machine 42
    1.7.3 Step 3: Defining State-Handler Functions.............................43
1.8 The Execution Model 48
    1.8.1 Simple Nonpreemptive "Vanilla" Scheduler 48
    1.8.2 The QK Preemptive Kernel..49
    1.8.3 Traditional OS/RTOS 50
1.9 Comparison to the Traditional Approach..50
1.10 Summary..52

Chapter 2: A Crash Course in UML State Machines *55*
2.1 The Oversimplification of the Event-Action Paradigm 56
2.2 Basic State Machine Concepts...59
    2.2.1 States 60
    2.2.2 State Diagrams 61

2.2.3   State Diagrams versus Flowcharts 61
2.2.4   Extended State Machines 63
2.2.5   Guard Conditions... 64
2.2.6   Events 66
2.2.7   Actions and Transitions 67
2.2.8   Run-to-Completion Execution Model 67
2.3   UML Extensions to the Traditional FSM Formalism 68
2.3.1   Reuse of Behavior in Reactive Systems 69
2.3.2   Hierarchically Nested States 69
2.3.3   Behavioral Inheritance 71
2.3.4   Liskov Substitution Principle for States....................... 73
2.3.5   Orthogonal Regions 74
2.3.6   Entry and Exit Actions 75
2.3.7   Internal Transitions 77
2.3.8   Transition Execution Sequence 78
2.3.9   Local versus External Transitions 81
2.3.10 Event Types in the UML... 82
2.3.11 Event Deferral 83
2.3.12 Pseudostates 83
2.3.13 UML Statecharts and Automatic Code Synthesis................... 85
2.3.14 The Limitations of the UML State Diagrams 86
2.3.15 UML State Machine Semantics: An Exhaustive Example 87
2.4   Designing A UML State Machine....................................91
2.4.1   Problem Specification 91
2.4.2   High-Level Design 92
2.4.3   Scavenging for Reuse 93
2.4.4   Elaborating Composite States 94
2.4.5   Refining the Behavior 95
2.4.6   Final Touches 96
2.5   Summary 96

Chapter 3: Standard State Machine Implementations..................101
3.1   The Time-Bomb Example 102
3.1.1   Executing the Example Code....................................104
3.2   A Generic State Machine Interface 105
3.2.1   Representing Events 106
3.3   Nested Switch Statement 108
3.3.1   Example Implementation 108
3.3.2   Consequences..112
3.3.3   Variations of the Technique...................................113
3.4   State Table..113
3.4.1   Generic State-Table Event Processor..........................114
3.4.2   Application-Specific Code 118

3.4.3   Consequences...122
3.4.4   Variations of the Technique.......................................123
3.5   Object-Oriented State Design Pattern..................................124
3.5.1   Example Implementation 126
3.5.2   Consequences...130
3.5.3   Variations of the Technique.......................................131
3.6   QEP FSM Implementation..132
3.6.1   Generic QEP Event Processor.....................................133
3.6.2   Application-Specific Code 137
3.6.3   Consequences...142
3.6.4   Variations of the Technique.......................................143
3.7   General Discussion of State Machine Implementations 144
3.7.1   Role of Pointers to Functions.....................................144
3.7.2   State Machines and C++ Exception Handling 145
3.7.3   Implementing Guards and Choice Pseudostates...........145
3.7.4   Implementing Entry and Exit Actions 146
3.8   Summary 146

Chapter 4: Hierarchical Event Processor Implementation..........149
4.1   Key Features of the QEP Event Processor 150
4.2   QEP Structure..152
4.2.1   QEP Source Code Organization 153
4.3   Events...154
4.3.1   Event Signal (QSignal)...154
4.3.2   QEvent Structure in C 155
4.3.3   QEvent Structure in C++ 157
4.4   Hierarchical State-Handler Functions....................................158
4.4.1   Designating the Superstate (Q_SUPER() Macro).........158
4.4.2   Hierarchical State-Handler Function Example in C ...158
4.4.3   Hierarchical State-Handler Function Example in C++ ...160
4.5   Hierarchical State Machine Class 161
4.5.1   Hierarchical State Machine in C (Structure QHsm)........162
4.5.2   Hierarchical State Machine in C++ (Class QHsm) 163
4.5.3   The Top State and the Initial Pseudostate 164
4.5.4   Entry/Exit Actions and Nested Initial Transitions.........166
4.5.5   Reserved Events and Helper Macros in QEP............168
4.5.6   Topmost Initial Transition (QHsm_init())....................170
4.5.7   Dispatching Events (QHsm_dispatch(), General Structure)..........174
4.5.8   Executing a Transition in the State Machine
          (QHsm_dispatch(), Transition).......................................177
4.6   Summary of Steps for Implementing HSMs with QEP.............183
4.6.1   Step 1: Enumerating Signals.......................................185
4.6.2   Step 2: Defining Events...185

        4.6.3   Step 3: Deriving the Specific State Machine 186
        4.6.4   Step 4: Defining the Initial Pseudostate............................188
        4.6.5   Step 5: Defining the State-Handler Functions....................188
        4.6.6   Coding Entry and Exit Actions 189
        4.6.7   Coding Initial Transitions 189
        4.6.8   Coding Internal Transitions...190
        4.6.9   Coding Regular Transitions..190
        4.6.10 Coding Guard Conditions...190
    4.7   Pitfalls to Avoid While Coding State Machines with QEP 191
        4.7.1   Incomplete State Handlers 192
        4.7.2   Ill-Formed State Handlers..193
        4.7.3   State Transition Inside Entry or Exit Action....................193
        4.7.4   Incorrect Casting of Event Pointers................................194
        4.7.5   Accessing Event Parameters in Entry/Exit Actions or Initial
                Transitions...194
        4.7.6   Targeting a Nonsubstate in the Initial Transition 195
        4.7.7   Code Outside the `switch` Statement 196
        4.7.8   Suboptimal Signal Granularity 197
        4.7.9   Violating the Run-to-Completion Semantics....................198
        4.7.10 Inadvertent Corruption of the Current Event 198
    4.8   Porting and Configuring QEP..199
    4.9   Summary 201

Chapter 5: State Patterns 203
    5.1   Ultimate Hook 205
        5.1.1   Intent 205
        5.1.2   Problem..205
        5.1.3   Solution...206
        5.1.4   Sample Code 207
        5.1.5   Consequences 211
    5.2   Reminder 211
        5.2.1   Intent 211
        5.2.2   Problem..212
        5.2.3   Solution...212
        5.2.4   Sample Code 213
        5.2.5   Consequences 218
    5.3   Deferred Event..219
        5.3.1   Intent 219
        5.3.2   Problem..219
        5.3.3   Solution...220
        5.3.4   Sample Code 222
        5.3.5   Consequences 229
        5.3.6   Known Uses...230

5.4   Orthogonal Component..230
    5.4.1   Intent 230
    5.4.2   Problem..230
    5.4.3   Solution..231
    5.4.4   Sample Code..234
    5.4.5   Consequences..243
    5.4.6   Known Uses...244
5.5   Transition to History 245
    5.5.1   Intent 245
    5.5.2   Problem..245
    5.5.3   Solution..245
    5.5.4   Sample Code..246
    5.5.5   Consequences..250
    5.5.6   Known Uses...251
5.6   Summary 251

Part II: Real-Time Framework 253

Chapter 6: Real-Time Framework Concepts 255
6.1   Inversion of Control 256
6.2   CPU Management 257
    6.2.1   Traditional Sequential Systems 257
    6.2.2   Traditional Multitasking Systems...259
    6.2.3   Traditional Event-Driven Systems..263
6.3   Active Object Computing Model..266
    6.3.1   System Structure 267
    6.3.2   Asynchronous Communication...269
    6.3.3   Run-to-Completion 269
    6.3.4   Encapsulation 269
    6.3.5   Support for State Machines...271
    6.3.6   Traditional Preemptive Kernel/RTOS......................................273
    6.3.7   Cooperative Vanilla Kernel...274
    6.3.8   Preemptive RTC Kernel 276
6.4   Event Delivery Mechanisms 279
    6.4.1   Direct Event Posting..280
    6.4.2   Publish-Subscribe 281
6.5   Event Memory Management 282
    6.5.1   Copying Entire Events...282
    6.5.2   Zero-Copy Event Delivery 284
    6.5.3   Static and Dynamic Events 286
    6.5.4   Multicasting Events and the Reference-Counting Algorithm...........286
    6.5.5   Automatic Garbage Collection 287
    6.5.6   Event Ownership 288

        6.5.7   Memory Pools 289
  6.6   Time Management 291
        6.6.1   Time Events 291
        6.6.2   System Clock Tick 293
  6.7   Error and Exception Handling 294
        6.7.1   Design by Contract...294
        6.7.2   Errors versus Exceptional Conditions...................................296
        6.7.3   Customizable Assertions in C and C++.................................297
        6.7.4   State-Based Handling of Exceptional Conditions..................300
        6.7.5   Shipping with Assertions 301
        6.7.6   Asserting Guaranteed Event Delivery...................................302
  6.8   Framework-Based Software Tracing 303
  6.9   Summary 304

Chapter 7: Real-Time Framework Implementation....................................**307**
  7.1   Key Features of the QF Real-Time Framework....................................308
        7.1.1   Source Code 309
        7.1.2   Portability..309
        7.1.3   Scalability 310
        7.1.4   Support for Modern State Machines......................................312
        7.1.5   Direct Event Posting and Publish-Subscribe Event Delivery 312
        7.1.6   Zero-Copy Event Memory Management.................................312
        7.1.7   Open-Ended Number of Time Events....................................312
        7.1.8   Native Event Queues 313
        7.1.9   Native Memory Pool 313
        7.1.10   Built-in "Vanilla" Scheduler 313
        7.1.11   Tight Integration with the QK Preemptive Kernel 313
        7.1.12   Low-Power Architecture 313
        7.1.13   Assertion-Based Error Handling...314
        7.1.14   Built-in Software Tracing Instrumentation............................314
  7.2   QF Structure...315
        7.2.1   QF Source Code Organization 316
  7.3   Critical Sections in QF 318
        7.3.1   Saving and Restoring the Interrupt Status............................319
        7.3.2   Unconditional Locking and Unlocking Interrupts 321
        7.3.3   Internal QF Macros for Interrupt Locking/Unlocking..................323
  7.4   Active Objects..324
        7.4.1   Internal State Machine of an Active Object..........................328
        7.4.2   Event Queue of an Active Object 328
        7.4.3   Thread of Execution and Active Object Priority...................330
  7.5   Event Management in QF..333
        7.5.1   Event Structure..333
        7.5.2   Dynamic Event Allocation 335

7.5.3   Automatic Garbage Collection 339
7.5.4   Deferring and Recalling Events...341
7.6   Event Delivery Mechanisms in QF 343
7.6.1   Direct Event Posting..343
7.6.2   Publish-Subscribe Event Delivery 344
7.7   Time Management..351
7.7.1   Time Event Structure and Interface......................................351
7.7.2   The System Clock Tick and the QF_tick() Function 354
7.7.3   Arming and Disarming a Time Event.....................................356
7.8   Native QF Event Queue 359
7.8.1   The QEQueue Structure 360
7.8.2   Initialization of QEQueue..362
7.8.3   The Native QF Active Object Queue....................................362
7.8.4   The "Raw" Thread-Safe Queue 367
7.9   Native QF Memory Pool 369
7.9.1   Initialization of the Native QF Memory Pool 372
7.9.2   Obtaining a Memory Block from the Pool..............................375
7.9.3   Recycling a Memory Block Back to the Pool..........................376
7.10   Native QF Priority Set...377
7.11   Native Cooperative "Vanilla" Kernel 379
7.11.1   The qvanilla.c Source File...380
7.11.2   The qvanilla.h Header File 384
7.12   QP Reference Manual...386
7.13   Summary..387

Chapter 8: Porting and Configuring QF 389
8.1   The QP Platform Abstraction Layer..390
8.1.1   Building QP Applications 390
8.1.2   Building QP Libraries..391
8.1.3   Directories and Files...392
8.1.4   The qep_port.h Header File 398
8.1.5   The qf_port.h Header File 400
Types of Platform-Specific QActive Data Members 402
Base Class for Derivation of QActive....................................402
The Maximum Number of Active Objects in the Application.........403
Various Object Sizes Within the QF Framework........................403
QF Critical Section Mechanism...404
Include Files Used by this QF Port.......................................404
Interface Used Only Inside QF, But Not in Applications...............405
Active Object Event Queue Operations..................................406
QF Event Pool Operations..406
8.1.6   The qf_port.c Source File 407
8.1.7   The qp_port.h Header File 411

8.1.8   Platform-Specific QF Callback Functions.................................412
8.1.9   System Clock Tick (Calling `QF_tick()`)..............................413
8.1.10   Building the QF Library 413
8.2   Porting the Cooperative "Vanilla" Kernel 414
8.2.1   The `qep_port.h` Header File 414
8.2.2   The `qf_port.h` Header File 415
8.2.3   The System Clock Tick (`QF_tick()`)..................................417
8.2.4   Idle Processing (`QF_onIdle()`).....................................418
8.3   QF Port to µC/OS-II (Conventional RTOS) 420
8.3.1   The `qep_port.h` Header File 422
8.3.2   The `qf_port.h` Header File 423
8.3.3   The `qf_port.c` Source File 425
8.3.4   Building the µC/OS-II Port 430
8.3.5   The System Clock Tick (`QF_tick()`)..................................430
8.3.6   Idle Processing 431
8.4   QF Port to Linux (Conventional POSIX-Compliant OS) 431
8.4.1   The `qep_port.h` Header File 432
8.4.2   The `qf_port.h` Header File 432
8.4.3   The `qf_port.c` Source File 435
8.5   Summary 441

Chapter 9: Developing QP Application.....................................443
9.1   Guidelines for Developing QP Applications...............................444
9.1.1   Rules...444
9.1.2   Heuristics 445
9.2   The Dining Philosopher Problem 446
9.2.1   Step 1: Requirements...447
9.2.2   Step 2: Sequence Diagrams 447
9.2.3   Step 3: Signals, Events, and Active Objects 449
9.2.4   Step 4: State Machines 451
9.2.5   Step 5: Initializing and Starting the Application 457
9.2.6   Step 6: Gracefully Terminating the Application 460
9.3   Running DPP on Various Platforms......................................461
9.3.1   "Vanilla" Kernel on DOS...461
9.3.2   "Vanilla" Kernel on Cortex-M3 465
9.3.3   µC/OS-II 469
9.3.4   Linux 472
9.4   Sizing Event Queues and Event Pools 476
9.4.1   In Sizing Event Queues 477
9.4.2   Sizing Event Pools 479
9.4.3   System Integration..480
9.5   Summary 480

Chapter 10: Preemptive Run-to-Completion Kernel*............................*483
10.1   Reasons for Choosing a Preemptive Kernel...483
10.2   Introduction to RTC Kernels 485
    10.2.1   Preemptive Multitasking with a Single Stack.........................486
    10.2.2   Nonblocking Kernel 487
    10.2.3   Synchronous and Asynchronous Preemptions........................487
    10.2.4   Stack Utilization 491
    10.2.5   Comparison to Traditional Preemptive Kernels 494
10.3   QK Implementation 496
    10.3.1   QK Source Code Organization 497
    10.3.2   The `qk.h` Header File...498
    10.3.3   Interrupt Processing..503
    10.3.4   The `qk_sched.c` Source File (QK Scheduler) 506
    10.3.5   The `qk.c` Source File (QK Startup and Idle Loop)..............511
10.4   Advanced QK Features..514
    10.4.1   Priority-Ceiling Mutex 515
    10.4.2   Thread-Local Storage 518
    10.4.3   Extended Context Switch (Coprocessor Support) 520
10.5   Porting QK..524
    10.5.1   The `qep_port.h` Header File..525
    10.5.2   The `qf_port.h` Header File..525
    10.5.3   The `qk_port.h` Header File..526
    10.5.4   Saving and Restoring FPU Context 531
10.6   Testing the QK Port 531
    10.6.1   Asynchronous Preemption Demonstration..............................531
    10.6.2   Priority-Ceiling Mutex Demonstration...................................535
    10.6.3   TLS Demonstration 536
    10.6.4   Extended Context Switch Demonstration 539
10.7   Summary...540

Chapter 11: Software Tracing for Event-Driven Systems *............................*541
11.1   Software Tracing Concepts..542
11.2   Quantum Spy Software-Tracing System...544
    11.2.1   Example of a Software-Tracing Session.................................545
    11.2.2   The Human-Readable Trace Output..547
11.3   QS Target Component...550
    11.3.1   QS Source Code Organization...552
    11.3.2   The QS Platform-Independent Header Files `qs.h` and `qs_dummy.h`..553
    11.3.3   QS Critical Section..560
    11.3.4   General Structure of QS Records..561
    11.3.5   QS Filters 562
          Global On/Off Filter...562

          Local Filters 564
  11.3.6   QS Data Protocol 566
          Transparency 567
          Endianness..568
  11.3.7   QS Trace Buffer..569
          Initializing the QS Trace Buffer `QS_initBuf()` 569
          Byte-Oriented Interface: `QS_getByte()` 571
          Block-Oriented Interface: `QS_getBlock()` 573
  11.3.8   Dictionary Trace Records..574
          Object Dictionaries 575
          Function Dictionaries...577
          Signal Dictionaries...577
  11.3.9   Application-Specific QS Trace Records 578
  11.3.10 Porting and Configuring QS 580
11.4 The QSPY Host Application...581
  11.4.1   Installing QSPY 582
  11.4.2   Building QSPY Application from Sources 584
          Building QSPY for Windows with Visual
          C++ 2005 584
          Building QSPY for Windows with MinGW...............................584
          Building QSPY for Linux 584
  11.4.3   Invoking QSPY 585
11.5 Exporting Trace Data to MATLAB 587
  11.5.1   Analyzing Trace Data with MATLAB......................................587
  11.5.2   MATLAB Output File...589
  11.5.3   MATLAB Script `qspy.m`...590
  11.5.4   MATLAB Matrices Generated by `qspy.m` 593
11.6 Adding QS Software Tracing to a QP Application 596
  11.6.1   Initializing QS and Setting Up the Filters 596
  11.6.2   Defining Platform-Specific QS Callbacks 598
  11.6.3   Generating QS Timestamps with the `QS_onGetTime()`
          Callback..601
  11.6.4   Generating QS Dictionary Records from
          Active Objects 604
  11.6.5   Adding Application-Specific Trace Records............................607
  11.6.6   "QSPY Reference Manual" 608
11.7 Summary..608

Chapter 12: QP-nano: How Small Can You Go?...................................611
12.1 Key Features of QP-nano 612
12.2 Implementing the "Fly 'n' Shoot" example with QP-nano........................614
  12.2.1   The `main()` function 615
  12.2.2   The `qpn_port.h` Header File...618

  12.2.3 Signals, Events, and Active Objects in the "Fly 'n' Shoot"
         Game..620
  12.2.4 Implementing the Ship Active Object in QP-nano 622
  12.2.5 Time Events in QP-nano...626
  12.2.6 Board Support Package for "Fly 'n' Shoot" Application in
         QP-nano 628
  12.2.7 Building the "Fly 'n' Shoot" QP-nano Application......................630
12.3 QP-nano Structure 631
  12.3.1 QP-nano Source Code, Examples, and Documentation..................633
  12.3.2 Critical Sections in QP-nano 634
         Task-Level Interrupt Locking 635
         ISR-Level Interrupt Locking.. 635
  12.3.3 State Machines in QP-nano...637
  12.3.4 Active Objects in QP-nano...640
  12.3.5 The System Clock Tick in QP-nano 642
12.4 Event Queues in QP-nano.. 644
  12.4.1 The Ready-Set in QP-nano (QF_readySet_)...............................645
  12.4.2 Posting Events from the Task Level (QActive_post())............646
  12.4.3 Posting Events from the ISR Level (QActive_postISR())........649
12.5 The Cooperative "Vanilla" Kernel in QP-nano.............................. 650
  12.5.1 Interrupt Processing Under the "Vanilla" Kernel 655
  12.5.2 Idle Processing under the "Vanilla" Kernel 655
12.6 The Preemptive Run-to-Completion QK-nano Kernel..................... 655
  12.6.1 QK-nano Interface qkn.h 656
  12.6.2 Starting Active Objects and the QK-nano Idle Loop.....................658
  12.6.3 The QK-nano Scheduler...660
  12.6.4 Interrupt Processing in QK-nano 665
  12.6.5 Priority Ceiling Mutex in QK-nano 666
12.7 The PELICAN Crossing Example.. 666
  12.7.1 PELICAN Crossing State Machine...668
  12.7.2 The Pedestrian Active Object...671
  12.7.3 QP-nano Port to MSP430 with QK-nano Kernel..........................672
  12.7.4 QP-nano Memory Usage 675
12.8 Summary.. 678

Appendix A. Licensing Policy for QP and QP-nano *679*
A.1 Open-Source Licensing.. 679
A.2 Closed-Source Licensing 680
A.3 Evaluating the Software... 680
A.4 NonProfits, Academic Institutions, and Private Individuals........................ 680
A.5 GNU General Public License Version 2... 681

Appendix B. Guide to Notation **685**
  B.1   Class Diagrams...685
  B.2   State Diagrams 688
  B.3   Sequence Diagrams 689
  B.4   Timing Diagrams..690

Bibliography **693**
Index...**699**

Preface

To create a usable piece of software, you have to fight for every fix, every feature, every little accommodation that will get one more person up the curve. There are no shortcuts. Luck is involved, but you don't win by being lucky, it happens because you fought for every inch.
—Dave Winer

For many years, I had been looking for a book or a magazine article that would describe a truly practical way of coding modern state machines (UML[1] statecharts) in a mainstream programming language such as C or C++. I have never found such a technique.

In 2002, I wrote *Practical Statecharts in C/C++: Quantum Programming for Embedded Systems* (*PSiCC*), which was the first book to provide what had been missing thus far: a compact, efficient, and highly maintainable implementation of UML state machines in C and C++ with full support for hierarchical nesting of states. *PSiCC* was also the first book to offer complete C and C++ source code of a generic, state machine-based, real-time application framework for embedded systems.

To my delight, *PSiCC* continues to be one of the most popular books about statecharts and event-driven programming for embedded systems. Within a year of its publication, *PSiCC* was translated into Chinese, and a year later into Korean. I've received and answered literally thousands of e-mails from readers who successfully used the published code in consumer, medical, industrial, wireless, networking, research, defense, robotics, automotive, space exploration, and many other applications worldwide. In 2003 I started to speak about the subject matter at

---

[1] UML stands for Unified Modeling Language and is the trademark of Object Management Group.

the Embedded Systems Conferences on both U.S. coasts. I also began to consult to companies. All this gave me additional numerous opportunities to find out firsthand how engineers actually use the published design techniques in a wide range of application areas.

What you're holding in your hands is the second edition of *PSiCC*. It is the direct result of the plentiful feedback I've received as well as five years of the "massive parallel testing" and scrutiny that has occurred in the trenches.

What's New in the Second Edition?

As promised in the first edition of *PSiCC*, I continued to advance the code and refine the design techniques. This completely revised second edition incorporates these advancements as well the numerous lessons learned from readers.

New Code

First of all, this book presents an entirely new version of the software, which is now called Quantum Platform (QP) and includes the hierarchical event processor (QEP) and the real-time framework (QF) as well as two new components. QP underwent several quantum leaps of improvement since the first publication six years ago. The enhancements introduced since the first edition of *PSiCC* are too numerous to list here, but the general areas of improvements include greater efficiency and testability and better portability across different processors, compilers, and operating systems. The two new QP components are the lightweight, *preemptive*, real-time kernel (QK) described in Chapter 10 and the software-tracing instrumentation (QS) covered in Chapter 11. Finally, I'm quite excited about the entirely new, ultralight, reduced-feature version of QP called QP-nano that scales the approach down to the lowest-end 8- and 16-bit MCUs. I describe QP-nano in Chapter 12.

Open Source and Dual Licensing

In 2004, I decided to release the entire QP code as open source under the terms of the GNU General Public License (GPL) version 2, as published by the Free Software Foundation. Independent of the open-source licensing, the QP source code is also available under the terms of traditional commercial licenses, which expressly supersede the GPL and are specifically designed for users interested in retaining the proprietary status of their applications based on QP. This increasingly popular strategy of combining open source with commercial licensing, called *dual licensing*, is explained in more detail in Appendix A.

C as the Primary Language of Exposition

Most of the code samples in the first edition of *PSiCC* pertained to the C++ implementation. However, as I found out in the field, many embedded software developers come from a hardware background (mostly EE) and are often unnecessarily intimidated by C++.

In this edition, I decided to exactly reverse the roles of C and C++. As before, the companion Website contains the complete source code for both C and C++ versions. But now, most of the code examples in the text refer to the C version, and the C++ code is discussed only when the differences between it and the C implementation become nontrivial and important.

As far as the C source code is concerned, I no longer use the C+ object-oriented extension that I've applied and documented in the first edition. The code is still compatible with C+, but the C+ macros are not used.

More Examples

Compared to the first edition, this book presents more examples of event-driven systems and the examples are more complete. I made a significant effort to come up with examples that are not utterly trivial yet don't obscure the general principles in too many details. I also chose examples that don't require any specific domain knowledge, so I don't need to waste space and your attention explaining the problem specification.

Preemptive Multitasking Support

An event-driven infrastructure such as QP can work with a variety of concurrency mechanisms, from a simple "superloop" to fully preemptive, priority-based multitasking. The previous version of QP supported the simple nonpreemptive scheduling natively but required an external RTOS to provide preemptive multitasking, if such capability was required.

In Chapter 10, I describe the new real-time kernel (QK) component that provides deterministic, fully preemptive, priority-based multitasking to QP. QK is a very special, super-simple, run-to-completion, single-stack kernel that perfectly matches the universally assumed run-to-completion semantics required for state machine execution.

Testing Support

A running application built of concurrently executing state machines is a highly structured affair where all important system interactions funnel through the event-driven framework that ties all the state machines together. By instrumenting just this tiny "funnel" code, you can gain unprecedented insight into the live system. In fact, the software trace data from an instrumented event-driven framework can tell you much more about the application than any traditional real-time operating system (RTOS) because the framework "knows" so much more about the application.

Chapter 11 describes the new QS ("spy") component that provides a comprehensive software-tracing instrumentation to the QP event-driven platform. The trace data produced by the QS component allows you to perform a live analysis of your running real-time embedded system with minimal target system resources and without stopping or significantly slowing down the code. Among other things, you can reconstruct complete sequence diagrams and detailed, timestamped state machine activities for all active objects in the system. You can monitor all event exchanges, event queues, event pools, time events (timers), and preemptions and context switches. You can also use QS to add your own instrumentation to the application-level code.

Ultra-Lightweight QP-nano Version

The event-driven approach with state machines scales down better than any conventional real-time kernel or RTOS. To address really small embedded systems, a reduced QP version called QP-nano implements a subset of features supported in QP/C or QP/C++. QP-nano has been specifically designed to enable event-driven programming with hierarchical state machines on low-end 8- and 16-bit microcontrollers (MCUs), such as AVR, MSP430, 8051, PICmicro, 68HC(S)08, M16C, and many others. Typically, QP-nano requires around 1-2KB of ROM and just a few bytes of RAM per state machine. I describe QP-nano in Chapter 12.

Removed Quantum Metaphor

In the first edition of *PSiCC*, I proposed a quantum-mechanical metaphor as a way of thinking about the event-driven software systems. Though I still believe that this analogy is remarkably accurate, it hasn't particularly caught on with readers, even though providing such a metaphor is one of the key practices of eXtreme Programming (XP) and other agile methods.

Respecting readers' feedback, I decided to remove the quantum metaphor from this edition. For historical reasons, the word *quantum* still appears in the names of the software components, and the prefix *Q* is consistently used in the code for type and function names to clearly distinguish the QP code from other code, but you don't need to read anything into these names.

What You Need to Use QP

Most of the code supplied with this book is highly portable C or C++, independent of any particular CPU, operating system, or compiler. However, to focus the discussion I provide *executable examples* that run in a DOS console under any variant of Windows. I've chosen the legacy 16-bit DOS as a demonstration platform because it allows programming a standard x86-based PC at the bare-metal level. Without leaving your desktop, you can work with interrupts, directly manipulate CPU registers, and directly access the I/O space. No other modern 32-bit development environment for the standard PC allows this much so easily.

The additional advantage of the legacy DOS platform is the availability of mature and free tools. To that end, I have compiled the examples with the legacy Borland Turbo C++ 1.01 toolset, which is available for a *free download* from Borland.

To demonstrate modern embedded systems programming with QP, I also provide examples for the inexpensive[2] ARM Corterx-M3-based Stellaris EV-LM3S811 evaluation kit form Luminary Micro. The Cortex-M3 examples use the exact same source code as the DOS counterparts and differ only in the board support package (BSP). The Cortex-M3 examples require the 32KB-limited KickStart edition of the IAR EWARM toolset, which is included in the Stellaris kit and is also available for a *free download* from IAR.

Finally, some examples in this book run on Linux as well as any other POSIX-compliant operating system such as BSD, QNX, Max OS X, or Solaris. You can also build the Linux examples on Windows under Cygwin.

---

[2] At the time of this writing, the EKIEV-LM3S811 kit was available for $49 (www.luminarymicro.com).

The companion Website to this book at www.quantum-leaps.com/psicc2 provides the links for downloading all the tools used in the book, as well as other resources. The Website also contains links to dozens of QP ports to various CPUs, operating systems, and compilers. Keep checking this Website; new ports are added frequently.

Intended Audience

This book is intended for the following software developers interested in event-driven programming and modern state machines:

- Embedded programmers and consultants will find a *complete*, ready-to-use, event-driven infrastructure to develop applications. The book describes both state machine coding strategies and, equally important, a compatible real-time framework for executing concurrent state machines. These two elements are synergistically complementary, and one cannot reach its full potential without the other.

- Embedded developers looking for a real-time kernel or RTOS will find that the QP event-driven platform can do everything one might expect from an RTOS and that, in fact, QP actually contains a fully preemptive real-time kernel as well as a simple cooperative scheduler.

- Designers of ultra low-power systems, such as wireless sensor networks, will find how to scale down the event-driven, state machine-based approach to fit the tiniest MCUs. The ultra-light QP-nano version (Chapter 12) combines a hierarchical event processor, a real-time framework, and either a cooperative or a fully preemptive kernel in just 1–2KB of ROM.

- On the opposite end of the complexity spectrum, designers of very large-scale, massively parallel server applications will find that the event-driven approach combined with hierarchical state machines scales up easily and is ideal for managing very large numbers of stateful components, such as client sessions. As it turns out, the "embedded" design philosophy of QP provides the critical per-component efficiency both in time and space.

- The open-source community will find that QP complements other open-source software, such as Linux or BSD. The QP port to Linux (and more generally to POSIX-compliant operating systems) is described in Chapter 8.

- GUI developers and computer game programmers using C or C++ will find that QP very nicely complements GUI libraries. QP provides the high-level "screen logic" based on hierarchical state machines, whereas the GUI libraries handle low-level widgets and rendering of the images on the screen.

- System architects might find in QP a lightweight alternative to heavyweight design automation tools.

- Users of design automation tools will gain deeper understanding of the inner workings of their tools. The glimpse "under the hood" will help them use the tools more efficiently and with greater confidence.

Due to the *code-centric approach*, this book will primarily appeal to software developers tasked with creating actual, working code, as opposed to just modeling. Many books about UML already do a good job of describing model-driven analysis and design as well as related issues, such as software development processes and modeling tools.

This book does *not* provide yet another CASE tool. Instead, this book is about practical, manual coding techniques for hierarchical state machines and about combining state machines into robust event-driven systems by means of a real-time framework.

To benefit from the book, you should be reasonably proficient in C or C++ and have a general understanding of computer architectures. I am not assuming that you have prior knowledge of UML state machines, and I introduce the underlying concepts in a crash course in Chapter 2. I also introduce the basic real-time concepts of multitasking, mutual exclusion, and blocking in Chapter 6.

The Companion Websites

This book has a companion Website at www.quantum-leaps.com/psicc2 that contains the following information:

- Source code downloads for QP/C, QP/C++, and QP-nano

- All QP ports and examples described in the book

- Reference manuals for QP/C, QP/C++, and QP-nano in HTML and CHM file formats

- Links for downloading compilers and other tools used in the book

- Selected reviews and reader feedback

- Errata

Additionally, the Quantum Leaps Website at `www.quantum-leaps.com` has been supporting the QP user community since the publication of the first edition of *PSiCC* in 2002. This Website offers the following resources:

- Latest QP downloads

- QP ports and development kits

- Programmer manuals

- Application notes

- Resources and goodies such as Visio stencils for drawing UML diagrams, design patterns, links to related books and articles, and more

- Commercial licensing and technical support information

- Consulting and training in the technology

- News and events

- Discussion forum

- Newsletter

- Blog

- Links to related Websites

- And more

Finally, QP is also present on SourceForge.net—the world's largest repository of open source code and applications. The QP project is located at `https://sourceforge. net/projects/qpc/`.

Acknowledgments

First and foremost, I'd like to thank my wonderful family for the unfading support over the years of creating the software and the two editions of this book.

I would also like to thank the team at Elsevier, which includes Rachel Roumeliotis and Heather Scherer, and John (Jay) Donahue.

Finally, I'm grateful to all the software developers who contacted me with thought-provoking questions, bug reports, and countless suggestions for improvements in the code and documentation. As a rule, a software system only gets better if it is used and scrutinized by many people in many different real-life projects.

Introduction

Almost all computer systems in general, and embedded systems in particular, are event-driven, which means that they continuously wait for the occurrence of some external or internal event such as a time tick, an arrival of a data packet, a button press, or a mouse click. After recognizing the event, such systems react by performing the appropriate computation that may include manipulating the hardware or generating "soft" events that trigger other internal software components. (That's why event-driven systems are alternatively called *reactive* systems.) Once the event handling is complete, the software goes back to waiting for the next event.

You are undoubtedly accustomed to the basic sequential control, in which a program waits for events in various places in its execution path by either actively polling for events or passively blocking on a semaphore or other such operating system mechanism. Though this approach to programming event-driven systems is functional in many situations, it doesn't work very well when there are multiple possible sources of events whose arrival times and order you cannot predict and where it is important to handle the events in a timely manner. The problem is that while a sequential program is waiting for one kind of event, it is not doing any other work and is not responsive to other events.

Clearly, what we need is a program structure that can respond to a multitude of possible events, any of which can arrive at unpredictable times and in an unpredictable sequence. Though this problem is very common in embedded systems such as home appliances, cell phones, industrial controllers, medical devices and many others, it is also very common in modern desktop computers. Think about using a Web browser, a word processor, or a spreadsheet. Most of these programs have a modern graphical user interface (GUI), which is clearly capable of handling multiple events. All developers of modern GUI systems, and many embedded applications, have adopted a common program structure that elegantly solves the problem of dealing with many asynchronous events in a timely manner. This program structure is generally called *event-driven programming*.

Inversion of Control

Event-driven programming requires a distinctly different way of thinking than conventional sequential programs, such as "superloops" or tasks in a traditional RTOS. Most modern event-driven systems are structured according to the *Hollywood principle*, which means "Don't call us, we'll call you." So an event-driven program is *not* in control while waiting for an event; in fact, it's not even active. Only once the event arrives, the program is called to process the event and then it quickly relinquishes the control again. This arrangement allows an event-driven system to wait for many events in parallel, so the system remains responsive to all events it needs to handle.

This scheme has three important consequences. First, it implies that an event-driven system is naturally divided into the application, which actually handles the events, and the supervisory event-driven infrastructure, which waits for events and dispatches them to the application. Second, the control resides in the event-driven infrastructure, so from the application standpoint the control is *inverted* compared to a traditional sequential program. And third, the event-driven application must return control after handling each event, so the execution context cannot be preserved in the stack-based variables and the program counter as it is in a sequential program. Instead, the event-driven application becomes a *state machine*, or actually a set of collaborating state machines that preserve the context from one event to the next in the static variables.

The Importance of the Event-Driven Framework

The *inversion of control*, so typical in all event-driven systems, gives the event-driven infrastructure all the defining characteristics of an *application framework* rather than a toolkit. When you use a toolkit, such as a traditional operating system or an RTOS, you write the main body of the application and call the toolkit code that you want to reuse. When you use a framework, you reuse the main body and write the code *it* calls.

Another important point is that an event-driven framework is actually necessary if you want to combine multiple event-driven state machines into systems. It really takes more than "just" an API, such as a traditional RTOS, to execute concurrent state machines.

State machines require an infrastructure (framework) that provides, at a minimum, run-to-completion (RTC) execution context for each state machine, queuing of events, and event-based timing services. This is really the pivotal point. State machines cannot operate in a vacuum and are not really practical without an event-driven framework.

Active Object Computing Model

This book brings together two most effective techniques of decomposing event-driven systems: hierarchical state machines and an event-driven framework. The combination of these two elements is known as the *active object computing model*. The term *active object* comes from the UML and denotes an autonomous object engaging other active objects asynchronously via events. The UML further proposes the UML variant of statecharts with which to model the behavior of event-driven active objects.

In this book, active objects are implemented by means of the event-driven framework called QF, which is the main component of the QP event-driven platform. The QF framework orderly executes active objects and handles all the details of thread-safe event exchange and processing within active objects. QF guarantees the universally assumed RTC semantics of state machine execution, by queuing events and dispatching them sequentially (one at a time) to the internal state machines of active objects.

The fundamental concepts of hierarchical state machines combined with an event-driven framework are not new. In fact, they have been in widespread use for at least two decades. Virtually all commercially successful design automation tools on the market today are based on hierarchical state machines (statecharts) and incorporate internally a variant of an event-driven, real-time framework similar to QF.

The Code-Centric Approach

The approach I assume in this book is *code-centric*, minimalist, and low-level. This characterization is not pejorative; it simply means that you'll learn how to map hierarchical state machines and active objects directly to C or C++ source code, without big tools. The issue here is not a tool—the issue is understanding.

The modern design automation tools are truly powerful, but they are not for everyone. For many developers the tool simply can't pull its own weight and gets abandoned. For such developers, the code-centric approach presented in this book can provide a lightweight alternative to the heavyweight tools.

Most important, though, no tool can replace conceptual understanding. For example, determining which exit and entry actions fire in which sequence in a nontrivial state transition is not something you should discover by running a tool-supported animation of your state machine. The answer should come from your understanding of the underlying state machine implementation (discussed in Chapters 3 and 4). Even if you later decide to use a design automation tool and even if that particular tool would use a different statechart implementation technique than discussed in this book, you will still apply the concepts with greater confidence and more efficiency because of your understanding of the fundamental mechanisms at a low level.

In spite of many pressures from existing users, I persisted in keeping the QP event-driven platform lean by directly implementing only the essential elements of the bulky UML specification and supporting the niceties as design patterns. Keeping the core implementation small and simple has real benefits. Programmers can learn and deploy QP quickly without large investments in tools and training. They can easily adapt and customize the framework's source code to the particular situation, including to severely resource-constrained embedded systems. They can understand, and indeed regularly use, all the provided features.

Focus on Real-Life Problems

You can't just look at state machines and the event-driven framework as a collection of features, because some of the features will make no sense in isolation. You can only use these powerful concepts effectively if you are thinking about design, not simply coding. And to understand state machines that way, you must understand the problems with event-driven programming in general.

This book discusses event-driven programming problems, why they are problems, and how state machines and active object computing model can help. Thus, I begin most chapters with the programming problems the chapter will address. In this way, I hope to move you, a little at a time, to the point where hierarchical state machines and the event-driven framework become a much more natural way of solving the problems than the traditional approaches such as deeply nested IFs and ELSEs for coding stateful behavior or passing events via semaphores or event flags of a traditional RTOS.

Object Orientation

Even though I use C as the primary programming language, I also extensively use object-oriented design principles. Like virtually all application frameworks, QP uses the basic concepts of encapsulation (classes) and single inheritance as the primary mechanisms of customizing, specializing, and extending the framework to a particular application. Don't worry if these concepts are new to you, especially in C. At the C language level, encapsulation and inheritance become just simple coding idioms, which I introduce in Chapter 1. I specifically avoid polymorphism in the C version because implementing late binding in C is a little more involved. Of course, the C++ version uses classes and inheritance directly and QP/C++ applications can use polymorphism.

More Fun

When you start using the techniques described in this book, your problems will change. You will no longer struggle with 15 levels of convoluted if–else statements, and you will stop worrying about semaphores or other such low-level RTOS mechanisms. Instead, you'll start thinking at a *higher level of abstraction* about state machines, events, and active objects. After you experience this quantum leap you will find, as I did, that programming can be much more *fun*. You will never want to go back to the "spaghetti" code or the raw RTOS.

How to Contact Me

If you have comments or questions about this book, the code, or event-driven programming in general, I'd be pleased to hear from you. Please e-mail me at miro@quantum-leaps.com.

"IT MAY NOT BE A PERFECT WHEEL, BUT IT'S
A STATE-OF-THE-ART WHEEL."

www.CartoonStock.com

PART I UML STATE MACHINES

State machines are the best-known formalism for specifying and implementing event-driven systems that must react to incoming events in a timely fashion. The advanced UML state machines represent the current state of the art in state machine theory and notation.

Part I of this book shows practical ways of using UML state machines in event-driven applications to help you produce efficient and maintainable software with well-understood behavior, rather than creating "spaghetti" code littered with convoluted IFs and ELSEs. Chapter 1 presents an overview of the method based on a working example.

Chapter 2 introduces state machine concepts and the UML notation. Chapter 3 shows the standard techniques of coding state machines, and Chapter 4 describes a generic hierarchical event processor. Part I concludes with Chapter 5, which presents a mini-catalogue of five state design patterns. You will learn that UML state machines are a powerful design method that you can use, even without complex code-synthesizing tools.

CHAPTER 1

Getting Started with UML State Machines and Event-Driven Programming

It is common sense to take a method and try it. If it fails, admit it frankly and try another. But above all, try something.
—Franklin D. Roosevelt

This chapter presents an example project implemented entirely with UML state machines and the event-driven paradigm. The example application is an interactive "Fly 'n' Shoot"-type game, which I decided to include early in the book so that you can start playing (literally) with the code as soon as possible. My aim in this chapter is to show the essential elements of the method in a real, nontrivial program, but without getting bogged down in details, rules, and exceptions. At this point, I am not trying to be complete or even precise, although this example as well as all other examples in the book is meant to show a good design and the recommended coding style. I don't assume that you know much about UML state machines, UML notation, or event-driven programming. I will either briefly introduce the concepts, as needed, or refer you to the later chapters of the book for more details.

The example "Fly 'n' Shoot" game is based on the Quickstart application provided in source code with the Stellaris EV-LM3S811 evaluation kit from Luminary Micro [Luminary 06]. I was trying to make the "Fly 'n' Shoot" example behave quite similarly to the original Luminary Micro Quickstart application so that you can directly compare the event-driven approach with the traditional solution to essentially the same problem specification.

1.1   Installing the Accompanying Code

The companion Website to this book at www.quantum-leaps.com/psicc2 contains the self-extracting archive with the complete source code of the QP event-driven platform and all executable examples described in this book; as well as documentation, development tools, resources, and more. You can uncompress the archive into any directory. The installation directory you choose will be referred henceforth as the QP Root Directory <qp>.

NOTE

Although in the text I mostly concentrate on the C implementation, the accompanying Website also contains the equivalent C++ version of virtually every element available in C. The C++ code is organized in exactly the same directory tree as the corresponding C code, except you need to look in the <qp>\qpcpp\... directory branch.

Specifically to the "Fly 'n' Shoot" example, the companion code contains two versions[1] of the game. I provide a DOS version for the standard Windows-based PC (see Figure 1.1) so that you don't need any special embedded board to play the game and experiment with the code.

NOTE

I've chosen the legacy 16-bit DOS platform because it allows programming a standard PC at the bare-metal level. Without leaving your desktop, you can work with interrupts, directly manipulate CPU registers, and directly access the I/O space. No other modern 32-bit development environment for the standard PC allows this much so easily. The ubiquitous PC running under DOS (or a DOS console within any variant of Windows) is as close as it gets to emulating embedded software development on the commodity 80x86 hardware. Additionally, you can use free, mature tools, such as the Borland C/C++ compiler.

I also provide an embedded version for the inexpensive[2] ARM Cortex-M3-based Stellaris EV-LM3S811 evaluation kit (see Figure 1.2). Both the PC and Cortex-M3

---

[1] The accompanying code actually contains many more versions of the "Fly 'n' Shoot" game, but they are not relevant at this point.

[2] At the time of this writing the EV-LM3S811 kit was available for $49 (www.luminarymicro.com).

versions use the exact same source code for all application components and differ only in the Board Support Package (BSP).

1.2 Let's Play

The following description of the "Fly 'n' Shoot" game serves the dual purpose of explaining how to play the game and as the problem specification for the purpose of designing and implementing the software later in the chapter. To accomplish these two goals I need to be quite detailed, so please bear with me.

Your objective in the game is to navigate a spaceship through an endless horizontal tunnel with mines. Any collision with the tunnel or the mine destroys the ship. You can move the ship up and down with Up-arrow and Down-arrow keys on the PC (see Figure 1.1) or via the potentiometer wheel on the EV-LM3S811 board (see Figure 1.2). You can also fire a missile to destroy the mines in the tunnel by pressing the Spacebar on the PC or the User button on the EV-LM3S811 board. Score accumulates for survival (at the rate of 30 points per second) and destroying the mines. The game lasts for only one ship.

The game starts in a demo mode, where the tunnel walls scroll at the normal pace from right to left and the "Press Button" text flashes in the middle of the screen. You need to generate the "fire missile" event for the game to begin (press Spacebar on the PC or the User button on the EV-LM3S811 board).

You can have only one missile in flight at a time, so trying to fire a missile while it is already flying has no effect. Hitting the tunnel wall with the missile brings you no points, but you earn extra points for destroying the mines.

The game has two types of mines with different behavior. In the original Luminary Quickstart application both types of mines behave the same, but I wanted to demonstrate how state machines can elegantly handle differently behaving mines.

Mine type 1 is small, but can be destroyed by hitting any of its pixels with the missile. You earn 25 points for destroying a mine type 1. Mine type 2 is bigger but is nastier in that the missile can destroy it only by hitting its center, not any of the "tentacles." Of course, the ship is vulnerable to the whole mine. You earn 45 points for destroying a mine type 2.

When you crash the ship, by either hitting a wall or a mine, the game ends and displays the flashing "Game Over" text as well as your final score. After 5 seconds of flashing, the "Game Over" screen changes back to the demo screen, where the game waits to be started again.

Additionally the application contains a screen saver because the OLED display of the original EV-LM3S811 board has burn-in characteristics similar to a CRT. The screen saver only becomes active if 20 seconds elapse in the demo mode without starting the game (i.e., the screen saver never appears during game play). The screen saver is a simple random pixel type rather than the "Game of Life" algorithm used in the original Luminary Quickstart application. I've decided to simplify this aspect of the implementation because the more elaborate pixel-mixing algorithm does not contribute any new or interesting behavior.

After a minute of running the screen saver, the display turns blank and only a single random pixel shows on the screen. Again, this is a little different from the original Quickstart application, which instead blanks the screen and starts flashing the User LED. I've changed this behavior because I have a better purpose for the User LED (to visualize the activity of the idle loop).

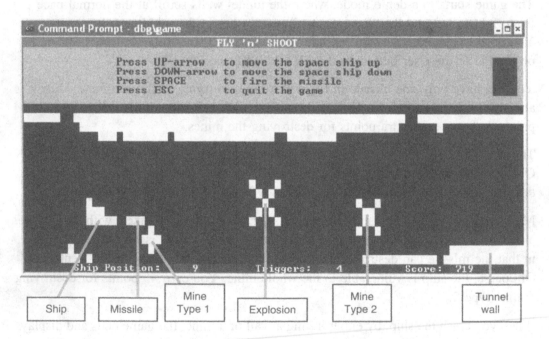

Figure 1.1: The "Fly 'n' Shoot" game running in a DOS window under Windows XP.

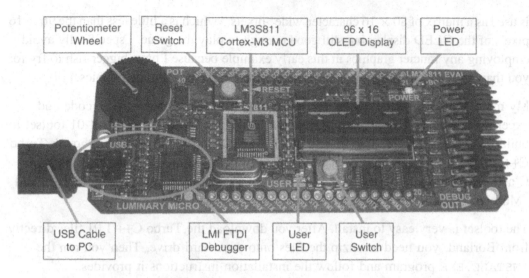

Figure 1.2: The "Fly 'n' Shoot" game running on the Stellaris EV-LM3S811 evaluation board.

1.2.1 Running the DOS Version

The "Fly 'n' Shoot" sample code for the DOS version (in C) is located in the `<qp>\qpc\examples\80x86\dos\tcpp101\l\game\` directory, where `<qp>` stands for the installation directory in which you chose to install the accompanying software.

The compiled executable is provided, so you can run the game on any Windows-based PC by simply double-clicking the executable `game.exe` located in the directory `<qp>\qpc\examples\80x86\dos\tcpp101\l\game\dbg\`. The first screen you see is the game running in the demo mode with the text "Push Button" flashing in the middle of the display. At the top of the display you see a legend of keystrokes recognized by the application. You need to hit the SPACEBAR to start playing the game. Press the ESC key to cleanly exit the application.

If you run "Fly 'n' Shoot" in a window under Microsoft Windows, the animation effects in the game might appear a little jumpy, especially compared to the Stellaris version of the same game. You can make the application execute significantly more smoothly if you switch to the full-screen mode by pressing and holding the Alt key and then pressing the Enter key. You go back to the window mode via the same Alt-Enter key combination.

As you can see in Figure 1.1, the DOS version uses simply the standard VGA text mode to emulate the OLED display of the EV-LM3S811 board. The lower part of the DOS screen is used as a matrix of 80×16 character-wide "pixels," which is a little less than the 96×16 pixels of the OLED display but still good enough to play the game. I specifically avoid employing any fancier graphics in this early example because I have bigger fish to fry for you than to worry about the irrelevant complexities of programming graphics.

My main goal is to make it easy for you to understand the event-driven code and experiment with it. To this end, I chose the legacy Borland Turbo C++ 1.01 toolset to build this example as well as several other examples in this book. Even though Turbo C++ 1.01 is an older compiler, it is adequate to demonstrate all features of both the C and C++ versions. Best of all, it is available for a free download from the Borland "Museum" at http://bdn.borland.com/article/0,1410,21751,00.html.

The toolset is very easy to install. After you download the Turbo C++ 1.01 files directly from Borland, you need to unzip the files onto your hard drive. Then you run the INSTALL.EXE program and follow the installation instructions it provides.

> **NOTE**
>
> I strongly recommend that you install the Turbo C++ 1.01 toolset into the directory C:\tools\tcpp101\. That way you will be able to directly use the provided project files and make scripts.

Perhaps the easiest way to experiment with the "Fly 'n' Shoot" code is to launch the Turbo C++ IDE (TC.EXE) and open the provided project file GAME-DBG.PRJ, which is located in the directory <qp>\qpc\examples\80x86\dos\tcpp101\1\game\. You can modify, recompile, execute, and debug the program directly from the IDE. However, you should avoid terminating the program stopped in the debugger, because this will not restore the standard DOS interrupt vectors for the time tick and keyboard interrupts. You should always cleanly exit the application by letting it freely run and pressing the Esc key.

The next section briefly describes how to run the embedded version of the game. If you are not interested in the Cortex-M3 version, feel free to skip to Section 1.3, where I start explaining the application code.

1.2.2 Running the Stellaris Version

In contrast to the "Fly 'n' Shoot" version for DOS running in the ancient real mode of the 80x86 processor, the exact same source code runs on one of the most modern processors in the industry: the ARM Cortex-M3.

The sample code for the Stellaris EV-LM3S811 board is located in the `<qp>\qpc\examples\cortex-m3\vanilla\iar\game-ev-lm3s811\` directory, where `<qp>` stands for the root directory in which you chose to install the accompanying software.

The code for the Stellaris kit has been compiled with the 32KB-limited Kickstart edition of the IAR Embedded Workbench for ARM (IAR EWARM) v 5.11, which is provided with the Stellaris EV-LM3S811 kit. You can also download this software free of charge directly from IAR Systems (www.iar.com) after filling out an online registration.

The installation of IAR EWARM is quite straightforward, since the software comes with the installation utility. You also need to install the USB drivers for the hardware debugger built into the EV-LM3S811 board, as described in the documentation of the Stellaris EV-LM3S811 kit.

> **NOTE**
>
> I strongly recommend that you install the IAR EWARM toolset into the directory `C:\tools\iar\arm_ks_5.11`. That way you will be able to directly use the provided EWARM workspace files and make scripts.

Before you program the "Fly 'n' Shoot" game to the EV-LM3S811 board, you might want to play a little with the original Quickstart application that comes preprogrammed with the EV-LM3S811 kit.

To program the "Fly 'n' Shoot" game to the Flash memory of the EV-LM3S811 board, you first connect the EV-LM3S811 board to your PC with the USB cable provided in the kit and make sure that the Power LED is on (see Figure 1.2). Next, you need to launch the IAR Embedded Workbench and open the workspace `game-ev-1m3s811.eww` located in the `<qp>\qpc\examples\cortex-m3\vanilla\iar\game-ev-1m3s811\` directory. At this point your screen should look similar to the screenshot shown in Figure 1.3.

The `game-ev-1m3s811` project is set up to use the LMI FTDI debugger, which is the piece of hardware integrated on the EV-LM3S811 board (see Figure 1.2). You can verify this setup by opening the "Options" dialog box via the Project | Options menu. Within the "Options" dialog box, you need to select the Debugger category in the panel on the left. While you're at it, you could also verify that the Flash loading is enabled by selecting the "Download" tab. The checked "Use flash loader(s)" check box means that the Flash loader application provided by IAR will be first loaded to the RAM of the MCU, and this application will program the Flash with the image of your application.

To start the Flash programming process, select the Project | Debug menu, or simply click the Debug button (see Figure 1.3) in the toolbar. The IAR Workbench should respond by showing the Flash programming progress bar for several seconds, as shown in Figure 1.3. Once the Flash programming completes, the IAR EWARM switches to the IAR C-Spy debugger and the program should stop at the entry to main(). You can start playing the game either by clicking the Go button in the debugger or you can close the debugger and reset the board by pressing the Reset button. Either way, the "Fly 'n' Shoot" game is now permanently programmed into the EV-LM3S811 board and will start automatically on every powerup.

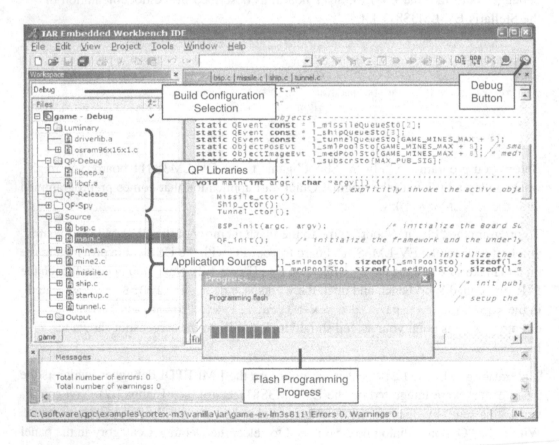

Figure 1.3: Loading the "Fly 'n' Shoot" game into the flash of LM3S811 MCU with IAR EWARM IDE.

The IAR Embedded Workbench environment allows you to experiment with the "Fly 'n' Shoot" code very easily. You can edit the files and recompile the application at a click of a button (F7). The only caveat is that the first time after the installation of the IAR toolset you need to build the Luminary Micro driver library for the LM3S811 MCU from the sources. You accomplish this by loading the workspace `ek-lm3s811.eww` located in the directory `<IAR-EWARM>\ARM\examples \Luminary\Stellaris\boards\ek-lm3s811`, where `<IAR-EWARM>` stands for the directory name where you've installed the IAR toolset. In the `ev-lm3s811.eww` workspace, you select the "`driverlib - Debug`" project from the drop-down list at the top of the Workspace panel and then press F7 to build the library.

1.3   The main() Function

Perhaps the best place to start the explanation of the "Fly 'n' Shoot" application code is the `main()` function, located in the file `main.c`. Unless indicated otherwise in this chapter, you can browse the code in either the DOS version or the EV-LM3S811 version, because the application source code is identical in both. The complete `main.c` file is shown in Listing 1.1.

NOTE

To explain code listings, I place numbers in parentheses at the interesting lines in the left margin of the listing. I then use these labels in the left margin of the explanation section that immediately follows the listing. Occasionally, to unambiguously refer to a line of a particular listing from sections of text other than the explanation section, I use the full reference consisting of the listing number followed by the label. For example, Listing 1.1(21) refers to the label (21) in Listing 1.1.

Listing 1.1 The file `main.c` of the "Fly 'n' Shoot" game application

```
(1)    #include "qp_port.h"                                    /* the QP port */
(2)    #include "bsp.h"                              /* Board Support Package */
(3)    #include "game.h"                                /* this application */

       /* Local-scope objects -----------------------------------------------*/
(4)    static QEvent const * l_missileQueueSto[2];                /* event queue */
(5)    static QEvent const * l_shipQueueSto[3];                   /* event queue */
(6)    static QEvent const * l_tunnelQueueSto[GAME_MINES_MAX + 5]; /* event queue */
```

Continued onto next page

```
(7)    static ObjectPosEvt   l_smlPoolSto[GAME_MINES_MAX + 8]; /* small-size pool */
(8)    static ObjectImageEvt l_medPoolSto[GAME_MINES_MAX + 8]; /* medium-size pool */
(9)    static QSubscrList     l_subscrSto[MAX_PUB_SIG];        /* publish-subscribe */

       /*..........................................................*/
       void main(int argc, char *argv[]) {
                                    /* explicitly invoke the active objects' ctors... */
(10)        Missile_ctor();
(11)        Ship_ctor();
(12)        Tunnel_ctor();

(13)        BSP_init(argc, argv);         /* initialize the Board Support Package */
(14)        QF_init();    /* initialize the framework and the underlying RT kernel */

                                          /* initialize the event pools... */
(15)        QF_poolInit(l_smlPoolSto, sizeof(l_smlPoolSto), sizeof(l_smlPoolSto[0]));
(16)        QF_poolInit(l_medPoolSto, sizeof(l_medPoolSto), sizeof(l_medPoolSto[0]));

(17)        QF_psInit(l_subscrSto, Q_DIM(l_subscrSto)); /* init publish-subscribe */

                                          /* start the active objects... */
(18)        QActive_start(AO_Missile, /* global pointer to the Missile active object */
                    1,                                        /* priority (lowest) */
                    l_missileQueueSto, Q_DIM(l_missileQueueSto), /* evt queue */
                    (void *)0, 0,                         /* no per-thread stack */
                    (QEvent *)0);                       /* no initialization event */
(19)        QActive_start(AO_Ship,     /* global pointer to the Ship active object */
                    2,                                               /* priority */
                    l_shipQueueSto,    Q_DIM(l_shipQueueSto),    /* evt queue */
                    (void *)0, 0,                         /* no per-thread stack */
                    (QEvent *)0);                       /* no initialization event */
(20)        QActive_start(AO_Tunnel, /* global pointer to the Tunnel active object */
                    3,                                               /* priority */
                    l_tunnelQueueSto, Q_DIM(l_tunnelQueueSto), /* evt queue */
                    (void *)0, 0,                         /* no per-thread stack */
                    (QEvent *)0);                       /* no initialization event */

(21)        QF_run();                                   /* run the QF application */
       }
```

(1)   The "Fly 'n' Shoot" game is an example of an application implemented with the
      QP event-driven platform. Every application C-file that uses QP must include the
      qp_port.h header file. This header file contains the specific adaptation of QP to
      the given processor, operating system, and compiler, which is called a *port*. Each
      QP port is located in a separate directory, and the C compiler finds the right
      qp_port.h header file through the include search path provided to the compiler

(typically via the –I compiler option). That way I don't need to change the application source code to recompile it for a different processor or compiler. I only need to instruct the compiler to look in a different QP port directory for the `qp_port.h` header file. For example, the DOS version includes the `qp_port.h` header file from the directory `<qp>\qpc\`**ports**`\80x86\dos \tcpp101\l\`, and the EV-LM3S811 version from the directory `<qp>\qpc \`**ports**`\cortex-m3\vanilla\iar\`.

(2)   The `bsp.h` header file contains the interface to the Board Support Package and is located in the application directory.

(3)   The `game.h` header file contains the declarations of events and other facilities shared among the components of the application. I will discuss this header file in the upcoming Section 1.7. This header file is located in the application directory.

The QP event-driven platform is a collection of components, such as the QEP event processor that executes state machines according to the UML semantics and the QF real-time framework that implements the active object computing model. Active objects in QF are encapsulated state machines (each with an event queue, a separate task context, and a unique priority) that communicate with one another asynchronously by sending and receiving events, whereas QF handles all the details of thread-safe event exchange and queuing. Within an active object, the events are processed by the QEP event processor sequentially in a run-to-completion (RTC) fashion, meaning that processing of one event must necessarily complete before processing the next event. (See also Section 6.3.3 in Chapter 6.)

(4-6)   The application must provide storage for the event queues of all active objects used in the application. Here the storage is provided at compile time through the statically allocated arrays of immutable (`const`) pointers to events, because QF event queues hold just pointers to events, not events themselves. Events are represented as instances of the `QEvent` structure declared in the `qp_port.h` header file. Each event queue of an active object can have a different size, and you need to decide this size based on your knowledge of the application. Event queues are discussed in Chapters 6 and 7.

(7,8)   The application must also provide storage for event pools that the framework uses for fast and deterministic dynamic allocation of events. Each event pool can provide only fixed-size memory blocks. To avoid wasting memory by using oversized blocks for small events, the QF framework can manage up to three event pools of different block sizes (for small, medium, and large events). The "Fly 'n' Shoot" application uses only two out of the three possible event pools (the small and medium pools).

The QF real-time framework supports two event delivery mechanisms: the simple direct event posting to active objects and the more advanced mechanism called *publish-subscribe* that decouples event producers from the consumers. In the publish-subscribe mechanism, active objects subscribe to events by the framework. Event producers publish the events to the framework. Upon each publication request, the framework delivers the event to all active objects that had subscribed to that event type. One obvious implication of publish-subscribe is that the framework must store the subscriber information, whereas it must be possible to handle multiple subscribers to any given event type. The event delivery mechanisms are described in Chapters 6 and 7.

(9)   The "Fly 'n' Shoot" application uses the publish-subscribe event delivery mechanism supported by QF, so it needs to provide the storage for the subscriber lists. The subscriber lists remember which active objects have subscribed to which events. The size of the subscriber database depends on both the number of published events, which is specified in the MAX_PUB_SIG constant found in the game.h header file, and the maximum number of active objects allowed in the system, which is determined by the QF configuration parameter QF_MAX_ACTIVE.

(10-12) These functions perform an early initialization of the active objects in the system. They play the role of static "constructors," which in C you need to invoke explicitly. (C++ calls such static constructors implicitly before entering main()).

(13)   The function BSP_init() initializes the board and is defined in the bsp.c file.

(14)   The function QF_init() initializes the QF component and the underlying RTOS/kernel, if such software is used. You need to call QF_init() before you invoke any QF services.

(15,16) The function QF_poolInit() initializes the event pools. The parameters of this function are the pointer to the event pool storage, the size of this storage, and the block-size of this pool. You can call this function up to three times to initialize up to three event pools. The subsequent calls to QF_poolInit() must be made in the increasing order of block size. For instance, the small block-size pool must be initialized before the medium block-size pool.

(17) The function QF_poolInit() initializes the publish-subscribe event delivery mechanism of QF. The parameters of this function are the pointer to the subscriber-list array and the dimension of this array.

The utility macro Q_DIM(a) provides the dimension of a one-dimensional array a[] computed as sizeof(a)/sizeof(a[0]), which is a compile-time constant. The use of this macro simplifies the code because it allows me to eliminate many #define constants that otherwise I would need to provide for the dimensions of various arrays. I can simply hard-code the dimension right in the definition of an array, which is the only place that I specify it. I then use the macro Q_DIM() whenever I need this dimension in the code.

(18-20) The function QActive_start() tells the QF framework to start managing an active object as part of the application. The function takes the following parameters: the pointer to the active object structure, the priority of the active object, the pointer to its event queue, the dimension (length) of that queue, and three other parameters that I explain in Chapter 7 (they are not relevant at this point). The active object priorities in QF are numbered from 1 to QF_MAX_ACTIVE, inclusive, where a higher-priority number denotes higher urgency of the active object. The constant QF_MAX_ACTIVE is defined in the QF port header file qf_port.h and currently cannot exceed 63.

I like to keep the code and data of every active object strictly encapsulated within its own C-file. For example, all code and data for the active object Ship are encapsulated in the file ship.c, with the external interface consisting of the function Ship_ctor() and the pointer AO_Ship.

(21) At this point, you have provided to the framework all the storage and information it needs to manage your application. The last thing you must do is call the function QF_run() to pass the control to the framework.

After the call to QF_run() the framework is in full control. The framework executes the application by calling your code, not the other way around. The function QF_run() never returns the control back to main(). In the DOS version of the

"Fly 'n' Shoot" game, you can terminate the application by pressing the Esc key, in which case QF_run() exits to DOS but not to main(). In an embedded system, such as the Stellaris board, QF_run() runs forever or till the power is removed, whichever comes first.

> **NOTE**
>
> For best cross-platform portability, the source code consistently uses the *UNIX end-of-line convention* (lines are terminated with LF only, 0xA character). This convention seems to be working for all C/C++ compilers and cross-compilers, including legacy DOS-era tools. In contrast, the DOS/Windows end-of-line convention (lines terminated with the CR,LF, or 0xD,0xA pair of characters) is known to cause problems on UNIX-like platforms, especially in the multiline preprocessor macros.

1.4   The Design of the "Fly 'n' Shoot" Game

To proceed further with the explanation of the "Fly 'n' Shoot" application, I need to step up to the design level. At this point I need to explain how the application has been decomposed into the active objects and how these objects exchange events to collectively deliver the functionality of the "Fly 'n' Shoot" game.

In general, the decomposition of a problem into active objects is not trivial. As usual in any decomposition, your goal is to achieve possibly loose coupling among the active object components (ideally no sharing of any resources), and you also strive for minimizing the communication in terms of the frequency and size of exchanged events.

In the case of the "Fly 'n' Shoot" game, I need to first identify all objects with reactive behavior (i.e., with a state machine). I applied the simplest object-oriented technique of identifying objects, which is to pick the frequently used nouns in the problem specification. From Section 1.2, I identified Ship, Missile, Mines, and Tunnel. However, not every state machine in the system needs to be an active object (with a separate task context, an event queue, and a unique priority level), and merging them is a valid option when performance or space is needed. As an example of this idea, I ended up merging the Mines into the Tunnel active object, whereas I preserved the Mines as independent state machine components of the Tunnel active object. By doing so I applied the "Orthogonal Component" design pattern described in Chapter 5.

The next step in the event-driven application design is assigning responsibilities and resources to the identified active objects. The general design strategy for avoiding sharing of resources is to encapsulate each resource inside a dedicated active object and to let that object manage the resource for the rest of the application. That way, instead of sharing the resource directly, the rest of the application shares the dedicated active object via events.

So, for example, I decided to put the Tunnel active object in charge of the display. Other active objects and state machine components, such as Ship, Missile, and Mines, don't draw on the display directly, but rather send events to the Tunnel object with the request to render the Ship, Missile, or Mine bitmaps at the provided (x, y) coordinates of the display.

With some understanding of the responsibilities and resource allocations to active objects I can move on to devising the various scenarios of event exchanges among the objects. Perhaps the best instrument to aid the thinking process at this stage is the UML *sequence diagram*, such as the diagram depicted in Figure 1.4. This particular sequence diagram shows the most common event exchange scenarios in the "Fly 'n' Shoot" game (the primary use cases, if you will). The explanation section immediately following the diagram illuminates the interesting points.

NOTE

A UML sequence diagram like Figure 1.4 has two dimensions. Horizontally arranged boxes represent the various objects participating in the scenario, whereas heavy borders indicate active objects. As usual in the UML, the object name is underlined. Time flows down the page along the vertical dashed lines descending from the objects. Events are represented as horizontal arrows originating from the sending object and terminating at the receiving object. Optionally, thin rectangles around instance lines indicate focus of control.

NOTE

To explain diagrams, I place numbers in parentheses at the interesting elements of the diagram. I then use these labels in the left margin of the explanation section that immediately follows the diagram. Occasionally, to unambiguously refer to a specific element of a particular diagram from sections of text other than the explanation section, I use the full reference consisting of the figure number followed by the label. For example, Figure 1.4(12) refers to the element (12) in Figure 1.4.

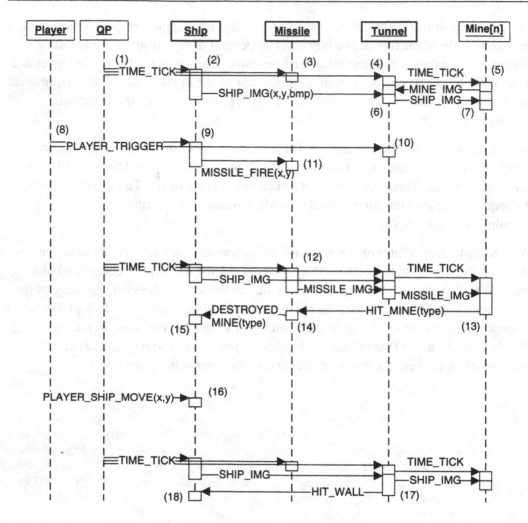

Figure 1.4: The sequence diagram of the "Fly 'n' Shoot" game.

(1) The TIME_TICK is the most important event in the game. This event is generated by the QF framework from the system time tick interrupt at a rate of 30 times per second, which is needed to drive a smooth animation of the display. Because the TIME_TICK event is of interest to virtually all objects in the application, it is published by the framework to all active objects. (The publish-subscribe event delivery in QF is described in Chapter 6.)

(2) Upon reception of the TIME_TICK event, the Ship object advances its position by one step and posts the event SHIP_IMG(x, y, bmp) to the Tunnel object. The

SHIP_IMG event has parameters *x* and *y*, which are the coordinates of the Ship on the display, as well as the bitmap number *bmp* to draw at these coordinates.

(3) The Missile object is not in flight yet, so it simply ignores the TIME_TICK event this time.

(4) The Tunnel object performs the heaviest lifting for the TIME_TICK event. First, Tunnel redraws the entire display from the current frame buffer. This action, performed 30 times per second, provides the illusion of animation of the display. Next, the Tunnel clears the frame buffer and starts filling it up again for the next time frame. The Tunnel advances the tunnel walls by one step and copies the walls to the frame buffer. The Tunnel also dispatches the TIME_TICK event to all its Mine state machine components.

(5) Each Mine advances its position by one step and posts the MINE_IMG(x, y, bmp) event to the Tunnel to render the appropriate Mine bitmap at the position (*x*, *y*) in the current frame buffer. Mines of type 1 send the bitmap number MINE1_BMP, whereas mines of type 2 send MINE2_BMP.

(6) Upon receipt of the SHIP_IMG(x, y, bmp) event from the Ship, the Tunnel object renders the specified bitmap in the frame buffer and checks for any collision between the ship bitmap and the tunnel walls. Tunnel also dispatches the original SHIP_IMG(x, y, bmp) event to all active Mines.

(7) Each Mine determines whether the Ship is in collision with that Mine.

(8) The PLAYER_TRIGGER event is generated when the Player reliably presses the button (button press is debounced). This event is published by the QF framework and is delivered to the Ship and Tunnel objects, which both subscribe to the PLAYER_TRIGGER event.

(9) Ship generates the MISSILE_FIRE(x, y) event to the Missile object. The parameters of this event are the current (*x*, *y*) coordinates of the Ship, which are the starting point for the Missile.

(10) Tunnel receives the published PLAYER_TRIGGER event as well because Tunnel occasionally needs to start the game or terminate the screen saver mode based on this stimulus.

(11) Missile reacts to the MISSILE_FIRE(x, y) event by starting to fly, whereas it sets its initial position from the (*x*, *y*) event parameters delivered from the Ship.

(12)   This time around, the TIME_TICK event arrives while Missile is in flight. Missile posts the MISSILE_IMG(x, y, bmp) event to the Table.

(13)   Table renders the Missile bitmap in the current frame buffer and dispatches the MISSILE_IMG(x, y, bmp) event to all the Mines to let the Mines test for the collision with the Missile. This determination depends on the type of the Mine. In this scenario a particular Mine[n] object detects a hit and posts the HIT_MINE (score) event to the Missile. The Mine provides the score earned for destroying this particular mine as the parameter of this event.

(14)   Missile handles the HIT_MINE(score) event by becoming immediately ready to launch again and lets the Mine do the exploding. Because I decided to make the Ship responsible for the scorekeeping, the Missile also generates the DESTROYED_MINE (score) event to the Ship, to report the score for destroying the Mine.

(15)   Upon reception of the DESTROYED_MINE(score) event, the Ship increments the score by the value received from the Missile.

(16)   The Ship object handles the PLAYER_SHIP_MOVE(x, y) event by updating its position from the event parameters.

(17)   When the Tunnel object handles the SHIP_IMG(x, y, bmp_id) event next time around, it detects a collision between the Ship and the tunnel wall. In that case it posts the event HIT_WALL to the Ship.

(18)   The Ship responds to the HIT_WALL event by transitioning to the "exploding" state.

Even though the sequence diagram in Figure 1.4 shows merely some selected scenarios of the "Fly 'n' Shoot" game, I hope that the explanations give you a big picture of how the application works. More important, you should start getting the general idea about the thinking process that goes into designing an event-driven system with active objects and events.

1.5   Active Objects in the "Fly 'n' Shoot" Game

I hope that the analysis of the sequence diagram in Figure 1.4 makes it clear that actions performed by an active object depend as much on the events it receives as on the internal mode of the object. For example, the Missile active object handles the TIME_TICK event very differently when the Missile is in flight (Figure 1.4(12)) compared to the time when it is not (Figure 1.4(3)).

The best-known mechanism for handling such modal behavior is through state machines because a state machine makes the behavior explicitly dependent on both the event and the state of an object. Chapter 2 introduces UML state machine concepts more thoroughly. In this section, I give a cursory explanation of the state machines associated with each object in the "Fly 'n' Shoot" game.

1.5.1   The Missile Active Object

I start with the Missile state machine shown in Figure 1.5 because it turns out to be the simplest one. The explanation section immediately following the diagram illuminates the interesting points.

NOTE

A UML state diagram like Figure 1.5 preserves the general form of the traditional state transition diagrams, where states are represented as nodes and transitions as arcs connecting the nodes. In the UML notation the state nodes are represented as rectangles with rounded corners. The name of the state appears in bold type in the name compartment at the top of the state. Optionally, right below the name, a state can have an internal transition compartment separated from the name by a horizontal line. The internal transition compartment can contain entry actions (actions following the reserved symbol "entry"), exit actions (actions following the reserved symbol "exit"), and other internal transitions (e.g., those triggered by TIME_TICK in Figure 1.5(3)). State transitions are represented as arrows originating at the boundary of the source state and pointing to the boundary of the target state. At a minimum, a transition must be labeled with the triggering event. Optionally, the trigger can be followed by event parameters, a guard, and a list of actions.

---

(1)   The state transition originating at the black ball is called the *initial transition*. Such transition designates the first active state after the state machine object is created. An initial transition can have associated actions, which in the UML notation are enlisted after the forward slash (/). In this particular case, the Missile state machine starts in the "armed" state and the actions executed upon the initialization consist of subscribing to the event TIME_TICK. Subscribing to an event means that the framework will deliver the specified event to the Missile active object every time the event is published to the framework. Chapter 7 describes the implementation of the publish-subscribe event delivery in QF.

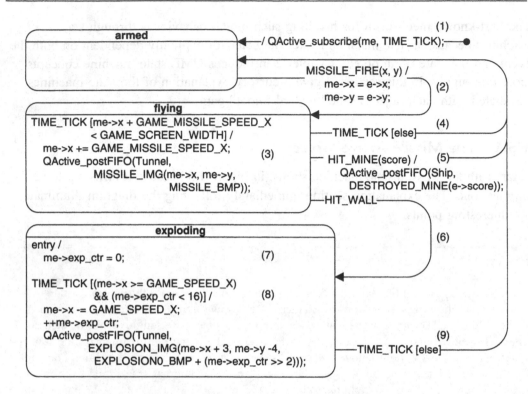

Figure 1.5: Missile state machine diagram.

(2) The arrow labeled with the MISSILE_FIRE(x, y) event denotes a state transition, that is, a change of state from "armed" to "flying." The MISSILE_FIRE(x, y) event is generated by the Ship object when the Player triggers the Missile (see the sequence diagram in Figure 1.4). In the MISSILE_FIRE event, Ship provides Missile with the initial coordinates in the event parameters (x, y).

NOTE

The UML intentionally does not specify the notation for actions. In practice, the actions are often written in the programming language used for coding the particular state machine. In all state diagrams in this book, I assume the C programming language. Furthermore, in the C expressions I refer to the data members associated with the state machine object through the "me->" prefix and to the event parameters through the "e->" prefix. For example, the action "me->x = e->x;" means that the internal data member x of the Missile active object is assigned the value of the event parameter x.

(3) The event name `TIME_TICK` enlisted in the compartment below the state name denotes an *internal transition*. Internal transitions are simple reactions to events performed without a change of state. An internal transition, as well as a regular transition, can have a guard condition, enclosed in square brackets. Guard condition is a Boolean expression evaluated at runtime. If the guard evaluates to TRUE, the transition is taken. Otherwise, the transition is not taken and no actions enlisted after the forward slash (/) are executed. In this particular case, the guard condition checks whether the *x*-coordinate propagated by the Missile speed is still visible on the screen. If so, the actions are executed. These actions include propagation of the Missile position by one step and posting the `MISSILE_IMG` event with the current Missile position and the `MISSILE_BMP` bitmap number to the Tunnel active object. Direct event posting to an active object is accomplished by the QF function `QActive_postFIFO()`, which I discuss in Chapter 7.

(4) The same event `TIME_TICK` with the `[else]` guard denotes a regular state transition with the guard condition complementary to the other occurrence of the `TIME_TICK` event in the same state. In this case, the `TIME_TICK` transition to "armed" is taken if the Missile object flies out of the screen.

(5) The event `HIT_MINE(score)` triggers another transition to the "armed" state. The action associated with this transition posts the `DESTROYED_MINE` event with the parameter `e->score` to the Ship object, to report destroying the mine.

(6) The event `HIT_WALL` triggers a transition to the "exploding" state, with the purpose of animating the explosion bitmaps on the display.

(7) The label "entry" denotes the *entry action* to be executed unconditionally upon the entry to the "exploding" state. This action consists of clearing the explosion counter (`me->exp_ctr`) member of the Missile object.

(8) The `TIME_TICK` internal transition is guarded by the condition that the explosion does not scroll off the screen and that the explosion counter is lower than 16. The actions executed include propagation of the explosion position and posting the `EXPLOSION_IMG` event to the Tunnel active object. Please note that the bitmap of the explosion changes as the explosion counter gets bigger.

(9) The `TIME_TICK` regular transition with the complementary guard changes the state back to the "armed" state. This transition is taken after the animation of the explosion completes.

1.5.2 The Ship Active Object

The state machine of the Ship active object is shown in Figure 1.6. This state machine introduces the profound concept of *hierarchical state nesting*. The power of state nesting derives from the fact that it is designed to eliminate repetitions that otherwise would have to occur.

One of the main responsibilities of the Ship active object is to maintain the current position of the Ship. On the original EV-LM3S811 board, this position is determined by the potentiometer wheel (see Figure 1.2). The PLAYER_SHIP_MOVE(x, y) event is generated whenever the wheel position changes, as shown in the sequence diagram (Figure 1.4). The Ship object must always keep track of the wheel position, which means that all states of the Ship state machine must handle the PLAYER_SHIP_MOVE(x, y) event.

In the traditional finite state machine (FSM) formalism, you would need to repeat the Ship position update from the PLAYER_SHIP_MOVE(x, y) event in every state. But such repetitions would bloat the state machine and, more important, would represent multiple points of maintenance both in the diagram and the code. Such repetitions go against the DRY (Don't Repeat Yourself) principle, which is vital for flexible and maintainable code [Hunt+ 00].

Hierarchical state nesting remedies the problem. Consider the state "active" that surrounds all other states in Figure 1.6. The high-level "active" state is called the *superstate* and is abstract in that the state machine cannot be in this state directly but only in one of the states nested within, which are called the *substates* of "active." The UML semantics associated with state nesting prescribe that any event is first handled in the context of the currently active substate. If the substate cannot handle the event, the state machine attempts to handle the event in the context of the next-level superstate. Of course, state nesting in UML is not limited to just one level and the simple rule of processing events applies recursively to any level of nesting.

Specifically to the Ship state machine diagram shown in Figure 1.6, suppose that the event PLAYER_SHIP_MOVE(x, y) arrives when the state machine is in the "parked" state. The "parked" state does not handle the PLAYER_SHIP_MOVE(x, y) event. In the traditional finite state machine this would be the end of the story—the PLAYER_SHIP_MOVE(x, y) event would be silently discarded. However, the state machine in Figure 1.6 has another layer of the "active" superstate. Per the semantics of state nesting, this higher-level superstate handles the PLAYER_SHIP_MOVE(x, y) event, which is exactly what's needed. The same exact argumentation applies for any other substate of the "active" superstate, such as "flying" or "exploding," because none of these substates handle the PLAYER_SHIP_MOVE(x, y) event. Instead, the "active" superstate handles the event in one single place, without repetitions.

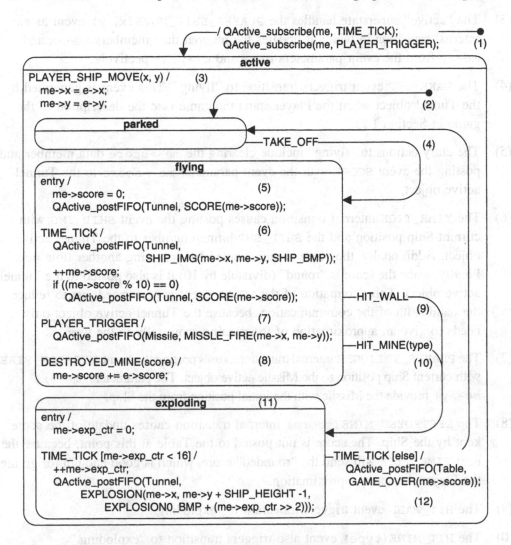

Figure 1.6: Ship state machine diagram.

(1) Upon the initial transition, the Ship state machine enters the "active" superstate and subscribes to events TIME_TICK and PLAYER_TRIGGER.

(2) At each level of nesting, a superstate can have a private initial transition that designates the active substate after the superstate is entered directly. Here the

initial transition of state "active" designates the substate "parked" as the initial active substate.

(3)   The "active" superstate handles the PLAYER_SHIP_MOVE(x, y) event as an internal transition in which it updates the internal data members me->x and me->y from the event parameters e->x and e->y, respectively.

(4)   The TAKE_OFF event triggers transition to "flying." This event is generated by the Tunnel object when the Player starts the game (see the description of the game in Section 1.2).

(5)   The entry actions to "flying" include clearing the me->score data member and posting the event SCORE with the event parameter me->score to the Tunnel active object.

(6)   The TIME_TICK internal transition causes posting the event SHIP_IMG with current Ship position and the SHIP_BMP bitmap number to the Tunnel active object. Additionally, the score is incremented for surviving another time tick. Finally, when the score is "round" (divisible by 10) it is also posted to the Tunnel active object. This decimation of the SCORE event is performed just to reduce the bandwidth of the communication, because the Tunnel active object only needs to give an approximation of the running score tally to the user.

(7)   The PLAYER_TRIIGGER internal transition causes posting the event MISSILE_FIRE with current Ship position to the Missile active object. The parameters (me->x, me->y) provide the Missile with the initial position from the Ship.

(8)   The DESTROYED_MINE(score) internal transition causes update of the score kept by the Ship. The score is not posted to the Table at this point, because the next TIME_TICK will send the "rounded" score, which is good enough for giving the Player the score approximation.

(9)   The HIT_WALL event triggers transition to "exploding."

(10)   The HIT_MINE(type) event also triggers transition to "exploding."

(11)   The "exploding" state of the Ship state machine is very similar to the "exploding" state of Missile (see Figure 1.5(7-9)).

(12)   The TIME_TICK[else] transition is taken when the Ship finishes exploding. Upon this transition, the Ship object posts the event GAME_OVER(me->score) to the Tunnel active object to terminate the game and display the final score to the Player.

1.5.3   The Tunnel Active Object

The Tunnel active object has the most complex state machine, which is shown in
Figure 1.7. Unlike the previous state diagrams, the diagram in Figure 1.7 shows only the
high level of abstraction and omits a lot of details such as most entry/exit actions,
internal transitions, guard conditions, or actions on transitions. Such a "zoomed out"
view is always legal in the UML because UML allows you to choose the level of detail
that you want to include in your diagram.

The Tunnel state machine uses state hierarchy more extensively than the Ship state
machine in Figure 1.6. The explanation section immediately following Figure 1.7
illuminates the new uses of state nesting as well as the new elements not explained yet
in the other state diagrams.

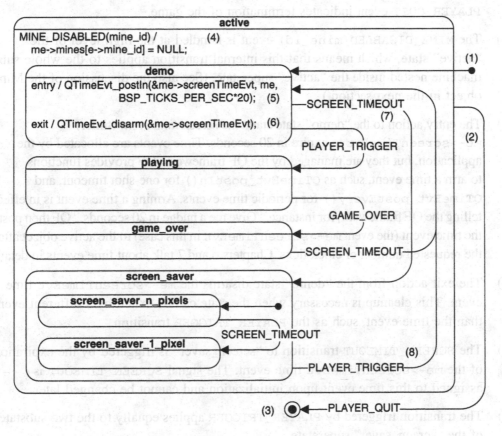

Figure 1.7: Tunnel state machine diagram.

(1)   An initial transition can target a substate at any level of state hierarchy, not necessarily just the next-lower level. Here the topmost initial transition goes down two levels to the substate "demo."

(2)   The superstate "active" handles the PLAYER_QUIT event as a transition to the final state (see explanation of element (3)). Please note that the PLAYER_QUIT transition applies to all substates directly or transitively nested in the "active" superstate. Because a state transition always involves execution of all exit actions from the states, the high-level PLAYER_QUIT transition guarantees the proper cleanup that is specific to the current state context, whichever substate happens to be active at the time when the PLAYER_QUIT event arrives.

(3)   The final state is indicated in the UML notation as the bull's-eye symbol and typically indicates destruction of the state machine object. In this case, the PLAYER_QUIT event indicates termination of the game.

(4)   The MINE_DISABLED(mine_id) event is handled at the high level of the "active" state, which means that this internal transition applies to the whole sub-machine nested inside the "active" superstate. (See also the discussion of the Mine object in the next section.)

(5)   The entry action to the "demo" state starts the screen *time event* (timer) me->screenTimeEvt to expire in 20 seconds. Time events are allocated by the application, but they are managed by the QF framework. QF provides functions to arm a time event, such as QTimeEvt_postIn() for one-shot timeout, and QTimeEvt_postEvery() for periodic time events. Arming a time event is in effect telling the QF framework, for instance, "Give me a nudge in 20 seconds." QF then posts the time event (the event me->screenTimeEvt in this case) to the active object after the requested number of clock ticks. Chapters 6 and 7 talk about time events in detail.

(6)   The exit action from the "demo" state disarms the me->screenTimeEvt time event. This cleanup is necessary when the state can be exited by a different event than the time event, such as the PLAYER_TRIGGER transition.

(7)   The SCREEN_TIMEOUT transition to "screen_saver" is triggered by the expiration of the me->screenTimeEvt time event. The signal SCREEN_TIMEOUT is assigned to this time event upon initialization and cannot be changed later.

(8)   The transition triggered by PLAYER_TRIGGER applies equally to the two substates of the "screen_saver" superstate.

1.5.4   The Mine Components

Mines are also modeled as hierarchical state machines, but are not active objects. Instead, Mines are components of the Tunnel active object and share its event queue and priority level. The Tunnel active object communicates with the Mine components *synchronously* by directly dispatching events to them via the function QHsm_dispatch(). Mines communicate with Tunnel and all other active objects *asynchronously* by posting events to their event queues via the function QActive_postFIFO().

NOTE

Active objects exchange events asynchronously, meaning that the sender of the event merely posts the event to the event queue of the recipient active object without waiting for the completion of the event processing. In contrast, synchronous event processing corresponds to a function call (e.g., QHsm_dispatch()), which processes the event in the caller's thread of execution.

As shown in Figure 1.8, Tunnel maintains the data member mines[], which is an array of pointers to hierarchical state machines (QHsm *). Each of these pointers can point either to a Mine1 object, a Mine2 object, or NULL, if the entry is unused. Note that Tunnel "knows" the Mines only as generic state machines (pointers to the QHsm structure defined in QP). Tunnel dispatches events to Mines uniformly, without differentiating between different types of Mines. Still, each Mine state machine handles the events in its specific way. For example, Mine type 2 checks for collision with the Missile differently than with the Ship, whereas Mine type 1 handles both identically.

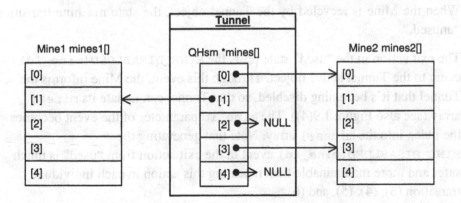

Figure 1.8: The Table active object manages two types of Mines.

> **NOTE**
>
> The last point is actually very interesting. Dispatching the same event to different Mine objects results in different behavior, specific to the type of the Mine, which in OOP is known as *polymorphism*. I'll have more to say about this in Chapter 3.

Each Mine object is fairly autonomous. The Mine maintains its own position and is responsible for informing the Tunnel object whenever the Mine gets destroyed or scrolls out of the display. This information is vital for the Tunnel object so that it can keep track of the unused Mines.

Figure 1.9 shows a hierarchical state machine of Mine2 state machine. Mine1 is very similar, except that it uses the same bitmap for testing collisions with the Missile and the Ship.

(1) The Mine starts in the "unused" state.

(2) The Tunnel object plants a Mine by dispatching the MINE_PLANT(x, y) event to the Mine. The Tunnel provides the (x, y) coordinates as the original position of the Mine.

(3) When the Mine scrolls off the display, the state machine transitions to "unused."

(4) When the Mine hits the Ship, the state machine transitions to "unused."

(5) When the Mine finishes exploding, the state machine transitions to "unused."

(6) When the Mine is recycled by the Tunnel object, the state machine transitions to "unused."

(7) The exit action in the "used" state posts the MINE_DISABLDED(mine_id) event to the Tunnel active object. Through this event, the Mine informs the Tunnel that it's becoming disabled, so that Tunnel can update its mines[] array (see also Figure 1.9(4)). The mine_id parameter of the event becomes the index into the mines[] array. Note that generating the MINE_DISABLDED(mine_id) event in the exit action from "used" is much safer and more maintainable than repeating this action in each individual transition (3), (4), (5), and (6).

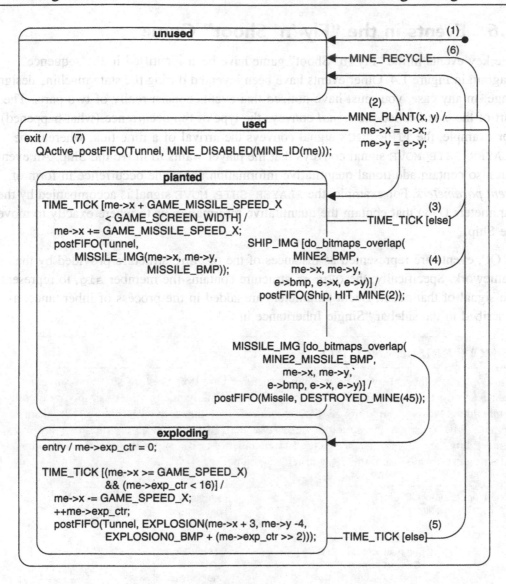

Figure 1.9: Mine2 state machine diagram.

1.6   Events in the "Fly 'n' Shoot" Game

The key events in the "Fly 'n' Shoot" game have been identified in the sequence diagram in Figure 1.4. Other events have been invented during the state machine design stage. In any case, you must have noticed that events consist really of two parts. The part of the event called the *signal* conveys the type of the occurrence (what happened). For example, the TIME_TICK signal conveys the arrival of a time tick, whereas the PLAYER_SHIP_MOVE signal conveys that the player wants to move the Ship. An event can also contain additional quantitative information about the occurrence in form of *event parameters*. For example, the PLAYER_SHIP_MOVE signal is accompanied by the parameters (x, y) that contain the quantitative information as to where exactly to move the Ship.

In QP, events are represented as instances of the QEvent structure provided by the framework. Specifically, the QEvent structure contains the member sig, to represent the signal of that event. Event parameters are added in the process of inheritance, as described in the sidebar "Single Inheritance in C."

SINGLE INHERITANCE IN C

Inheritance is the ability to derive new structures based on existing structures in order to reuse and organize code. You can implement single inheritance in C very simply by literally embedding the base structure as the first member of the derived structure. For example, Figure 1.10(A) shows the structure ScoreEvt derived from the base structure QEvent by embedding the QEvent instance as the first member of ScoreEvt. To make this idiom better stand out, I always name the base structure member super.

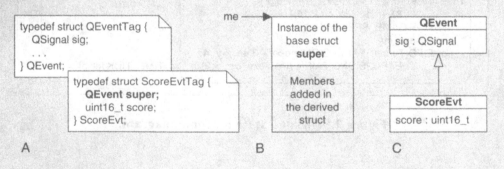

A   B   C

Figure 1.10: (A) Derivation of structures in C, (B) memory alignment, and (C) the UML class diagram.

As shown in Figure 1.10(B), such nesting of structures always aligns the data member super at the beginning of every instance of the derived structure, which is actually guaranteed by the C standard. Specifically, WG14/N1124 Section 6.7.2.1.13 says: "... A pointer to a structure object, suitably converted, points to its initial member. There may be unnamed padding within a structure object, but not at its beginning" [ISO/IEC 9899:TC2]. The alignment lets you treat a pointer to the derived ScoreEvt structure as a pointer to the QEvent base structure. All this is legal, portable, and guaranteed by the C standard. Consequently, you can always safely pass a pointer to ScoreEvt to any C function that expects a pointer to QEvent. (To be strictly correct in C, you should explicitly cast this pointer. In OOP such casting is called *upcasting* and is always safe.) Therefore, all functions designed for the QEvent structure are automatically available to the ScoreEvt structure as well as other structures derived from QEvent. Figure 1.10(C) shows the UML class diagram depicting the inheritance relationship between ScoreEvt and QEvent structures.

QP uses single inheritance quite extensively not just for derivation of events with parameters, but also for derivation of state machines and active objects. Of course, the C++ version of QP uses the native C++ support for class inheritance rather than "derivation of structures." You'll see more examples of inheritance later in this chapter and throughout the book.

Because events are explicitly shared among most of the application components, it is convenient to declare them in the separate header file game.h shown in Listing 1.2. The explanation section immediately following the listing illuminates the interesting points.

Listing 1.2 Signals, event structures, and active object interfaces defined in file game.h

```
(1)   enum GameSignals {                              /* signals used in the game */
(2)       TIME_TICK_SIG = Q_USER_SIG,                /* published from tick ISR */
          PLAYER_TRIGGER_SIG, /* published by Player (ISR) to trigger the Missile */
          PLAYER_QUIT_SIG,          /* published by Player (ISR) to quit the game */
          GAME_OVER_SIG,        /* published by Ship when it finishes exploding */
          /* insert other published signals here ... */
(3)       MAX_PUB_SIG,                               /* the last published signal */

          PLAYER_SHIP_MOVE_SIG,  /* posted by Player (ISR) to the Ship to move it */
          BLINK_TIMEOUT_SIG,          /* signal for Tunnel's blink timeout event */
          SCREEN_TIMEOUT_SIG,        /* signal for Tunnel's screen timeout event */
          TAKE_OFF_SIG,     /* from Tunnel to Ship to grant permission to take off */
          HIT_WALL_SIG,            /* from Tunnel to Ship when Ship hits the wall */
          HIT_MINE_SIG,      /* from Mine to Ship or Missile when it hits the mine */
          SHIP_IMG_SIG,        /* from Ship to the Tunnel to draw and check for hits */
          MISSILE_IMG_SIG,   /* from Missile the Tunnel to draw and check for hits */
          MINE_IMG_SIG,               /* sent by Mine to the Tunnel to draw the mine */
```

Continued onto next page

```
            MISSILE_FIRE_SIG,                    /* sent by Ship to the Missile to fire */
            DESTROYED_MINE_SIG,   /* from Missile to Ship when Missile destroyed Mine */
            EXPLOSION_SIG,     /* from any exploding object to render the explosion */
            MINE_PLANT_SIG,                   /* from Tunnel to the Mine to plant it */
            MINE_DISABLED_SIG,      /* from Mine to Tunnel when it becomes disabled */
            MINE_RECYCLE_SIG,          /* sent by Tunnel to Mine to recycle the mine */
            SCORE_SIG,     /* from Ship to Tunnel to adjust game level based on score */
            /* insert other signals here ... */
(4)         MAX_SIG                          /* the last signal (keep always last) */
        };


(5)     typedef struct ObjectPosEvtTag {
(6)         QEvent super;                              /* extend the QEvent class */
(7)         uint8_t x;                              /* the x-position of the object */
(8)         uint8_t y;                              /* new y-position of the object */
        } ObjectPosEvt;


        typedef struct ObjectImageEvtTag {
            QEvent super;                              /* extend the QEvent class */
            uint8_t x;                              /* the x-position of the object */
            int8_t  y;                              /* the y-position of the object */
            uint8_t bmp;                    /* the bitmap ID representing the object */
        } ObjectImageEvt;


        typedef struct MineEvtTag {
            QEvent super;                              /* extend the QEvent class */
            uint8_t id;                                     /* the ID of the Mine */
        } MineEvt;


        typedef struct ScoreEvtTag {
            QEvent super;                              /* extend the QEvent class */
            uint16_t score;                               /* the current score */
        } ScoreEvt;


        /* opaque pointers to active objects in the application */
(9)     extern QActive * const AO_Tunnel;
(10)    extern QActive * const AO_Ship;
(11)    extern QActive * const AO_Missile;


        /* active objects' "constructors" */
(12)    void Tunnel_ctor(void);
(13)    void Ship_ctor(void);
(14)    void Missile_ctor(void);
```

(1) In QP, signals of events are simply enumerated constants. Placing all signals in a single enumeration is particularly convenient to avoid inadvertent overlap in the numerical values of different signals.

(2)   The application-level signals do not start from zero but rather are offset by the constant Q_USER_SIG. This is because QP reserves the lowest few signals for the internal use and provides the constant Q_USER_SIG as an offset from which user-level signals can start. Also note that by convention, I attach the suffix _SIG to all signals so that I can easily distinguish signals from other constants. I drop the suffix _SIG in the state diagrams to reduce the clutter.

(3)   The constant MAX_PUB_SIG delimits the published signals from the rest. The publish-subscribe event delivery mechanism consumes some RAM, which is proportional to the number of published signals. I save some RAM by providing the lower limit of published signals to QP (MAX_PUB_SIG) rather than the maximum of all signals used in the application. (See also Listing 1.1(9)).

(4)   The last enumeration MAX_SIG indicates the maximum of all signals used in the application.

(5)   The event structure ObjectPosEvt defines a "class" of events that convey the object's position on the display in the event parameters.

(6)   The structure ObjectPosEvt derives from the base structure QEvent, as explained in the sidebar "Single Inheritance in C."

(7,8) The structure ObjectPosEvt adds parameters *x* and *y*, which are coordinates of the object on the display.

NOTE

Throughout this book I use the following standard exact-width integer types (WG14/N843 C99 Standard, Section 7.18.1.1) [ISO/IEC 9899:TC2]:

Exact Size	Unsigned	Signed
8-bits	uint8_t	int8_t
16-bits	uint16_t	int16_t
32-bits	uint32_t	int32_t

If your (pre-standard) compiler does not provide the <stdint.h> header file, you can always typedef the exact-width integer types using the standard C data types such as signed/unsigned char, short, int, and long.

(9-11)   These global pointers represent active objects in the application and are used for posting events directly to active objects. Because the pointers can be initialized at compile time, I like to declare them const, so that they can be placed in ROM. The active object pointers are "opaque" because they cannot access the whole active object, only the part inherited from the QActive structure. I'll have more to say about this in the next section.

(12-14)   These functions perform an early initialization of the active objects in the system. They play the role of static "constructors," which in C you need to call explicitly, typically at the beginning of main(). (See also Listing 1.1 (10-12).)

1.6.1   Generating, Posting, and Publishing Events

The QF framework supports two types of asynchronous event exchange:

1.   The simple mechanism of direct event posting supported through the functions QActive_postFIFO() and QActive_postLIFO(), where the producer of an event directly posts the event to the event queue of the consumer active object.

2.   A more sophisticated publish-subscribe event delivery mechanism supported through the functions QF_publish() and QActive_subscribe(), where the producers of the events "publish" them to the framework, and the framework then delivers the events to all active objects that had "subscribed" to these events.

In QF, any part of the system, not necessarily only the active objects, can produce events. For example, interrupt service routines (ISRs) or device drivers can also produce events. On the other hand, only active objects can consume events, because only active objects have event queues.

NOTE

QF also provides "raw" thread-safe event queues (struct QEQueue), which can consume events as well. These "raw" thread-safe queues cannot block and are intended to deliver events to ISRs or device drivers. Refer to Chapter 7 for more details.

The most important characteristic of event management in QF is that the framework passes around only pointers to events, not the events themselves. QF never copies the events by value ("zero-copy" policy); even in case of publishing events that often involves multicasting the same event to multiple subscribers. The actual event instances are either constant events statically allocated at compile time or dynamic events allocated at runtime from one of the event pools that the framework manages. Listing 1.3 provides examples of publishing static events and posting dynamic events from the ISRs of the "Fly 'n' Shoot" version for the Stellaris board (file `<qp>\qpc\examples \cortex-m3\vanilla\iar\game-ev-1m3s811\bsp.c`). In Section 1.7.3 you will see other examples of event posting from active objects in the state machine code.

Listing 1.3 Generating, posting, and publishing events from the ISRs in `bsp.c` for the Stellaris board

```
(1)   void ISR_SysTick(void) {
(2)       static QEvent const tickEvt = { TIME_TICK_SIG, 0 };
(3)       QF_publish(&tickEvt);      /* publish the tick event to all subscribers */
(4)       QF_tick();                           /* process all armed time events */
      }
      /*.......................................................................*/
(5)   void ISR_ADC(void) {
          static uint32_t adcLPS = 0;               /* Low-Pass-Filtered ADC reading */
          static uint32_t wheel = 0;                    /* the last wheel position */
          unsigned long tmp;

          ADCIntClear(ADC_BASE, 3);                     /* clear the ADC interrupt */
(6)       ADCSequenceDataGet(ADC_BASE, 3, &tmp);   /* read the data from the ADC */

          /* 1st order low-pass filter: time constant ~= 2^n samples
           * TF = (1/2^n)/(z-((2^n - 1)/2^n)),
           * e.g., n=3, y(k+1) = y(k) - y(k)/8 + x(k)/8 => y += (x - y)/8
           */
(7)       adcLPS += (((int)tmp - (int)adcLPS + 4) >> 3);       /* Low-Pass-Filter */

          /* compute the next position of the wheel */
(8)       tmp = (((1 << 10) - adcLPS)*(BSP_SCREEN_HEIGHT - 2)) >> 10;

          if (tmp != wheel) {                     /* did the wheel position change? */
(9)           ObjectPosEvt *ope = Q_NEW(ObjectPosEvt, PLAYER_SHIP_MOVE_SIG);
(10)          ope->x = (uint8_t)GAME_SHIP_X;             /* x-position is fixed */
(11)          ope->y = (uint8_t)tmp;
(12)          QActive_postFIFO(AO_ship, (QEvent *)ope);   /* post to the Ship AO */
              wheel = tmp;                      /* save the last position of the wheel */
          }
          . . .
      }
```

(1)   In the case of the Stellaris board, the function `ISR_SysTick()` services the system clock tick ISR generated by the Cortex-M3 system tick timer.

(2)   The `TIME_TICK` event never changes, so it can be statically allocated just once. This event is declared as `const`, which means that it can be placed in ROM. The initializer list for this event consists of the signal `TIME_TICK_SIG` followed by zero. This zero informs the QF framework that this event is static and should never be recycled to an event pool.

(3)   The ISR calls the framework function `QF_publish()`, which takes the pointer to the `tickEvt` event to deliver to all subscribers.

(4)   The ISR calls the function `QF_tick()`, in which the framework manages the armed time events.

(5)   The function `ISR_ADC()` services the ADC conversions, which ultimately deliver the position of the Ship.

(6)   The ISR reads the data from the ADC.

(7,8)   A low-pass filter is applied to the raw ADC reading and the potentiometer wheel position is computed.

(9)   The QF macro `Q_NEW(ObjectPosEvt, PLAYER_SHIP_MOVE_SIG)` dynamically allocates an instance of the `ObjectPosEvt` event from an event pool managed by QF. The macro also performs the association between the signal `PLAYER_SHIP_MOVE_SIG` and the allocated event. The `Q_NEW()` macro returns the pointer to the allocated event.

NOTE

The `PLAYER_SHIP_MOVE(x, y)` event is an example of an event with changing parameters. In general, such an event cannot be allocated statically (like the `TIME_TICK` event at label (2)) because it can change asynchronously next time the ISR executes. Some active objects in the system might still be referring to the event via a pointer, so the event should not be changing. Dynamic event allocation of QF solves all such concurrency issues because every time a new event is allocated. QF then recycles the dynamic events after it determines that all active objects are done with accessing the events.

(10,11)   The x and y parameters of the event are assigned.

(12)   The dynamic event is posted directly to the Ship active object.

1.7   Coding Hierarchical State Machines

Contrary to widespread misconceptions, you don't need big design automation tools to translate hierarchical state machines (UML statecharts) into efficient and highly maintainable C or C++. This section explains how to hand-code the Ship state machine from Figure 1.6 with the help of the QF real-time framework and the QEP hierarchical processor, which is also part of the QP event-driven platform. Once you know how to code this state machine, you know how to code them all.

The source code for the Ship state machine is found in the file `ship.c` located either in the DOS version or the Stellaris version of the "Fly 'n' Shoot" game. I break the explanation of this file into three steps.

1.7.1   Step 1: Defining the Ship Structure

In the first step you define the `Ship` data structure. Just as in the case of events, you use inheritance to derive the `Ship` structure from the framework structure `QActive` (see the sidebar "Single Inheritance in C"). Creating this inheritance relationship ties the `Ship` structure to the QF framework.

The main responsibility of the `QActive` base structure is to store the information about the current active state of the state machine as well as the event queue and priority level of the Ship active object. In fact, `QActive` itself derives from a simpler QEP structure `QHsm` that represents just the current active state of a hierarchical state machine. On top of that information, almost every state machine must also store other "extended-state" information. For example, the Ship object is responsible for maintaining the Ship position as well as the score accumulated in the game. You supply this additional information by means of data members enlisted after the base structure member `super`, as shown in Listing 1.4.

Listing 1.4 Deriving the Ship structure in file `ship.c`

```
(1)  #include "qp_port.h"                                    /* the QP port */
(2)  #include "bsp.h"                               /* Board Support Package */
(3)  #include "game.h"                                  /* this application */
     /* local objects ------------------------------------------------------*/
(4)  typedef struct ShipTag {
(5)      QActive super;                         /* derive from the QActive struct */
(6)      uint8_t x;              /* x-coordinate of the Ship position on the display */
(7)      uint8_t y;              /* y-coordinate of the Ship position on the display */
(8)      uint8_t exp_ctr;       /* explosion counter, used to animate explosions */
```

Continued onto next page

```
 (9)      uint16_t score;                               /* running score of the game */
(10)   } Ship;                          /* the typedef-ed name for the Ship struct */

                                               /* state handler functions... */
(11)   static QState Ship_active    (Ship *me, QEvent const *e);
(12)   static QState Ship_parked    (Ship *me, QEvent const *e);
(13)   static QState Ship_flying    (Ship *me, QEvent const *e);
(14)   static QState Ship_exploding(Ship *me, QEvent const *e);

(15)   static QState Ship_initial   (Ship *me, QEvent const *e);

(16)   static Ship l_ship;              /* the sole instance of the Ship active object */

       /* global objects ------------------------------------------------------------*/
(17)   QActive * const AO_ship = (QActive *)&l_ship;   /* opaque pointer to Ship AO */
```

(1) Every application-level C file that uses the QP platform must include the
     qp_port.h header file.

(2) The bsp.h header file contains the interface to the Board Support Package.

(3) The game.h header file contains the declarations of events and other facilities
     shared among the components of the application (see Listing 1.2).

(4) This structure defines the Ship active object.

NOTE

I like to keep active objects, and indeed all state machine objects (such as Mines), strictly
encapsulated. Therefore, I don't put the state machine structure definitions in header files; rather,
I define them right in the implementation file, such as ship.c. That way I can be sure that the
internal data members of the Ship structure are not known to any other parts of the application.

(5) The Ship active object structure derives from the framework structure
     QActive, as described in the sidebar "Single Inheritance in C."

(6,7) The x and y data members represent the position of the Ship on the display.

(8) The exp_ctr member is used for pacing the explosion animation (see also the
     "exploding" state in the Ship state diagram in Figure 1.6).

(9) The score member stores the accumulated score in the game.

(10)   I use the `typedef` to define the shorter name `Ship` equivalent to `struct ShipTag`.

(11-14) These four functions are called *state-handler functions* because they correspond one to one to the states of the Ship state machine shown in Figure 1.6. For example, the `Ship_active()` function represents the "active" state. The QEP event processor calls the state-handler functions to realize the UML semantics of state machine execution. All state-handler functions have the same signature. A state-handler function takes the state machine pointer and the event pointer as arguments and returns the status of the operation back to the event processor—for example whether the event was handled or not. The return type `QState` of state-handler functions is `typedef`-ed to `uint8_t` as `QState` in the header file `<qp>\qpc\include\qep.h`.

NOTE

I use a simple naming convention to strengthen the association between the structures and the functions designed to operate on these structures. First, I name the functions by combining the `typedef`'ed structure name with the name of the operation (e.g., `Ship_active`). Second, I always place the pointer to the structure as the first argument of the associated function, and I always name this argument "me" (e.g., `Ship_active(Ship *me, ...)`).

---

(15)   In addition to state-handler functions, every state machine must declare the initial pseudostate, which QEP invokes to execute the topmost initial transition (see Figure 1.6(1)). The initial pseudostate handler has a signature identical to the regular state-handler function.

(16)   In this line I statically allocate the storage for the `Ship` active object. Note that the object `l_ship` is defined as `static` so that it is accessible only locally at the file scope of the `ship.c` file.

(17)   In this line I define and initialize the global pointer `AO_Ship` to the Ship active object (see also Listing 1.2(10)). This pointer is "opaque" because it treats the Ship object as the generic `QActive` base structure rather than the specific `Ship` structure. The power of an "opaque" pointer is that it allows me to completely hide the definition of the `Ship` structure and make it inaccessible to the rest of the application. Still, the other application components can access the Ship object to post events directly to it via the `QActive_postFIFO(QActive *me, QEvent const *e)` function.

1.7.2 Step 2: Initializing the State Machine

The state machine initialization is divided into the following two steps for increased flexibility and better control of the initialization timeline:

1. The state machine "constructor"; and

2. The top-most initial transition.

The state machine "constructor," such as `Ship_ctor()`, intentionally does not execute the topmost initial transition defined in the initial pseudostate because at that time some vital objects can be missing and critical hardware might not be properly initialized yet.[3] Instead, the state machine "constructor" merely puts the state machine in the initial pseudostate. Later, the user code must trigger the topmost initial transition explicitly, which happens actually inside the function `QActive_start()` (see Listing 1.1(18-20)). Listing 1.5 shows the instantiation (the "constructor" function) and initialization (the initial pseudostate) of the Ship active object.

Listing 1.5 Instantiation and initialization of the Ship active object in `ship.c`

```
(1)  void Ship_ctor(void) {                                  /* instantiation */
(2)      Ship *me = &l_ship;
(3)      QActive_ctor(&me->super, (QStateHandler)&Ship_initial);
(4)      me->x = GAME_SHIP_X;
(5)      me->y = GAME_SHIP_Y;
     }
     /*..................................................................*/
(6)  QState Ship_initial(Ship *me, QEvent const *e) {      /* initialization */
(7)      QActive_subscribe((QActive *)me, TIME_TICK_SIG);
(8)      QActive_subscribe((QActive *)me, PLAYER_TRIGGER_SIG);

(9)      return Q_TRAN(&Ship_active);         /* top-most initial transition */
     }
```

---

(1) The global function `Ship_ctor()` is prototyped in `game.h` and called at the beginning of `main()`.

(2) The "me" pointer points to the statically allocated Ship object (see Listing 1.4(16)).

---

[3] In C++, the static constructors run even before `main()`.

(3) Every derived structure is responsible for initializing the part inherited from the base structure. The "constructor" `QActive_ctor()` puts the state machine in the initial pseudostate `&Ship_initial` (see Listing 1.4(15)).

(4,5) The Ship position is initialized.

(6) The `Ship_initial()` function defines the topmost initial transition in the Ship state machine (see Figure 1.6(1)).

(7,8) The Ship active object subscribes to signals `TIME_TICK_SIG` and `PLAYER_TRIGGER_SIG`, as specified in the state diagram in Figure 1.6(1).

(9) The initial state "active" is specified by returning the QP macro `Q_TRAN()`.

NOTE

The macro `Q_TRAN()` must always follow the `return` statement.

1.7.3   Step 3: Defining State-Handler Functions

In the last step, you actually code the Ship state machine by implementing one state at a time as a state-handler function. To determine what elements belong to any given state-handler function, you follow around the state's boundary in the diagram (Figure 1.6). You need to implement all transitions originating at the boundary, any entry and exit actions defined in the state, and all internal transitions enlisted directly in the state. Additionally, if there is an initial transition embedded directly in the state, you need to implement it as well.

Take for example the state "flying" shown in Figure 1.6. This state has an entry action and two transitions originating at its boundary: `HIT_WALL` and `HIT_MINE(type)` as well as three internal transitions `TIME_TICK`, `PLAYER_TRIGGER`, and `DESTROYED_MINE(score)`. The "flying" state nests inside the "active" superstate.

Listing 1.6 shows two state-handler functions of the Ship state machine from Figure 1.6. The state-handler functions correspond to the states "active" and "flying," respectively. The explanation section immediately following the listing highlights the important implementation techniques.

Listing 1.6 State-handler functions for states "active" and "flying" in `ship.c`

```
(1)   QState Ship_active(Ship *me, QEvent const *e) {
(2)       switch (e->sig) {
(3)           case Q_INIT_SIG: {                          /* nested initial transition */
(4)               /* any actions associated with the initial transition */
(5)               return Q_TRAN(&Ship_parked);
              }
(6)           case PLAYER_SHIP_MOVE_SIG: {
(7)               me->x = ((ObjectPosEvt const *)e)->x;
(8)               me->y = ((ObjectPosEvt const *)e)->y;
(9)               return Q_HANDLED();
              }
          }
(10)      return Q_SUPER(&QHsm_top);                      /* return the superstate */
      }
      /*..........................................................................*/
      QState Ship_flying(Ship *me, QEvent const *e) {
          switch (e->sig) {
(11)          case Q_ENTRY_SIG: {
(12)              ScoreEvt *sev;

                  me->score = 0;                          /* reset the score */
(13)              sev = Q_NEW(ScoreEvt, SCORE_SIG);
(14)              sev->score = me->score;
(15)              QActive_postFIFO(AO_Tunnel, (QEvent *)sev);
(16)              return Q_HANDLED();
              }
              case TIME_TICK_SIG: {
                  /* tell the Tunnel to draw the Ship and test for hits */
                  ObjectImageEvt *oie = Q_NEW(ObjectImageEvt, SHIP_IMG_SIG);
                  oie->x   = me->x;
                  oie->y   = me->y;
                  oie->bmp = SHIP_BMP;
                  QActive_postFIFO(AO_Tunnel, (QEvent *)oie);

                  ++me->score;   /* increment the score for surviving another tick */

                  if ((me->score % 10) == 0) {            /* is the score "round"? */
                      ScoreEvt *sev = Q_NEW(ScoreEvt, SCORE_SIG);
                      sev->score = me->score;
                      QActive_postFIFO(AO_Tunnel, (QEvent *)sev);
                  }
                  return Q_HANDLED();
              }
              case PLAYER_TRIGGER_SIG: {                  /* trigger the Missile */
                  ObjectPosEvt *ope = Q_NEW(ObjectPosEvt, MISSILE_FIRE_SIG);
                  ope->x = me->x;
                  ope->y = me->y + SHIP_HEIGHT - 1;
```

```
                    QActive_postFIFO(AO_Missile, (QEvent *)ope);
                    return Q_HANDLED();
                }
                case DESTROYED_MINE_SIG: {
                    me->score += ((ScoreEvt const *)e)->score;
                    /* the score will be sent to the Tunnel by the next TIME_TICK */
                    return Q_HANDLED();
                }
(17)            case HIT_WALL_SIG:
(18)            case HIT_MINE_SIG: {
(19)                /* any actions associated with the transition */
(20)                return Q_TRAN(&Ship_exploding);
                }
            }
(21)    return Q_SUPER(&Ship_active);                       /* return the superstate */
        }
```

(1)   Each state handler must have the same signature, that is, it must take two parameters: the state machine pointer "me" and the pointer to QEvent. The keyword const before the * in the event pointer declaration means that the event pointed to by that pointer cannot be changed inside the state-handler function (i.e., the event is read-only). A state-handler function must return QState, which conveys the status of the event handling to the QEP event processor.

(2)   Typically, every state handler is structured as a switch statement that discriminates based on the signal of the event e->sig.

(3)   This line of code pertains to the nested *initial transition* Figure 1.6(2). QEP provides a reserved signal Q_INIT_SIG that the framework passes to the state-handler function when it wants to execute the initial transition.

(4)   You can enlist any actions associated with this initial transition (none in this particular case).

(5)   You designate the target substate with the Q_TRAN() macro. This macro must always follow the return statement, through which the state-handler function informs the QEP event processor that the transition has been taken.

> **NOTE**
>
> The initial transition must necessarily target a direct or transitive substate of a given state. An initial transition cannot target a peer state or go up in state hierarchy to higher-level states, which in the UML would represent a "malformed" state machine.

(6) This line of code pertains to the internal transition `PLAYER_SHIP_MOVE_SIG (x, y)` in Figure 1.6(3).

(7,8) You access the data members of the Ship state machine via the "me" argument of the state-handler function. You access the event parameters via the "e" argument. You need to cast the event pointer from the generic `QEvent` base class to the specific event structure expected for the `PLAYER_SHIP_MOVE_SIG`, which is `ObjectPosEvt` in this case.

> **NOTE**
>
> The association between the event signal and event structure (event parameters) is established at the time the event is generated. All recipients of that event must know about this association to perform the cast to the correct event structure.

(9) You terminate the case statement with "`return QHandled ()`", which informs the QEP processor that the event has been handled but no transition has been taken.

(10) The final `return` from a state-handler function designates the *superstate* of that state by means of the QEP macro `Q_SUPER()`. The final `return` statement from a state-handler function represents the single point of maintenance for changing the nesting level of a given state. The state "active" in Figure 1.6 has no explicit superstate, which means that it is implicitly nested in the "top" state. The "top" state is a UML concept that denotes the ultimate root of the state hierarchy in a hierarchical state machine. QEP provides the "top" state as a state-handler function `QHsm_top()`, and therefore the `Ship_active()` state handler uses the pointer `&QHsm_top` as the argument of the macro `Q_SUPER()`.

NOTE

In C and C++, a pointer-to-function `QHsm_top()` can be written either as `QHsm_top` or `&QHsm_top`. Even though the notation `QHsm_top` is more succinct, I prefer adding the ampersand explicitly, to leave absolutely no doubt that I mean a pointer-to-function `&QHsm_top`.

(11) This line of code pertains to the *entry action* into state "flying" (Figure 1.6(5)). QEP provides a reserved signal `Q_ENTRY_SIG` that the framework passes to the state-handler function when it wants to execute the entry actions.

(12) The entry action to "flying" posts the `SCORE` event to the Tunnel active object (Figure 1.6(5)). This line defines a temporary pointer to the event structure `ScoreEvt`.

(13) The QF macro `Q_NEW(ScoreEvt, SCORE_SIG)` dynamically allocates an instance of the `ScoreEvt` from an event pool managed by QF. The macro also performs the association between the signal `SCORE_SIG` and the allocated event. The `Q_NEW()` macro returns the pointer to the allocated event.

(14) The `score` parameter of the `ScoreEvt` is set from the state machine member `me->score`.

(15) The `sev` event is posted directly to the Tunnel active object by means of the QP function `QActive_postFIFO()`. The arguments of this function are the recipient active object (`AO_Tunnel` in this case) and the pointer to the event (the temporary pointer `sev` in this case).

(16) You terminate the case statement with `return Q_Handled()`, which informs QEP that the entry actions have been handled.

(17,18) These two lines of code pertain to the *state transitions* from "flying" to "exploding" (Figure 1.6(9, 10)).

(19) You can enlist any actions associated with the transition (none in this particular case).

(20) You designate the target of the transition with the `Q_TRAN()` macro.

(21)   The final return from a state-handler function designates the *superstate* of that state. The state "flying" in Figure 1.6 nests in the state "active," so the state handler `Ship_flying()` returns the pointer `&Ship_active`.

When implementing state-handler functions you need to keep in mind that the QEP event processor is in charge here rather than your code. QEP will invoke a state-handler function for various reasons: for hierarchical event processing, for execution of entry and exit actions, for triggering initial transitions, or even just to elicit the superstate of a given state handler. Therefore, you should not assume that a state handler would be invoked only for processing signals enlisted in the `case` statements. You should avoid any code outside the `switch` statement, especially code that would have side effects.

1.8   The Execution Model

As you saw in Listing 1.1(21), the `main()` function eventually gives control to the event-driven framework by calling `QF_run()` to execute the application. In this section, I briefly explain how QF allocates the CPU cycles to various tasks within the system and what options you have in choosing the execution model.

1.8.1   Simple Nonpreemptive "Vanilla" Scheduler

The "Fly 'n' Shoot" example uses the simplest QF configuration, in which QF runs on a bare-metal target processor without any underlying operating system or kernel.[4] I call such a QF configuration "plain vanilla" or just "vanilla."

QF includes a simple nonpreemptive "vanilla" kernel, which executes one active object at a time in the infinite loop (similar to the "superloop"). The "vanilla" kernel is engaged after each event is processed in the run-to-completion (RTC) fashion to choose the next highest-priority active object ready to process the next event. The "vanilla" scheduler is cooperative, which means that all active objects cooperate to share a single CPU and implicitly yield to each other after every RTC step. The kernel is nonpreemptive, meaning that every active object must completely process an event before any other active object can start processing another event.

---

[4] The 80×86 version of the "Fly 'n' Shoot" game runs on top of DOS, but DOS does not provide any multitasking support.

The ISRs can preempt the execution of active objects at any time, but due to the simplistic nature of the "vanilla" kernel, every ISR returns to exactly the preemption point. If the ISR posts or publishes an event to any active object, the processing of this event won't start until the current RTC step completes. The maximum time an event for the highest-priority active object can be delayed this way is called the *task-level response*. With the nonpreemptive "vanilla" kernel, the task-level response is equal to the longest RTC step of all active objects in the system. Note that the task-level response of the "vanilla" kernel is still a lot better than the traditional "superloop" (a.k.a. main+ISRs) architecture. I'll have more to say about this in the upcoming Section 1.9, where I compare the event-driven "Fly 'n' Shoot" example to the traditionally structured Quickstart application.

The task-level response of the simple "vanilla" kernel turns out to be adequate for surprisingly many applications because state machines by nature handle events quickly without a need to busy-wait for events. (A state machine simply runs to completion and becomes dormant until another event arrives.) Also note that often you can make the task-level response as fast as you need by breaking up longer RTC steps into shorter ones (e.g., by using the "Reminder" state pattern described in Chapter 5).

1.8.2   The QK Preemptive Kernel

In some cases, breaking up long RTC steps into short enough pieces might be very difficult, and consequently the task-level response of the nonpreemptive "vanilla" kernel might be too long. An example system could be a GPS receiver. Such a receiver performs a lot of floating-point number crunching on a fixed-point CPU to calculate the GPS position. At the same time, the GPS receiver must track the GPS satellite signals, which involves closing control loops in submillisecond intervals. It turns out that it's not easy to break up the position-fix computation into short enough RTC steps to allow reliable signal tracking.

But the RTC semantics of state machine execution do not mean that a state machine has to monopolize the CPU for the duration of the RTC step. A preemptive kernel can perform a context switch in the middle of the long RTC step to allow a higher-priority active object to run. As long as the active objects don't share resources, they can run concurrently and complete their RTC steps independently (see also Section 6.3.3 in Chapter 6).

The QP event-driven platform includes a tiny, fully **preemptive**, priority-based real-time kernel component called QK, which is specifically designed for processing events in the RTC fashion. Configuring QP to use the preemptive QK kernel is very easy, but as with any fully preemptive kernel you must be very careful with any resources shared among active objects.[5] The "Fly 'n' Shoot" example has been purposely designed to avoid any resource sharing among active objects, so the application code does not need to change at all to run on top of the QK preemptive kernel, or any other preemptive kernel or RTOS for that matter. The accompanying code contains the "Fly 'n' Shoot" example with QK in the following directory: `<qp>\qpc\examples\80x86\qk\tcpp101\l\game\`. You can execute this example in a DOS-console on any standard Windows-based PC.

1.8.3 Traditional OS/RTOS

QP can also work with a traditional operating system (OS), such as Windows or Linux, or virtually any real-time operating system (RTOS) to take advantage of the existing device drivers, communication stacks, and other middleware.

QP contains a platform abstraction layer (PAL), which makes adapting QP to virtually any operating system easy. The carefully designed PAL allows tight integration with the underlying OS/RTOS by reusing any provided facilities for interrupt management, message queues, and memory partitions. I cover porting QP in Chapter 8.

1.9   Comparison to the Traditional Approach

The "Fly 'n' Shoot" game behaves intentionally almost identically to the Quickstart application provided in source code with the Luminary Micro Stellaris EV-LM3S811 evaluation kit [Luminary 06]. In this section I'd like to compare the traditional approach represented by the Quickstart application with the state machine-based solution exemplified in the "Fly 'n' Shoot" game.

Figure 1.11(A) shows schematically the flowchart of the Quickstart application; Figure 1.11(B) shows the flowchart of the "Fly 'n' Shoot" game running on top of the cooperative "vanilla" kernel. At the highest level, the flowcharts are similar in that they both consist of an endless loop surrounding the entire processing. But the internal structure of the main loop is very different in the two cases. As indicated by the heavy

---

[5] QK provides a mutex facility for enforcing a mutually exclusive access to shared resources. The QK mutex uses the priority-ceiling protocol to avoid priority inversions. Refer to Chapter 10 for more information.

lines in the flowcharts, the Quickstart application spends most of its time in the tight "event loops" designed to busy-wait for certain events, such as the screen update event. In contrast, the "Fly 'n' Shoot" application spends most of its time right in the main loop. The QP framework dispatches any available event to the appropriate state machine that handles the event and returns quickly to the main loop without ever waiting for events internally.

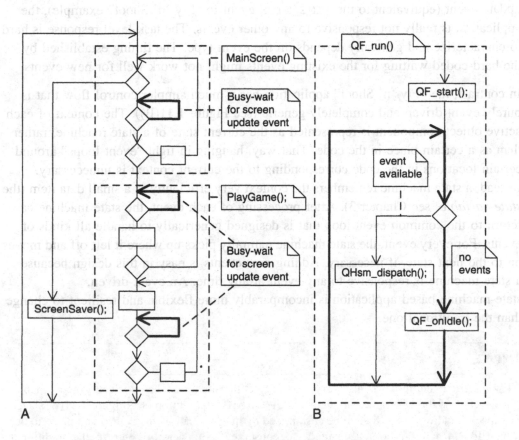

Figure 1.11: The control flow in the Quickstart application (A) and the "Fly 'n' Shoot" example (B). The heavy lines represent the most frequently exercised paths through the code.

The Quickstart application has much more convoluted flow of control than the "Fly 'n' Shoot" example because the traditional solution is very specific to the problem at hand whereas the state-machine approach is generic. The Quickstart application is structured very much like a traditional sequential program that tries to stay in control from the beginning to the end. From time to time, the application pauses to busy-wait for a certain event, whereas the code is generally not ready to handle any other events than the one it chooses to wait for. All this contributes to the inflexibility of the design. Adding new events is hard because the whole structure of the intervening code is designed to accept only very specific events and would need to change dramatically to accommodate new events. Also, while busy-waiting for the screen update event (equivalent to the TIME_TICK event in "Fly 'n' Shoot" example), the application is really not responsive to any other events. The task-level response is hard to characterize and generally depends on the event type. The timing established by the hard-coded waiting for the existing events might not work well for new events.

In contrast, the "Fly 'n' Shoot" application has a much simpler control flow that is purely event-driven and completely generic (see Figure 1.11(B)). The context of each active object component is represented as the current state of a state machine, rather than as a certain place in the code. That way, hanging in tight "event loops" around certain locations in the code corresponding to the current context is unnecessary. Instead, a state machine remembers the context very efficiently as a small data item (the *state-variable;* see Chapter 3). After processing of each event, the state machine can return to the common event loop that is designed generically to handle all kinds of events. For every event, the state machine naturally picks up where it left off and moves on to the next state, if necessary. Adding new events is easy in this design because a state machine is responsive to any event at any time. An event-driven, state-machine-based application is incomparably more flexible and resilient to change than the traditional one.

NOTE

The generic event loop can also very easily detect the situation when no events are available, in which case the QP framework calls the QF_onIdle() function (see Figure 1.11(B)). This callback function is designed to be customized by the application and is the ideal place to put the CPU in a low-power sleep mode to conserve the power. In contrast, the traditional approach does not offer any single place to transition to the low-power sleep mode and consequently is much less friendly for implementing truly low-power designs.

1.10   Summary

If you've never done event-driven programming before, the internal structure of the "Fly 'n' Shoot" game must certainly represent a big paradigm shift for you. In fact,

I hope that it actually blows your mind, because otherwise I'm not sure that you really appreciate the complete reversal of control of an event-driven program compared to the traditional sequential code. This reversal of control, known as the Hollywood Principle (don't call us, we'll call *you*), baffles many newcomers, who often find it "mind-boggling," "backward," or "weird."

The "Fly 'n' Shoot" game is by no means a big application, but at the same time it is definitely *not* trivial, either. You shouldn't worry if you don't fully understand it at the first reading. In the upcoming chapters, I will provide a closer look at the state machine design and coding techniques. In Part II, I discuss the features, implementation, and porting of the QF real-time framework.

My main goal in this chapter was just to introduce you to the event-driven paradigm and the modern state machines to convince you that these powerful concepts aren't particularly hard to implement directly in C or C++. Indeed, I hope you noticed that the actual coding of the nontrivial "Fly 'n' Shoot" game wasn't a big deal at all. All you needed to know was just a few cookie-cutter rules for coding state machines and familiarity with a few framework services for implementing the actions.

Wile the coding turned out to be essentially a nonissue; the bulk of the programming effort was spent on the *design* of the application. At this point, I hope that the "Fly 'n' Shoot" example helps you get the big picture of how the method works. Under the event-driven model, the program structure is divided into two rough groups: events and state machine components (active objects). An event represents the occurrence of something interesting. A state machine codifies the reactions to the events, which generally depend both on the nature of the event and on the state of the component. While events often originate from the outside of your program, such as time ticks or button presses in the "Fly 'n' Shoot" game, events can also be generated internally by the program itself. For example, the Mine components generate notification events when they detect a collision with the Missile or the Ship.

An event-driven program executes by constantly checking for possible events and, when an event is detected, dispatching the event to the appropriate state machine component (see Figure 1.11(B)). For this approach to work, the events must be checked continuously and frequently. This implies that the state machines must execute quickly so that the program can get back to checking for events. To meet this requirement, a state machine cannot go into a condition where it is busy-waiting for some long or indeterminate time. The most common example of this would be a `while` loop inside a state-handler function, where the condition for termination was not under program control—for instance, the button press. This kind of program structure, an indefinite loop, is referred to as "blocking" code,[6] and you saw examples of it in the Quickstart application (see Figure 1.11(A)). For the event-driven programming model to work, you must only write "nonblocking" code [Carryer 05].

Finally, the "Fly 'n' Shoot" example demonstrates the use of the event-driven platform called QP, which is a collection of components for building event-driven applications. The QF real-time framework component embodies the Hollywood Principle by calling the application code, not the other way around. Such an arrangement is very typical for event-driven systems and application frameworks similar to QF are at the heart of virtually every design automation tool on the market today.

The QF framework operates in the "Fly 'n' Shoot" game in its simplest configuration, in which QF runs on a bare-metal target processor without any operating system. QF can also be configured to work with the build-in preemptive real-time kernel called QK (see Chapter 10) or can be easily ported to almost any traditional OS or RTOS (see Chapter 8). In fact, you can view the QF framework itself as a high-level, event-driven, real-time operating system.

---

[6] In the context of a multitasking operating system the "blocking" code corresponds to waiting on a semaphore, event flag, message mailbox, or other such operating system primitive.

CHAPTER 2

A Crash Course in UML State Machines

One place we could really use help is in optimizing IF-THEN-ELSE constructs. Most programs start out fairly well structured. As bugs are found and features are grafted on, IFs and ELSEs are added until no human being really has a good idea how data flows through a function. Pretty printing helps, but does not reduce the complexity of 15 nested IF statements.
—Jack Ganssle, "Break Points," ESP Magazine, January 1991

Traditional, sequential programs can be structured as a single flow of control, using standard constructs such as loops and nested function calls. Such programs represent most of the execution context in the location of the program counter, in the procedure call tree, and in the temporary variables allocated on the stack.

Event-driven programs, in contrast, require a series of fine-granularity event-handler functions for handling events. These event-handler functions must execute quickly and always return to the main event-loop, so no context can be preserved in the call tree and the program counter. In addition, all stack variables disappear across calls to the separate event-handlers. Thus, event-driven programs rely heavily on static variables to preserve the execution context from one event-handler invocation to the next.

Consequently, one of the biggest challenges of event-driven programming lies in managing the execution context represented as data. The main problem here is that the context data must somehow feed back into the control flow of the event-handler code so that each event handler can execute only the actions appropriate in the current context. Traditionally, this dependence on the context very often leads to deeply nested if-else constructs that direct the flow of control based on the context data.

If you could eliminate even a fraction of these conditional branches (a.k.a. "spaghetti" code), the software would be much easier to understand, test, and maintain, and the sheer number of convoluted execution paths through the code would drop radically, typically by orders of magnitude. Techniques based on *state machines* are capable of achieving exactly this—a dramatic reduction of the different paths through the code and simplification of the conditions tested at each branching point.

In this chapter I briefly introduce UML state machines that represent the current state of the art in the long evolution of these techniques. My intention is not to give a complete, formal discussion of UML state machines, which the official OMG specification [OMG 07] covers comprehensively and with formality. Rather, my goal in this chapter is to lay a foundation by establishing basic terminology, introducing basic notation,[1] and clarifying semantics. This chapter is restricted to only a subset of those state machine features that are arguably most fundamental. The emphasis is on the role of UML state machines in practical, everyday programming rather than mathematical abstractions.

2.1   The Oversimplification of the Event-Action Paradigm

The currently dominating approach to structuring event-driven software is the ubiquitous "event-action" paradigm, in which events are directly mapped to the code that is supposed to be executed in response. The event-action paradigm is an important stepping stone for understanding state machines, so in this section I briefly describe how it works in practice.

I will use an example from the graphical user interface (GUI) domain, given that GUIs make exemplary event-driven systems. In the book *Constructing the User Interface with Statecharts* [Horrocks 99], Ian Horrocks discusses a simple GUI calculator application distributed in millions of copies as a sample program with Microsoft Visual Basic, in which he found a number of serious problems. As Horrocks notes, the point of this analysis is not to criticize this particular program but to identify the shortcomings of the general principles used in its construction.

When you launch the Visual Basic calculator (available from the companion Website, in the directory `<qp>\resources\vb\calc.exe`), you will certainly find out that most of the time it correctly adds, subtracts, multiplies, and divides (see Figure 2.1(A)).

---

[1] Appendix B contains a comprehensive summary of the notation.

What's not to like? However, play with the program for a while longer and you can discover many corner cases in which the calculator provides misleading results, freezes, or crashes altogether.

NOTE

Ian Horrocks found 10 serious errors in the Visual Basic calculator after only an hour of testing. Try to find at least half of them.

For example, the Visual Basic calculator often has problems with the "–" event; just try the following sequence of operations: 2, –, –, –, 2, =. The application crashes with a runtime error (see Figure 2.1(B)). This is because the same button (–) is used to negate a number and to enter the subtraction operator. The correct interpretation of the "–" button-click event, therefore, depends on the context, or mode, in which it occurs. Likewise, the CE (Cancel Entry) button occasionally works erroneously—try 2, x, CE, 3, =, and observe that CE had no effect, even though it appears to cancel the 2 entry from the display. Again, CE should behave differently when canceling an operand than canceling an operator. As it turns out, the application handles the CE event always the same way, regardless of the context. At this point, you probably have noticed an

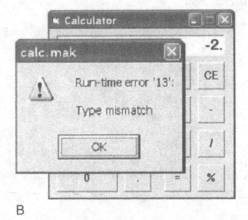

Figure 2.1: Visual Basic calculator before the crash (A) and after a crash with a runtime error (B).

emerging pattern. The application is especially vulnerable to events that require different handling depending on the *context*.

This is not to say that the Visual Basic calculator does not attempt to handle the context. Quite the contrary, if you look at the calculator code (available from the companion Website, in the directory `<qp>\resources\vb\calc.frm`), you'll notice that managing the context is in fact the *main* concern of this application. The code is littered with a multitude of global variables and flags that serve only one purpose: handling the context. For example, `DecimalFlag` indicates that a decimal point has been entered, `OpFlag` represents a pending operation, `LastInput` indicates the type of the last button press event, `NumOps` denotes the number of operands, and so on. With this representation, the context of the computation is represented ambiguously, so it is difficult to tell precisely in which mode the application is at any given time. Actually, the application has no notion of any single mode of operation but rather a bunch of tightly coupled and overlapping conditions of operation determined by values of the global variables and flags.

Listing 2.1 shows the conditional logic in which the event-handler procedure for the operator events (+, − , *, and /) attempts to determine whether the − (minus) button-click should be treated as negation or subtraction.

Listing 2.1 Fragment of Visual Basic code that attempts to determine whether the − (minus) button-click event should be treated as negation or subtraction

```
Private Sub Operator_Click(Index As Integer)
    . . .
    Select Case NumOps
        Case 0
        If Operator(Index).Caption = "-" And LastInput <> "NEG" Then
            ReadOut = "-" & ReadOut
            LastInput = "NEG"
        End If
        Case 1
            Op1 = ReadOut
            If Operator(Index).Caption = "-" And LastInput <> "NUMS" And
                                        OpFlag <> "=" Then
            ReadOut = "-"
            LastInput = "NEG"
        End If
    . . .
```

The approach exemplified in Listing 2.1 is a fertile ground for the "corner case" behavior (a.k.a. bugs) for at least three reasons:

- It always leads to convoluted conditional logic (a.k.a. "spaghetti" code).

- Each branching point requires evaluation of a complex expression.

- Switching between different modes requires modifying many variables, which can easily lead to inconsistencies.

Convoluted conditional expressions like the one shown in Listing 2.1, scattered throughout the code, are unnecessarily complex and expensive to evaluate at runtime. They are also notoriously difficult to get right, even by experienced programmers, as the bugs still lurking in the Visual Basic calculator attest. This approach is insidious because it appears to work fine initially, but doesn't scale up as the problem grows in complexity. Apparently, the calculator application (overall only seven event handlers and some 140 lines of Visual Basic code including comments) is just complex enough to be difficult to get right with this approach.

The faults just outlined are rooted in the oversimplification of the event-action paradigm. The Visual Basic calculator example makes it clear, I hope, that an event alone does *not* determine the actions to be executed in response to that event. The *current context* is at least equally important. The prevalent event-action paradigm, however, recognizes only the dependency on the event-type and leaves the handling of the context to largely ad hoc techniques that all too easily degenerate into spaghetti code.

2.2 Basic State Machine Concepts

The event-action paradigm can be extended to explicitly include the dependency on the execution context. As it turns out, the behavior of most event-driven systems can be divided into a relatively small number of chunks, where event responses within each individual chunk indeed depend only on the current event-type but no longer on the sequence of past events (the context). In other words, the event-action paradigm is still applied, but only *locally* within each individual chunk.

A common and straightforward way of modeling behavior based on this idea is through a *finite state machine* (FSM). In this formalism, "chunks of behavior" are called *states*, and change of behavior (i.e., change in response to any event) corresponds to change of state and is called a *state transition*. An FSM is an efficient way to specify *constraints* of the overall behavior of a system. Being in a state means that the system responds only to a subset of all allowed events, produces only a subset of possible responses, and changes state directly to only a subset of all possible states.

The concept of an FSM is important in programming because it makes the event handling explicitly dependent on both the event-type and on the execution context (state) of the system. When used correctly, a state machine becomes a powerful "spaghetti reducer" that drastically cuts down the number of execution paths through the code, simplifies the conditions tested at each branching point, and simplifies the transitions between different modes of execution.

2.2.1 States

A state captures the relevant aspects of the system's history very efficiently. For example, when you strike a key on a keyboard, the character code generated will be either an uppercase or a lowercase character, depending on whether the Caps Lock is active.[2] Therefore, the keyboard's behavior can be divided into two chunks (states): the "default" state and the "caps_locked" state. (Most keyboards actually have an LED that indicates that the keyboard is in the "caps_locked" state.) The behavior of a keyboard depends only on certain aspects of its history, namely whether the Caps Lock key has been pressed, but not, for example, on how many and exactly which other keys have been pressed previously. A state can abstract away all possible (but irrelevant) event sequences and capture only the relevant ones.

To relate this concept to programming, this means that instead of recording the event history in a multitude of variables, flags, and convoluted logic, you rely mainly on just one *state variable* that can assume only a limited number of a *priori* determined values (e.g., two values in case of the keyboard). The value of the state variable crisply defines the current state of the system at any given time. The concept of state reduces the problem of identifying the execution context in the code to testing just the state variable instead of many variables, thus eliminating a lot of conditional logic. Actually, in all but the most basic state machine implementation techniques, such as the "nested-switch statement" technique discussed in Chapter 3, even the explicit testing of the state variable disappears from the code, which reduces the "spaghetti" further still (you will experience this effect later in Chapters 3 and 4). Moreover, switching between different states is vastly simplified as well, because you need to reassign just one state variable instead of changing multiple variables in a self-consistent manner.

---

[2] Ignore at this print the effects of the Shift, Ctrl, and Alt keys.

2.2.2   State Diagrams

FSMs have an expressive graphical representation in the form of *state diagrams*. These diagrams are directed graphs in which nodes denote states and connectors denote state transitions.[3]

For example, Figure 2.2 shows a UML state transition diagram corresponding to the computer keyboard state machine. In UML, states are represented as rounded rectangles labeled with state names. The transitions, represented as arrows, are labeled with the triggering events followed optionally by the list of executed actions. The initial transition originates from the solid circle and specifies the starting state when the system first begins. Every state diagram should have such a transition, which should not be labeled, since it is not triggered by an event. The initial transition can have associated actions.

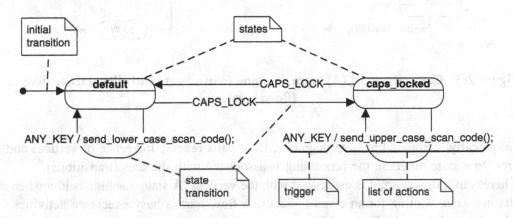

Figure 2.2: UML state diagram representing the computer keyboard state machine.

2.2.3   State Diagrams versus Flowcharts

Newcomers to the state machine formalism often confuse state diagrams with flowcharts. The UML specification [OMG 07] isn't helping in this respect because it lumps activity graphs in the state machine package. Activity graphs are essentially elaborate flowcharts.

---

[3] Appendix B contains a succinct summary of the graphical notations used throughout the book, including state transition diagrams.

Figure 2.3 shows a comparison of a state diagram with a flowchart. A state machine (panel (A)) performs actions in response to explicit triggers. In contrast, the flowchart (panel (B)) does not need explicit triggers but rather transitions from node to node in its graph automatically upon completion of activities.

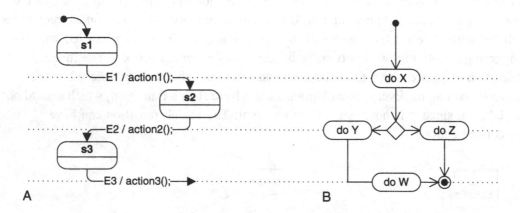

Figure 2.3: Comparison of (A) state machine (statechart) with (B) activity diagram (flowchart).

Graphically, compared to state diagrams, flowcharts reverse the sense of vertices and arcs. In a state diagram, the processing is associated with the arcs (transitions), whereas in a flowchart, it is associated with the vertices. A state machine is idle when it sits in a state waiting for an event to occur. A flowchart is busy executing activities when it sits in a node. Figure 2.3 attempts to show that reversal of roles by aligning the arcs of the statecharts with the processing stages of the flowchart.

You can compare a flowchart to an assembly line in manufacturing because the flowchart describes the progression of some task from beginning to end (e.g., transforming source code input into object code output by a compiler). A state machine generally has no notion of such a progression. A computer keyboard, for example, is not in a more advanced stage when it is in the "caps_locked" state, compared to being in the "default" state; it simply reacts differently to events. A state in a state machine is an efficient way of specifying a particular behavior, rather than a stage of processing.

The distinction between state machines and flowcharts is especially important because these two concepts represent two diametrically opposed programming paradigms: event-driven programming (state machines) and transformational programming (flowcharts). You cannot devise effective state machines without constantly thinking about the available events. In contrast, events are only a secondary concern (if at all) for flowcharts.

2.2.4   Extended State Machines

One possible interpretation of state for software systems is that each state represents one distinct set of valid values of the whole program memory. Even for simple programs with only a few elementary variables, this interpretation leads to an astronomical number of states. For example, a single 32-bit integer could contribute to over 4 billion different states. Clearly, this interpretation is not practical, so program variables are commonly dissociated from states. Rather, the complete condition of the system (called the *extended state*) is the combination of a qualitative aspect (the state) and the quantitative aspects (the extended state variables). In this interpretation, a change of variable does not always imply a change of the qualitative aspects of the system behavior and therefore does not lead to a change of state [Selic+ 94].

State machines supplemented with variables are called *extended state machines*. Extended state machines can apply the underlying formalism to much more complex problems than is practical without including extended state variables. For instance, suppose the behavior of the keyboard depends on the number of characters typed on it so far and that after, say, 1,000 keystrokes, the keyboard breaks down and enters the final state. To model this behavior in a state machine without memory, you would need to introduce 1,000 states (e.g., pressing a key in state stroke123 would lead to state stroke124, and so on), which is clearly an impractical proposition. Alternatively, you could construct an extended state machine with a key_count down-counter variable. The counter would be initialized to 1,000 and decremented by every keystroke without changing state. When the counter reached zero, the state machine would enter the final state.

The state diagram from Figure 2.4 is an example of an *extended state machine*, in which the complete condition of the system (called the *extended state*) is the combination of a qualitative aspect—the "state"—and the quantitative aspects—the extended state variables (such as the down-counter key_count). In extended state machines, a change of a variable does not always imply a change of the qualitative aspects of the system behavior and therefore does not always lead to a change of state.

The obvious advantage of extended state machines is flexibility. For example, extending the lifespan of the "cheap keyboard" from 1,000 to 10,000 keystrokes would not complicate the extended state machine at all. The only modification required would be changing the initialization value of the key_count down-counter in the initial transition.

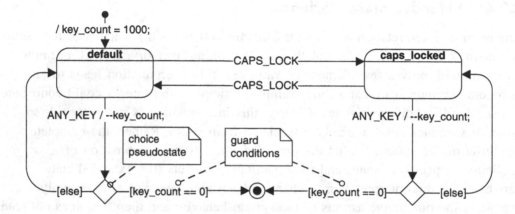

Figure 2.4: **Extended state machine of "cheap keyboard" with extended state variable** key_count **and various guard conditions.**

2.2.5   Guard Conditions

This flexibility of extended state machines comes with a price, however, because of the complex coupling between the "qualitative" and the "quantitative" aspects of the extended state. The coupling occurs through the *guard conditions* attached to transitions, as shown in Figure 2.4.

Guard conditions (or simply guards) are Boolean expressions evaluated dynamically based on the value of extended state variables and event parameters (see the discussion of events and event parameters in the next section). Guard conditions affect the behavior of a state machine by enabling actions or transitions only when they evaluate to TRUE and disabling them when they evaluate to FALSE. In the UML notation, guard conditions are shown in square brackets (e.g., [key_count == 0]).

The need for guards is the immediate consequence of adding memory (extended state variables) to the state machine formalism. Used sparingly, extended state variables and guards make up an incredibly powerful mechanism that can immensely simplify designs. But don't let the fancy name ("guard") and the concise UML notation fool you. When you actually code an extended state machine, the guards become the same IFs and ELSEs that you wanted to eliminate by using the state machine in the first place. Too many of them, and you'll find yourself back in square one ("spaghetti"), where the guards effectively take over handling of all the relevant conditions in the system.

Indeed, abuse of extended state variables and guards is the primary mechanism of architectural decay in designs based on state machines. Usually, in the day-to-day battle, it seems very tempting, especially to programmers new to state machine formalism, to add yet another extended state variable and yet another guard condition (another `if` or an `else`) rather than to factor out the related behavior into a new qualitative aspect of the system—the state. From my experience in the trenches, the likelihood of such an architectural decay is directly proportional to the overhead (actual or perceived) involved in adding or removing states. (That's why I don't particularly like the popular state-table technique of implementing state machines that I describe in Chapter 3, because adding a new state requires adding and initializing a whole new column in the table.)

One of the main challenges in becoming an effective state machine designer is to develop a sense for which parts of the behavior should be captured as the "qualitative" aspects (the "state") and which elements are better left as the "quantitative" aspects (extended state variables). In general, you should actively look for opportunities to capture the event history (what happened) as the "state" of the system, instead of storing this information in extended state variables. For example, the Visual Basic calculator uses an extended state variable `DecimalFlag` to remember that the user entered the decimal point to avoid entering multiple decimal points in the same number. However, a better solution is to observe that entering a decimal point really leads to a distinct state "entering_the_fractional_part_of_a_number," in which the calculator ignores decimal points. This solution is superior for a number of reasons. The lesser reason is that it eliminates one extended state variable and the need to initialize and test it. The more important reason is that the state-based solution is more robust because the context information is used very locally (only in this particular state) and is discarded as soon as it becomes irrelevant. Once the number is correctly entered, it doesn't really matter for the subsequent operation of the calculator whether that number had a decimal point. The state machine moves on to another state and automatically "forgets" the previous context. The `DecimalFlag` extended state variable, on the other hand, "lays around" well past the time the information becomes irrelevant (and perhaps outdated!).

Worse, you must not forget to reset `DecimalFlag` before entering another number or the flag will incorrectly indicate that indeed the user once entered the decimal point, but perhaps this happened in the context of the previous number.

Capturing behavior as the quantitative "state" has its disadvantages and limitations, too. First, the state and transition topology in a state machine must be static and fixed at compile time, which can be too limiting and inflexible. Sure, you can easily devise "state machines" that would modify themselves at runtime (this is what often actually happens when you try to recode "spaghetti" as a state machine). However, this is like writing self-modifying code, which indeed was done in the early days of programming but was quickly dismissed as a generally *bad* idea. Consequently, "state" can capture only static aspects of the behavior that are known *a priori* and are unlikely to change in the future.

For example, it's fine to capture the entry of a decimal point in the calculator as a separate state "entering_the_fractional_part_of_a_number," because a number can have only one fractional part, which is both known *a priori* and is not likely to change in the future. However, implementing the "cheap keyboard" without extended state variables and guard conditions would be practically impossible. This example points to the main weakness of the quantitative "state," which simply cannot store too much information (such as the wide range of keystroke counts). Extended state variables and guards are thus a mechanism for adding extra runtime flexibility to state machines.

2.2.6   Events

In the most general terms, an *event* is an occurrence in time and space that has significance to the system. Strictly speaking, in the UML specification the term *event* refers to the type of occurrence rather than to any concrete instance of that occurrence [OMG 07]. For example, Keystroke is an event for the keyboard, but each press of a key is not an event but a concrete *instance* of the Keystroke event. Another event of interest for the keyboard might be Power-on, but turning the power on tomorrow at 10:05:36 will be just an instance of the Power-on event.

An event can have associated parameters, allowing the event instance to convey not only the occurrence of some interesting incident but also quantitative information regarding that occurrence. For example, the Keystroke event generated by pressing a key on a computer keyboard has associated parameters that convey the character scan code as well as the status of the Shift, Ctrl, and Alt keys.

An event instance outlives the instantaneous occurrence that generated it and might convey this occurrence to one or more state machines. Once generated, the event instance goes through a processing life cycle that can consist of up to three stages. First, the event instance is received when it is accepted and waiting for processing (e.g., it is placed on the event queue). Later, the event instance is dispatched to the state machine, at which point it becomes the current event. Finally, it is consumed when the state machine finishes processing the event instance. A consumed event instance is no longer available for processing.

2.2.7 Actions and Transitions

When an event instance is dispatched, the state machine responds by performing actions, such as changing a variable, performing I/O, invoking a function, generating another event instance, or changing to another state. Any parameter values associated with the current event are available to all actions directly caused by that event.

Switching from one state to another is called *state transition*, and the event that causes it is called the *triggering event*, or simply the *trigger*. In the keyboard example, if the keyboard is in the "default" state when the Caps Lock key is pressed, the keyboard will enter the "caps_locked" state. However, if the keyboard is already in the "caps_locked" state, pressing Caps Lock will cause a different transition—from the "caps_locked" to the "default" state. In both cases, pressing Caps Lock is the triggering event.

In extended state machines, a transition can have a guard, which means that the transition can "fire" only if the guard evaluates to TRUE. A state can have many transitions in response to the same trigger, as long as they have nonoverlapping guards; however, this situation could create problems in the sequence of evaluation of the guards when the common trigger occurs. The UML specification intentionally does not stipulate any particular order; rather, UML puts the burden on the designer to devise guards in such a way that the order of their evaluation does not matter. Practically, this means that guard expressions should have no side effects, at least none that would alter evaluation of other guards having the same trigger.

2.2.8 Run-to-Completion Execution Model

All state machine formalisms, including UML statecharts, universally assume that a state machine completes processing of each event before it can start processing the next event. This model of execution is called *run to completion*, or RTC.

In the RTC model, the system processes events in discrete, indivisible RTC steps. New incoming events cannot interrupt the processing of the current event and must be stored (typically in an event queue) until the state machine becomes idle again. These semantics completely avoid any internal concurrency issues within a single state machine. The RTC model also gets around the conceptual problem of processing actions associated with transitions,[4] where the state machine is not in a well-defined state (is between two states) for the duration of the action. During event processing, the system is unresponsive (unobservable), so the ill-defined state during that time has no practical significance.

Note, however, that RTC does not mean that a state machine has to monopolize the CPU until the RTC step is complete. The preemption restriction only applies to the task context of the state machine that is already busy processing events. In a multitasking environment, other tasks (not related to the task context of the busy state machine) can be running, possibly preempting the currently executing state machine. As long as other state machines do not share variables or other resources with each other, there are no concurrency hazards.

The key advantage of RTC processing is simplicity. Its biggest disadvantage is that the responsiveness of a state machine is determined by its longest RTC step.[5] Achieving short RTC steps can often significantly complicate real-time designs.

2.3   UML Extensions to the Traditional FSM Formalism

Though the traditional FSMs are an excellent tool for tackling smaller problems, it's also generally known that they tend to become unmanageable, even for moderately involved systems. Due to the phenomenon known as *state explosion,* the complexity of a traditional FSM tends to grow much faster than the complexity of the reactive system it describes. This happens because the traditional state machine formalism inflicts repetitions. For example, if you try to represent the behavior of the Visual Basic calculator introduced in Section 2.1 with a traditional FSM, you'll immediately notice that many events (e.g., the Clear event) are handled identically in many states. A conventional FSM, however, has no means of capturing such a commonality and requires repeating the same actions and transitions in many states. What's missing in the traditional state machines is the mechanism for factoring out the common behavior in order to share it across many states.

---

[4] State machines that associate actions with transitions are classified as Mealy machines.

[5] A state machine can improve responsiveness by breaking up the CPU-intensive processing into sufficiently short RTC steps (see also the "Reminder" state pattern in Chapter 5).

The formalism of statecharts, invented by David Harel in the 1980s [Harel 87], addresses exactly this shortcoming of the conventional FSMs. Statecharts provide a very efficient way of sharing behavior so that the complexity of a statechart no longer explodes but tends to faithfully represent the complexity of the reactive system it describes. Obviously, formalism like this is a godsend to embedded systems programmers (or any programmers working with event-driven systems) because it makes the state machine approach truly applicable to real-life problems.

UML state machines, known also as UML statecharts [OMG 07], are object-based variants of Harel statecharts and incorporate several concepts defined in ROOMcharts, a variant of the statechart defined in the Real-time Object-Oriented Modeling (ROOM) language [Selic+ 94]. UML statecharts are extended state machines with characteristics of both Mealy and Moore automata. In statecharts, actions generally depend on both the state of the system and the triggering event, as in a Mealy automaton. Additionally, UML statecharts provide optional entry and exit actions, which are associated with states rather than transitions, as in a Moore automaton.

2.3.1   Reuse of Behavior in Reactive Systems

All reactive systems seem to reuse behavior in a similar way. For example, the characteristic look and feel of all GUIs results from the same pattern, which the Windows guru Charles Petzold calls the "Ultimate Hook" [Petzold 96]. The pattern is brilliantly simple: A GUI system dispatches every event first to the application (e.g., Windows calls a specific function inside the application, passing the event as an argument). If not handled by the application, the event flows back to the system. This establishes a hierarchical order of event processing. The application, which is conceptually at a lower level of the hierarchy, has the first shot at every event; thus the application can choose to react in any way it likes. At the same time, all unhandled events flow back to the higher level (i.e., to the GUI system), where they are processed according to the standard look and feel. This is an example of *programming by difference* because the application programmer needs to code only the differences from the standard system behavior.

2.3.2   Hierarchically Nested States

Harel statecharts bring the "Ultimate Hook" pattern to the logical conclusion by combining it with the state machine formalism. The most important innovation of statecharts over the classical FSMs is the introduction of *hierarchically nested states* (that's why statecharts are also called *hierarchical state machines*, or HSMs). The semantics associated with state nesting are as follows (see Figure 2.5(A)): If a system is in the nested state "s11" (called the *substate*), it also (implicitly) is in the surrounding state "s1" (called the *superstate*). This state machine will attempt to handle any event in the context of state "s11," which conceptually is at the lower level of the hierarchy. However, if state "s11" does not prescribe how to handle the event, the event is not quietly discarded as in a traditional "flat" state machine; rather, it is automatically handled at the higher level context of the superstate "s1." This is what is meant by the system being in state "s11" as well as "s1." Of course, state nesting is not limited to one level only, and the simple rule of event processing applies recursively to any level of nesting.

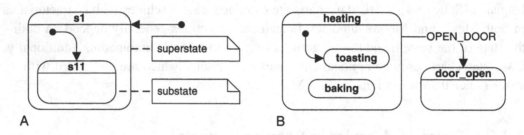

Figure 2.5: UML notation for hierarchically nested states (A), and a state model of a toaster oven in which states "toasting" and "baking" share the common transition from state "heating" to "door_open" (B).

States that contain other states are called *composite states*; conversely, states without internal structure are called *simple states* or *leaf states*. A nested state is called a *direct substate* when it is not contained by any other state; otherwise, it is referred to as a *transitively nested substate*.

Because the internal structure of a composite state can be arbitrarily complex, any hierarchical state machine can be viewed as an internal structure of some (higher-level) composite state. It is conceptually convenient to define one composite state as the ultimate root of state machine hierarchy. In the UML specification, every state machine has a *top state* (the abstract root of every state machine hierarchy), which contains all the other elements of the entire state machine. The graphical rendering of this all-enclosing top state is optional [OMG 07].

As you can see, the semantics of hierarchical state decomposition are designed to facilitate sharing of behavior through the direct support for the "Ultimate Hook" pattern. The substates (nested states) need only define the differences from the superstates (surrounding states). A substate can easily reuse the common behavior from its superstate(s) by simply ignoring commonly handled events, which are then automatically handled by higher-level states. In this manner, the substates can share all aspects of behavior with their superstates. For example, in a state model of a toaster oven shown in Figure 2.5(B), states "toasting" and "baking" share a common transition DOOR_OPEN to the "door_open" state, defined in their common superstate "hating."

The aspect of state hierarchy emphasized most often is *abstraction*—an old and powerful technique for coping with complexity. Instead of facing all aspects of a complex system at the same time, it is often possible to ignore (abstract away) some parts of the system. Hierarchical states are an ideal mechanism for hiding internal details because the designer can easily zoom out or zoom in to hide or show nested states. Although abstraction by itself does not reduce overall system complexity, it is valuable because it reduces the amount of detail you need to deal with at one time. As Grady Booch [Booch 94] notes:

. . . we are still constrained by the number of things that we can comprehend at one time, but through abstraction, we use chunks of information with increasingly greater semantic content.

However valuable abstraction in itself might be, you cannot cheat your way out of complexity simply by hiding it inside composite states. However, the composite states don't simply hide complexity, they also actively reduce it through the powerful mechanism of reuse (the "Ultimate Hook" pattern). Without such reuse, even a moderate increase in system complexity often leads to an explosive increase in the number of states and transitions. For example, if you transform the statechart from Figure 2.5(B) to a classical flat state machine,[6] you must repeat one transition (from heating to "door_open") in two places: as a transition from "toasting" to "door_open" and from "baking" to "door_open." Avoiding such repetitions allows HSMs to grow proportionally to system complexity. As the modeled system grows, the opportunity for reuse also increases and thus counteracts the explosive increase in states and transitions typical for traditional FSMs.

2.3.3 Behavioral Inheritance

Hierarchical states are more than merely the "grouping of [nested] state machines together without additional semantics" [Mellor 00]. In fact, hierarchical states have

---

[6] Such a transformation is always possible because HSMs are mathematically equivalent to classical FSMs.

simple but profound semantics. Nested states are also more than just "great diagrammatic simplification when a set of events applies to several substates" [Douglass 99]. The savings in the number of states and transitions are real and go far beyond less cluttered diagrams. In other words, simpler diagrams are just a side effect of behavioral reuse enabled by state nesting.

The fundamental character of state nesting comes from the combination of abstraction and hierarchy, which is a traditional approach to reducing complexity and is otherwise known in software as *inheritance*. In OOP, the concept of class inheritance describes relations between classes of objects. Class inheritance describes the "is a ..." relationship among classes. For example, class `Bird` might derive from class `Animal`. If an object *is a* bird (instance of the `Bird` class), it automatically is *an* animal, because all operations that apply to animals (e.g., eating, eliminating, reproducing) also apply to birds. But birds are more specialized, since they have operations that are not applicable to animals in general. For example, `flying()` applies to birds but not to fish.

The benefits of class inheritance are concisely summarized by Gamma and colleagues [Gamma+ 95]:

Inheritance lets you define a new kind of class rapidly in terms of an old one, by reusing functionality from parent classes. It allows new classes to be specified by difference rather than created from scratch each time. It lets you get new implementations almost for free, inheriting most of what is common from the ancestor classes.

As you saw in the previous section, all these basic characteristics of inheritance apply equally well to nested states (just replace the word *class* with *state*), which is not surprising because state nesting is based on the same fundamental "is a ..." classification as object-oriented class inheritance. For example, in a state model of a toaster oven, state "toasting" nests inside state "heating." If the toaster *is in* the "toasting" state, it automatically *is in* the "heating" state because all behavior pertaining to "heating" applies also to "toasting" (e.g., the heater must be turned on). But "toasting" is more specialized because it has behaviors not applicable to "heating" in general. For example, setting toast color (light or dark) applies to "toasting" but not to "baking."

In the case of nested states, the "is a ..." (is-a-kind-of) relationship merely needs to be replaced by the "is in ..." (is-in-a-state) relationship; otherwise, it is the same fundamental classification. State nesting allows a substate to inherit state behavior from its ancestors (superstates); therefore, it's called *behavioral inheritance*.

> **NOTE**
>
> The term "behavioral inheritance" does not come from the UML specification. Note too that behavioral inheritance describes the relationship between substates and superstates, and you should not confuse it with traditional (class) inheritance applied to entire state machines.

The concept of inheritance is fundamental in software construction. Class inheritance is essential for better software organization and for code reuse, which makes it a cornerstone of OOP. In the same way, behavioral inheritance is essential for efficient use of HSMs and for behavior reuse, which makes it a cornerstone of event-driven programming. In Chapter 5, a mini-catalog of state patterns shows ways to structure HSMs to solve recurring problems. Not surprisingly, behavioral inheritance plays the central role in all these patterns.

2.3.4   Liskov Substitution Principle for States

Identifying the relationship among substates and superstates as inheritance has many practical implications. Perhaps the most important is the Liskov Substitution Principle (LSP) applied to state hierarchy. In its traditional formulation for classes, LSP requires that a subclass can be freely substituted for its superclass. This means that every instance of the subclass should be compatible with the instance of the superclass and that any code designed to work with the instance of the superclass should continue to work correctly if an instance of the subclass is used instead.

Because behavioral inheritance is just a special kind of inheritance, LSP can be applied to nested states as well as classes. LSP generalized for states means that the behavior of a substate should be consistent with the superstate. For example, all states nested inside the "heating" state of the toaster oven (e.g., "toasting" or "baking") should share the same basic characteristics of the "heating" state. In particular, if being in the "heating" state means that the heater is turned on, then none of the substates should turn the heater off (without transitioning out of the "heating" state). Turning the heater off and staying in the "toasting" or "baking" state would be inconsistent with being in the "heating" state and would indicate poor design (violation of the LSP).

Compliance with the LSP allows you to build better (more correct) state hierarchies and make efficient use of abstraction. For example, in an LSP-compliant state hierarchy, you can safely zoom out and work at the higher level of the "heating" state (thus abstracting away the specifics of "toasting" and "baking"). As long as all the substates are consistent with their superstate, such abstraction is meaningful. On the other hand, if the substates violate basic assumptions of being in the superstate, zooming out and ignoring the specifics of the substates will be incorrect.

2.3.5   Orthogonal Regions

Hierarchical state decomposition can be viewed as exclusive-OR operation applied to states. For example, if a system is in the "heating" superstate (Figure 2.5(B)), it means that it's either in "toasting" substate OR the "baking" substate. That is why the "heating" superstate is called an OR-state.

UML statecharts also introduce the complementary AND-decomposition. Such decomposition means that a composite state can contain two or more *orthogonal regions* (*orthogonal* means independent in this context) and that being in such a composite state entails being in all its orthogonal regions simultaneously [Harel+ 98].

Orthogonal regions address the frequent problem of a combinatorial increase in the number of states when the behavior of a system is fragmented into independent, concurrently active parts. For example, apart from the main keypad, a computer keyboard has an independent numeric keypad. From the previous discussion, recall the two states of the main keypad already identified: "default" and "caps_locked" (Figure 2.2). The numeric keypad also can be in two states—"numbers" and "arrows"— depending on whether Num Lock is active. The complete state space of the keyboard in the standard decomposition is the cross-product of the two components (main keypad and numeric keypad) and consists of four states: "default–numbers," "default–arrows," "caps_locked–numbers," and "caps_locked–arrows." However, this is unnatural because the behavior of the numeric keypad does not depend on the state of the main keypad and vice versa. Orthogonal regions allow you to avoid mixing the independent behaviors as a cross-product and, instead, to keep them separate, as shown in Figure 2.6.

Note that if the orthogonal regions are fully independent of each other, their combined complexity is simply additive, which means that the number of independent states needed to model the system is simply the sum $k + l + m + \ldots$, where $k, l, m, \ldots$ denote numbers of OR-states in each orthogonal region. The general case of mutual dependency, on the other hand, results in multiplicative complexity, so in general, the number of states needed is the product $k \times l \times m \times \ldots$.

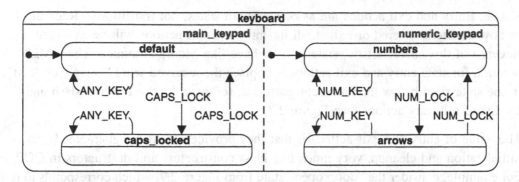

Figure 2.6: Two orthogonal regions (main keypad and numeric keypad) of a computer keyboard.

In most real-life situations, however, orthogonal regions are only approximately orthogonal (i.e., they are not independent). Therefore, UML statecharts provide a number of ways for orthogonal regions to communicate and synchronize their behaviors. From these rich sets of (sometimes complex) mechanisms, perhaps the most important is that orthogonal regions can coordinate their behaviors by sending event instances to each other.

Even though orthogonal regions imply independence of execution (i.e., some kind of concurrency), the UML specification does not require that a separate thread of execution be assigned to each orthogonal region (although it can be implemented that way). In fact, most commonly, orthogonal regions execute within the same thread. The UML specification only requires that the designer not rely on any particular order in which an event instance will be dispatched to the involved orthogonal regions.

NOTE

The HSM implementation described in this book (see Chapter 4) does not directly support orthogonal regions. Chapter 5 describes the "Orthogonal Component" state pattern, which emulates orthogonal regions by composition of HSMs.

2.3.6 Entry and Exit Actions

Every state in a UML statechart can have optional *entry actions*, which are executed upon entry to a state, as well as optional *exit actions*, which are executed upon exit from a state. Entry and exit actions are associated with states, not transitions.[7] Regardless of how a state is entered or exited, all its entry and exit actions will be executed. Because of this characteristic, statecharts behave like Moore automata. The UML notation for state entry and exit actions is to place the reserved word "entry" (or "exit") in the state right below the name compartment, followed by the forward slash and the list of arbitrary actions (see Figure 2.7).

The value of entry and exit actions is that they provide means for guaranteed initialization and cleanup, very much like class constructors and destructors in OOP. For example, consider the "door_open" state from Figure 2.7, which corresponds to the toaster oven behavior while the door is open. This state has a very important safety-critical requirement: Always disable the heater when the door is open.[8] Additionally, while the door is open, the internal lamp illuminating the oven should light up.

Of course, you could model such behavior by adding appropriate actions (disabling the heater and turning on the light) to every transition path leading to the "door_open" state (the user may open the door at any time during "baking" or "toasting" or when the oven is not used at all). You also should not forget to extinguish the internal lamp

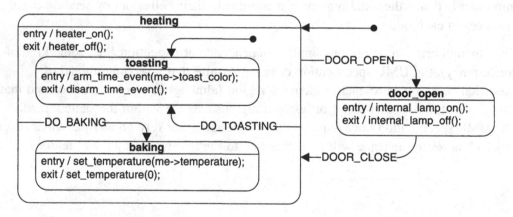

Figure 2.7: Toaster oven state machine with entry and exit actions.

---

[7] State machines are classified as Mealy machines if actions are associated with transitions and as Moore machines if actions are associated with states. UML state machines have characteristics of both Mealy machines and Moore machines.

[8] Commonly, such a safety-critical function is (and should be) redundantly safeguarded by mechanical interlocks, but for the sake of this discussion, suppose you need to implement it entirely in software.

with every transition leaving the "door_open" state. However, such a solution would cause the repetition of actions in many transitions. More important, such an approach is error-prone in view of changes to the state machine (e.g., the next programmer working on a new feature, such as top-browning, might simply forget to disable the heater on transition to "door_open").

Entry and exit actions allow you to implement the desired behavior in a much safer, simpler, and more intuitive way. As shown in Figure 2.7, you could specify that the exit action from "heating" disables the heater, the entry action to "door_open" lights up the oven lamp, and the exit action from "door_open" extinguishes the lamp. The use of entry and exit action is superior to placing actions on transitions because it avoids repetitions of those actions on transitions and eliminates the basic safety hazard of leaving the heater on while the door is open. The semantics of exit actions guarantees that, regardless of the transition path, the heater will be disabled when the toaster is not in the "heating" state.

Because entry actions are executed automatically whenever an associated state is entered, they often determine the conditions of operation or the identity of the state, very much as a class constructor determines the identity of the object being constructed. For example, the identity of the "heating" state is determined by the fact that the heater is turned on. This condition must be established before entering any substate of "heating" because entry actions to a substate of "heating," like "toasting," rely on proper initialization of the "heating" superstate and perform only the differences from this initialization. Consequently, the order of execution of entry actions must always proceed from the outermost state to the innermost state.

Not surprisingly, this order is analogous to the order in which class constructors are invoked. Construction of a class always starts at the very root of the class hierarchy and follows through all inheritance levels down to the class being instantiated. The execution of exit actions, which corresponds to destructor invocation, proceeds in the exact reverse order, starting from the innermost state (corresponding to the most derived class).

2.3.7 Internal Transitions

Very commonly, an event causes only some internal actions to execute but does not lead to a change of state (state transition). In this case, all actions executed comprise the *internal transition*. For example, when you type on your keyboard, it responds by generating different character codes. However, unless you hit the Caps Lock key, the state of the keyboard does not change (no state transition occurs). In UML, this situation should be modeled with internal transitions, as shown in Figure 2.8. The UML notation for internal transitions follows the general syntax used for exit (or entry) actions, except instead of the word *entry* (or *exit*) the internal transition is labeled with the triggering event (e.g., see the internal transition triggered by the ANY_KEY event in Figure 2.8).

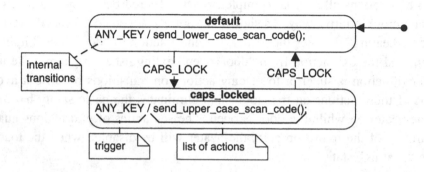

Figure 2.8: UML state diagram of the keyboard state machine with internal transitions.

In the absence of entry and exit actions, internal transitions would be identical to self-transitions (transitions in which the target state is the same as the source state). In fact, in a classical Mealy automaton, actions are associated exclusively with state transitions, so the only way to execute actions without changing state is through a self-transition (depicted as a directed loop in Figure 2.2). However, in the presence of entry and exit actions, as in UML statecharts, a self-transition involves the execution of exit and entry actions and therefore it is distinctively different from an internal transition.

In contrast to a self-transition, no entry or exit actions are *ever* executed as a result of an internal transition, even if the internal transition is inherited from a higher level of the hierarchy than the currently active state. Internal transitions inherited from superstates at any level of nesting act as if they were defined directly in the currently active state.

2.3.8   Transition Execution Sequence

State nesting combined with entry and exit actions significantly complicates the state transition semantics in HSMs compared to the traditional FSMs. When dealing with hierarchically nested states and orthogonal regions, the simple term *current state* can be quite confusing. In an HSM, more than one state can be active at once. If the state machine is in a leaf state that is contained in a composite state (which is possibly contained in a higher-level composite state, and so on), all the composite states that either directly or transitively contain the leaf state are also active. Furthermore, because some of the composite states in this hierarchy might have orthogonal regions, the current active state is actually represented by a tree of states starting with the single top state at the root down to individual simple states at the leaves. The UML specification refers to such a state tree as *state configuration* [OMG 07].

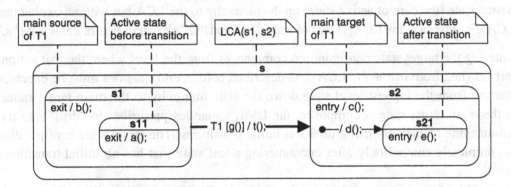

Figure 2.9: State roles in a state transition.

In UML, a state transition can directly connect any two states. These two states, which may be composite, are designated as the *main source* and the *main target* of a transition. Figure 2.9 shows a simple transition example and explains the state roles in that transition. The UML specification prescribes that taking a state transition involves executing the following actions in the following sequence [OMG 07, Section 15.3.13]:

1. Evaluate the guard condition associated with the transition and perform the following steps only if the guard evaluates to TRUE.

2. Exit the source state configuration.

3. Execute the actions associated with the transition.

4. Enter the target state configuration.

The transition sequence is easy to interpret in the simple case of both the main source and the main target nesting at the same level. For example, transition T1 shown in Figure 2.9 causes the evaluation of the guard g(); followed by the sequence of actions: a(); b(); t(); c(); d(); and e(), assuming that the guard g() evaluates to TRUE.

However, in the general case of source and target states nested at different levels of the state hierarchy, it might not be immediately obvious how many levels of nesting need to be exited. The UML specification prescribes that a transition involves exiting all nested states from the current active state (which might be a direct or transitive substate of the main source state) up to, but not including, the *least common ancestor* (LCA) state of the main source and main target states. As the name indicates, the LCA is the lowest composite state that is simultaneously a superstate (ancestor) of both the source and the target states. As described before, the order of execution of exit actions is always from the most deeply nested state (the current active state) up the hierarchy to the LCA but without exiting the LCA. For instance, the LCA(s1, s2) of states "s1" and "s2" shown in Figure 2.9 is state "s."

Entering the target state configuration commences from the level where the exit actions left off (i.e., from inside the LCA). As described before, entry actions must be executed starting from the highest-level state down the state hierarchy to the main target state. If the main target state is composite, the UML semantics prescribes to "drill" into its submachine recursively using the local initial transitions. The target state configuration is completely entered only after encountering a leaf state that has no initial transitions.

NOTE

The HSM implementation described in this book (see Chapter 4) preserves the essential order of exiting the source configuration followed by entering the target state configuration, but executes the actions associated with the transition entirely in the context of the source state, that is, *before* exiting the source state configuration. Specifically, the implemented transition sequence is as follows:

1. Evaluate the guard condition associated with the transition and perform the following steps only if the guard evaluates to TRUE.

2. Execute the actions associated with the transition.

3. *Atomically* exit the source state configuration and enter the target state configuration.

For example, the transition T1 shown in Figure 2.9 will cause the evaluation of the guard g(); followed by the sequence of actions: t(); a(); b(); c(); d(); and e(), assuming that the guard g() evaluates to TRUE.

One big problem with the UML transition sequence is that it requires executing actions associated with the transition *after* destroying the source state configuration but before creating the target state configuration. In the analogy between exit actions in state machines and destructors in OOP, this situation corresponds to executing a class method after partially destroying an object. Of course, such action is illegal in OOP. As it turns out, it is also particularly awkward to implement for state machines.

Executing actions associated with a transition is much more natural in the context of the source state—the same context in which the guard condition is evaluated. Only after the guard and the transition actions execute, the source state configuration is exited and the target state configuration is entered *atomically*. That way the state machine is observable only in a stable state configuration, either before or after the transition, but not in the middle.

2.3.9   Local versus External Transitions

Before UML 2, the only transition semantics in use was the *external transition*, in which the main source of the transition is always exited and the main target of the transition is always entered. UML 2 preserved the "external transition" semantics for backward compatibility, but also introduced a new kind of transition called *local transition* [OMG 07, Section 15.3.15]. For many transition topologies, external and local transitions are actually identical. However, a local transition doesn't cause exit from the main source state if the main target state is a substate of the main source. In addition, local state transition doesn't cause exit and reentry to the target state if the main target is a superstate of the main source state.

Figure 2.10 contrasts local (A) and external (B) transitions. In the top row, you see the case of the main source containing the target. The local transition does not cause exit

Local transition   External transitions

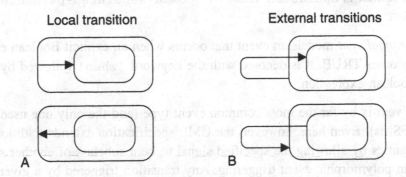

A   B

Figure 2.10: Local (A) versus external transitions (B). QP implements only the local transitions.

from the source, while the external transition causes exit and re-entry to the source. In the bottom row of Figure 2.10, you see the case of the target containing the source. The local transition does not cause entry to the target, whereas the external transition causes exit and reentry to the target.

NOTE

The HSM implementation described in Chapter 4 of this book (as well as the HSM implementation described in the first edition) supports exclusively the *local* state transition semantics.

2.3.10   Event Types in the UML

The UML specification defines four kinds of events, each one distinguished by a specific notation.

- *SignalEvent* represents the reception of a particular (asynchronous) signal. Its format is `signal-name '(' comma-separated-parameter-list ')'`.

- *TimeEvent* models the expiration of a specific deadline. It is denoted with the keyword `'after'` followed by an expression specifying the amount of time. The time is measured from the entry to the state in which the TimeEvent is used as a trigger.

- *CallEvent* represents the request to synchronously invoke a specific operation. Its format is `operation-name '(' comma-separated-parameter-list ')'`.

- *ChangeEvent* models an event that occurs when an explicit Boolean expression becomes TRUE. It is denoted with the keyword `'when'` followed by a Boolean expression.

A SignalEvent is by far the most common event type (and the only one used in classical FSMs). Even here, however, the UML specification extends traditional FSM semantics by allowing the specified signal to be a subclass of another signal, resulting in polymorphic event triggering. Any transition triggered by a given signal event is also triggered by any subevent derived directly or indirectly from the original event.

> **NOTE**
>
> The HSM implementation described in this book (see Chapter 4) supports only the SignalEvent type. The real-time framework described in Part II also adds support for the TimeEvent type, but TimeEvents in QF require explicit arming and disarming, which is not compatible with the UML 'after' notation. The polymorphic event triggering for SignalEvents is not supported, due to its inherent complexity and very high performance costs.

2.3.11   Event Deferral

Sometimes an event arrives at a particularly inconvenient time, when a state machine is in a state that cannot handle the event. In many cases, the nature of the event is such that it can be postponed (within limits) until the system enters another state, in which it is much better prepared to handle the original event.

UML state machines provide a special mechanism for deferring events in states. In every state, you can include a clause 'deferred / [event list]'. If an event in the current state's deferred event list occurs, the event will be saved (deferred) for future processing until a state is entered that does not list the event in its deferred event list. Upon entry to such state, the UML state machine will automatically recall any saved event(s) that are no longer deferred and process them as if they have just arrived.

> **NOTE**
>
> The HSM implementation described in this book (see Chapter 4) does not directly support the UML-style event deferral. However, the "Deferred Event" state pattern presented in Chapter 5 shows how to approximate this feature in a much less expensive way by explicitly deferring and recalling events.

2.3.12   Pseudostates

Because statecharts started as a visual formalism [Harel 87], some nodes in the diagrams other than the regular states turned out to be useful for implementing various features (or simply as a shorthand notation). The various "plumbing gear" nodes are collectively called *pseudostates*. More formally, a pseudostate is an abstraction that encompasses different types of transient vertices (nodes) in the state machine graph. The UML specification [OMG 07] defines the following kinds of pseudostates:

- The initial pseudostate (shown as a black dot) represents a source for initial transition. There can be, at most, one initial pseudostate in a composite state. The outgoing transition from the initial pseudostate may have actions but not a trigger or guard.

- The choice pseudostate (shown as a diamond or an empty circle) is used for dynamic conditional branches. It allows the splitting of transitions into multiple outgoing paths, so the decision as to which path to take could depend on the results of prior actions performed in the same RTC step.

- The shallow-history pseudostate (shown as a circled letter *H*) is a shorthand notation that represents the most recent active direct substate of its containing state. A transition coming into the shallow-history vertex (called a *transition to history*) is equivalent to a transition coming into the most recent active substate of a state. A transition can originate from the history connector to designate a state to be entered in case a composite state has no history yet (has never been active before).

- The deep-history pseudostate (shown as a circled *H**) is similar to shallow-history except it represents the whole, most recent state configuration of the composite state that directly contains the pseudostate.

- The junction pseudostate (shown as a black dot) is a semantics-free vertex that chains together multiple transitions. A junction is like a Swiss Army knife: It performs various functions. Junctions can be used both to merge multiple incoming transitions (from the same concurrent region) and to split an incoming transition into multiple outgoing transition segments with different guards. The latter case realizes a static conditional branch because the use of a junction imposes static evaluation of all guards before the transition is taken.

- The join pseudostate (shown as a vertical bar) serves to merge several transitions emanating from source vertices in different orthogonal regions.

- The fork pseudostate (represented identically as a join) serves to split an incoming transition into two or more transitions terminating in different orthogonal regions.

2.3.13 UML Statecharts and Automatic Code Synthesis

UML statecharts provide sufficiently well-defined semantics for building executable state models. Indeed, several design automation tools on the market support various versions of statecharts (see the sidebar "Design Automation Tools Supporting Statecharts"). The commercially available design automation tools typically not only automatically generate code from statecharts but also enable debugging and testing of the state models at the graphical level [Douglass 99].

But what does automatic code generation really mean? And more important, what kind of code is actually generated by such statechart-based tools?

Many people understand automatic code synthesis as the generation of a program to solve a problem from a statement of the problem specification. Statechart-based tools cannot provide this because a statechart is just a higher-level (mostly visual) solution rather than the statement of the problem.

As far as the automatically generated code is concerned, the statechart-based tools can autonomously generate only so-called "housekeeping code" [Douglass 99]. The modeler explicitly must provide all the application-specific code, such as action and guard expressions, to the tool. The role of housekeeping code is to "glue" the various action and guard expressions together to ensure proper state machine execution in accordance with the statechart semantics. For example, synthesized code typically handles event queuing, event dispatching, guard evaluation, or transition chain execution (including exit and entry of appropriate states). Almost universally, the tools also encompass some kind of real-time framework (see Part II of this book) that integrates tightly with the underlying operating system.

DESIGN AUTOMATION TOOLS SUPPORTING STATECHARTS

Some of the computer-aided software-engineering (CASE) tools with support for statecharts currently available on the market are (see also Queens University CASE tool index [Queens 07]):

- Telelogic Statemate, www.telelogic.com (the tool originally developed by I-Logix, Inc. acquired by Telelogic in 2006, which in turn is in the process of being acquired by IBM)

- Telelogic Rhapsody, www.telelogic.com

Continued onto next page

- Rational Suite Development Studio Real-Time, Rational Software Corporation, www. ibm.com/software/rational (Rational was acquired by IBM in 2006)

- ARTiSAN Studio, ARTiSAN Software Tools, Ltd., www.artisansw.com

- Stateflow, The Mathworks, www.mathworks.com

- VisualState, IAR Systems, www.iar.com

2.3.14   The Limitations of the UML State Diagrams

Statecharts have been invented as "a visual formalism for complex systems" [Harel 87], so from their inception, they have been inseparably associated with graphical representation in the form of state diagrams. However, it is important to understand that the concept of HSMs transcends any particular notation, graphical or textual. The UML specification [OMG 07] makes this distinction apparent by clearly separating state machine semantics from the notation.

However, the notation of UML statecharts is not purely visual. Any nontrivial state machine requires a large amount of textual information (e.g., the specification of actions and guards). The exact syntax of action and guard expressions isn't defined in the UML specification, so many people use either structured English or, more formally, expressions in an implementation language such as C, C++, or Java [Douglass 99b]. In practice, this means that UML statechart notation depends heavily on the specific programming language.

Nevertheless, most of the statecharts semantics are heavily biased toward graphical notation. For example, state diagrams poorly represent the sequence of processing, be it order of evaluation of guards or order of dispatching events to orthogonal regions. The UML specification sidesteps these problems by putting the burden on the designer not to rely on any particular sequencing. But, as you will see in Chapters 3 and 4, when you actually implement UML state machines, you will always have full control over the order of execution, so the restrictions imposed by UML semantics will be unnecessarily restrictive. Similarly, statechart diagrams require a lot of plumbing gear (pseudostates, like joins, forks, junctions, choicepoints, etc.) to represent the flow of control graphically. These elements are nothing but the old flowchart in disguise, which structured programming techniques proved far less significant[9] a long time ago. In other words, the graphical notation does not add much value in representing flow of control as compared to plain structured code.

---

[9] You can find a critique of flowcharts in Brooks [Brooks 95].

This is not to criticize the graphical notation of statecharts. In fact, it is remarkably expressive and can scarcely be improved. Rather, I want merely to point out some shortcomings and limitations of the pen-and-paper diagrams.

The UML notation and semantics are really geared toward computerized design automation tools. A UML state machine, as represented in a tool, is a not just the state diagram, but rather a mixture of graphical and textual representation that precisely captures both the state topology and the actions. The users of the tool can get several complementary views of the same state machine, both visual and textual, whereas the generated code is just one of the many available views.

2.3.15   UML State Machine Semantics: An Exhaustive Example

The very rich UML state machine semantics might be quite confusing to newcomers, and even to fairly experienced designers. Wouldn't it be great if you could generate the exact sequence of actions for every possible transition so that you know *for sure* what actions get executed and in which order?

In this section, I present an executable example of a hierarchical state machine shown in Figure 2.11 that contains all possible transition topologies up to four levels of state nesting. The state machine contains six states: "s," "s1," "s11," "s2," "s21," and "s211." The state machine recognizes nine events A through I, which you can generate by typing either uppercase or lowercase letters on your keyboard. All the actions of these state machines consist only of `printf()` statements that report the status of the state machine to the screen. The executable console application for Windows is located in the directory `<qp>\qpc\examples\80x86\dos\tcpp101\l\qhsmtst \dbg\`. The name of the application is `QHSMTST.EXE`.

Figure 2.12 shows an example run of the `QHSMTST.EXE` application. Note the line numbers in parentheses at the left edge of the window, added for reference. Line (1) shows the effect of the topmost initial transition. Note the sequence of entry actions and initial transitions ending at the "s211-ENTRY" printout. Injecting events into the state machine begins in line (2). Every generated event (shown on a gray background) is followed by the sequence of exit actions from the source state configuration followed by entry actions and initial transitions entering the target state configuration. From these printouts you can always determine the order of transition processing as well as the active state, which is the last state entered. For instance, the active state before injecting event G in line (2) is "s211" because this is the last state entered in the previous line.

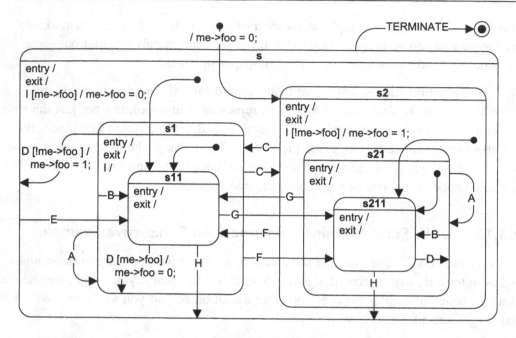

Figure 2.11: Hypothetical state machine that contains all possible state transition topologies up to four levels of state nesting.

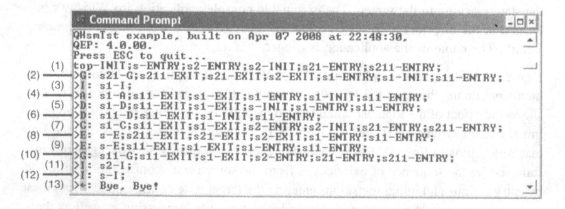

Figure 2.12: QHSMTST.EXE example application running in the command window. The line numbers in brackets to the right are added for reference.

Per the semantics of UML sate machines, the event G injected in line (2) is handled in the following way. First, the active state ("s211") attempts to handle the event. However, as you can see in the state diagram in Figure 2.11, state "s211" does not prescribe how to handle event G. Therefore the event is passed on to the next higher-level state in the hierarchy, which is state "s21." The superstate "s21" prescribes how to handle event G because it has a state transition triggered by G. State "s21" executes the actions associated with the transition (printout "s21-G;" in line (2) of Figure 2.12). Next the state executes the transition chain that exits the source state configuration and enters the target state configuration. The transition chain starts with execution of exit actions from the active state through higher and higher levels of hierarchy. Next the target state configuration is entered in exact opposite order, that is, starting from highest levels of state hierarchy down to the lowest. The transition G from state "s21" terminates at state "s1." However, the transition chain does not end at the direct target of the transition but continues via the initial transition defined in the direct target state "s21." Finally, a new active state is established by entering state "s11" that has no initial transition.

In line (3) of Figure 2.12, you see how the statechart handles an internal transition. Event I is injected while state machine is in state "s11." Again, the active state does not prescribe how to handle event I, so it is passed on to the next level of state hierarchy, that is, to state "s1." State "s1" has an internal transition triggered by I defined in its internal transition compartment; therefore "s1" handles the event (printout "s1-I;" in line (3)). And at this point the processing ends. No change of state ever occurs in the internal transition, even if such transition is inherited from higher levels of state hierarchy.

In the UML state machines internal transitions are different from self-transitions. Line (4) of Figure 2.12 demonstrates the difference. The state machine is in state "s11" when event A is injected. As in the case of the internal transition, the active state "s11" does not prescribe how to handle event A, so the event is passed on to the superstate "s1." The superstate has a self-transition triggered by A and so it executes actions associated with the transition (printout "s1-A;" in line (4)). This time, however, a regular transition is taken, which requires exiting the source state configuration and entering the target state configuration.

UML statecharts are extended state machines, meaning that in general, the actions executed by the state machine can depend also on the value of the extended state variables. Consider for example event D injected in line (5). The active state "s11" has transition D, but this transition has a guard `[me->foo]`. The variable `me->foo` is an extended state variable of the state machine from Figure 2.11. You can see that `me->foo` is initialized to 0 on the topmost initial transition. Therefore the guard `[me->foo]`, which is a test of `me->foo` against 0, evaluates to FALSE. The guard condition temporarily disables the transition D in state "s11," which is handled as though state "s11" did not define the transition in the first place. Therefore, the event D is passed on to the next higher level, that is, to state "s1." State "s1" has transition D with a complementary guard `[!me->foo]`. This time, the guard evaluates to TRUE and the transition D in state "s1" is taken. As indicated in the diagram, the transition action changes the value of the extended state variable `me->foo` to 1. Therefore when another event D is injected again in line (6), the guard condition `[me->foo]` on transition D in state "s11" evaluates to TRUE and this transition is taken, as indicated in line (6) of Figure 2.12.

Line (7) of Figure 2.12 demonstrates that all exit and entry actions are always executed, regardless of the exit and entry path. The main target of transition C from "s1" is "s2." The initial transition in the main target state goes "over" the substate "s21" all the way to the substate "s211." However, the entry actions don't skip the entry to "s21." This example demonstrates the powerful concept of guaranteed cleanup of the source state configuration and guaranteed initialization of the target state configuration, regardless of the complexity of the exit or entry path.

Interestingly, in hierarchical state machines, the same transition can cause different sequences of exit actions, depending on which state inherits the transition. For example, in lines (8) and (9) of Figure 2.12, event E triggers the exact same state transition defined in the superstate "s." However, the responses in lines (8) and (9) are different because transition E fires from different state configurations—once when "s211" is active (line (8)) and next when "s11" is active (line (9)).

Finally, lines (11) and (12) of Figure 2.12 demonstrate that guard conditions can also be used on internal transitions. States "s2" and *d* both define internal transition I with complementary guard conditions. In line (11), the guard `[!me->foo]` enables internal transition in state "s2." In line (12) the same guard disables the internal transition in "s2," and therefore the internal transition defined in the superstate *d* is executed.

You can learn much more about the semantics of UML state machines by injecting various events to the QHSMTST.EXE application and studying its output. Because the state machine from Figure 2.11 has been specifically designed to contain all possible state transition configurations up to level 4 of nesting, this example can "answer" virtually all your questions regarding the semantics of statecharts. Moreover, you can use the source code that actually implements the state machine (located in `<qp>\qpc\examples\ 80x86\dos\tcpp101\l\qhsmtst\qhsmtst.c`) as a template for implementing your own statecharts. Simply look up applicable fragments in the diagram from Figure 2.11 and check how they have been implemented in `qhsmtst.c`.

2.4 Designing A UML State Machine

Designing a UML state machine, as any design, is not a strict science. Typically it is an iterative and incremental process: You design a little, code a little, test a little, and so on. In that manner you may converge at a correct design in many different ways, and typically also, more than one correct HSM design satisfies a given problem specification. To focus the discussion, here I walk you through a design of an UML state machine that implements correctly the behavior of a simple calculator similar to the Visual Basic calculator used at the beginning of this chapter. Obviously, the presented solution is just one of the many possible.

Figure 2.13: A simple electronic calculator used as a model for the statechart example.

2.4.1 Problem Specification

The calculator (see Figure 2.13) operates broadly as follows: a user enters an operand, then an operator, then another operand, and finally clicks the equals button to get a result. From the programming perspective, this means that the calculator needs to parse numerical expressions, defined formally by the following BNF grammar:

```
expression ::= operand1 operator operand2 '='
operand1   ::= expression | ['+' | '-'] number
operand2   ::= ['+' | '-'] number
number     ::= {'0' | '1' | ... '9'}* ['.' {'0' | '1' | ... '9'}*]
operator   ::= '+' | '-' | '*' | '/'
```

The problem is not only to correctly parse numerical expressions, but also to do it interactively ("on the fly"). The user can provide any symbol at any time, not necessarily only the symbols allowed by the grammar in the current context. It is up to the application to ignore such symbols. (This particular application ignores invalid inputs. Often an even better approach is to actively prevent generation of the invalid inputs in the first place by disabling invalid options, for example.) In addition, the application must handle inputs not related to parsing expressions, for example Cancel (C) or Cancel Entry (CE). All this adds up to a nontrivial problem, which is difficult to tackle with the traditional event-action paradigm (see Section 2.1) or even with the traditional (nonhierarchical) FSM.

2.4.2   High-Level Design

Figure 2.14 shows first steps in elaborating the calculator statechart. In the very first step (panel (a)), the state machine attempts to realize the primary function of the system (the primary use case), which is to compute expressions: operand1 operator operand2 equals... The state machine starts in the "operand1" state, whose function is to ensure that the user can only enter a valid operand. This state obviously needs some internal submachine to accomplish this goal, but we ignore it for now. The criterion for transitioning out of "operand1" is entering an operator (+, −, *, or /). The statechart then enters the "opEntered" state, in which the calculator waits for the second operand. When the user clicks a digit (0 .. 9) or a decimal point, the state machine transitions to the "operand2" state, which is similar to "operand1." Finally, the user clicks =, at which point the calculator computes and displays the result. It then transitions back to the "operand1" state to get ready for another computation.

The simple state model from Figure 2.14(A) has a major problem, however. When the user clicks = in the last step, the state machine cannot transition directly to "operand1" because this would erase the result from the display (to get ready for the first operand). We need another state "result" in which the calculator pauses to display

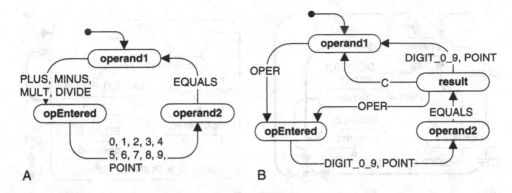

Figure 2.14: The first two steps in elaborating the calculator statechart[10].

the result (Figure 2.14(B)). Three things can happen in the "result" state: (1) the user may click an operator button to use the result as the first operand of a new computation (see the recursive production in line 2 of the calculator grammar), (2) the user may click Cancel (C) to start a completely new computation, or (3) the user may enter a number or a decimal point to start entering the first operand.

TIP

Figure 2.14(B) illustrates a trick worth remembering: the consolidation of signals PLUS, MINUS, MULTIPLY, and DIVIDE into a higher-level signal OPER (operand). This transformation avoids repetition of the same group of triggers on two transitions (from "operand1" to "opEntered" and from "result" to "opEntered"). Although most events are generated externally to the statechart, in many situations it is still possible to perform simple transformations before dispatching them (e.g., a transformation of raw button presses into the calculator events). Such transformations often simplify designs more than the trickiest state and transition topologies.

2.4.3 Scavenging for Reuse

The state machine from Figure 2.14(B) accepts the C (Cancel) command only in the result state. However, the user expects to be able to cancel and start over at any time. Similarly, the user expects to be able to turn the calculator off at any time. Statechart in Figure 2.15(A) adds these features in a naïve way. A better solution is to factor

---

[10] In this section, I am using a shorthand notation to represent many transitions with the same source and target as just one transition arrow with a multiple triggers.

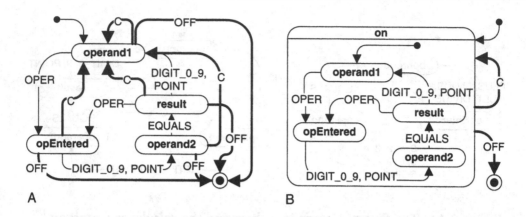

Figure 2.15: Applying state nesting to factorize out the common Cancel transition (C).

out the common transition into a higher-level state named On and let all substates reuse the Cancel and Off transitions through behavioral inheritance, as shown in Figure 2.15(B).

2.4.4   Elaborating Composite States

The states "operand1" and "operand2" need submachines to parse floating-point numbers. Figure 2.16 refers to both these states simultaneously as "operandX" state.

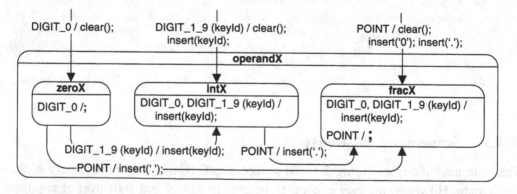

Figure 2.16: Internal submachine of states "operand1" and "operand2."

These submachines consist of three substates. The "zero" substate is entered when the user clicks 0. Its function is to ignore additional zeros that the user may try to enter (so that the calculator displays only one 0). Note my notation for explicitly ignoring an event. I use the internal transition (DIGIT_0 in this case) followed by an explicitly empty list of actions (a semicolon in C).

The function of the "int" substate is to parse integer part of a number. This state is entered either from outside or from the "zero" peer substate (when the user clicks 1 through 9). Finally, the substate "frac" parses the fractional part of the number. It is entered from either outside or both peer substates when the user clicks a decimal point (.). Again, note that the "frac" substate explicitly ignores the decimal point POINT event, so that the user cannot enter multiple decimal points in the fractional part of a number.

2.4.5 Refining the Behavior

The last step brings the calculator statechart to the point at which it can actually compute expressions. However, it can handle only positive numbers. In the next step, I will add handling of negative numbers. This turns out to be perhaps the toughest problem in this design because the same button, – (minus), represents in some contexts the binary operator of subtraction and sometimes the unary operator of negation.

There are only two possible contexts in which – can unambiguously represent the negation rather than the subtraction: (1) in the "opEntered" state (as in the expression: 2 * −2 =), and (2) at the beginning of a new computation (as in the expression: −2 * 2 =).

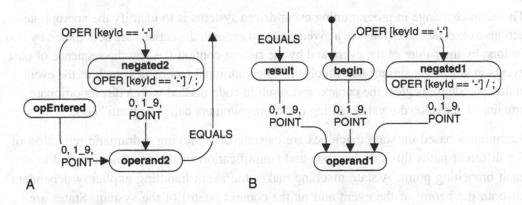

Figure 2.17: Two cases of handling negative numbers.

The solution to the first case (shown in Figure 2.17(A)) is simpler. We need one more state "negated2," which is entered when the operator is MINUS (note the use of the guard). Upon entry, this state sets up the display to show –0 and subsequently does not clear the display when transitioning to the "operand2" state. This is a different behavior from "opEntered" because in this latter state the display must be cleared to prepare for entering of the second operand.

The second case in which – represents the negation is trickier because the specification "beginning of new computation" is much more subtle. Here it indicates the situation just after launching the application or after the user clicks Cancel but not when the calculator displays the result from the previous computation. Figure 2.17(B) shows the solution. A new state "begin" is created to capture the behavior specific to the "beginning of new computation" (note the initial transition pointing now to "begin" rather than to "operand1"). The rest of the solution is analogous as in the first case, except now the state "begin" plays the role of "opEntered."

2.4.6   Final Touches

The calculator is almost ready now. The final touches (which I leave as an exercise) include adding Cancel-Entry transitions in appropriate contexts and adding an "error" state to capture overflows and division by zero. Figure 2.18 shows the final calculator state diagram. The actual C implementation of this state machine will be described in Chapter 4.

2.5   Summary

The main challenge in programming event-driven systems is to identify the appropriate actions to execute in response to a given event. In general, the actions are determined by two factors: by the nature of the event and by the current context (i.e., by the sequence of past events in which the system was involved). The traditional techniques, such as the event-action paradigm, neglect the context and result in code riddled with a disproportionate amount of convoluted conditional logic that programmers call "spaghetti" code.

Techniques based on state machines are capable of achieving a dramatic reduction of the different paths through the code and simplification of the conditions tested at each branching point. A state machine makes the event handling explicitly dependent on both the nature of the event and on the context (state) of the system. States are "chunks of behavior," whereas the event-action paradigm is applied locally within each state. The concept of state is a very useful abstraction of system history, capable of capturing only relevant sequences of stimuli (and ignoring all irrelevant ones). In extended state machines (state machines with "memory"), state corresponds to qualitative aspects of system behavior, whereas extended state variables (program memory) correspond to the quantitative aspects.

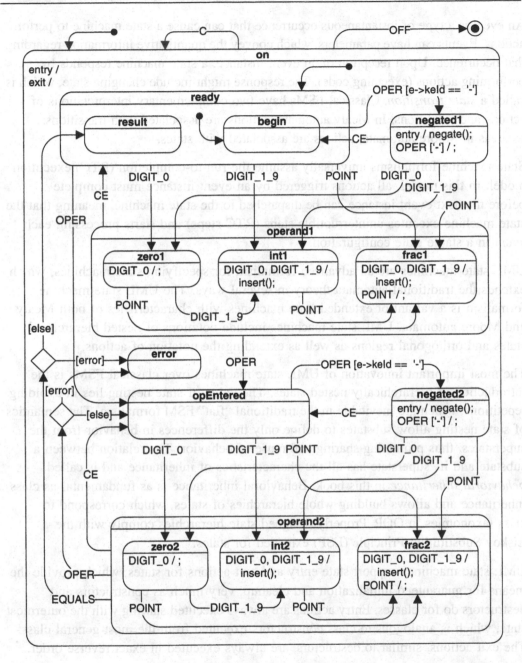

Figure 2.18: The final calculator statechart.

An event is a type of instantaneous occurrence that can cause a state machine to perform actions. Events can have parameters, which convey the quantitative information regarding that occurrence. Upon reception of an event instance, a state machine responds by performing actions (executing code). The response might include changing state, which is called a *state transition*. Classical FSMs have two complementary interpretations of actions and transitions. In Mealy automata, actions are associated with transitions, whereas in Moore automata, actions are associated with states.

State machine formalisms universally assume the run-to-completion (RTC) execution model. In this model, all actions triggered by an event instance must complete before the next event instance can be dispatched to the state machine, meaning that the state machine executes uninterruptible steps (RTC steps) and starts processing each event in a stable state configuration.

UML state machines are an advanced formalism for specifying state machines, which extends the traditional automata theory in several ways. The UML state machine formalism is a variant of extended state machines with characteristics of both Mealy and Moore automata. UML state machines include notations of nested hierarchical states and orthogonal regions as well as extending the notation of actions.

The most important innovation of UML state machines over classical FSMs is the introduction of hierarchically nested states. The value of state nesting lies in avoiding repetitions, which are inevitable in the traditional "flat" FSM formalism. The semantics of state nesting allow substates to define only the differences in behavior from the superstates, thus promoting sharing and reuse of behavior. The relation between a substate and its superstate has all the characteristics of inheritance and is called *behavioral inheritance* in this book. Behavioral inheritance is as fundamental as class inheritance and allows building whole hierarchies of states, which correspond to class taxonomies in OOP. Properly designed state hierarchies comply with the Liskov Substitution Principle (LSP) extended for states.

UML state machines support state entry and exit actions for states, which provide the means for guaranteed initialization and cleanup, very much as constructors and destructors do for classes. Entry actions are always executed starting with the outermost state, which is analogous to class constructors executed from the most general class. The exit actions, similar to destructors, are always executed in exact reverse order.

Entry and exit actions combined with state nesting complicate transition sequence. The precise semantics of state transitions can be confusing. The exhaustive example

QHSMTST.EXE discussed in this chapter can precisely "answer" virtually all your questions regarding the semantics of state machine execution.

Statecharts were first invented as a visual formalism; therefore, they are heavily biased toward graphical representation. However, it is important to distinguish the underlying concept of the HSM from the graphical notation. It is also important to distinguish between statecharts and flowcharts.

Designing effective UML state machines is not trivial and, as with most designs, typically requires incremental, iterative process. Reuse does not come automatically, but you must actively look for it. Chapter 5 presents a mini-catalog of proven, effective state machine designs called *state patterns*.

CHAPTER 3

Standard State Machine Implementations

An expert is a man who has made all the mistakes which can be made, in a narrow field.
— Niels Bohr

This chapter discusses standard state machine implementation techniques, which you can find in the literature or in the working code. They are mostly applicable to the traditional nonhierarchical extended finite state machines (FSMs) because hardly any standard implementations of hierarchical state machines (HSMs) are intended for manual coding.[1]

Typical implementations of state machines in high-level programming languages, such as C or C++, include:

- The nested `switch` statement

- The state table

- The object-oriented State design pattern

- Other techniques that are often a combination of the above

---

[1] Design automation tools often use standard state machine implementation techniques to generate code for hierarchical state machines. The resulting code, however, is typically intended not for manual maintenance but rather to be regenerated every time you change the state diagram inside the tool.

3.1   The Time-Bomb Example

To focus the discussion and allow meaningful comparisons, in all following state machine implementation techniques I'll use the same time-bomb example. As shown Figure 3.1, the time bomb has a control panel with an LCD that shows the current value of the timeout and three buttons: UP, DOWN, and ARM. The user begins with setting up the time bomb using the UP and DOWN buttons to adjust the timeout in one-second steps. Once the desired timeout is selected, the user can arm the bomb by pressing the ARM button. When armed, the bomb starts decrementing the timeout every second and explodes when the timeout reaches zero. An additional safety feature is the option to defuse an armed bomb by entering a secret code. The secret defuse code is a certain combination of the UP and DOWN buttons terminated with the ARM button press. Of course, the defuse code must be correctly entered before the bomb times out.

Figure 3.1: Time bomb controller user interface.

Figure 3.2 shows FSM that models the time-bomb behavior. The following explanation section illuminates the interesting points.

(1)   The initial transition initializes the `me->timeout` extended state variable and enters the "setting" state.

(2)   If the timeout is below the 60-second limit, the internal transition UP in state "setting" increments the `me->timeout` variable and displays it on the LCD.

(3)   If the timeout is above the 1 second limit, the internal transition DOWN in state "setting" decrements the `me->timeout` variable and displays it on the LCD.

(4)   The transition ARM to state "timing" clears the `me->code` extended state variable to make sure that the code for defusing the bomb is wiped clean before entering the "timing" state.

(5)   The internal transition UP in state "timing" shifts the me->code variable and inserts the 1-bit into the least-significant-bit position.

(6)   The internal transition DOWN in state "timing" just shifts the me->code variable and leaves the least-significant-bit at zero.

(7)   If the entered defuse code me->code matches exactly the secret code me->defuse given to the bomb in the constructor, the transition ARM to state "setting" disarms the ticking bomb. Note that the "setting" state does not handle the TICK event, which means that TICK is ignored in this state.

(8)   The handling of the most important TICK event in state "timing" is the most complex. To make the time-bomb example a little more interesting, I decided to generate the TICK event 10 times per second and to include an event parameter fine_time with the TICK event. The fine_time parameter contains the fractional part of a second in 1/10 s increments, so it cycles through 0, 1, .., 9 and back to 0. The guard [e->fine_time == 0] checks for the full-second rollover condition, at which time the me->timeout variable is decremented and displayed on the LCD.

(9)   The choice pseudostate is evaluated only if the first segment of the TICK transition fires. If the me->timeout variable is zero, the transition segment is executed that calls the BSP_boom() function to trigger the destruction of the bomb (transition to the final state).

(10)   Otherwise the choice pseudostate selects the transition back to "timing."

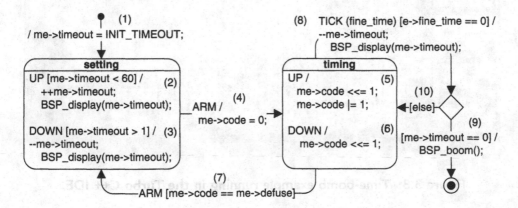

Figure 3.2: UML state diagram representing the time-bomb state machine.

3.1.1 Executing the Example Code

Every state machine implementation technique described in this chapter comes with the complete source code for the time-bomb example and the precompiled executable program. The C version of the code is located in the directory <qp>\qpc\examples\80x86\dos\tcpp101\bomb\, whereas the C++ version is found in <qp>\qpcpp\examples\80x86\dos\tcpp101\bomb\.

The executable program for each coding technique is a simple console application compiled with the free Turbo C++ 1.01 compiler (see Section 1.1 in Chapter 1). The Turbo C++ project files are included with the application source code. Figure 3.3 shows an example run of the time-bomb application.

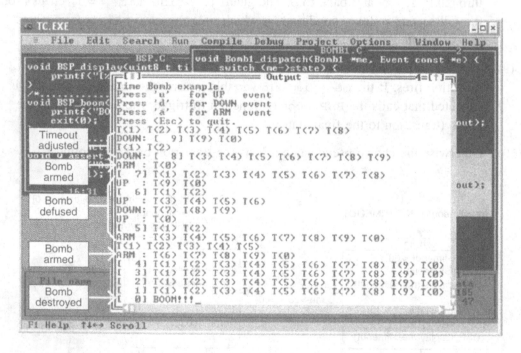

Figure 3.3: Time-bomb example running in the Turbo C++ IDE.

> **NOTE**
>
> Because all the examples in this chapter are simple console applications, they can easily be compiled with just about any C/C++ compiler for your favorite desktop workstation, including Linux.[2] The code accompanying this book provides the project files for the legacy Turbo C++ 1.01 compiler — the same one that I've used in Chapter 1. You can run the generated executables in any variant of Windows.

You generate events by pressing appropriate keys on the keyboard (see the legend of recognized keypresses at the top of the output window in Figure 3.3.). The time bomb responds by printing every event it receives (the TICK event is represented as *T*) as well as every update to the LCD, which is shown in square brackets. The application terminates automatically when the bomb explodes. You can also quit the program at any time by pressing the Esc key.

3.2 A Generic State Machine Interface

The majority of published state machine code presents state machines intimately intertwined with a specific concurrency model and a particular event-passing method. For example, embedded systems engineers[3] often present their state machines inside polling loops or interrupt service routines (ISRs) that extract events and event parameters directly from hardware registers or global variables. GUI programmers are typically more disciplined in this respect because the GUI system already provides a consistent interface, but then again GUI programmers seldom use state machines, as demonstrated in the Visual Basic Calculator example in Chapter 2.

However, it is far better to separate the state machine code from a particular concurrency model and to provide a flexible and uniform way of passing events with arbitrary parameters. Therefore, implementations in this chapter use a simple and

---

[2] The Linux platform requires slightly different techniques for interacting with the console, but the state machine code can be exactly the same.

[3] Judging by 20 years of articles (1988–2008) published on the subject in *Embedded Systems Design* magazine (formerly *Embedded Systems Programming*).

generally applicable interface to a state machine.[4] The interface I propose consists of just two functions: init(), to trigger the top-level initial transition in a state machine, and dispatch(), to dispatch an event to the state machine. In this simple model, a state machine is externally driven by invoking init() once and dispatch() repetitively, for each event.

3.2.1   Representing Events

To nail down the signature of the dispatch() function, we need a uniform representation of events. Again, this is where the standard approaches vary the most. For example, the GUI systems, such as Microsoft Windows, provide only events with fixed sets of event parameters that are passed to the WinMain() function and thus are not generally applicable outside the GUI domain (after all, the most complex event parameters that a GUI needs to handle are the parameters of the mouse-click event).

As described in Chapter 2, events consist really of two parts: the *signal* of the event conveys the type of the occurrence (such as arrival of the time tick), and *event parameters* convey the quantitative information about the occurrence (such as the fractional part of a second in the time tick event). In event-driven systems, event instances are frequently passed around, placed in event queues, and eventually consumed by state machines. Consequently, it is very convenient to represent events as *event objects* that combine the signal and the event parameters into one entity.

The following Event structure represents event objects. The scalar data member sig contains the signal information, which is an integer number that identifies the event, such as UP, DOWN, ARM, or TICK:

```
typedef struct EventTag {
    uint16_t sig;                                /* signal of the event */
    /* add event parameters by derivation from the Event structure... */
} Event;
```

---

[4] Because of the special nature, the object-oriented State design pattern uses a different interface for dispatching events.

> **NOTE**
>
> Throughout this book I use the following standard exact-width integer types (WG14/N843 C99 Standard, Section 7.18.1.1) [ISO/IEC 9899:TC2]:
>
> | Exact Size | Unsigned | Signed |
> |---|---|---|
> | 8 bits | uint8_t | int8_t |
> | 16 bits | uint16_t | int16_t |
> | 32 bits | uint32_t | int32_t |
>
> If your (prestandard) compiler does not provide the `<stdint.h>` header file, you can always `typedef` the exact-width integer types using the standard C data types such as `signed`/`unsigned char`, `short`, `int` and `long`.

You can add arbitrary event parameters to an event in the process of "derivation of structures" described in the sidebar "Single Inheritance in C" (Chapter 1, Section 1.6). For example, the following `TickEvt` structure declares an event with the parameter `fine_time` used in the `TICK(fine_time)` event (see Figure 3.2(8)).

```
typedef struct TickEvtTag {
    Event super;                /* derive from the Event structure */
    uint8_t fine_time;              /* the fine 1/10 s counter */
} TickEvt;
```

As shown in Figure 3.4, such nesting of structures always aligns the data member `super` at the beginning of every instance of the derived structure. In particular, this alignment lets you treat a pointer to the derived `TickEvt` structure as a pointer to the `Event` base structure. Consequently, you can always safely pass a pointer to `TickEvt` to any C function that expects a pointer to `Event`.

With this representation of events, the signature of the `dispatch()` function looks as follows:

```
void dispatch(StateMachine *me, Event const *e);
```

The first argument 'StateMachine *me' is the pointer to the state machine object. Different state machine implementation techniques will define the StateMachine structure differently. The second argument 'Event const *e' is the pointer to the event object, which might point to a structure derived from Event and thus might contain arbitrary event parameters (see Figure 3.4(B)). This interface becomes clearer when you see how it is used in the concrete implementation techniques.

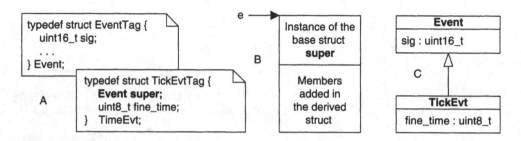

Figure 3.4: Adding parameters to events in the process of derivation of structures (inheritance).

3.3   Nested Switch Statement

Perhaps the most popular and straightforward technique of implementing state machines is the nested switch statement, with a scalar state variable used as the discriminator in the first level of the switch and the signal of the event used in the second level.

3.3.1   Example Implementation

Listing 3.1 shows a typical implementation of the time bomb FSM from Figure 3.2. The explanation section immediately following the listing illuminates the interesting points.

Listing 3.1 Time bomb state machine implemented using the nested switch statement technique (see file bomb1.c)

```
(1)    enum BombSignals {                        /* all signals for the Bomb FSM */
           UP_SIG,
           DOWN_SIG,
           ARM_SIG,
           TICK_SIG
       };
       /*..............................................................*/
```

```
(2)     enum BombStates {                          /* all states for the Bomb FSM */
            SETTING_STATE,
            TIMING_STATE
        };
        /*.........................................................................*/
(3)     typedef struct EventTag {
(4)         uint16_t sig;                                       /* signal of the event */
            /* add event parameters by derivation from the Event structure... */
        } Event;

(5)     typedef struct TickEvtTag {
(6)         Event super;                            /* derive from the Event structure */
(7)         uint8_t fine_time;                         /* the fine 1/10 s counter */
        } TickEvt;
        /*.........................................................................*/
(8)     typedef struct Bomb1Tag {                              /* the Bomb FSM */
(9)         uint8_t state;                            /* the scalar state-variable */
(10)        uint8_t timeout;                   /* number of seconds till explosion */
(11)        uint8_t code;              /* currently entered code to disarm the bomb */
(12)        uint8_t defuse;               /* secret defuse code to disarm the bomb */
        } Bomb1;
                                            /* macro for taking a state transition */
(13)    #define TRAN(target_) (me->state = (uint8_t)(target_))
        /*.........................................................................*/
(14)    void Bomb1_ctor(Bomb1 *me, uint8_t defuse) {          /* the "constructor" */
(15)        me->defuse = defuse;    /* the defuse code is assigned at instantiation */
        }
        /*.........................................................................*/
(16)    void Bomb1_init(Bomb1 *me) {                       /* initial transition */
(17)        me->timeout = INIT_TIMEOUT;/* timeout is initialized in initial tran. */
(18)        TRAN(SETTING_STATE);
        }
        /*.........................................................................*/
(19)    void Bomb1_dispatch(Bomb1 *me, Event const *e) {        /* dispatching */
(20)        switch (me->state) {
(21)            case SETTING_STATE: {
(22)                switch (e->sig) {
(23)                    case UP_SIG: {         /* internal transition with a guard */
(24)                        if (me->timeout < 60) {            /* guard condition */
(25)                            ++me->timeout;
(26)                            BSP_display(me->timeout);
                            }
(27)                        break;
                        }
                    case DOWN_SIG: {       /* internal transition with a guard */
                        if (me->timeout > 1) {
                            --me->timeout;
```

Continued onto next page

```
                                    BSP_display(me->timeout);
                        }
                        break;
                    }
                    case ARM_SIG: {                    /* regular transition */
                        me->code = 0;                  /* transition action */
(28)                    TRAN(TIMING_STATE);        /* transition to "timing" */
                        break;
                    }
                }
(29)        break;
            }
            case TIMING_STATE: {
                switch (e->sig) {
                    case UP_SIG: {
                        me->code <<= 1;
                        me->code |= 1;
                        break;
                    }
                    case DOWN_SIG: {
                        me->code <<= 1;
                        break;
                    }
                    case ARM_SIG: {        /* regular transition with a guard */
(30)                    if (me->code == me->defuse) {
(31)                        TRAN(SETTING_STATE);   /* transition to "setting" */
                        }
                        break;
                    }
                    case TICK_SIG: {
(32)                    if (((TickEvt const *)e)->fine_time == 0) {
                            --me->timeout;
                            BSP_display(me->timeout);
(33)                        if (me->timeout == 0) {
                                BSP_boom();                /* destroy the bomb */
                            }
                            else {
                                TRAN(TIMING_STATE);
                            }
                        }
                        break;
                    }
                }
            break;
            }
        }
    }
```

(1) Event signals are typically represented as an enumeration.

(2) States are also typically represented as an enumeration.

(3) The Event structure represents signal events without parameters.

(4) The scalar data member sig holds the signal. Here it's declared as the C99-standard 16-bit unsigned integer with the dynamic range of 64K signals.

(5) The TickEvt structure represents TICK events with the fine_time parameter described in the explanation to Figure 3.2(8).

(6) The TickEvt structure derives from Event structure, as described in the Sidebar "Single Inheritance in C" in Chapter 1. By convention, I name the base structure member super.

(7) The event parameter(s) are added after the member super.

(8) The Bomb1 structure represents the time-bomb state machine implemented with the nested switch statement technique.

(9) The data member state is the scalar *state variable* in this implementation. Here it's declared as the C99-standard 8-bit unsigned integer with the dynamic range of 256 states. You can adjust it to suit your needs.

(10-12) The data members timeout, defuse, and code are the extended state variables used in the state diagram shown in Figure 3.2.

(13) The TRAN() macro encapsulates the transition, which in this method consists of reassigning the state variable state.

(14) The state machine "constructor" performs just a basic initialization but does not trigger the initial transition. In C, you need to call the "constructor" explicitly at the beginning of main().

(15) Here, the "constructor" initializes the secret defuse code, which is assigned to the bomb at instantiation.

(16) This is the init() function of the generic state machine interface. Calling this function triggers the initial transition in the state machine.

(17) The initial transition initializes the me->timeout extended state variable, as prescribed in the diagram in Figure 3.2(1).

(18)   The initial transition changes the state to the "setting" state by means of the `TRAN()` macro.

(19)   This is the `dispatch()` function of the generic state machine interface. Calling this function dispatches one event to the state machine.

(20)   The first level of `switch` statement discriminates based on the scalar state variable `me->state`.

(21)   Each state corresponds to one `case` statement in the first level of the `switch`.

(22)   Within each state the second level of `switch` discriminates based on the event signal `e->sig`.

(23)   For example, the internal transition UP is coded as a nested `case` statement.

(24)   The guard condition (see Figure 3.2(2)) is coded by means of the `if` statement.

(25,26)   The actions associated with the transition are coded directly.

(27)   Handling of each event case must be terminated with the `break` statement.

(28)   A state transition is coded by reassigning the state variable, here achieved by the `TRAN()` macro.

(29)   Handling of each state case must be terminated with the `break` statement.

(30,31)   A regular transition with a guard is coded with an `if` statement and `TRAN()` macro.

(32)   Events with parameters, such as the TICK event, require explicit casting from the generic base structure `Event` to the specific derived structure `TickEvt`, in this case.

(33)   The choice pseudostate is coded as an `if` statement that tests all the outgoing guards of the choice point.

3.3.2   Consequences

The nested `switch` statement implementation has the following consequences:

- It is simple.
- It requires enumerating both signals and states.
- It has a small memory footprint, since only one small scalar state variable is necessary to represent the current state of a state machine.

- It does not promote code reuse because all elements of a state machine must be coded specifically for the problem at hand.

- The whole state machine is coded as one monolithic function, which easily can grow too large.

- Event dispatching time is not constant but depends on the performance of the two levels of `switch` statements, which degrade with increasing number of cases (typically as O(log n), where *n* is the number of cases).

- The implementation is not hierarchical. You could manually code entry/exit actions directly in every transition, but this would be prone to error and difficult to maintain in view of changes in the state machine topology. This is mainly because the code pertaining to one state (e.g., an entry action) would become distributed and repeated in many places (on every transition leading to this state).

- The latter property is not a problem for code-synthesizing tools, which often use a nested `switch` statement type of implementation.

3.3.3 Variations of the Technique

The variations of this method include eliminating the second level of the `switch`, if the state machine handles only one type of event. For example, parser state machines often receive identical characters from the input stream. In addition, signal-processing state machines often receive identical time samples of the signal under control. The "Fly 'n' Shoot" game introduced in Chapter 1 provides an example of a simple switch-debouncing state machine coded with the `switch` statement technique (see file `<qp>\qpc\examples\cortex-m3\dos\iar\game\bsp.c`).

3.4 State Table

Another common approach to implementing state machines is based on state table representation of a state machine. The most popular is the two-dimensional state table that lists events along the horizontal dimension and states along the vertical dimension. The contents of the cells are transitions represented as `{action, next-state}` pairs. For example, Table 3.1 shows the two-dimensional state table corresponding to the time-bomb state diagram from Figure 3.2.

Table 3.1: Two-dimensional state table for the time bomb

		Events →			
		UP	DOWN	ARM	TICK
States →	Setting	setting_UP(), setting	setting_DOWN(), setting	setting_ARM(), timing	empty(), setting
	Timing	timing_UP(), timing	timing_DOWN(), timing	timing_ARM(), setting(*)	timing_TICK(), timing(**)

Notes:
(*) The transition to "setting" is taken only when (me->code == me->defuse).
(**) The self-transition to "timing" is taken only when (e->fine_time == 0) and (me->timeout != 0).

3.4.1 Generic State-Table Event Processor

One of the most interesting aspects of the state-table approach is that it represents a state machine as a very regular data structure (the table). This allows writing a simple and generic piece of software called an *event processor* that can execute any state machine specified in the tabular form.

As shown in Figure 3.5, the generic event processor consists of the StateTable structure that manages an external array of transitions and the Event structure for

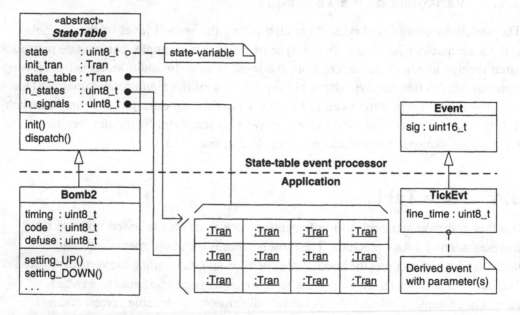

Figure 3.5: The structure of a generic state table-based event processor.

derivation of events with parameters or is used as is for events without parameters. Additionally, the StateTable structure has two functions associated with it. The init() function triggers the initial transition, and dispatch() dispatches an event to the state machine. The StateTable structure is abstract, meaning that it is not intended for direct instantiation but rather only for derivation of concrete[5] state machine structures, such as Bomb2.

In this implementation variant, the state table contains just pointers to transition functions instead of {action, next-state} pairs. This pushes the responsibility of changing the state to the transition function, but in the end is much more flexible because the transition function can evaluate guard conditions and change state only conditionally. Listing 3.2 shows the header file; Listing 3.3 shows the implementation of the generic event processor depicted in Figure 3.5.

Listing 3.2 Generic state-table event processor interface (file statetbl.h**)**

```
(1)   typedef struct EventTag {
          uint16_t sig;                                        /* signal of the event */
          /* add event parameters by derivation from the Event structure */
      } Event;

(2)   struct StateTableTag;                                    /* forward declaration */

(3)   typedef void (*Tran)(struct StateTableTag *me, Event const *e);

(4)   typedef struct StateTableTag {
(5)       Tran const *state_table;                             /* the State-Table */
(6)       uint8_t n_states;                                    /* number of states */
(7)       uint8_t n_signals;                                   /* number of signals */
(8)       uint8_t state;                                       /* the current active state */
(9)       Tran initial;                                        /* the initial transition */
      } StateTable;

(10)  void StateTable_ctor(StateTable *me,
                  Tran const *table, uint8_t n_states, uint8_t n_signals,
                  Tran initial);
(11)  void StateTable_init(StateTable *me);                    /* init method */
(12)  void StateTable_dispatch(StateTable *me, Event const *e); /* dispatch method */
```

Continued onto next page

---

[5] *Concrete class* is an OOP term and denotes a class that can be instantiated because it has no abstract (partially defined) operations or protected constructors.

```
(13)   void StateTable_empty(StateTable *me, Event const *e);      /* empty action */

             /* macro for taking a state transition inside a transition function */
(14)   #define TRAN(target_)  (((StateTable *)me)->state = (uint8_t)(target_))
```

(1)   The Event structure represents signal events (see also Listing 3.1(3-4)).

(2)   This forward declaration is used in the following definition of a pointer-to-function type.

(3)   This typedef defines Tran type as a pointer to transition function that takes the pointer to the StateTable struct and a pointer to the Event struct as arguments and returns uint8_t. The value returned from the transition function represents the *next state* for the state machine after executing the transition. The pivotal aspect of this design is that the transition functions can be used with respect to structures derived (inheriting) from the StateTable.

(4-7) The StateTable structure does not physically contain the state table but rather manages an arbitrary table state_table of transitions with an arbitrary number of states n_states and signals n_signals.

(8)   The data member state is the scalar *state variable* in this implementation.

(9)   The StateTable structure also contains a pointer to the initial transition.

(10) The state table "constructor" performs just a basic initialization but does not trigger the initial transition. In C, you need to call the "constructor" explicitly at the beginning of main().

(11) This is the init() function of the generic state machine interface. Calling this function triggers the initial transition in the state machine.

(12) This is the dispatch() function of the generic state machine interface. Calling this function dispatches one event to the state machine.

(13) The StateTable_empty() function is the default empty action useful for initializing the empty cells of the state table.

(14) The TRAN() macro encapsulates the transition, which in this method consist of re-assigning the state-variable state. Note the explicit cast (upcast) of the me pointer, which typically points to a structure derived from StateTable, rather than StateTable directly.

Listing 3.3 Generic state-table event processor implementation (file `statetbl.c`)

```
      #include "statetbl.h"
(1)   #include <assert.h>
      /*.................................................................*/
(2)   void StateTable_ctor(StateTable *me,
                          Tran const *table, uint8_t n_states, uint8_t n_signals,
                          Tran initial)
      {
          me->state_table = table;
          me->n_states    = n_states;
          me->n_signals   = n_signals;
          me->initial     = initial;
(3)       me->state       = n_states;               /* initialize state out of range */
      }
      /*.................................................................*/
      void StateTable_init(StateTable *me) {
(4)       (*me->initial)(me, (Event *)0);           /* top-most initial transition */

(5)       assert(me->state < me->n_states);  /* the initial tran. must change state */
      /*.................................................................*/
      void StateTable_dispatch(StateTable *me, Event const *e) {
          Tran t;

(6)       assert(e->sig < me->n_signals);            /* require the signal in range */

(7)       t = me->state_table[me->state*me->n_signals + e->sig];
(8)       (*t)(me, e);                               /* execute the transition function */

(9)       assert(me->state < me->n_states);    /* ensure that state stays in range */
      }
      /*.................................................................*/
      void StateTable_empty(StateTable *me, Event const *e) {
          (void)me;                  /* void compiler warning about unused parameter */
          (void)e;                   /* void compiler warning about unused parameter */
      }
```

(1) The event processor implementation uses assertions to prevent incorrect execution of the externally defined state machine. Here the standard assertions are used. See Section 6.7.3 in Chapter 6 for the implementation of customizable assertions in C and C++.

(2)   The state table "constructor" initializes the `state_table` pointer, the table geometry, and the initial transition.

(3)   The state variable is initially set outside the valid range.

(4)   The `init()` function calls the initial transition via the pointer to transition function.

(5)   The state variable must be in range after the initial transition (see also (3)).

(6)   The signal of the event dispatched to the state machine must be in range.

(7)   The transition function pointer corresponding to the current state and current event is obtained by indexing into the external state table array.

(8)   The transition function is invoked via the pointer to transition function obtained in the previous step.

(9)   The state variable must be in range after the transition.

3.4.2   Application-Specific Code

The application-specific part of the implementation provides (1) enumerated signals and states, (2) the state machine structure derived from `StateTable` that includes all the extended state variables, (3) all the transition functions, and (4) the state table initialized with the pointers to the transition functions. Listing 3.4 shows all these elements.

Listing 3.4 Time bomb state machine implemented using the state-table technique (file `bomb2.c`)

```
      #include "statetbl.h"          /* the generic state table event processor */

(1)   enum BombSignals {                       /* all signals for the Bomb FSM */
          UP_SIG,
          DOWN_SIG,
          ARM_SIG,
          TICK_SIG,
(2)       MAX_SIG                               /* the number of signals */
      };

(3)   enum BombStates {                         /* all states for the Bomb FSM */
          SETTING_STATE,
          TIMING_STATE,
(4)       MAX_STATE                             /* the number of states */
      };
(5)   typedef struct TickEvtTag {
(6)       Event super;                          /* derive from the Event structure */
```

```
(7)       uint8_t fine_time;                          /* the fine 1/10 s counter */
      } TickEvt;

(8)  typedef struct Bomb2Tag {                              /* the Bomb FSM */
(9)       StateTable super;          /* derive from the StateTable structure */
(10)      uint8_t timeout;                /* number of seconds till explosion */
(11)      uint8_t defuse;            /* secret defuse code to disarm the bomb */
(12)      uint8_t code;          /* currently entered code to disarm the bomb */
      } Bomb2;

(13) void Bomb2_ctor(Bomb2 *me, uint8_t defuse);          /* the "constructor" */
(14) void Bomb2_initial    (Bomb2 *me);   /* the initial transition function */
(15) void Bomb2_setting_UP   (Bomb2 *me, Event const *e); /* transition function */
     void Bomb2_setting_DOWN (Bomb2 *me, Event const *e); /* transition function */
     void Bomb2_setting_ARM  (Bomb2 *me, Event const *e); /* transition function */
     void Bomb2_timing_UP    (Bomb2 *me, Event const *e); /* transition function */
     void Bomb2_timing_DOWN  (Bomb2 *me, Event const *e); /* transition function */
     void Bomb2_timing_ARM   (Bomb2 *me, Event const *e); /* transition function */
     void Bomb2_timing_TICK  (Bomb2 *me, Event const *e); /* transition function */

                                       /* the initial value of the timeout */
     #define INIT_TIMEOUT  10

     /*..............................................................*/
     void Bomb2_ctor(Bomb2 *me, uint8_t defuse) {
         /* state table for Bomb state machine */
(16)     static const Tran bomb2_state_table[MAX_STATE][MAX_SIG] = {
             { (Tran)&Bomb2_setting_UP, (Tran) &Bomb2_setting_DOWN,
             (Tran)&Bomb2_setting_ARM,        &StateTable_empty   },
             { (Tran)&Bomb2_timing_UP,  (Tran) &Bomb2_timing_DOWN,
             (Tran)&Bomb2_timing_ARM, (Tran) &Bomb2_timing_TICK }
         };
(17)     StateTable_ctor(&me->super,
                     &bomb2_state_table[0][0], MAX_STATE, MAX_SIG,
                     (Tran)&Bomb2_initial); /* construct the superclass */
(18)     me->defuse = defuse;                     /* set the secret defuse code */
     }
     /*..............................................................*/
     void Bomb2_initial(Bomb2 *me) {
(19)     me->timeout = INIT_TIMEOUT;
(20)     TRAN(SETTING_STATE);
     }
     /*..............................................................*/
     void Bomb2_setting_UP(Bomb2 *me, Event const *e) {
         (void)e;          /* avoid compiler warning about unused parameter */
(21)     if (me->timeout < 60) {
             ++me->timeout;
             BSP_display(me->timeout);
```

Continued onto next page

```
            }
        }
        /*..........................................................*/
        void Bomb2_setting_DOWN(Bomb2 *me, Event const *e) {
            (void)e;          /* avoid compiler warning about unused parameter */
            if (me->timeout > 1) {
                --me->timeout;
                BSP_display(me->timeout);
            }
        }
        /*..........................................................*/
        void Bomb2_setting_ARM(Bomb2 *me, Event const *e) {
            (void)e;          /* avoid compiler warning about unused parameter */
            me->code = 0;
(22)        TRAN(TIMING_STATE);                      /* transition to "timing" */
        }
        /*..........................................................*/
        void Bomb2_timing_UP(Bomb2 *me, Event const *e) {
            (void)e;          /* avoid compiler warning about unused parameter */
            me->code <<= 1;
            me->code |= 1;
        }
        /*..........................................................*/
        void Bomb2_timing_DOWN(Bomb2 *me, Event const *e) {
            (void)e;          /* avoid compiler warning about unused parameter */
            me->code <<= 1;
        }
        /*..........................................................*/
        void Bomb2_timing_ARM(Bomb2 *me, Event const *e) {
            (void)e;          /* avoid compiler warning about unused parameter */
            if (me->code == me->defuse) {
                TRAN(SETTING_STATE);                 /* transition to "setting" */
            }
        }
        /*..........................................................*/
        void Bomb2_timing_TICK(Bomb2 *me, Event const *e) {
(23)        if (((TickEvt const *)e)->fine_time == 0) {
                --me->timeout;
                BSP_display(me->timeout);
(24)            if (me->timeout == 0) {
                    BSP_boom();                       /* destroy the bomb */
                }
            }
        }
```

(1) Event signals are typically represented as an enumeration.

(2) The extra enumeration added at the end corresponds to the total number of signals, which you need to know to size the state table array.

(3) States are also typically represented as an enumeration.

(4) The extra enumeration added at the end corresponds to the total number of signals, which you need to know to size the state table array.

(5) The `TickEvt` structure represents TICK events with the `fine_time` parameter described in the explanation to Figure 3.2(8).

(6) The `TickEvt` structure derives from `Event` structure, as described in the Sidebar "Single Inheritance in C" in Chapter 1. By convention, I always name the base structure member `super`.

(7) The event parameter(s) are added after the member `super`.

(8) The `Bomb2` structure represents the time-bomb state machine implemented with the state table technique.

(9) The `Bomb2` structure derives from `StateTable` structure, as described in the Sidebar "Single Inheritance in C" in Chapter 1. By convention, I always name the base structure member `super`.

(10-12) The data members `timeout`, `defuse`, and `code` are the extended state variables used in the state diagram shown in Figure 3.2.

(13) The state machine "constructor" performs just the basic initialization, but does not trigger the initial transition. In C, you need to call the "constructor" explicitly at the beginning of `main()`.

(14) The initial transition function performs the actions of the initial transition (see Figure 3.2(1)), and initializes the state variable to the default state.

(15) The transition functions are specified for each implemented state-signal combination. For example, `Bomb2_setting_UP()` corresponds to transition UP in state "setting."

(16) The state table array specifies the structure of the state machine. The table is known and initialized at compile time, so it can be declared `const`. The table is initialized with the pointers to the `Bomb2` transition functions. Typically, you need to explicitly cast these pointers to function on the `Tran` type because they refer to `Bomb2` subclass of `StateTable` rather than directly to `StateTable`.

(17) The sate table is passed to the `StateTable` event processor constructor, along with the dimensions of the table and the initial transition.

(18)   The Bomb "constructor" also initializes the secret defuse code, which is assigned to the bomb at instantiation.

(19)   The initial transition initializes the `me->timeout` extended state variable, as prescribed in the diagram in Figure 3.2(1).

(20)   The initial transition changes the state to the "setting" state by means of the `TRAN()` macro defined in the event processor interface (file `statetbl.h`).

(21)   The transition functions are responsible for testing the guards, which are implemented as `if` statements.

(22)   The transition functions are responsible for changing the current active state by means of the `TRAN()` macro.

(23)   Events with parameters, such as the TICK event, require explicit casting from the generic base structure `Event` to the specific derived structure `TickEvt`, in this case.

(24)   The choice pseudostate is coded as an `if` statement that tests all the outgoing guards of the choice point.

3.4.3   Consequences

The state table implementation technique has the following consequences.

1.   It maps directly to the highly regular state table representation of a state machine.

2.   It requires the enumeration of states and signals that are used as indexes into the state table.

3.   Because states and signals are used as indexes into an array, they must both be contiguous and start with zero.

4.   It provides relatively good and deterministic performance for event dispatching (O(const), not taking into account action execution).

5.   It promotes code reuse of the generic event processor, which is typically small.

6.   It requires a large state table, which is typically sparse. However, because the state table is constant, it often can be stored in ROM rather than RAM.

7.   It requires a complicated initialization of the state table that must implicitly match the enumerated states and signals. Manual maintenance of this initialization, in view

of changes in the state machine topology, is tedious and prone to error. For instance, adding a new state requires adding and initializing a whole row in the state table.

> **NOTE**
>
> Because of the complex initialization and rapid growth of the state table, programmers often perceive adding new states or events as expensive. This perception often discourages programmers from evolving the state machine. Instead, they tend to misuse extended state variables and guard conditions.

8. It requires a large number of fine-granularity functions representing actions.

9. It typically relies heavily on pointers to functions when implemented in C/C++ (see Section 3.7.1) because state tables typically contain large numbers of such pointers to functions.

10. It is not hierarchical. Although the state table can be extended to implement state nesting, entry/exit actions, and transition guards, these extensions require hardcoding whole transition chains into transition action functions, which is prone to error and inflexible.

3.4.4 Variations of the Technique

There seem to be two main variations on state table implementation in C/C++. Concrete state machines can either derive from the generic state table event processor (inheritance) or contain a state table processor (aggregation). The technique presented here falls into the inheritance category. However, the aggregation approach seems to be quite popular as well (e.g., see [Douglass 99, 01]). Aggregation introduces the indirection layer of a context class—that is, a structure containing the extended state variables on behalf of which the aggregated state table event processor executes actions. Inheritance eliminates this indirection because the `StateTable` class plays the role of the context class (state machine class) simultaneously. In other words, by virtue of inheritance, every derived state machine (like `Bomb2`) also simultaneously "is a" `StateTable`.

The main shortcoming of the two-dimensional state-table representation is the difficulty of showing guard conditions and transitions leading to different states based on different guards, as indicated by the notes added to Table 3.1. Therefore, some authors use a one-dimensional state transition table, shown in Table 3.2

Table 3.2: One-dimensional state transition table for the time bomb

Current State	Event (Parameters)	[Guard]	Next State	Actions
setting	UP	[me->timeout < 60]	setting	++me->timeout; BSP_display (me->timeout);
	DOWN	[me->timeout > 1]	setting	--me->timeout; BSP_display (me->timeout);
	ARM		timing	me->code = 0;
	TICK		setting	
timing	UP		timing	me->code <<=1; me->code |= 1;
	DOWN		timing	me->code <<= 1;
	ARM	[me->code == me-> defuse]	setting	
	TICK (fine_time)	[e->fine_time == 0]	choice	--me->timeout; BSP_display (me->timeout);
		[me->timeout == 0]	final	BSP_boom();
		[else]	timing	

[Diaz-Herrera 93]. In this case, the table can explicitly incorporate event parameters, guard conditions, and actions.

In the direct implementation of the one-dimensional state table, the transitions are more complex objects that contain:

- Pointer to the guard function
- Next state
- A list of pointers to action functions

3.5   Object-Oriented State Design Pattern

The object-oriented approach to implementing state machines is known as the *State design pattern* [Gamma+ 95]. The intent of the pattern is to make a state machine object appear to change its class at runtime as it transitions from state to state. An instance of the State pattern applied to the time-bomb state machine is shown as a UML class diagram[6] in Figure 3.6.

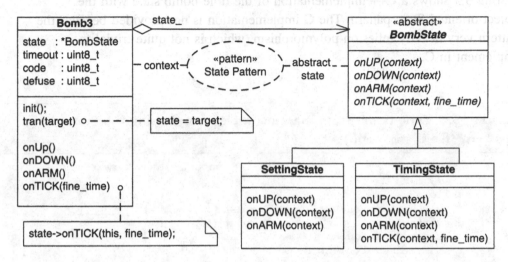

Figure 3.6: Object-oriented State design pattern applied to the time-bomb state machine [Gamma+ 95].

The key idea in this pattern is to introduce an abstract class BombState to represent the states of the time bomb. The BombState declares an interface common to all states, where each operation corresponds to an event. Subclasses of BombState, such as SettingState and TimingState, implement state-specific behavior by overriding the operations inherited from BombState. For example, SettingState handles the UP event in its own specific way by defining the SettingState::onUP() operation. Adding new events requires adding new operations to the abstract BombState class, and adding new states requires adding new subclasses of BombState.

The context class Bomb3 maintains the state as a pointer to a subclass of BombState (see the state data member). Bomb3 also contains all the extended state variables that the time bomb uses, such as timeout, code, and defuse. The class Bomb3 provides an interface identical to the BombState, that is, each handled event corresponds to an operation. The event-handler operations in Bomb3 delegate all state-specific requests to the current state object via the state pointer. In this technique change of state corresponds to changing the current state object, which is accomplished in the context class operation tran().

---

[6] Appendix B contains a quick summary of the UML notation, which includes class diagrams.

3.5.1 Example Implementation

Listing 3.5 shows a C++ implementation of the time-bomb state with the object-oriented State pattern. The C implementation is not provided because the pattern very heavily relies on polymorphism, which is not quite trivial to implement in C.

Listing 3.5 Time-bomb state machine implemented using the State design pattern (file `bomb3.cpp`**)**

```
(1)   class Bomb3;                          // context class, forward declaration
(2)   class BombState {
      public:
(3)       virtual void onUP   (Bomb3 *) const {}
          virtual void onDOWN (Bomb3 *) const {}
          virtual void onARM  (Bomb3 *) const {}
(4)       virtual void onTICK (Bomb3 *, uint8_t) const {}
      };

(5)   class SettingState : public BombState {
      public:
          virtual void onUP   (Bomb3 *context) const;
          virtual void onDOWN (Bomb3 *context) const;
          virtual void onARM  (Bomb3 *context) const;
      };

(6)   class TimingState : public BombState {
      public:
          virtual void onUP   (Bomb3 *context) const;
          virtual void onDOWN (Bomb3 *context) const;
          virtual void onARM  (Bomb3 *context) const;
          virtual void onTICK (Bomb3 *context, uint8_t fine_time) const;
      };

(7)   class Bomb3 {
      public:
(8)       Bomb3 (uint8_t defuse) : m_defuse(defuse) {}

(9)       void init();                                  // the init() FSM interface

(10)      void onUP   ()                { m_state->onUP   (this); }
          void onDOWN ()                { m_state->onDOWN (this); }
          void onARM  ()                { m_state->onARM  (this); }
```

```
(11)          void onTICK(uint8_t fine_time) { m_state->onTICK(this, fine_time); }

     private:
(12)          void tran(BombState const *target) { m_state = target; }

     private:
(13)          BombState const *m_state;                        // the state variable
(14)          uint8_t m_timeout;                  // number of seconds till explosion
              uint8_t m_code;          // currently entered code to disarm the bomb
              uint8_t m_defuse;              // secret defuse code to disarm the bomb


     private:
(15)          static SettingState const setting;
(16)          static TimingState  const timing;

(17)          friend class SettingState;
(18)          friend class TimingState;
     };


     //...............................................................
                                       // the initial value of the timeout
     #define INIT_TIMEOUT  10

(19) SettingState const Bomb3::setting;
(20) TimingState  const Bomb3::timing;

     void Bomb3::init() {
(21)     m_timeout = INIT_TIMEOUT;
(22)     tran(&Bomb3::setting);
     }
     //...............................................................
     void SettingState::onUP(Bomb3 *context) const {
(23)     if (context->m_timeout < 60) {
             ++context->m_timeout;
             BSP_display(context->m_timeout);
         }
     }
     void SettingState::onDOWN(Bomb3 *context) const {
         if (context->m_timeout > 1) {
             --context->m_timeout;
             BSP_display(context->m_timeout);
         }
     }
     void SettingState::onARM(Bomb3 *context) const {
         context->m_code = 0;
(24)     context->tran(&Bomb3::timing);             // transition to "timing"
     }
     //...............................................................
```

Continued onto next page

```
      void TimingState::onUP(Bomb3 *context) const {
          context->m_code <<= 1;
          context->m_code |= 1;
      }
      void TimingState::onDOWN(Bomb3 *context) const {
          context->m_code <<= 1;
      }
      void TimingState::onARM(Bomb3 *context) const {
          if (context->m_code == context->m_defuse) {
              context->tran(&Bomb3::setting);      // transition to "setting"
          }
      }
      void TimingState::onTICK(Bomb3 *context, uint8_t fine_time) const {
(25)      if (fine_time == 0) {
              --context->m_timeout;
              BSP_display(context->m_timeout);
(26)          if (context->m_timeout == 0) {
                  BSP_boom();                                // destroy the bomb
              }
          }
      }
```

(1)   The context class Bomb3 needs to be forward-declared because it is used in the signatures of the operations inside the state classes.

(2)   The BombState abstract class declares an interface common to all time-bomb states, where each operation corresponds to an event. The class provides default empty implementations for all event handlers.

(3)   The onUP() operation handles the UP event without parameters.

(4)   The onTICK() operation handles the TICK(fine_time) event with a parameter. Note that the signature of the event operation contains strongly typed event parameters.

(5)   The class SettingState derives from the abstract BombState and overrides all events handled in the "setting" state. Note that this state does not handle the TICK event, so the SettingState class defaults to the empty implementation inherited from the BombState superclass.

(6)   The class TimingState derives from the abstract BombState and overrides all events handled in the "timing" state.

(7)   The context class Bomb3 keeps track of the current state and contains all extended state variables.

(8)   The constructor initializes selected extended-state variables but does not take the initial transition.

(9)   This is the `init()` function of the generic state machine interface. Calling this function triggers the initial transition in the state machine.

(10)   The context class `Bomb3` duplicates the event interface from the abstract state class `BombState`. All state-dependent behavior is delegated to the current state object.

NOTE

The standard State design pattern does not use the `dispatch()` method for dispatching events to the state machine. Instead, for every signal event, the context class provides a specific (type-safe) event handler operation.

---

(11)   The signatures of event operations contain strongly typed event parameters.

(12)   The private `tran()` operation changes the state by reassigning the state variable.

(13)   The state variable in this technique is a pointer to the subclass of the abstract state class `BombState`.

(14)   The context class `Bomb3` contains also all extended-state variables.

(15,16)   The state objects are static and constant members of the context class `Bomb3`. Note that the state objects contain only operations but no data and therefore can be safely shared among all instances of the context class.

(17,18)   The context class `Bomb3` declares friendship with all state classes because the event-handler operations in the state classes must be able to access the private members of the context class.

(19,20)   The constant state objects must be defined.

(21)   The initial transition performs the actions specified in the state diagram in Figure 3.2(1).

(22)   The default state is specified by means of the `tran()` operation.

(23)   The guard condition is coded as an `if` statement. Note that the state class must access the extended-state variables via the `context` argument.

(24) The transition is achieved by calling the `tran()` operation on behalf of the context state machine object.

(25) The event parameters are available directly to state classes because they are arguments of the event operations.

(26) The choice pseudostate is coded as an `if`-statement.

3.5.2 Consequences

The object-oriented State design pattern has the following consequences:

- It relies heavily on polymorphism and requires an object-oriented language like C++.

- It partitions state-specific behavior and localizes it in separate classes.

- It makes state transitions efficient (reassigning one pointer).

- It provides very good performance for event dispatching through the late binding mechanism (O(const), not taking into account action execution). This performance is generally better than indexing into a state table plus invoking a method via a function pointer, as used in the state table technique. However, such performance is only possible because the selection of the appropriate event handler is not taken into account. Indeed, clients typically will use a `switch` statement to perform such selections. (See the `main()` function in `bomb3.cpp`.)

- It allows you to customize the signature of each event handler. Event parameters are explicit, and the typing system of the language verifies the appropriate type of all parameters at compile time (e.g., `onTICK()` takes a parameter of type `uint8_t`).

- The implementation is memory efficient. If the concrete state objects don't have attributes (only operations), they can be shared (as in the `Bomb3` example).

- It does not require enumerating states.

- It does not require enumerating events.

- It compromises the encapsulation of the context class, which typically requires granting friendship to all state classes.

- It enforces indirect access to the context's parameters from the methods of the concrete state subclasses (via the `context` pointer).

- Adding states requires subclassing the abstract state class.

- Handling new events requires adding event handlers to the abstract state class interface.

- The event handlers are typically of fine granularity, as in the state table approach.

- The pattern is not hierarchical.

3.5.3   Variations of the Technique

The State pattern can be augmented to support entry and exit actions to states. As shown in Figure 3.7, the changes include adding operations onEntry() and onExit() to the abstract state class BombState. Additionally, as shown in the note to operation onTick(fine_time), each event handler in the context class must detect the state change and invoke the onExit() operation to exit the source state and onEntry() to enter the target state.

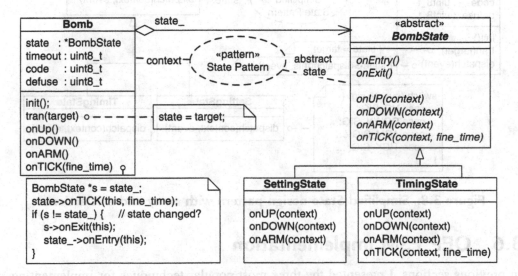

Figure 3.7: Object-oriented State design pattern augmented with entry and exit actions.

The standard State design pattern can also be simplified to provide the standard state machine interface consisting of operations init() and dispatch() with the event representation described in Section 3.2.1. A generic dispatch() operation of the abstract state class handles all events in a state and thus becomes a generic *state-handler* operation. As shown in Figure 3.8, the abstract state class then also becomes generic since it no longer depends on specific event signatures. Additionally, each state handler must perform explicit demultiplexing of events (based on the signal), which typically involves one level of `switch` statement, as shown in the note to the `SettingState::dispatch()` operation.

> **NOTE**
>
> The generic `dispatch(Event const *e)` operation is weakly typed because it accepts generic `Event` superclass and must perform explicit downcasting to the `Event` subclasses based on the signal.

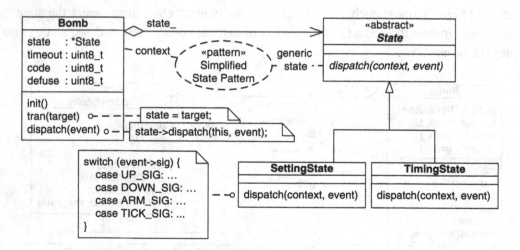

Figure 3.8: Simplified State design pattern with entry and exit actions.

3.6   QEP FSM Implementation

In previous sections, I presented the three most popular techniques for implementing FSMs. From my experience, though, none of these techniques in its pure form is truly optimal. However, one particular *combination* of these techniques repeatedly proved to be the most succinct and efficient implementation of the traditional nonhierarchical FSMs. This technique is part of the QEP event processor that I introduced in Chapter 1. The QEP FSM implementation is object-based, but unlike the State pattern, does not depend on polymorphism and is easy to implement in C.

3.6.1 Generic QEP Event Processor

The QEP support for the basic FSMs combines elements from the nested `switch` statement, state table, and the simplified State design pattern, but it also adds some original ideas. The design is based on a generic event processor (the QEP), similar in functionality to the state-table event processor discussed in Section 3.4.1. The novelty of the QEP design comes from mapping states directly to *state-handler* functions that handle all events in the state they represent. As shown in Figure 3.9, the central element of the QEP event processor is the `QFsm` structure that keeps track of the current state by means of a pointer to a state-handler function. The `QFsm` structure also provides the standard state machine interface functions `init()` and `dispatch()`. The `QFsm` structure is abstract, meaning that it is not intended for direct instantiation but rather only for derivation of concrete state machine structures, such as `Bomb4`. The derived state machine structure adds its own extended state variables such as `timeout`, `code`, and `defuse`, as well as all state-handler functions.

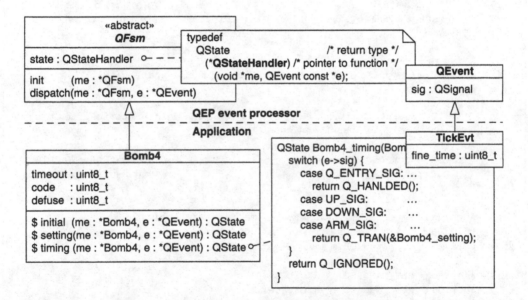

Figure 3.9: The structure of the QEP event processor support for traditional FSMs.

The QEP event processor supports both simple FSMs and hierarchical state machines (HSMs), which I will discuss in Chapter 4. Even though the basic FSMs are a strict subset of HSMs, the QEP provides a separate FSM implementation as an optimization compared to the full-featured HSM. You can use this optimization for performance-critical portions of your code, such as inside interrupt service routines or device drivers. Furthermore, you can use FSMs in extremely resource-constrained embedded systems that simply cannot fit the full-featured HSMs. The QEP support for FSMs requires typically about 120 bytes of code (ROM). For comparison, the support for HSMs requires about 600 bytes of code space for the hierarchical event processor on ARM Cortex-M3. Both FSM and HSM require just one pointer to function in RAM per state machine.

Listings 3.6 and 3.7 show fragments of the QEP files that pertain to the FSM implementation. The header files are located in the directory `<qp>\qpc\include\`, and the implementation files are found in the directory `<qp>\qpc\qep\source\`.

Listing 3.6 QEP FSM event processor interface (fragments of the files `qevent.h` and `qep.h`)

```
        /* qevent.h --------------------------------------------------------*/
(1)     typedef struct QEventTag {                          /* the event structure */
(2)         QSignal sig;                                    /* signal of the event */
(3)         uint8_t dynamic_;        /* dynamic attribute of the event (0 for static) */
        } QEvent;

        /* qep.h -----------------------------------------------------------*/
(4)     typedef uint8_t QState;    /* status returned from a state-handler function */
(5)     typedef                              /* pointer to function type definition */
            QState                                          /* return type */
                (*QStateHandler)          /* name of the pointer-to-function type */
                    (void *me, QEvent const *e);            /* argument list */

(6)     typedef struct QFsmTag {                            /* Finite State Machine */
(7)         QStateHandler state;                            /* current active state */
        } QFsm;

(8)     #define QFsm_ctor (me_, initial_) ((me_)->state = (initial_))
(9)     void QFsm_init    (QFsm *me, QEvent const *e);
(10)    void QFsm_dispatch(QFsm *me, QEvent const *e);

(11)    #define Q_RET_HANDLED    ((QState)0)
(12)    #define Q_RET_IGNORED    ((QState)1)
(13)    #define Q_RET_TRAN       ((QState)2)

(14)    #define Q_HANDLED()      (Q_RET_HANDLED)
(15)    #define Q_IGNORED()      (Q_RET_IGNORED)
(16)    #define Q_TRAN(target_) \
            (((QFsm *)me)->state = (QStateHandler)(target_), Q_RET_TRAN)
```

```
(17)   enum QReservedSignals {
           Q_ENTRY_SIG = 1,                      /* signal for coding entry actions */
           Q_EXIT_SIG,                            /* signal for coding exit actions */
           Q_INIT_SIG,                  /* signal for coding initial transitions */
(18)       Q_USER_SIG         /* first signal that can be used in user applications */
       };
```

(1) The structure QEvent represents events in QEP. Arbitrary event parameters can be added in the process of derivation of structures (see also Section 3.2.1).

(2) The scalar data member sig holds the signal on an event. The data type QSignal is an unsigned integer that can be configured to be 1 byte, 2 bytes, or 4 bytes wide.

(3) The byte-wide member dynamic_ is used by the QF real-time framework to manage dynamically allocated events. The QEP event processor does not use this data member and you can ignore it for now. I'll explain dynamic event allocation in Chapters 6 and 7.

(4) This typedef defines QState as a byte that conveys the status of the event handling to the event processor (see also lines (11-13)).

(5) This typedef defines QStateHandler type as a pointer to state-handler function that takes the pointer to a generic state machine and a pointer to QEvent as arguments and returns QState. The pivotal aspect of this design is that the state-handler functions signature can be used with respect to structures derived (inheriting) from QFsm.

(6) The structure QFsm is the base for derivation of state machine structures.

(7) The data member state is a pointer to a state-handler function. This is the *state variable* in this implementation.

(8) The "constructor" function-like macro initializes the state variable to the initial-pseudostate function that defines the initial transition. Note that the initial transition is not actually executed at this point.

(9) This is the init() function of the generic state machine interface. Calling this function triggers the initial transition in the state machine.

(10) This is the dispatch() function of the generic state machine interface. Calling this function dispatches one event to the state machine.

(11-13)   These constants define the status returned from state-handler functions to the event processor.

(14)   A state-handler function returns the macro Q_HANDLED() whenever it handles the current event.

(15)   A state-handler function returns the macro Q_IGNORED() whenever it ignores (does not handle) the current event.

(16)   The Q_TRAN() macro encapsulates the transition, which in this state machine implementation technique consist of reassigning the state variable state. The Q_TRAN() macro is defined using the *comma expression*. A comma expression is evaluated from left to right, whereas the type and value of the whole expression is the right-most operand. The right-most operand is in this case the status of the operation (transition), which is returned from the state-handler function. The pivotal aspect of this design is that the Q_TRAN() macro can be used with respect to structures derived (inheriting) from QFsm, which in C requires explicit casting (upcasting) to the QFsm base structure (see the sidebar "Single Inheritance in C" in Chapter 1).

(17)   The QEP event processor reserves these few lowest signals for internal use.

(18)   The signal Q_USER_SIG is the first signal available for the users. In other words, the user signals must necessarily be offset from zero by Q_USER_SIG to avoid overlapping the reserved QEP signals.

Listing 3.7 QEP FSM event processor implementation (files qfsm_ini.c and qfsm_dis.c)

```
        /* file qfsm_ini.c ------------------------------------------------*/
        #include "qep_port.h"                    /* the port of the QEP event processor */
        #include "qassert.h"                     /* embedded systems-friendly assertions */

        void QFsm_init(QFsm *me, QEvent const *e) {
(1)         (*me->state)(me, e);                 /* execute the top-most initial transition */

                                                 /* enter the target */
(2)         (void)(*me->state)(me, &QEP_reservedEvt_[Q_ENTRY_SIG]);
        }
        /* file qfsm_dis.c------------------------------------------------*/
        void QFsm_dispatch(QFsm *me, QEvent const *e) {
(3)         QStateHandler s = me->state;                    /* save the current state */
```

```
(4)        QState r = (*s)(me, e);                           /* call the event handler */
(5)        if (r == Q_RET_TRAN) {                              /* transition taken? */
(6)            (void)(*s)(me, &QEP_reservedEvt_[Q_EXIT_SIG]);      /* exit the source */
(7)            (void)(*me->state)(me, &QEP_reservedEvt_[Q_EXIT_SIG]); /*enter target*/
        }
    }
```

(1)   The initial transition is invoked via the pointer to function. The initial state handler changes the current state me->state to the default state by calling the Q_TRAN() macro.

(2)   The default state is entered by sending the reserved signal Q_ENTRY_SIG to its state handler.

NOTE

QEP maintains internally a constant array of reserved events QEP_reservedEvt_[]. This array is indexed by the reserved signals enumerated in Listing 3.6(17).

---

(3)   The QFsm_dispatch() function saves the current state in a temporary stack variable s.

(4)   The current state-handler function is invoked via the pointer to function.

(5)   If the status returned from the state-handler function indicates that a transition has been taken (via the Q_TRAN() macro), then . . .

(6)   The source of the transition is exited by sending the reserved signal Q_EXIT_SIG to the source state handler.

(7)   The target of the transition (the new current state) is entered by sending the reserved signal Q_ENTRY_SIG to its state handler.

3.6.2   Application-Specific Code

The application-specific part of the implementation provides the elements shown below the dashed line in Figure 3.9. These elements are (1) events with parameters derived from the QEvent structure, (2) the state machine structure derived from QFsm that includes all the extended state variables, and (3) all the state-handler functions. Listing 3.8 shows the application-level code.

NOTE

The time-bomb state machine from Figure 3.2 has been slightly modified for the QEP FSM implementation to demonstrate the usage of entry actions. The action of clearing of the defuse code (me->code = 0) has been moved from the transition ARM in state "setting," to entry action in state "timing."

Listing 3.8 Time-bomb state machine implemented using the optimal FSM technique (file bomb4.c)

```
(1)    #include "qep_port.h"              /* the port of the QEP event processor */
(2)    #include "bsp.h"                         /* board support package */

(3)    enum BombSignals {                       /* all signals for the Bomb FSM */
(4)        UP_SIG = Q_USER_SIG,
           DOWN_SIG,
           ARM_SIG,
           TICK_SIG
       };

(5)    typedef struct TickEvtTag {
(6)        QEvent super;                         /* derive from the QEvent structure */
(7)        uint8_t fine_time;                    /* the fine 1/10 s counter */
       } TickEvt;

(8)    typedef struct Bomb4Tag {
(9)        QFsm super;                                   /* derive from QFsm */
(10)       uint8_t timeout;                    /* number of seconds till explosion */
(11)       uint8_t code;             /* currently entered code to disarm the bomb */
(12)       uint8_t defuse;              /* secret defuse code to disarm the bomb */
       } Bomb4;

(13)   void  Bomb4_ctor    (Bomb4 *me, uint8_t defuse);
(14)   QState Bomb4_initial(Bomb4 *me, QEvent const *e);
(15)   QState Bomb4_setting(Bomb4 *me, QEvent const *e);
(16)   QState Bomb4_timing  (Bomb4 *me, QEvent const *e);

       /*-----------------------------------------------------------------*/
                                            /* the initial value of the timeout */
```

```
       #define INIT_TIMEOUT  10

       /*......................................................*/
       void Bomb4_ctor(Bomb4 *me, uint8_t defuse) {
(17)       QFsm_ctor(&me->super, (QStateHandler)&Bomb4_initial);
(18)       me->defuse = defuse;     /* the defuse code is assigned at instantiation */
       }
       /*......................................................*/
       QState Bomb4_initial(Bomb4 *me, QEvent const *e) {
           (void)e;
(19)       me->timeout = INIT_TIMEOUT;
(20)       return Q_TRAN(&Bomb4_setting);
       }
       /*......................................................*/
(21)   QState Bomb4_setting(Bomb4 *me, QEvent const *e) {
(22)       switch (e->sig) {
(23)           case UP_SIG: {
(24)               if (me->timeout < 60) {
                       ++me->timeout;
                       BSP_display(me->timeout);
                   }
(25)               return Q_HANDLED();
               }
               case DOWN_SIG: {
                   if (me->timeout > 1) {
                       --me->timeout;
                       BSP_display(me->timeout);
                   }
                   return Q_HANDLED();
               }
               case ARM_SIG: {
(26)               return Q_TRAN(&Bomb4_timing);        /* transition to "timing" */
               }
           }
(27)       return Q_IGNORED();
       }
       /*......................................................*/
       void Bomb4_timing(Bomb4 *me, QEvent const *e) {
           switch (e->sig) {
(28)           case Q_ENTRY_SIG: {
(29)               me->code = 0;                          /* clear the defuse code */
(30)               return Q_HANDLED();
               }
               case UP_SIG: {
                   me->code <<= 1;
                   me->code |= 1;
                   return Q_HANDLED();
               }
```

Continued onto next page

```
        case DOWN_SIG: {
            me->code <<= 1;
            return Q_HANDLED();
        }
        case ARM_SIG: {
            if (me->code == me->defuse) {
                return Q_TRAN(&Bomb4_setting);
            }
            return Q_HANDLED();
        }
        case TICK_SIG: {
            if (((TickEvt const *)e)->fine_time == 0) {
                --me->timeout;
                BSP_display(me->timeout);
                if (me->timeout == 0) {
                    BSP_boom();                      /* destroy the bomb */
                }
            }
            return Q_HANDLED();
        }
    }
    return Q_IGNORED();
}
```

(1)   Every application C file that uses the QP framework must include the qp_port.h header file. This header file contains the specific adaptation of QP to the given processor, operating system, and compiler, which is called a *port*—in this case, the "qp_port.h" header file, located in the directory <qp>\qpc\ports \80x86\dos\tcpp101\l\.

(2)   The application also includes the board support package.

(3)   All signals used in the application are enumerated.

(4)   The user signals must be offset by Q_USER_SIG, to avoid overlapping the reserved signals.

(5)   The TickEvt structure represents TICK events with the fine_time parameter described in the explanation to Figure 3.2(8).

(6)   The TickEvt structure derives from QEvent structure, as described in the sidebar "Single Inheritance in C" in Chapter 1. By convention, I name the base structure member super.

(7)   The event parameter(s) are added after the member super.

(8)   The Bomb4 structure represents the time bomb state machine implemented with the QEP event processor.

(9)   The Bomb4 structure derives from QFsm structure, as described in the sidebar "Single Inheritance in C" in Chapter 1. By convention, I always name the base structure member super.

(10-12)   The data members timeout, defuse, and code are the extended state variables used in the state diagram shown in Figure 3.2.

(13)   The state machine "constructor" performs just the basic initialization but does not trigger the initial transition. In C, you need to call the "constructor" explicitly at the beginning of main().

(14)   The initial transition function performs the actions of the initial transition (see Figure 3.1(1)), and initializes the state variable to the default state.

(15,16)   The state-handler functions are specified for each state.

(17)   The constructor of the Bomb4 state machine is responsible for invoking the constructor of the base structure QFsm_ctor(), which requires the initial pseudostate handler.

(18)   The Bomb4 constructor can also initialize any extended state variables.

(19)   The initial transition initializes the me->timeout extended state variable, as prescribed in the diagram in Figure 3.2(1).

(20)   The initial transition designates "setting" as the default state by means of the Q_TRAN() macro returned to the event processor (file qep.h).

(21)   This state-handler function corresponds to the state "setting."

(22)   Each state-handler function is typically structured as a switch statement that discriminates based on the signal of the event.

(23)   Each event is handled in a separate case labeled with the enumerated signal of the event.

(24)   A guard condition is coded as an if statement.

(25)   After handling of an event, the state-handler function returns Q_HANDLED() to the event processor.

(26)   A state transition is coded by means of the `Q_TRAN()` macro (see Listing 3.6(16)).

(27)   The final return statement is reached only when no case statements have handled the event. The state-handler function returns `Q_IGNORED()` to the event processor.

NOTE

The `QFsm_dispatch()` function shown in Listing 3.7 cares only whether a state transition has been taken but does not check to see whether the event has been handled or ignored. However, Listing 3.7 does not show the software tracing instrumentation built into the QEP event processor, which indeed makes use of each status value reported by state-handler functions. I discuss the software-tracing instrumentation in Chapter 11.

---

(28)   The entry action is coded as a response to the reserved event `Q_ENTRY_SIG`, which the event processor dispatches to the state-handler function when the state needs to be entered (see Listing 3.6(17)).

(29)   Entry actions are coded directly.

(30)   As all other case statements, entry actions are terminated with "`return Q_HANDLED()`" macro.

NOTE

You should *never* return `Q_TRAN()` from entry or exit actions!

3.6.3 Consequences

The QEP FSM implementation has the following consequences:

- It is simple and can easily be coded in C.

- It partitions state-specific behavior and localizes it in separate *state-handler* functions. These functions have just about the right granularity—neither too fine (as the action functions in the state table or event operations in the State pattern) nor monolithic (as in the nested `switch` statement technique).

- It provides direct and efficient access to extended state variables from state-handler functions (via the "me" pointer) and does not require compromising the encapsulation of the state machine structure.

- It has a small footprint in RAM because only one state variable (a pointer to function) is necessary to represent a state machine instance (see Listing 3.6(7)). No data space is required for states.

- It promotes code reuse of a small and generic QEP event processor that takes typically around 120 bytes of code space (ROM) for the non-hierarchical FSM implementation on ARM Cortex-M3.

- It makes state transitions efficient (the Q_TRAN() macro reassigns just one pointer to function).

- It provides good performance for event dispatching by eliminating one level of switch from the nested switch statement technique and replacing it with a very efficient pointer to function dereferencing. In typical implementations, state handlers still need one level of a switch statement to discriminate events based on the signal, which has performance dependent on the number of cases (typically O(log n), where *n* is the number of cases). The switch statement can be replaced by a one-dimensional lookup table in selected (time-critical) state handlers.

- It is scalable, flexible, maintainable, and *traceable*. It is easy to add both states and events, as well as to change state machine topology, even late in the development cycle, because every state machine element is represented in the code exactly once.

- It requires enumerating events.

- It does not require enumerating states.

- It is not hierarchical, but can be extended to include state hierarchy without sacrificing its good characteristics, as described in Chapter 4.

3.6.4 Variations of the Technique

In the literature, you often find techniques that apply pointers to functions in a very similar way as the QP FSM implementation but still use a scalar state variable to resolve the state handler through a lookup table (e.g., see [Gomez 00]). This approach has several weaknesses:

- It requires enumerating states, which are used as indexes into the call table.

- It requires allocating and initializing the call table.

- The indirection level of the call table degrades performance because of additional steps required by table lookup on top of dereferencing a pointer to function.

3.7 General Discussion of State Machine Implementations

3.7.1 Role of Pointers to Functions

Except for the nested `switch` statement, all other state machine implementation techniques in C/C++ rely heavily on pointers to functions. I also intentionally include here the object-oriented State pattern because it too ultimately resolves event-handler operations via virtual tables that are nothing else than call tables of pointers to functions.

To understand why pointers to functions are so popular in implementing state machines, it is very instructive to step down to the machine code level. Listing 3.9 shows disassembled instructions of a state-handler function called via a pointer in the QEP event processor (see Listing 3.7(4)).

Listing 3.9 Disassembled machine code for a function call via a pointer to function (x86 instruction set, 16-bit real mode, Turbo C++ compiler)

```
#QFSM_DIS#39:   (*s)(me, e);
cs:0101 FF760C   push   word ptr [bp+0C] ; push the "me" far pointer
cs:0104 FF760A   push   word ptr [bp+0A]
cs:0107 FF7608   push   word ptr [bp+08] ; push the "e" far pointer
cs:010A FF7606   push   word ptr [bp+06]
cs:010D FF5EFC   call   far [bp-04]       ; de-reference pointer-to-function
cs:0110 83CEFC   add    sp,0008           ; cleanup after the call
```

As shown in boldface in Listing 3.9, the actual function call via a pointer takes just one machine instruction! The four preceding instructions push the function arguments to the stack (the "me" pointer and the event pointer "e") and are needed no matter what technique you use. As it turns out, a pointer to function maps directly to the architecture of most CPUs and results in unbeatably small and fast code.

The point to remember from this discussion is that pointers to functions are the fastest mechanism for implementing state machines in C/C++. State machines are the "killer applications" for pointers to functions.

3.7.2   State Machines and C++ Exception Handling

Throwing and catching exceptions in C++ is fundamentally incompatible with the run-to-completion (RTC) semantics of state machines. An exception thrown somewhere in the middle of an RTC step typically corrupts a state machine by leaving the extended-state variables inconsistent with the main state variable or with each other. Therefore, in general, an RTC step of a state machine must be considered as one indivisible transaction that either atomically succeeds or entirely fails.

Note that the stack-unwinding process occurring when a thrown exception propagates up the call stack has much less value in the event-driven systems than traditional data-processing programs. As described in Chapter 2, event-driven systems rely much less on representing the context in the call tree and stack variables and instead capture the context in nonstack variables. The problem with exceptions is that they are specialized for cleaning up the stack but know nothing about the static data.

Therefore, you should be wary of the C++ exception handling in state machines, or more generally, in event-driven systems. If you cannot avoid the mechanism altogether (e.g., you rely on a library that throws exceptions), you should be careful to catch all exceptions in the same RTC step and before a thrown exception can cause any inconsistencies. This rule, of course, largely defeats the benefits of throwing exceptions in the first place.

However, state machines offer a better, language-independent way of handling exceptions. A state machine associated with an event-driven subsystem can represent all conditions of the subsystem, including fault conditions. Instead of throwing an exception, an action should generate an *exception event*, which then triggers a state-based exception handling. Section 6.7.4 in Chapter 6 describes the state-based exception handling in more detail.

3.7.3   Implementing Guards and Choice Pseudostates

As described in Chapter 2, guard conditions and choice pseudostates are elements of flowcharts that the UML statecharts simply reuse. As such, these elements are not specific to hierarchical state machines and can be applied equally well in classical flat-state machines.

If you know how to code a flowchart, you already know how to implement guards and choice pseudostates. Flowcharts map easily to plain structured code and are therefore straightforward to implement in those techniques that give you explicit choice of the target of a state transition, such as the nested `switch` statement, the State design pattern, and the QP FSM implementation. Conditional execution is harder to use in the traditional state-table representation because the rigidly structured state table explicitly specifies the targets of state transitions. The solution presented in Section 3.4.1 solves this problem by removing the "next-state" from the table and pushing the responsibility for changing states into the action functions.

A guard specified in the UML expression `[guard]/action` ... maps simply to the `if` statement `if (guard()) { action(); ...}`. A choice pseudostate has one incoming transition segment and many outgoing segments guarded by nonoverlapping guard expressions. This construct maps simply to chained `if-else` statements: `if (guard1()) { action1(); } else if (guard2()) { action2(); }` and so on.

3.7.4   Implementing Entry and Exit Actions

The traditional nonhierarchical FSMs can also reap the benefits of a guaranteed initialization of the state context through entry actions and a guaranteed cleanup in the exit actions. The lack of hierarchy vastly simplifies the problem, but at the same time it makes the feature much less powerful.

One way of implementing entry and exit actions is to dispatch reserved signals (e.g., `Q_ENTRY_SIG` and `Q_EXIT_SIG`) to the state machine. As shown in Listing 3.7, upon detecting a state transition, the state machine `dispatch()` operation sends the `Q_EXIT_SIG` signal to the source state and then sends the `Q_ENTRY_SIG` signal to the target state.

3.8   Summary

The standard implementation techniques and their variations discussed in this chapter can be freely mixed and matched to provide a continuum of possible trade-offs. Indeed, most of the implementations of state machines that you can find in the literature seem to be variations or combinations of the three fundamental techniques: the nested `switch` statement, the state table, and the object-oriented State design pattern. In this chapter, I provided concrete, executable code, and for each fundamental technique, I discussed the consequences of its use as well as some of the most common variations.

One particular combination of techniques, which is part of the QP framework, deserves special attention because it offers an optimal combination of good performance and a small memory footprint. As you will see in Chapter 4, it can be extended to hierarchical state machines (HSMs).

In all techniques, state machines tend to eliminate many conditional statements from your code. By crisply defining the state of the system at any given time, state machines require that you test only one variable (the state variable) instead of many variables to determine the mode of operation (recall the Visual Basic calculator example from Chapter 2). In all but the most basic approach of the nested `switch` statement, even this explicit test of the state variable disappears as a conditional statement. This coding aspect is similar to the effect of polymorphism in OOP, which eliminates many tests based on the type of the object and replaces them with more efficient (and extensible) late binding.

Hierarchical Event Processor Implementation

. . . the cost of adding a feature isn't just the time it takes to code it. The cost also includes the addition of an obstacle to future expansion. . . . The trick is to pick the features that don't fight each other.
— *John Carmack*

Chapter 2 introduced UML statecharts as a very effective way of getting around the state-explosion problem that plagues the traditional "flat" state machines. The particularly valuable innovation of UML state machines in this respect is the concept of *state nesting*, because it allows *reusing behavior* across many states instead of repeating the same actions and transitions over and over again. Hierarchical nesting of states lets you get new behavior almost for free by *inheriting* all of what is common from the superstates. It lets you define new states rapidly *by difference* from existing states rather than create every state from scratch each time. Needless to say, formalism like this is a godsend to the developers of event-driven software, because only state hierarchy makes the whole state machine approach truly applicable to real-life problems.

That is, the concept of the hierarchical state machine (HSM) is a true blessing only if it is easy enough to implement in a mainstream programming language, which for embedded systems developers means C. As a visual formalism, HSMs have been intended primarily for automatic code generation by specialized CASE tools (see Section 2.3.13). However, direct manual coding of HSMs isn't really any harder than coding the traditional nonhierarchical FSMs, especially when you use a generic *hierarchical event processor* that transparently handles all the intricacies of the UML state machine execution semantics.

This chapter describes such a generic hierarchical event processor called QEP, which is part of the QP event-driven framework. We already used the QEP event processor in Section 3.6 of Chapter 3 for implementing traditional "flat" FSMs. Here I describe how this technique can be generalized to support HSMs without sacrificing its good characteristics.

I begin with describing the structure of QEP, explaining both the C and C++ versions. Later in the chapter, I summarize the steps required to implement the calculator HSM designed in Chapter 2. I then provide some guidelines for using the QEP event processor in practice to avoid common pitfalls. I conclude with the instructions for porting and configuring QEP for various processors and compilers.

4.1   Key Features of the QEP Event Processor

QEP is a generic, efficient, and highly portable hierarchical event processor that you can use in any event-driven environment, such as GUI systems, computer games, or real-time embedded (RTE) systems. QEP makes every effort to be compliant with the UML specification [OMG 07], but it cannot really implement the entire bulky UML state machine package. Instead, the QEP design strategy is to supply just enough (but not more) of truly essential elements to allow building basic UML-compliant state machines directly and support the higher-level UML concepts only as design patterns. The main features of QEP are:

- Full support for hierarchical state nesting.

- Guaranteed entry/exit action execution on arbitrary state transition topology.

- Full support of nested initial transitions.

- Highly *maintainable* and *traceable* boilerplate state machine representation in C or C++, in which every state machine element is mapped to code precisely, unambiguously, and exactly once. This is in contrast to many automatic code generation techniques that often "flatten" the state hierarchy, which breaks traceability by repeating the same transitions and actions in many states.

> **NOTE**
>
> The direct, precise, and unambiguous mapping of every state machine element to code contribute to the excellent *traceability* of the QEP HSM implementation technique. The traceability from requirements through design to code is essential for mission-critical systems, such as medical devices or avionic systems.

- Extremely small RAM/ROM footprint. A state machine object requires only one function pointer in RAM. On the ARM Cortex-M3 processor, the hierarchical state machine code requires about 600 bytes whereas the simpler "flat" finite state machine takes only about 120 bytes of code space (ROM).

- No RAM required for representing states and transitions—the number of states is limited only by code space (ROM).

- Fully reentrant event processor code with minimal stack requirements.

- Support for events with arbitrary parameters.

- Easy to integrate with any event queuing and dispatching mechanism—for example, simple event-loop, a GUI system like Windows, or an event-driven framework like QP.

- Very clean source code passing strict static analysis with PC-Lint.

- Source code 98 percent compliant with the Motor Industry Software Reliability Association (MISRA) Guidelines for the Use of the C Language in Vehicle-Based Software [MISRA 98].

- Documentation, application examples, and ports to various compilers are available online.

- Q-SPY software tracing instrumentation for unprecedented observability, controllability, and testability (see Chapter 11).

> **NOTE**
>
> In Chapter 5, you will see how to realize event deferral, orthogonal regions, and transitions to history as *state design patterns* that build on top of the basic QEP implementation described in this chapter.

4.2   QEP Structure

Figure 4.1 shows the overall structure of QEP and its relation to the application-specific code, such as the calculator HSM from Figure 2.18. QEP consists of the QHsm class[1] for derivation of state machines and the QEvent class for derivation of events with parameters, or used as is for events without parameters.

The QHsm class keeps track of the current state and provides the standard state machine interface[2] operations init() and dispatch(). This class is abstract, which means that it

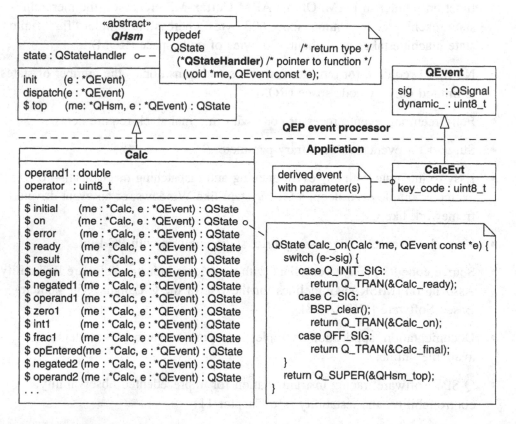

Figure 4.1: QEP event processor and the calculator HSM derived from it. The $ in front of the Calc operations denotes static member functions (see Appendix B).

---

[1] The concept of a "class" in C and derivation of classes is explained in the sidebar "Single Inheritance in C" in Chapter 1.

[2] Section 3.2 in Chapter 3 introduces the standard state machine interface.

is not intended for direct instantiation but rather only for derivation of concrete[3] state machine classes, such as `Calc`, shown in Figure 4.1. The derived state machine class defines all extended state variables as data members and provides state-handler functions for all states (e.g., see the note attached to the `Calc::on()` member function in Figure 4.1). The following sections explain the implementation of all the QEP elements in C and C++.

4.2.1 QEP Source Code Organization

Listing 4.1 shows the directories and files comprising the QEP event processor in C. The structure of the C++ version is identical, except that the implementation files have the `.cpp` extension.

```
Listing 4.1  QEP event processor source code organization

qpc\                    - QP/C root directory (qpcpp for QP/C++)
  |
  +-include\            - QP platform-independent include files (*.H files)
  | +-qassert.h         - QP embedded systems-friendly assertions
  | +-qep.h             - QEP interface
  | . . .
  |
  +-ports\              - QP platform-specific ports
  | +-80x88\            - QP ports to the 80x86 CPU
  | | +-dos\            - ports for DOS (the non-preemptive "vanilla" scheduler)
  | | | +-tcpp101\      - the Turbo C++ 1.01 compiler
  | | | | +-l\          - large memory model
  | | | | | +-dbg\      - build directory for the Debug configuration
  | | | | | | +-qep.lib - QEP library
  | | | | | +-qep_port.h - QEP port header file
  | | | | | +-make.bat  - make script for building the QP libraries
  | |
  | +-. . .             - QP ports to other CPUs ...
  |
  +-qep\                - QEP event processor component
  | |
  | +-source\           - QEP platform-independent source code (*.C files)
  | | +-qep_pkg.h       - internal, packet-scope interface of QEP
  | | +-qep.c           - contains definition of reserved signals
  | | +-qfsm_ini.c      - contains definition of QFsm_init()
  | | +-qfsm_dis.c      - contains definition of QFsm_dispatch()
```

Continued onto next page

---

[3] *Concrete class* is the OOP term and denotes a class that has no abstract operations or protected constructors. Concrete class can be instantiated, as opposed to *abstract class*, which cannot be instantiated.

```
|  | +-qhsm_ini.c - contains definition of QHsm_init()
|  | +-qhsm_dis.c - contains definition of QHsm_dispatch()
|  | +-qhsm_top.c - contains definition of QHsm_top()
|  | +-qhsm_in.c  - contains definition of QHsm_isIn()
|  |
| +-lint\
|  | +-opt_qep.lnt - specific PC-Lint options for linting QEP
```

The QEP source files contain typically just one function or a data structure definition per file. This design aims at deploying QEP as a fine-granularity library that you statically link with your applications. Fine granularity means that the QEP library consists of several small, loosely coupled modules (object files) rather than a single module that contains all functionality. For example, a separate module qhsm_in.c implements the QHsm_isIn() function; therefore, if your application never calls this function, the linker will not pull in the qhsm_in module. This strategy puts the burden on the linker to do the heavy lifting of eliminating any unused code automatically at link time, rather than on the application programmer to configure the QEP code for each application at compile time.

> **NOTE**
>
> The QEP code is instrumented with Q-SPY macros to generate software trace output from state machine execution. However, the instrumentation is disabled by default and will not be shown in the listings discussed in this chapter, for better clarity. Refer to Chapter 11 for more information about Q-SPY software tracing.

4.3   Events

The event representation for HSMs is in QEP exactly the same as for FSMs (see Section 3.6 in Chapter 3). Event instances are implemented as *event objects* that combine the signal and the event parameters into one entity. As shown in Figure 4.1, QEP provides the QEvent base class for direct instantiation of events without parameters or for derivation of events with arbitrary parameters.

4.3.1   Event Signal (QSignal)

A *signal* in UML is the specification of an asynchronous stimulus that triggers reactions [OMG 07] and as such is an essential part of an event. The signal conveys the type of the occurrence—what happened. Signals are typically enumerated constants. The following fragment of QEvent.h header file defines the type of the signal QSignal to be either 8 bits (uint8_t), 16 bits (uint16_t), or 32 bits wide (uint32_t), depending on the configuration macro Q_SIGNAL_SIZE. If you don't define this macro, a default of 8 bits is assumed.

```
#ifndef QEP_SIGNAL_SIZE
    #define QEP_SIGNAL_SIZE 1
#endif

#if (QEP_SIGNAL_SIZE == 1)
    typedef uint8_t QSignal;
#elif (QEP_SIGNAL_SIZE == 2)
    typedef uint16_t QSignal;
#elif (QEP_SIGNAL_SIZE == 4)
    typedef uint32_t QSignal;
#else
    #error "QEP_SIGNAL_SIZE defined incorrectly, expected 1, 2, or 4"
#endif
```

NOTE

All components of the QP framework, including QEP, use the following standard exact-width integer types (WG14/N843 C99 Standard, Section 7.18.1.1):

Exact Size	Unsigned	Signed
8 bits	uint8_t	int8_t
16 bits	uint16_t	int16_t
32 bits	uint32_t	int32_t

4.3.2   QEvent **Structure in C**

Listing 4.2 shows the definition of the QEvent structure in C. The member sig of type QSignal represents the signal. The byte-wide member dynamic_ is used by the QP framework to manage dynamically allocated events. You should never need to access this member from the application-level code. (I'll explain dynamic event allocation in Chapter 7.)

156

Chapter 4

Listing 4.2 QEvent **structure in C (file** `<qp>\qpc\include\qevent.h`**)**

```
typedef struct QEventTag {
    QSignal sig;                               /* signal of the event */
    uint8_t dynamic_; /* attributes of a dynamic event (0 for static event) */
    /* add event parameters by derivation from the QEvent structure... */
} QEvent;                                      /* the QEvent type */
```

The QEvent structure can be used as is for events without parameters or can serve as the base structure for derivation of events with arbitrary parameters. The following C code snippet shows how to derive the calculator event CalcEvt that contains the key-code parameter (also see the sidebar "Single Inheritance in C" in Chapter 1):

```
typedef struct CalcEvtTag {
    QEvent super;                              /* derives from QEvent */
    uint8_t key_code;                          /* code of the key */
} CalcEvt;                                     /* the CalcEvt type */
```

Having the common base structure QEvent for all events ensures that every event object contains the signal at the same offset within the event. This allows using a pointer to a derived event as a parameter to any function that expects a generic QEvent *e pointer to the base structure. Any such function can always access the sig data member (as e->sig) to determine what kind of derived event structure is used. The function can then perform an explicit downcast[4] to the derived event structure to get the event parameters. For example, to get the key_code parameter, a generic QEvent *e pointer needs to be cast to CalcEvt as follows: ((CalcEvt *)e)->key_code.

The point here is that the sig data member has double responsibility. It obviously must convey the occurrence (what happened?). But in addition, the signal must also uniquely identify the derived event structure so that state-handler functions can explicitly downcast to this derived structure based only on the sig value. This second responsibility will became clearer in the upcoming Section 4.4.2, where I provide examples of state handler functions.

---

[4] Casting from the subclass to the superclass is called in OOP downcasting because the cast goes down a traditionally drawn inheritance relationship in a class diagram, such as Figure 4.1.

4.3.3   `QEvent` **Structure in** C++

In C++, the `QEvent` structure can be defined without the ugly `typedef`, as shown in Listing 4.3. Please note that in C++ `struct` is exactly equivalent to `class`, except in `struct` the default protection level is public and in `class` it is private.

Listing 4.3 `QEvent` **structure in** C++ **(file** `<qp>\qpcpp\include\qevent.h`**)**

```
struct QEvent {
    QSignal sig;                    // signal of the event instance
    uint8_t dynamic_;   // attributes of a dynamic event (0 for static event)
    // add event parameters by inheriting from QEvent
};
```

Event instances are used primarily as "bags" for passing around signals and event parameters. To generate events efficiently, it's often convenient to use statically preallocated, constant event objects initialized with an initializer list. To allow such initialization in C++, a class must be an aggregate; that is, it must not have private or protected members, constructors, base classes, and virtual functions [Stroustrup 00]. For that reason, `QEvent` is declared as `struct` in Listing 4.3, without any private members or constructors. (An obvious constructor would take one argument to initialize the `sig` attribute.)

The following C++ code snippet shows how to derive the calculator event `CalcEvt` class that contains the key-code parameter:

```
struct CalcEvt : public QEvent {
    uint8_t key_code;                        // code of the key
};
```

When you derive from `QEvent`, the subclass is obviously no longer an aggregate. However, I recommend that you still keep your event classes simple and lightweight. I like to define the `QEvent` subclasses using the `struct` keyword as a reminder that they are lightweight. In particular, I avoid private members, constructors, or virtual functions in the derived event classes. As you will see in Chapter 7, events generally do not go through conventional instantiation (the standard operator `new` isn't used to create dynamic events), so the constructors aren't invoked and the virtual pointers aren't set up.

4.4   Hierarchical State-Handler Functions

In QEP, states are represented as *state-handler functions* that handle all events in the state they implement. The hierarchical state-handler functions use exactly the same signature QStateHandler as nonhierarchical state handler functions, as described in Section 3.6 of Chapter 3. The only extension to the nonhierarchical implementation technique discussed before is that a hierarchical state handler must additionally inform the event processor about the nesting level of the state. When the hierarchical state handler does *not* handle the event, the handler must provide the superstate so that the event processor can invoke the superstate handler function, per the semantics of state nesting (see Section 2.3.2). The hierarchical state-handler function provides this additional information to the event processor very similarly as it informs the event processor about a state transition. A state handler sets the state variable to the superstate handler and returns a special status information that distinguishes this situation from a state transition.

4.4.1   Designating the Superstate (Q_SUPER() Macro)

When a hierarchical state handler function does not handle the current event, it returns the macro Q_SUPER() to the event processor, which is defined as follows:

```
#define Q_RET_SUPER       ((QState)3)
#define Q_SUPER(super_) \
    (((QHsm *)me)->state = (QStateHandler)(super_), Q_RET_SUPER)
```

The Q_SUPER() macro is defined using the *comma expression*. A comma expression is evaluated from left to right, whereas the type and value of the whole expression is the rightmost operand. The rightmost operand is in this case the status of the operation (superstate), which is returned from the state-handler function. The pivotal aspect of this design is that the Q_SUPER() macro can be used with respect to structures derived (inheriting) from QHsm, which in C requires explicit casting (upcasting) to the QHsm base structure (see the sidebar "Single Inheritance in C" in Chapter 1).

4.4.2   Hierarchical State-Handler Function Example in C

Listing 4.4 shows an example of a hierarchical state-handler function that corresponds to the state "int1" in the calculator statechart in Figure 2.18. State "int1" controls entering the integer part of the first operand.

Listing 4.4 Example of a hierarchical state-handler function in C (file `calc.c`)

```
(1)  QState Calc_int1(Calc *me, QEvent const *e) {
(2)      switch (e->sig) {
(3)          case DIGIT_0_SIG:                 /* intentionally fall through */
(4)          case DIGIT_1_9_SIG: {
(5)              BSP_insert(((CalcEvt const *)e)->key_code);
(6)              return Q_HANDLED();
             }
             case POINT_SIG: {
                 BSP_insert(((CalcEvt const *)e)->key_code);
(7)              return Q_TRAN(&Calc_frac1);
             }
         }
(8)      return Q_SUPER(&Calc_operand1);
     }
```

(1) Each state handler takes two parameters: the state machine pointer "me" and the constant pointer "e" to QEvent. It returns QState, which conveys the status of the event handling to the event processor.

> **NOTE**
>
> The event pointer is declared as `const` to prevent modifying the event inside the state-handler function. In other words, the state handler is granted read-only access to the event.

(2) Generally, every state handler is structured as a single `switch` that discriminates based on the signal of the event `e->sig`.

(3,4) Each `case` is labeled by the event signal. Signals are typically enumerated constants.

(5) To get to the event parameters, a state handler must perform an explicit downcast from the generic QEvent const* pointer to the specific derived event pointer, such as `Calc const*` in this case.

> **NOTE**
>
> At this point it becomes apparent that the signature of the state-handler function is really weakly typed with respect to the event parameter. The compiler knows only that every event is passed as the generic QEvent* pointer, but the compiler does not know the specific type of the event, such

Continued onto next page

NOTE—CONT'D

as `CalcEvt`. The application programmer is ultimately responsible for performing a correct downcast to the derived event based on the signal (`e->sig`). The point to remember is that you need to be careful because the compiler cannot prevent an incorrect downcast.

---

(6) Returning `Q_HANLDED()` from a hierarchical state handler informs the QEP event processor that the particular event has been handled.

(7) A state transition is accomplished by returning the macro `Q_TRAN()` that requires the target of the transition as parameter.

(8) If no `case` executes, the state handler returns the `Q_SUPER()` macro, which designates the superstate and informs the event processor about it.

4.4.3 Hierarchical State-Handler Function Example in C++

Listing 4.5 shows an example of a hierarchical state-handler function that corresponds to the state "operand1" in the calculator statechart in Figure 2.18.

Listing 4.5 Example of a hierarchical state-handler function in C++ (file `calc.cpp`)

```
       QState Calc::int1(Calc *me, QEvent const *e) {
           switch (e->sig) {
               case DIGIT_0_SIG:                     // intentionally fall through
               case DIGIT_1_9_SIG: {
(1)                BSP_insert(((static_cast<CalcEvt const *>(e)))->key_code);
                   return Q_HANDLED();
               }
               case POINT_SIG: {
                   BSP_insert(((static_cast<CalcEvt const *>(e)))->key_code);
                   return Q_TRAN(&Calc::frac1);
               }
           }
           return Q_SUPER(&Calc::operand1);
       }
```

Apart from the trivial syntactic differences (such as the ": :" scope resolution operator instead of an underscore), the structure of a hierarchical state-handler function in C++ is identical to the C version from Listing 4.4. The only interesting difference is the downcast of the generic event pointer to the specific subclass in Listing 4.5(1). Here, I've used the new-style `static_cast<>` operator because the cast converts between types related by inheritance. Of course, you can also use the C-style cast if your older C++ compiler does not support the new-style casts.

POINTERS TO MEMBER FUNCTIONS IN C++

The C++ state-handler function takes the "`me`" pointer of its own class type, through which it accesses the state machine data members and member functions (e.g., `me->operand1 = ...`). This is because the state-handler functions are *static members* of the QHsm subclass, such as the calculator state machine class `Calc` (see Section 4.6.3).

An obvious and more elegant alternative would be to make the state-handler functions regular, nonstatic class members, which would allow them to access the class members much more naturally through the implicit "`this`" pointer.

Indeed this much more elegant alternative has been used in the earlier QEP/C++ version published in the first edition of this book. However, this alternative requires using pointers to *member* functions instead of simple pointers to functions, which turned out to be a problem in practice.

Even though the earlier C++ version of QEP used pointers to member functions in a rather standard way, the embedded developers have filed a number of alarming reports from the trenches, where the elegant approach either had very lousy performance or did not work at all. For example, some embedded C++ compilers used over 30 machine instructions to de-reference a pointer to member function and only three to de-reference a regular pointer to function. Needless to say, three machine instructions should do the job (see also Section 3.7.1 in Chapter 3).

As it turns out, too many C++ compilers simply don't support pointers to member functions well due to interference from other language features, such as multiple inheritance and virtual base classes. As eloquently explained in the online article "Member Function Pointers and the Fastest Possible C++ Delegates" [Clugston 07], even such widespread and important frameworks as the MFC actually use pointers to member functions in a nonstandard way by subverting the normal C++ type checking.

To avoid inefficiencies and portability issues, the current C++ version of QEP does not use pointers to member functions but simply plain pointers to functions to static member functions that don't have the "`this`" pointer and therefore are not affected by polymorphism or multiple inheritance. Note that the explicit "`me`" pointer required by static class members plays the same role as the "`context`" pointer required by the object-oriented State design pattern (see Section 3.5.1 in Chapter 3).

4.5   Hierarchical State Machine Class

As shown in Figure 4.1, the QHsm base class is the central element of the QEP design. The QHsm class is abstract, which means that it is not intended for direct instantiation but only for derivation of hierarchical state machines, such as the `Calc` state machine in Figure 4.1. The main responsibility of the QHsm class is keeping track of the current active state. In QEP, the state variable is a pointer to the state-handler function QStateHandler defined previously.

The QHsm class also provides the standard state machine interface functions init() and dispatch() as well as the constructor and the top state handler. The following sections explain these elements, first in the C version and later in C++.

4.5.1   Hierarchical State Machine in C (Structure QHsm)

In C, HSMs are derived from the QHsm base structure, shown in Listing 4.6.

Listing 4.6 QHsm **structure and related functions (file** `<qp>\qpc\include\qep.h`**)**

```
      typedef struct QHsmTag {
(1)       QStateHandler state;   /* current active state (state-variable) */
      } QHsm;

(2)  #define QHsm_ctor(me_, initial_) ((me_)->state = (initial_))

(3)  void QHsm_init    (QHsm *me, QEvent const *e);
(4)  void QHsm_dispatch(QHsm *me, QEvent const *e);
(5)  uint8_t QHsm_isIn (QHsm *me, QHsmState state);
(6)  QState QHsm_top    (QHsm *me, QEvent const *e);
```

(1)   The QHsm structure stores the state-variable state, which is a pointer to a state-handler function. Typically, the QHsm structure requires just 2 or 4 bytes of RAM, depending on the size of the pointer to function for a given CPU and C compiler options.

(2)   The QHsm "constructor" function-like macro initializes the state variable to the initial-pseudostate function that defines the initial transition. Note that the initial transition is not actually executed at this point.

(3)   The QHsm_init() function triggers the initial transition in the state machine. The function takes an initialization event argument. You can use this event to pass any parameters to initialize the state machine.

(4)   The QHsm_dispatch() function dispatches one event to the state machine.

(5)   The QHsm_isIn() function tests whether the HSM "is in" a given state. Note that an HSM simultaneously "is in" all superstates of the currently active state,

and `QHsm_isIn()` tests for it. The `QHsm_isIn()` function returns 1 (TRUE) if the HSM "is in" a given state (in the hierarchical sense). Otherwise the function returns 0 (FALSE).

(6) The `QHsm_top()` function is the hierarchical state handler for the *top state*. The top state is the UML concept that denotes the ultimate root of the state hierarchy. The top state handler "handles" every event by silently ignoring it, which is the default policy in the UML (see also Section 4.5.3).

NOTE

The application-level state-handler functions that don't explicitly nest in any other state return the `&QHsm_top` pointer to the event processor.

4.5.2 Hierarchical State Machine in C++ (Class `QHsm`)

In C++, HSMs are derived from the `QHsm` abstract base class, shown in Listing 4.7.

Listing 4.7 `QHsm` **class (file** `<qp>\qpcpp\include\qep.h`**)**

```
      class QHsm {
      protected:
(1)       QStateHandler m_state;    // current active state (state-variable)

      public:
(2)       void init     (QEvent const *e = (QEvent const *)0);
(3)       void dispatch(QEvent const *e);
(4)       uint8_t isIn  (QHsmState state);

      protected:
(5)       QHsm(QStateHandler initial) : m_state(initial) {} // protected ctor
(6)       static QState top(QHsm *me, QEvent const *e);
      };
```

---

(1) The `QHsm` class stores the state-variable state, which is a pointer to the hierarchical state-handler function. The state variable `m_state` is protected so that the concrete state machine classes derived from `QHsm` can access it through the macro `Q_TRAN()`.

(2) The `init()` member function triggers the initial transition in the state machine. The function takes an optional initialization event argument. You can use this event to pass any parameters to initialize the state machine.

(3) The `dispatch()` member function dispatches one event to the state machine.

(4) The `isIn()` member function tests whether the HSM "is in" a given state. Note that if an HSM is in a substate, it recursively also "is in" all the superstates. The `isIn()` function returns 1 (TRUE) if the HSM "is in" a given state (in the hierarchical sense). Otherwise the function returns 0 (FALSE).

(5) The constructor is protected to prevent direct instantiation of `QHsm` class, as it is abstract. The constructor initializes the state variable to the initial-pseudostate function that defines the initial transition. Note that the initial transition is not actually executed at this point.

(6) The `top()` *static* member function is the hierarchical state handler for the *top state*. The top state is in UML the ultimate root of the state hierarchy. The top state handler "handles" every event by silently ignoring it, which is the default policy in the UML (see also Section 4.5.3).

NOTE

The application-level state-handler functions that don't explicitly nest in any other state return the `&QHsm::top` pointer to the event processor. It is crucial in this design that `QHsm::top()` is a *static* member, because static member functions can be referenced by the simple pointers-to-functions, whereas regular member functions would require pointers to member functions (also see the sidebar "Pointers to Member Functions in C++").

4.5.3 The Top State and the Initial Pseudostate

Every HSM has the (typically implicit) *top state*, which surrounds all the other elements of the entire state machine, as depicted in Figure 4.2.

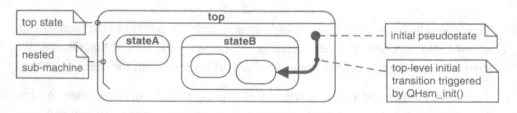

Figure 4.2: The top state and the initial pseudostate.

The QHsm class guarantees that the top state is available to every derived state machine by providing the QHsm_top() hierarchical state handler subsequently inherited by the subclasses. The QHsm_top() hierarchical state-handler function is defined as follows:

```
QState QHsm_top(QHsm *me, QEvent const *e) {
    (void)me;        /* avoid the compiler warning about unused parameter */
    (void)e;         /* avoid the compiler warning about unused parameter */
    return Q_IGNORED();              /* the top state ignores all events */
}
```

By the UML semantics, the top state has no superstates and silently ignores all events, so it always returns Q_IGNORED() to the event processor (see Listing 3.6(15) in Chapter 3). The only purpose, and legitimate use, of the top state is to provide the ultimate root of a state hierarchy so that the highest-level state handlers can return &QHsm_top as their superstate. In particular, you should never target the top state in a state transition.

The state machine initialization is intentionally divided into two steps. The QHsm constructor merely initializes the state variable to the initial pseudostate. Later, the application code must trigger the initial transition explicitly by invoking QHsm_init() (described in the upcoming Section 4.5.6). This design separates instantiation of the state machine from initialization, giving the applications full control over the sequence of initializations in the system. The following code shows an example of an initial pseudostate handler for the calculator state machine:

```
QState Calc_initial(Calc *me, QEvent const *e) {
    (void)e;           /* avoid the compiler warning about unused parameter */
    BSP_clear();                        /* clear the calculator display */
    return Q_TRAN(&Calc_on);           /* designate the default state */
}
```

Note that the topmost initial transition can fire only once (actually, exactly once), because after you leave the top state, you cannot transition back. In other words, your state machine cannot reuse the initial pseudostate in its life cycle.

4.5.4   Entry/Exit Actions and Nested Initial Transitions

In Chapter 2, you saw that UML state machines support elements of Moore automata such as entry and exit actions as well as nested initial transitions. These elements are sole characteristics of the state in which they are defined and do not depend, in particular, on the transition path through which the state has been reached. As described in Chapter 3, state-handler functions in QEP can (optionally) define state-specific behavior by responding to the following reserved signals defined in the qep.h header file:

```
enum QReservedSignals {
    Q_ENTRY_SIG = 1,                    /* signal for coding entry actions */
    Q_EXIT_SIG,                          /* signal for coding exit actions */
    Q_INIT_SIG,                   /* signal for coding initial transitions */

    Q_USER_SIG              /* first signal that can be used in user code */
};
```

A state handler can handle these signals by using them as case labels in the usual switch statement. A state handler is free to execute any actions in response to those signals, but it should not take any state transitions in entry/exit actions. Conversely, the response to the Q_INIT_SIG signal must always include the Q_TRAN() macro to designate the default substate of the current state.

NOTE

The target of a nested initial transition specified in the Q_TRAN() macro must be the direct or transitive substate of the composite state in which the initial transition is defined. In other words, the nested initial transition must "drill into" the state hierarchy but cannot "go up" to target superstates or "sideways" to target peer states. The QEP event processor does not check for such incorrect initial transition targets, which really violate the UML semantics and correspond to malformed state machines and could crash the QEP event processor.

Listing 4.8 provides an example of using an entry action, an exit action, and a nested initial transition.

Listing 4.8 Definition of the `Calc_on()` **state-handler function with entry and exit actions and an initial transition**

```
QState Calc_on(Calc *me, QEvent const *e) {
    switch (e->sig) {
        case Q_ENTRY_SIG: {                           /* entry action */
            BSP_message("on-ENTRY;");
            return Q_HANDLED();
        }
        case Q_EXIT_SIG: {                            /* exit action */
            BSP_message("on-EXIT;");
            return Q_HANDLED();
        }
        case Q_INIT_SIG: {                /* nested initial transition */
            BSP_message("on-INIT;");
            return Q_TRAN(&Calc_ready);
        }
        case C_SIG: {
            BSP_clear();
            return Q_TRAN(&Calc_on);
        }
        case OFF_SIG: {
            return Q_TRAN(&Calc_final);
        }
    }
    return Q_SUPER(&QHsm_top);
}
```

The reserved signals take up the lowest signal values (0..3), which are thus not available for the applications. For convenience, the public HSM interface contains the signal `Q_USER_SIG`, which indicates the first signal free for the users. A typical way of defining application-level signals is to use an enumeration. In this case, `Q_USER_SIG` can be used to offset the values of the entire enumeration, as shown in Listing 4.9.

Listing 4.9 Enumerating signals for the Calc state machine

```
enum CalcSignals {
    C_SIG = Q_USER_SIG,
    CE_SIG,
```

Continued onto next page

```
    DIGIT_0_SIG,
    DIGIT_1_9_SIG,
    POINT_SIG,
    OPER_SIG,
    EQUALS_SIG,
    OFF_SIG
};
```

NOTE

The reserved signals Q_ENTRY_SIG, Q_EXIT_SIG, and Q_INIT_SIG should cause no side effects in state-handler functions that do not have entry actions, exit actions, or initial transitions. The signal 0 (defined internally in the QEP as QEQ_EMPTY_SIG_) is reserved as well and should cause a state-handler function to always return the superstate without any side effects.

4.5.5   Reserved Events and Helper Macros in QEP

To execute entry/exit actions and initial transitions, the QEP event processor needs to invoke the various state-handler functions and pass to them pointers to event objects containing the reserved signals. For example, to trigger an entry action in the Calc_on() state handler, the event processor calls the Calc_on() function with a pointer to an event that has the signal equal to Q_ENTRY_SIG. To do this efficiently, QEP uses the QEP_reservedEvt_[] constant array of reserved events, defined as follows:

```
QEvent const QEP_reservedEvt_[] = {
    { (QSignal)QEP_EMPTY_SIG_, (uint8_t)0 },
    { (QSignal)Q_ENTRY_SIG,    (uint8_t)0 },
    { (QSignal)Q_EXIT_SIG,     (uint8_t)0 },
    { (QSignal)Q_INIT_SIG,     (uint8_t)0 }
};
```

NOTE

The reserved signal zero, enumerated as QEP_EMPTY_SIG_, is used only internally in QEP and therefore is not defined in the public QEP interface qep.h with the rest of the reserved signals.

The array `QEP_reservedEvt_[]` is designed to be indexed by the reserved signals. For example, `&QEP_reservedEvt_[Q_ENTRY_SIG]` represents a pointer to an event with the reserved signal `Q_ENTRY_SIG`.

The following three helper macros are then used extensively inside the QEP implementation (file `<qp>\qpc\qep\source\qep_pkg.h`):

```
    /** helper macro to trigger reserved event in an HSM */
#define QEP_TRIG_(state_, sig_) \
    ((*(state_))(me, &QEP_reservedEvt_[sig_]))

/** helper macro to trigger entry action in an HSM */
#define QEP_EXIT_(state_) \
    if (QEP_TRIG_(state_, Q_EXIT_SIG) == Q_RET_HANDLED) { \
        /* QS software tracing instrumentation for state entry */\
    }

/** helper macro to trigger exit action in an HSM */
#define QEP_ENTER_(state_) \
    if (QEP_TRIG_(state_, Q_ENTRY_SIG) == Q_RET_HANDLED) { \
        /* QS software tracing instrumentation for state exit */\
    }
```

For example, the macro `QEP_TRIG_()` calls a given state-handler function `state_` with the reserved event pointer argument `&QEP_reservedEvt_[sig_]`, where `sig_` is one of these: `QEP_EMPTY_SIG_`, `Q_ENTRY_SIG`, `Q_EXIT_SIG`, or `Q_INIT_SIG`. Note the characteristic syntax of the function call based on a pointer to function `(*(state_))(...)`.

DESIGN BY CONTRACT IN C AND C++

All components of the QP framework, including QEP, apply the elements of the Design by Contract[5] (DbC) philosophy, which is a method of programming based on precisely defined specifications of the various software components' mutual obligations (*contracts*). The central idea of this method is to inherently embed the contracts in the code and validate them automatically at runtime [Meyer 97].

Continued onto next page

---

[5] Design by Contract is a registered trademark of Interactive Software Engineering (ISE).

DESIGN BY CONTRACT IN C AND C++—CONT'D

In C or C++, you can implement the most important aspects of DbC with assertions (see [Myrphy 01a, 01b, Samek 03d]). Throughout this book, I use customized assertions defined in the header file qassert.h, located in directories <qp>\qpc\include\ as well as <qp>\qpcpp\include\. The qassert.h header file provides a number of macros, which include:

- Q_REQUIRE(), to assert a precondition
- Q_ENSURE(), to assert a postcondition
- Q_INVARIANT(), to assert an invariant
- Q_ASSERT(), to assert a general contract of another type
- Q_ALLEGE, to assert a general contract and always evaluate the condition, even when assertions are disabled at compile time

Each of these macro works similarly as the standard library macro assert(), and their different names serve only to document the purpose of the contract. Section 6.7.3 in Chapter 6 covers DbC and qassert.h in more detail.

4.5.6   Topmost Initial Transition (QHsm_init())

State nesting adds a lot of complexity to the topmost initial transition in an HSM compared to a nonhierarchical FSM. The initial transition in an HSM might be complex because UML semantics require "drilling" into the state hierarchy with the nested initial transitions until the leaf state is reached. For example, the topmost initial transition in the calculator example (Figure 2.18 in Chapter 2) involves the following six steps:

1. Execution of actions associated with the topmost initial transition

2. Execution of entry actions to the "on" state

3. Execution of the actions associated with the initial transition defined in the "on" state

4. Execution of the entry actions to the "result" state

5. Execution of the actions associated with the initial transition defined in the "result" state

6. Execution of the entry actions to the "begin" state; at this point the transition is done because "begin" is a leaf state with no nested initial transition

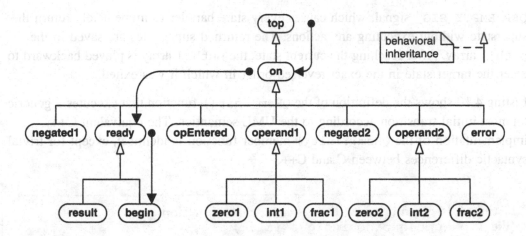

Figure 4.3: The inheritance tree of the calculator HSM with the nested initial transitions.

Figure 4.3 shows the inheritance tree of the states comprising the calculator statechart. The UML specification requires that higher-level states must be entered before entering lower-level states. Unfortunately, this is exactly the opposite order to the natural direction of navigation through the state handlers denoted by the behavioral inheritance arrow in Figure 4.3. As you recall from Section 4.4, a hierarchical state-handler function provides the superstate, so it's easy to traverse the state hierarchy from lower- to higher-level states. Although this order is very convenient for the efficient implementation of the most frequently used QHsm_dispatch() function, entering states is harder.

The solution implemented in QEP is to use a temporary array path[] to record the exit path from the target state of the initial transition without executing any actions (see Figure 4.4). This is achieved by calling the state handlers with the reserved

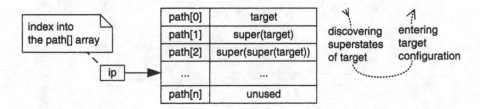

Figure 4.4: The use of the `path[]` **array to enter the target state configuration in the correct order.**

QEP_EMPTY_SIG_ signal, which causes every state handler to immediately return the superstate without executing any actions. The returned superstates are saved in the path[] array. After reaching the current state, the path[] array is played backward to enter the target state in the exact reversed order in which it was exited.

Listing 4.10 shows the definition of the QHsm_init() function that executes a generic topmost initial transition according to the UML semantics. The equivalent C++ implementation of the QHsm::init() member function is identical, except for trivial syntactic differences between C and C++.

Listing 4.10 Definition of the QHsm_init() function (file <qp>\qpc\qep\source\qep_ini.c)

```
(1)   void QHsm_init(QHsm *me, QEvent const *e) {
          QStateHandler t;

                                    /* the top-most initial transition must be taken */
(2)       Q_ALLEGE((*me->state)(me, e) == Q_RET_TRAN);

(3)       t = (QStateHandler)&QHsm_top;          /* HSM starts in the top state */
(4)       do {                                   /* drill into the target... */
(5)           QStateHandler path[QEP_MAX_NEST_DEPTH_];
(6)           int8_t ip = (int8_t)0;             /* transition entry path index */

(7)           path[0] = me->state;      /* save the target of the initial transition */
(8)           (void)QEP_TRIG_(me->state, QEP_EMPTY_SIG_);
(9)           while (me->state != t) {
(10)              path[++ip] = me->state;
                  (void)QEP_TRIG_(me->state, QEP_EMPTY_SIG_);
              }
(11)          me->state = path[0];            /* restore the target of the initial tran. */
                                              /* entry path must not overflow */
(12)          Q_ASSERT(ip < (int8_t)QEP_MAX_NEST_DEPTH_);

              do {           /* retrace the entry path in reverse (desired) order... */
(13)              QEP_ENTER_(path[ip]);                     /* enter path[ip] */
(14)          } while ((--ip) >= (int8_t)0);

(15)          t = path[0];                   /* current state becomes the new source */
(16)      } while (QEP_TRIG_(t, Q_INIT_SIG) == Q_RET_TRAN);
(17)      me->state = t;
      }
```

(1)   The QHsm_init() implements the QHsm "class operation" and therefore takes the "me" pointer. The event pointer parameter e can be used to provide additional initialization parameters to the state machine.

(2)   Inside the Q_ALLEGE() macro (see the sidebar "Design by Contract in C and C++") the initial pseudostate handler function is called via the pointer to function stored in the state variable me->state. The initial pseudostate handler returns the Q_TRAN() macro, in which it designates the target of the initial transition. The Q_ALLEGE() macro makes sure that the initial pseudostate always takes the initial transition. I use the Q_ALLEGE() macro because the initial transition must be executed even when assertions are disabled at compile time.

(3)   The temporary variable t holds the source state of the transition. The first source is the top state.

(4)   The do-loop performs the recursive execution of any nested initial transitions until a leaf state is reached.

(5)   The temporary array path[] stores the pointers to state-handler functions in the exit path from the target state of the initial transition (see Figure 4.4).

(6)   The temporary variable ip is used as the index into the path[] array.

(7)   The path[0] entry is initialized to hold the target state, which is placed in me->state by the Q_TRAN() macro called in the state-handler function.

(8)   The superstate of the current state is discovered by calling the state handler with the reserved QEP_EMPTY_SIG_ signal. The hierarchical state handler never handles the QEP_EMPTY_SIG_ signal, so it returns Q_SUPER() macro, which sets me->state to the superstate of the given state.

(9)   The discovery of superstates continues until the current source state is reached.

NOTE

It is crucial at this point that the target state of the initial transition indeed is a substate of the source.

(10)   The exit path from the target is stored in the `path[]` array.

(11)   The current state `me->state` is restored to the original target of the initial transition.

(12)   This assertion makes sure that the `path[]` array does not overflow (see the sidebar "Design by Contract in C and C++").

(13)   All states stored in the `path[]` array are entered in the correct order.

(14)   The entry to the target state configuration continues until index 0, which points to the target itself.

(15)   The current state becomes the new source.

(16)   The loop continues as long as the current state handler reports that it handled the initial transition. Otherwise, it is a leaf state and the job of the `QHsm_init()` function is done.

(17)   The current state is set to the final leaf state.

4.5.7   Dispatching Events (`QHsm_dispatch()`, General Structure)

Dispatching an event to an HSM requires implementing the UML state nesting semantics, that is, propagating the event through all the levels of nesting until it is handled or reaches the top state. This functionality as well as executing of transitions is implemented in the `QHsm_dispatch()` function.

The `QHsm_dispatch()` function is the most complicated in the QEP event processor.[6] I will break up the discussion of the implementation into two steps. First, in this section I explain how the function handles hierarchical event processing. In the next section, I take a closer look at the transition execution algorithm.

Listing 4.11 shows the general structure of the `QHsm_dispatch()` function. The implementation uses many elements already described for `QHsm_init()`. The code carefully optimizes the number and size of temporary stack variables to minimize the stack use. The equivalent C++ implementation of the `QHsm::dispatch()` member function is identical, except for trivial syntactic differences between C and C++.

---

[6] `QHsm_dispatch()` is not broken up in to smaller functions to conserve the stack space.

Listing 4.11 General structure of the `QHsm_dispatch()` **function (file** `<qp>\qpc\qep\source\qep_dis.c`**)**

```
(1)    void QHsm_dispatch(QHsm *me, QEvent const *e) {
(2)        QStateHandler path[QEP_MAX_NEST_DEPTH_];
           QStateHandler s;
           QStateHandler t;
           QState r;

(3)        t = me->state;                            /* save the current state */

(4)        do {                          /* process the event hierarchically...*/
(5)            s = me->state;
(6)            r = (*s)(me, e);                       /* invoke state handler s */
(7)        } while (r == Q_RET_SUPER);

(8)        if (r == Q_RET_TRAN) {                     /* transition taken? */
               int8_t ip = (int8_t)(-1);       /* transition entry path index */
               int8_t iq;                  /* helper transition entry path index */

(9)            path[0] = me->state;         /* save the target of the transition */
(10)           path[1] = t;

(11)           while (t != s) {       /* exit current state to transition source s...*/
(12)               if (QEP_TRIG_(t, Q_EXIT_SIG) == Q_RET_HANDLED) { /*exit handled? */
(13)                   (void)QEP_TRIG_(t, QEP_EMPTY_SIG_); /* find superstate of t */
                   }
(14)               t = me->state;                  /* me->state holds the superstate */
               }

(15)           . . .

(16)       me->state = t;         /* set new state or restore the current state */
       }
```

(1) The `QHsm_dispatch()` implements the `QHsm` "class operation" and therefore it takes the "`me`" pointer. The job of this function is to process the event e according to the UML semantics.

(2) The temporary array `path[]` stores the pointers to state-handler functions in the exit path from the target state of the initial transition. To conserve the stack space, this array is reused for other purposes as well.

(3) The current state is saved temporarily into '`t`'.

(4)   This do-loop processes the events hierarchically starting from the current state.

(5)   The current state is saved in the temporary 's' in case a transition is taken, which overwrites me->state. In that case, the variable 's' holds the source of the transition.

(6)   At each level of state nesting, the state-handler function is called. The status of the event handling reported from the state handler is stored in to the temporary variable 'r.'

(7)   The do-loop continues as long as state handler functions return superstates (via the Q_SUPER() macro). Please note that the top state handler ignores all events, so the loop must terminate even if the event is not explicitly handled at any level of state nesting.

(8)   If the returned status is Q_RET_TRAN, the last called state handler must have taken a state transition by executing the Q_TRAN() macro (see Listing 3.6(16) in Chapter 3).

(9)   The current state me->state holds the target of the transition, which is now placed in path[0] (see Figure 4.4).

(10)   The current state before the transition is copied to path[1], so that the variable 't' can be reused.

(11)   This while-loop executes the transition segment from the current state to the explicit source of the transition. This step covers the case of an inherited state transition—that is, the transition defined at a level higher than the currently active state.

For example, assume that the state "result" is the current active state of the calculator state machine in Figure 4.5 while the user presses one of the operator keys (+, −, *, or /). When the function QHsm_dispatch() receives the OPER event, it calls the currently active state handler first, which is the Calc_result() state handler. This state handler doesn't handle the OPER event, so it returns the superstate Calc_ready(). The QHsm_dispatch() function then calls the Calc_ready() state handler, which handles the OPER event by taking a state transition Q_TRAN(&Calc_opEntered). However, the correct exit of the current state configuration must include exiting "result" before exiting "ready." This transition segment is shown in grey in Figure 4.5.

(12)   The exit action is triggered in the state.

(13)   If the exit action is handled, I need to discover the superstate by calling the state handler with the empty event. If the exit action is unhandled, the state handler returned the Q_SUPER() macro, so me->state already contains the superstate.

(14)   The superstate is stored in the temporary variable 't' to be compared with the transition source.

(15)   The omitted part contains the state transition algorithm shown in Listing 4.12 and explained in the next section.

(16)   The current state is restored from the variable 't,' where it has been stored in step (3).

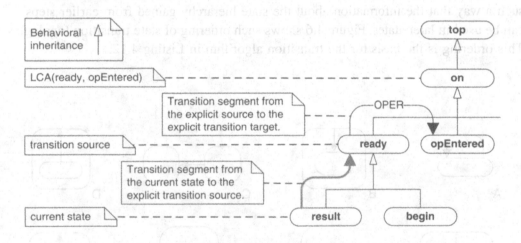

Figure 4.5: Two segments of an inherited state transition.

4.5.8   Executing a Transition in the State Machine
(QHsm_dispatch(), **Transition**)

Executing a generic state transition in a HSM is by far the most complex part of the QEP implementation. The challenge is to quickly find the least common ancestor (LCA) state of the source and target states. (The LCA is the lowest-hierarchy state that is simultaneously the superstate of the source and the target states.) The transition sequence involves the exit of all states up to the LCA (but without exiting the LCA itself), followed by the recursive entry into the target state, followed by "drilling" into the target state configuration with the initial transitions until a leaf state is reached.

Listing 4.12 shows the omitted part of the QHsm_dispatch() function that implements the general case of a state transition. A large part of the complexity of this part of the code results from the optimization of the workload required to efficiently execute the most frequently used types of state transitions. The optimization criterion used in the transition algorithm is to minimize the number of invocations of state-handler functions, in particular the "empty" invocations (with the reserved QEP_EMPTY_SIG_), which serve only for eliciting the superstate of a given state handler. The strategy is to order the possible source-target state combinations in such a way that the information about the state hierarchy gained from earlier steps can be used in later states. Figure 4.6 shows such ordering of state transition topologies. This ordering is the basis for the transition algorithm in Listing 4.12.

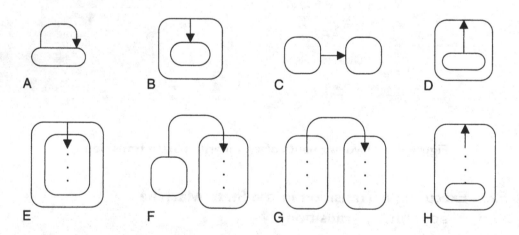

Figure 4.6: Ordering of transition types from simplest to progressively more complex in the transition algorithm (Listing 4.12).

Listing 4.12 Transition algorithm implementation in the `QHsm_dispatch()` **function (file** `<qp>\qpc\qep\source\qep_dis.c`**)**

```
/* NOTE: continued from Listing 4.11 */

        t = path[0];                                        /* target of the transition */

(1)     if (s == t) {                   /* (a) check source==target (transition to self) */
            QEP_EXIT_(s)                                             /* exit the source */
            ip = (int8_t)0;                                         /* enter the target */
        }
        else {
            (void)QEP_TRIG_(t, QEP_EMPTY_SIG_);              /* superstate of target */
            t = me->state;
(2)         if (s == t) {                          /* (b) check source==target->super */
                ip = (int8_t)0;                                     /* enter the target */
            }
            else {
(3)             (void)QEP_TRIG_(s, QEP_EMPTY_SIG_);          /* superstate of src */
                                    /* (c) check source->super==target->super */
(4)             if (me->state == t) {
                    QEP_EXIT_(s)                                    /* exit the source */
                    ip = (int8_t)0;                                /* enter the target */
                }
                else {
                                          /* (d) check source->super==target */
(5)                 if (me->state == path[0]) {
                        QEP_EXIT_(s)                               /* exit the source */
                    }
                    else {       /* (e) check rest of source==target->super->super..
                                  * and store the entry path along the way
                                  */
                        iq = (int8_t)0;              /* indicate that LCA not found */
                        ip = (int8_t)1;          /* enter target and its superstate */
                        path[1] = t;              /* save the superstate of target */
                        t = me->state;                         /* save source->super */
                                                        /* find target->super->super */
                        r = QEP_TRIG_(path[1], QEP_EMPTY_SIG_);
(6)                     while (r == Q_RET_SUPER) {
                            path[++ip] = me->state;         /* store the entry path */
                            if (me->state == s) {              /* is it the source? */
                                iq = (int8_t)1;         /* indicate that LCA found */
                                                   /* entry path must not overflow */
                            Q_ASSERT(ip < (int8_t)QEP_MAX_NEST_DEPTH_);
```

Continued onto next page

```
                                --ip;                      /* do not enter the source */
                                r = Q_RET_HANDLED;               /* terminate the loop */
                            }
                            else {              /* it is not the source, keep going up */
                                r = QEP_TRIG_(me->state, QEP_EMPTY_SIG_);
                            }
                        }
                        if (iq == (int8_t)0) {               /* the LCA not found yet? */

                                                        /* entry path must not overflow */
(7)                         Q_ASSERT(ip < (int8_t)QEP_MAX_NEST_DEPTH_);

(8)                         QEP_EXIT_(s)                         /* exit the source */

                            /* (f) check the rest of source->super
                             *                          == target->super->super...
                             */
                            iq = ip;
                            r = Q_RET_IGNORED;          /* indicate LCA NOT found */
                            do {
(9)                             if (t == path[iq] ) {               /* is this the LCA? */
                                    r = Q_RET_HANDLED;          /* indicate LCA found */
                                    ip = (int8_t)(iq - 1);          /* do not enter LCA */
                                    iq = (int8_t)(-1);          /* terminate the loop */
                                }
                                else {
                                    --iq;            /* try lower superstate of target */
                                }
                            } while (iq >= (int8_t)0);

                            if (r != Q_RET_HANDLED) {          /* LCA not found yet? */
                                /* (g) check each source->super->...
                                 * for each target->super...
                                 */
                                r = Q_RET_IGNORED;                  /* keep looping */
                                do {
                                                                /* exit t unhandled? */
(10)                                if (QEP_TRIG_(t, Q_EXIT_SIG)
                                        == Q_RET_HANDLED)
                                    {
                                        (void)QEP_TRIG_(t, QEP_EMPTY_SIG_);
                                    }
                                    t = me->state;               /* set to super of t */
                                    iq = ip;
                                    do {
(11)                                    if (t == path[iq] ) {               /* is this LCA? */
                                                                /* do not enter LCA */
                                            ip = (int8_t)(iq - 1);
```

```
                                         iq = (int8_t)(-1);              /*break inner */
                                         r = Q_RET_HANDLED;             /*break outer */
                                      }
                                      else {
                                         --iq;
                                      }
                                } while (iq >= (int8_t)0);
                           } while (r != Q_RET_HANDLED);
                        }
                     }
                 }
              }
           }
        }
                          /* retrace the entry path in reverse (desired) order... */
(12)    for (; ip >= (int8_t)0; --ip) {
            QEP_ENTER_(path[ip])                               /* enter path[ip] */
        }
        t = path[0];                             /* stick the target into register */
        me->state = t;                                /* update the current state */

                                         /* drill into the target hierarchy... */
(13)    while (QEP_TRIG_(t, Q_INIT_SIG) == Q_RET_TRAN) {
            ip = (int8_t)0;
            path[0] = me->state;
            (void)QEP_TRIG_(me->state, QEP_EMPTY_SIG_);          /* find superstate */
            while (me->state != t) {
                path[++ip] = me->state;
                (void)QEP_TRIG_(me->state, QEP_EMPTY_SIG_);     /* find superstate */
            }
            me->state = path[0];
                                              /* entry path must not overflow */
            Q_ASSERT(ip < (int8_t)QEP_MAX_NEST_DEPTH_);

            do {                      /* retrace the entry path in reverse (correct) order... */
                QEP_ENTER_(path[ip])                           /* enter path[ip] */
            } while ((--ip) >= (int8_t)0);

            t = path[0];
        }
    }
```

(1) From the point of view of reducing the number of state handler calls, the simplest transition type is transition to self (Figure 4.6(A)) because this transition type can be determined immediately by testing (source == target), that is, no state-handler invocations are necessary to check for this transition type. The

self-transition requires exiting the source and entering the target. The exit from the source can be performed right away by means of the helper macro QEP_EXIT_() introduced in Section 4.5.5.

(2)   The next transition type is the topology shown in Figure 4.6(B). The check for this transition type is (source == super(target)) and requires determining the superstate of the target. This transition type requires only entry to the target but no exit from the source.

(3)   To proceed further, the algorithm checks for the superstate of the source state. The information about superstates of both the target and the source collected so far is subsequently used to determine two transition types shown in Figure 4.6 (C) and (D).

(4)   The transition topology in Figure 4.6(C) is the peer-to-peer transition shown (perhaps the most common type). This transition topology can be determined by checking the condition (super(source) == super(target)) and requires exit from the source and entry to the target. The exit is performed right away by means of the helper macro QEP_EXIT_().

(5)   The topology shown in Figure 4.6(D) requires testing the condition (super(source) == target) and involves only entry to the target but no exit from the source.

(6)   The topology shown in Figure 4.6(E) requires probing all superstates of the target until a match with the source is found or until the top state is reached. The target state hierarchy determined in this part of the algorithm is stored in the temporary array path[] and is subsequently reused to perform the entry to the target state configuration in the desired order. Note that the entry path[0] already holds the target itself, and path[1] holds the superstate of target discovered in the prior steps.

(7)   The macro Q_ASSERT() implements a customizable, embedded-system friendly assertion. I discuss the use of assertions in the QP event-driven platform in Section 6.7 of Chapter 6.

(8)   The transition topology from Figure 4.6(E) is the last that might not require exiting the source state, so if the processed transition does not fall into the (E) category, the source state must be exited.

(9)   The topology shown in Figure 4.6(F) requires traversal of the target state
hierarchy stored in the array path[] to find the match with the superstate of
source still kept in the temporary variable s.

(10) The topologies shown in Figure 4.6(G) and (H) require traversal of the target
state hierarchy stored in the array path[] to find the match with any of the
superstates of the source.

(11) Because every scan for a match with a given superstate of the source
exhausts all possible matches for the LCA, the source's superstate can be
safely exited.

The transition types shown in Figure 4.6(A) through Figure 4.6(H) represent all valid
transition topologies, and every well-formed transition should be recognized as one of
the cases (A) through (H). Once QHsm_dispatch() detects the type of the transition
and executes all necessary exit actions up to the LCA, it must enter the target state
configuration.

(12) The entry to the target state configuration is straightforward and involves just a
simple for loop that scans through the array path[] in the reversed order as it
was filled.

(13) The target state can be composite and can have an initial transition as well.
Therefore, the while loop performs the "drilling" into the target until it detects
the leaf state. This part of the algorithm is very similar as the QHsm_init()
function explained in Listing 4.10.

4.6   Summary of Steps for Implementing HSMs
with QEP

Implementing state machines with the QEP event processor is quite a mechanical
process consisting of just a few simple steps. You already went through the process
at least once in Section 1.7 of Chapter 1, where I explained the implementation of the
Ship state machine in the "Fly 'n' Shoot" game. Here I present the implementation
of the calculator state machine, which was designed in Section 2.4 of Chapter 2 and also
served as an example throughout this chapter.

When you know how to code one HSM, you know how to code them all. Therefore, this section will necessarily repeat some of the descriptions from Chapter 1. However, I feel that having all the HSM coding steps conveniently summarized in one section will be helpful for daily programming work. In addition, to reduce the repetition, I describe here the state machine implementation in C++. If you are a C programmer, I hope that by now you are getting more familiar with the concepts of a "class" and "inheritance" and know how to code them in C.

The C++ source code for the calculator state machine is located in the directory `<qp>\qpcpp\examples\80x86\dos\tcpp101\l\calc\` and consists of the following files:

- `calc.h` contains the declaration of signals, events, and the global pointer to the `Calc` state machine.

- `calc.cpp` contains declaration of the `Calc` state machine structure and the implementation of the state-handler functions.

- `bsp.h` contains the board support package interface.

- `bsp.cpp` contains the implementation of the board-specific functions.

- `main.cpp` contains the `main()` function and the event loop.

- `CALC.PRJ` is the Turbo C++ project file for building the application.

As always, the code I provide is executable and I encourage you to try it out. You can run the example on any Windows PC by double-clicking on the executable located in the directory `<qp>\qpcpp\examples\80x86\dos\tcpp101\l\calc\dbg\CALC.EXE`.

The calculator example is interactive and you can perform computations with it. You use the keyboard to send keypress events to the application; the state of the calculator display is shown at the command prompt. The calculator recognizes keys: 0, 1, . . ., 9, ., +, −, *, /, =, C, and E (cancel entry, CE). The Esc key terminates the application. All other keys are ignored.

Figure 4.7 shows a screen shot in which you can see how the calculator handles the expression 2, −, −, −, 2, = that has crashed the Visual Basic calculator in Chapter 2. I'd like to challenge you to crash the state machine-based calculator. The calculator starts with displaying zero aligned at the right edge of the display [0]. To the right of the display, you can see the key sent to the calculator. For example, the first key is 2.

The key event is followed by the sequence of actions that the calculator HSM performs in response to the key event. I recommend that you correlate this output with the calculator state diagram from Figure 2.18.

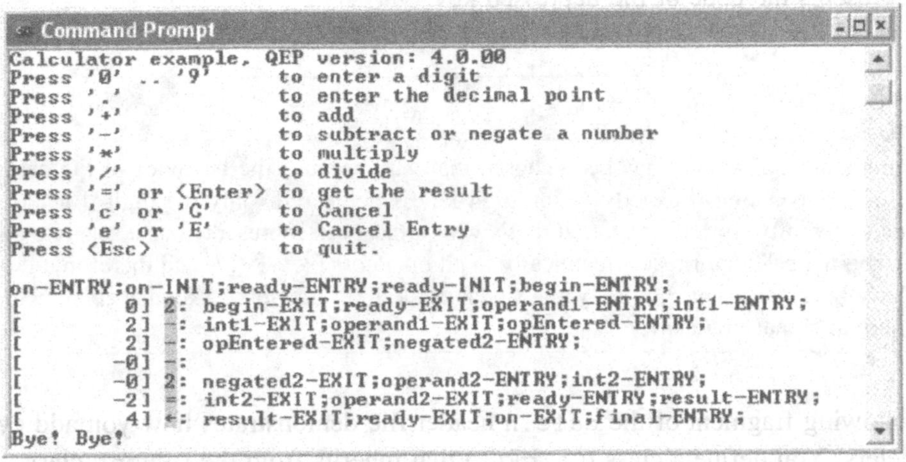

```
Command Prompt                                                    _ □ x
Calculator example, QEP version: 4.0.00
Press '0' .. '9'        to enter a digit
Press '.'               to enter the decimal point
Press '+'               to add
Press '-'               to subtract or negate a number
Press '*'               to multiply
Press '/'               to divide
Press '=' or <Enter> to get the result
Press 'c' or 'C'        to Cancel
Press 'e' or 'E'        to Cancel Entry
Press <Esc>             to quit.

on-ENTRY;on-INIT;ready-ENTRY;ready-INIT;begin-ENTRY;
[      0] 2:  begin-EXIT;ready-EXIT;operand1-ENTRY;int1-ENTRY;
[      2] =:  int1-EXIT;operand1-EXIT;opEntered-ENTRY;
[      2] -:  opEntered-EXIT;negated2-ENTRY;
[     -0] -:
[     -0] 2:  negated2-EXIT;operand2-ENTRY;int2-ENTRY;
[     -2] =:  int2-EXIT;operand2-EXIT;ready-ENTRY;result-ENTRY;
[      4] *:  result-EXIT;ready-EXIT;on-EXIT;final-ENTRY;
Bye! Bye!
```

Figure 4.7: The calculator HSM running in a Windows console.

4.6.1   Step 1: Enumerating Signals

The first step of the implementation consists of enumerating all signals recognized by the state machine shown in the state diagram (Figure 2.18 in Chapter 2), such as C, CE, DIGIT_0, DIGIT_1_9, and so on.

Listing 4.9 shows the enumeration of all signals recognized by the calculator state machine. Note that the user-level signals do not start from zero but rather are offset by the constant Q_USER_SIG. Also note that by QEP convention, all signals have the suffix _SIG to easily distinguish signals from other constants. The suffix _SIG is omitted in the state diagram to reduce the clutter.

4.6.2   Step 2: Defining Events

Many events consist only of the signal and don't need any additional parameters. You can represent such events directly as instances of the QEvent structure provided in the header file <qp>\qpc\include\qevent.h.

However, some events require parameters. For example, the calculator signal DIGIT_1_9_SIG communicates only that one of the digit keys 1..9 has been depressed, but the signal alone does not inform us as to which digit key this was. The missing information is added to the event in the form of the key_code parameter that represents the code of the depressed key.

NOTE

The granularity of signals has been chosen that way because the behavior of the calculator really is independent of exactly which digit key is depressed (only the digit 0 needs to be treated differently from the rest, and that's why it has been represented as a separate signal). Similarly, the calculator reacts identically to all operators (+, −, *, /) and therefore all operators have been represented by only one signal OPER_SIG. Section 4.7.8 talks about achieving the optimal signal granularity.

The following fragment of the calc.h header file demonstrates how you add event parameters. You define a class (CalcEvt) that inherits from the QEvent class. You then add arbitrary parameters as data members:

```
struct CalcEvt : public QEvent {
    uint8_t key_code;                              // code of the key
};
```

4.6.3   Step 3: Deriving the Specific State Machine

Hierarchical state machines are represented in QEP as subclasses of the QHsm abstract base class, which is defined in the header file <qp>\qpcpp\include\qep.h. Listing 4.13 demonstrates how you derive the Calc (calculator) class from QHsm.

NOTE

You should not be confused by the fact that the Ship state machine example in Chapter 1 derived from the QActive base class rather than QHsm. As you will see in Chapter 7, QActive is a subclass of QHsm, so the Ship state machine is in fact derived from QHsm, albeit not directly.

Listing 4.13 Deriving the `Calc` **class from** `QHsm`

```
(1) class Calc : public QHsm {
    private:
(2)     double  m_operand1;      // the value of operand 1 (extended state variable)
        uint8_t m_operator;      // operator key entered (extended state variable)

    public:
(3)     Calc() : QHsm((QStateHandler)&Calc::initial) {    // ctor
        }

    protected:
(4)     static QState initial   (Calc *me, QEvent const *e); // initial pseudostate
(5)     static QState on         (Calc *me, QEvent const *e);      // state handler
        static QState error      (Calc *me, QEvent const *e);      // state handler
        static QState ready      (Calc *me, QEvent const *e);      // state handler
        static QState result     (Calc *me, QEvent const *e);      // state handler
        static QState begin      (Calc *me, QEvent const *e);      // state handler
        static QState negated1   (Calc *me, QEvent const *e);      // state handler
        static QState operand1   (Calc *me, QEvent const *e);      // state handler
        static QState zero1      (Calc *me, QEvent const *e);      // state handler
        static QState int1       (Calc *me, QEvent const *e);      // state handler
        static QState frac1      (Calc *me, QEvent const *e);      // state handler
        static QState opEntered  (Calc *me, QEvent const *e);      // state handler
        static QState negated2   (Calc *me, QEvent const *e);      // state handler
        static QState operand2   (Calc *me, QEvent const *e);      // state handler
        static QState zero2      (Calc *me, QEvent const *e);      // state handler
        static QState int2       (Calc *me, QEvent const *e);      // state handler
        static QState frac2      (Calc *me, QEvent const *e);      // state handler
        static QState final      (Calc *me, QEvent const *e);      // state handler
};
```

(1) You define a class (`Calc`) that inherits the `QHsm` base class.

(2) You add arbitrary extended-state variables as data members to the derived class.

(3) You typically provide the default constructor (constructor without parameters) that conveniently encapsulates the initial pseudostate pointer passed to the `QHsm` constructor.

(4) You provide the initial pseudostate as a static member function with the shown signature.

(5) You provide the all state handlers as a static member functions with the shown signature.

4.6.4   Step 4: Defining the Initial Pseudostate

The initial pseudostate `Calc::initial()` shown here takes the "me" pointer to its own class (`Calc *`) as the first argument and an event pointer as the second parameter. This particular initial pseudostate ignores the event, but sometimes such an initialization event can be helpful to provide additional information required to initialize extended-state variables of the state machine.

```
QState Calc::initial(Calc *me, QEvent const * /* e */) {
    BSP_clear();
    return Q_TRAN(&Calc::on);
}
```

The initial pseudostate can initialize the extended state variables and perform any other actions, but its most important job is to set the default state of the state machine with the `Q_TRAN()` macro, as shown.

4.6.5   Step 5: Defining the State-Handler Functions

Earlier in this chapter, in Listing 4.5 you saw an example of the `Calc::int1()` state handler function. Typically, every state handler function consists of a `switch` statement that discriminates based on the event signal `e->sig`. Each `case` is labeled by a signal and terminates either with "`return Q_HANDLED()`" or "`return Q_TRAN(...)`." Either one of these return statements informs the QEP event processor that the particular event has been handled. On the other hand, if no `case` executes, the state handler exits through the final "`return Q_SUPER(...)`" statement, which informs the QEP event processor that the event needs to be handled by the designated superstate.

Highest-level states without explicit superstate (e.g., the "on" state in the calculator example) nest implicitly in the top state. Such states disignate `&QHsm::top` as the argument to the `Q_SUPER()` macro.

> **NOTE**
>
> The final `return` statement from a state handler function is the only place where you specify the hierarchy of states. Therefore, this one line of code represents the single point of maintenance for changing the nesting level of a given state.

While coding state-handler functions, you need to keep in mind that QEP will invoke them for various reasons: for hierarchical event processing, for execution of entry and exit actions, for triggering initial transitions, or even just to elicit the superstate of a given state handler. Therefore, you should not assume that a state handler would be invoked only for processing events enlisted in the `case` statements. You should also avoid any code outside the `switch` statement, especially code that would have side effects.

4.6.6   Coding Entry and Exit Actions

The `qep.h` header file provides two reserved signals `Q_ENTRY_SIG` and `Q_EXIT_SIG` that the QEP event processor passes to the appropriate state-handler function to execute the state entry actions or exit actions, respectively.

Therefore, as shown in Listing 4.8 earlier in this chapter, to code an entry action, you provide a `case` statement labeled with signal `Q_ENTRY_SIG`, enlist all the actions you want to execute upon the entry to the state, and terminate the lists with "`return Q_HANDLED()`," which informs the QEP that the entry actions have been handled.

Coding the exit actions is identical, except that you provide a `case` statement labeled with the signal `Q_EXIT_SIG`, call the actions you want to execute upon the exit from the state, and terminate the lists with "`return Q_HANDLED()`," which informs the QEP that the exit action has been handled.

4.6.7   Coding Initial Transitions

Every composite state (a state with substates) can have its own initial transition, which in the diagram is represented as an arrow originating from a black ball. For example, the calculator state "on" in Figure 2.18 has such a transition to substate "ready."

The QEP provides a reserved signal `Q_INIT_SIG` that the event processor passes to the appropriate state-handler function to execute the initial transition.

Therefore, as shown Listing 4.8 earlier in this chapter, to code an initial transition, you provide a `case` statement labeled with signal `Q_INIT_SIG`, enlist all the actions you want to execute upon the initial transition, and then designate the target substate with the `Q_TRAN()` macro. The status returned from the `Q_TRAN()` macro informs QEP that the initial transition has been handled.

The UML specification requires that the target of the initial transition is a direct or indirect substate of the source state. An initial transition to a nonsubstate (e.g., a peer state, or a superstate) corresponds to a malformed state machine and may even crash the event processor. Note that initial transitions cannot have guard conditions.

4.6.8   Coding Internal Transitions

Internal transitions are simple reactions to events that never lead to change of state and consequently never cause execution of exit actions, entry actions, or initial transitions.

To code an internal transition, you provide a `case` statement labeled with the triggering signal, enlist the actions, and terminate the list with "`return Q_HANDLED()`" to inform QEP that the event has been handled.

4.6.9   Coding Regular Transitions

State-handler `Calc::int1()` from Listing 4.5 provides two examples of regular state transitions. To code a regular transition, you provide a `case` statement labeled with the triggering signal (e.g., `POINT_SIG`), enlist the actions, and then designate the target state with the `Q_TRAN()` macro. The status returned from the `Q_TRAN()` macro informs QEP that a transition has been taken.

The `Q_TRAN()` macro can accept any target state at any level of nesting, such as a peer state, a substate, a superstate, or even the same state as the source of the transition (transition to self).

NOTE

The QEP hierarchical event processor automatically handles execution of appropriate exit and entry actions during arbitrary state transitions (in the `QHsm::dispatch()` function). Consequently, any change in state machine topology (change in state transitions or state nesting) requires only recompiling the state-handler functions. QEP automatically takes care of figuring out the correct sequence of exit/entry actions and initial transitions to execute for every state transition.

4.6.10   Coding Guard Conditions

Guard conditions (or simply guards) are Boolean expressions evaluated dynamically based on the value of event parameters and/or the variables associated with the state machine (extended-state variables). The following definition of the `Cacl::begin()` state-handler function shows an example of a state transition with a guard.

```
QState Calc::begin(Calc *me, QEvent const *e) {
    switch (e->sig) {
        case OPER_SIG: {
            if ((static_cast<CalcEvt const *>(e))->key_code == KEY_MINUS) {
                return Q_TRAN(&Calc::negated1);
            }
            break;
        }
    }
    return Q_SUPER(&Calc::ready);
}
```

The guard condition maps simply to an `if`-statement that conditionally executes actions. Note that only the TRUE branch of the `if` contains the "`return Q_TRAN()`" statement, meaning that only the TRUE branch reports that the event has been handled. If the TRUE branch is not taken, the `break` statement causes a jump to the final `return` that informs the QEP that the event has *not* been handled. This is in compliance with the UML semantics, which require treating an event as unhandled in case the guard evaluates to FALSE. In that case, the event should be propagated up to the higher levels of hierarchy (to the superstate).

Guard conditions are allowed not just for regular state transitions but for the internal transitions as well. In this case a guard maps to the `if` statement that contains "`return Q_HANDLED()`" only in the TRUE branch. The only difference for the internal transition is that you return the `Q_HANDLED()` macro instead of `Q_TRAN()`.

4.7 Pitfalls to Avoid While Coding State Machines with QEP

The QEP hierarchical event processor enables building efficient and maintainable state machine implementations in C and C++. However, it is also possible to use QEP incorrectly because the direct manual-coding approach leaves you a lot of freedom in structuring your state machine code. This section summarizes the main pitfalls that various QEP users have fallen into over the years and provides some guidelines on how to benefit the most from QEP.

4.7.1 Incomplete State Handlers

You should construct only *complete* state handlers, that is, state-handler functions that directly include all state machine elements pertaining to a given state (such as all actions, all transitions, and all guards), so that you or anyone else could at any time unambiguously draw the state in a diagram using only the state-handler function.

The key is the way you break up the code. Instead of thinking in terms of individual C statements, you should think at a higher level of abstraction, in terms of the idioms defined in Sections 4.6.5-4.6.10 for coding states, transitions, entry/exit actions, initial transitions, and guards.

Consider the following problematic implementation of the "on" state handler of the calculator state machine shown before, in Section 4.5.4:

```
QState Calc_on(Calc *me, QEvent const *e) {
    switch (e->sig) {
        . . .
        case C_SIG: {
            return Calc_onClear(me);          /* handle the Clear event */
        }
        . . .
    }
    return Q_SUPER(&QHsm_top);
}
. . .
QState Calc_onClear(Calc *me) {
    BSP_clear();
    return Q_TRAN(&Calc_on);                   /* transition to "on" */
}
```

This `Calc_on()` state-handler function differs from the original implementation discussed in Section 4.5.4 only in the way it handles the `C_SIG` signal. Though the problematic implementation is in principle equivalent to the original and would perform exactly the same way, the problematic state handler is *incomplete* because it does not follow the idiom for coding state transition from Section 4.6.9. In particular, the state handler hides the state transition to self triggered by `C_SIG` and from such an incomplete state handler alone you would not be able to correctly draw the state in the diagram.

In summary, perhaps the most important principle to keep in mind while coding state machines with QEP is that the code is as much an implementation as it is a *specification* of a state machine. This perspective on coding state machines with QEP will help you (and others) readily see the state machine structure right from the code and easily and unambiguously map the code back to state diagrams. Conversely, state machine code structured arbitrarily, even if working correctly, might be misleading and therefore difficult to maintain (see also [Samek 03f]).

4.7.2   Ill-Formed State Handlers

All nontrivial, semantically rich formalisms, including UML state machines, allow building ill-formed constructs. An ill-formed state machine is inherently wrong, not one that just happens to incorrectly represent behavior. For example, you could draw a UML state diagram with initial transitions targeting peer states rather than substates or conflicting transitions with overlapping guards. The specific state machine implementation technique, such as QEP, introduces additional opportunities of "shooting yourself in the foot." It is possible, for example, to code a state handler that would nest inside itself (the state-handler function would return a pointer to self). Such a state machine cannot be even drawn in a state diagram but is quite easy to code with QEP.

This section examines more of such situations that result with ill-formed state machines. Often, ill-formed state machines cause an assertion violation within the QEP code (see the sidebar "Design by Contract in C and C++"). However, some pathological cases, such as circular state nesting, could crash the QEP event processor.

4.7.3   State Transition Inside Entry or Exit Action

Novice QEP users sometimes try to code a transition inside the entry action to a state by using the Q_TRAN() macro. This happens typically when a developer confuses a statechart with a flowchart (see Section 2.2.3) and thinks of the entered state as just a stage of processing that automatically progresses to the next stage upon completion of the entry actions.

The UML does not allow transitions in entry or exit actions. The typical intention in coding a state transition in an entry action is to enter a given state only under some condition and transition to a different state under some other condition. The correct way of handling this situation is to explicitly code two transitions with complementary guards and with different target states.

4.7.4 Incorrect Casting of Event Pointers

As described in Section 4.3, event parameters are added to the QEvent structure in the process of class inheritance. However, events are uniformly passed to state-handler functions as the generic QEvent* pointer, even though they point to various derived event classes. That's why you need to *downcast* the generic QEvent* pointer onto the pointer to the specific event subclass as shown, for instance, in Listing 4.4(5).

However, to perform the downcast correctly, you need to know what derived event to cast to. The only information you have at this point is the signal of the event (e->sig), and therefore the signal alone must unambiguously identify the derived event structure. The problem arises if you use one signal with multiple event classes, because then you could cast incorrectly on the wrong event class.

4.7.5 Accessing Event Parameters in Entry/Exit Actions or Initial Transitions

A specific case of incorrect event casting is an attempt to access event parameters when handling entry/exit actions or initial transitions. For example, if a state has only one incoming transition that is triggered with an event with parameters, novice QEP users sometimes try to access these parameters in the entry action to this state. Consider the following hypothetical code example:

```
QState MyHSM_stateA(MyHSM *me, QEvent const *e) {
    switch (e->sig) {
        case EVTB_SIG: {
            . . .
            /* the only way to transition to stateB */
            return Q_TRAN(&MYHSM_state B);
        }
    }
    return Q_SUPER(&QHsm_top);
}
QState MyHSM_stateB(MyHSM *me, QEvent const *e) {
    switch (e->sig) {
        case Q_ENTRY_SIG: {
            EvtB const *evtB = (EvtB const *)e;        /* INCORRECT cast */
            if (evtB->foo != ) . . .                   /* INCORRECT access */
            . . .
        }
    }
    return Q_SUPER(&QHsm_top);
}
```

The transition in "stateA" triggered by EVTB_SIG is the only way to get to "stateB." In the entry action to "stateB" the programmer realizes that some parameter foo in the EvtB structure, associated with signal EVTB_SIG, is needed and casts the generic event on (EvtB const *). This is an incorrect cast, however, because by the time "stateB" is entered the original triggering event EvtB is no longer accessible. Instead, the entry action is triggered by a reserved signal Q_ENTRY_SIG, which is not associated with the EvtB event structure. This is actually logical because a state can be entered in many different transitions, each one triggered by a different event, and none of them are accessible by the time entry action is processed.

The correct way of handling this situation is to perform actions dependent on the event parameters directly on the transition triggered by this event rather than in the entry action. Alternatively, the event parameters can be stored in the extended-state variables (members of the state machine structure that you access through the "me" pointer). The extended-state variables are accessible all the time, so they can be used also in the entry/exit actions or initial transitions.

4.7.6 Targeting a Nonsubstate in the Initial Transition

All initial transitions must target direct or indirect substates of the state in which the initial transition is defined. Figure 4.8(A) shows several examples of correct initial transitions. Note that an initial transition can span more than one level of state hierarchy, but it must always target a direct or indirect substate of a given state. Figure 4.8(B) shows one example

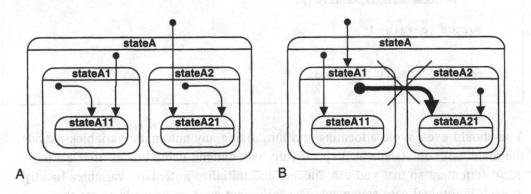

Figure 4.8: Correct (A) and incorrect (B) initial transitions.

of the highlighted initial transition in "stateA1" that targets "stateA21." The problem is that "stateA21" is not a substate of "stateA1" and therefore the state machine in Figure 4.8(B) is ill-formed according to the UML semantics. Coding such an initial transition in QEP will crash the event processor.

4.7.7   Code Outside the switch Statement

As the QEP user, you need to understand that each event that dispatched the state machine through the function QHsm_dispatch() might potentially cause invocation of many state-handler functions and some of them might be called more than once. This is because the event processor needs to call state-handler functions to perform hierarchical event processing, to handle entry/exit actions and initial transitions, or simply to discover the nesting level of a given state. Therefore, you should not assume that a state-handler function would be called exactly once for a given event, so you should avoid any code outside the main switch statement dedicated to events, especially if the code has side effects.

For example, the following state-handler function is certainly inefficient, and probably incorrect, because the for loop executes every time the state handler is invoked, which is not just for the events enlisted as cases in the switch statement.

```
QState MyHSM_stateA(MyHSM *me, QEvent const *e) {
    for (i = 0; i < N; ++i) { /* PROBLEMATIC: expensive loop outside switch */
        doSomethingExpensive();
    }
    switch (e->sig) {
        . . .
    }
    return Q_SUPER(&QHsm_top);
}
```

You should even avoid allocating and initializing any automatic variables outside the main switch statement. I specifically recommend using braces after each case statement so that you can allocate and initialize automatic variables locally in each individual case statement. The following code snippet illustrates the situation:

```
QState MyHSM_stateB(MyHSM *me, QEvent const *e) {
    uint32_t tmp = 0x12345678;    /* initialization occurring every time */
    switch (e->sig) {
        . . .
        case MY_EVT_SIG: {
            uint32_t tmp = 0x12345678; /* initialization only in this case */
            . . .
        }
    }
    return Q_SUPER(&QHsm_top);
}
```

4.7.8 Suboptimal Signal Granularity

Nothing affects state machine complexity and efficiency as much as the right granularity and semantics of events. The optimal granularity of signals falls somewhere between the two extremes of too fine and too coarse.

The granularity of signals is too fine if you repeatedly find the same groups of signals handled in the same way. For example, recall the calculator example (Section 2.4 in Chapter 2). The calculator HSM handles all numerals 1 through 9 in the same way. Therefore, introducing a separate signal for each numeral would lead to a signal granularity that is too fine, which would unnecessarily bloat the state-handler functions (you would see long lists of cases handled identically). Instead, the calculator statechart represents the whole group of numerals 1 through 9 as just one signal, IDC_1_9_SIG (see Figure 2.18).

The granularity of signals is too coarse if you find yourself frequently using guard conditions that test event parameters. In this case, event parameters are the *de facto* signals. Consider the Windows message WM_COMMAND, frequently used in Windows GUI applications for all buttons and menus of the application. This signal is too coarse because Windows applications typically must test the wParam parameter associated with the WM_COMMAND to determine what actually happened. In other words, values of wParam are the *de facto* signals. In this case, the too coarse signal granularity results in a suboptimal (and not very elegant) additional switch statement based on wParam nested within the WM_COMMAND case. When you encounter signals that are too coarse, the first thing you should try is to redefine or remap signals to the right level of granularity before dispatching them to the state machine. However, if you cannot do this, you should include all the *de facto* signals directly in your state handlers.

All too often, the additional layer of signal dispatching (such as the `switch` based on `wParam`) end up in a separate function, which makes state handlers incomplete in the sense discussed in Section 4.7.1.

4.7.9 Violating the Run-to-Completion Semantics

All state machine formalisms, including UML statecharts, universally assume run-to-completion (RTC) semantics of processing events. RTC means that a state machine must always complete processing of the previous event before it can start processing the next. The RTC restriction comes from the fact that a state machine must always go from one *stable* state configuration all the way to another *stable* state configuration in one indivisible step (RTC step). A state machine cannot accept events before reaching a stable state configuration.

The RTC semantics is implicit in the QEP implementation because each invocation of the function `QHsm_dispatch()` represents one RTC step. In single-threaded systems, such as all the examples discussed in this chapter, the RTC semantics cannot be violated because each function must return before it can be called again. However, in multitasking environments, even as simple as the superloop (main+ISRs), the RTC semantics can be easily violated by attempts to dispatch an event to a state machine from an ISR while the same state machine in the background loop is still busy processing the previous event.

4.7.10 Inadvertent Corruption of the Current Event

A very nasty and difficult-to-debug violation of the RTC semantics is an inadvertent corruption of the current event before the RTC step completes. Recall that the state-handler functions in QEP take just pointers to events, not the copies of the entire event objects. It is therefore possible that the memory pointed to by the event pointer will get corrupted before the current RTC step completes.

For example, consider once more the superloop (main+ISRs) architecture. An ISR produces an event and sets a global flag to trigger a state machine running in the background. The background loop starts processing the event, but before it completes, another interrupt preempts it. The ISR produces another event by overwriting the memory used previously for the first event. The RTC semantics are violated even though the ISR merely sets a flag instead of calling the state machine directly.

The general solution to guarantee RTC semantics in multitasking systems is to use
event queues to store events while a state machine is busy. The mechanisms for
a *thread-safe* event queuing and dispatching to multiple concurrently executing state
machines can be generalized and reused rather than being reinvented from scratch for
each application. Virtually all GUI systems (such as Microsoft Windows, X Windows,
and others) are examples of such reusable architectures. QEP can be used with
virtually any such event-driven infrastructure. In particular, QEP can be combined with
the QF event-driven framework design specifically for the domain of real-time
embedded systems. I introduce the QF component in Chapter 6.

4.8 Porting and Configuring QEP

Adapting the QEP software to a particular CPU and compiler is called *porting*. You port
and configure the QEP event processor by providing the qep_port.h header file,
which is included in all source files comprising QEP (see Listing 4.1). Listing 4.14
shows an example of qep_port.h for 80×86 CPU.

Listing 4.14 The qep_port.h **header file for the 80x86 QEP port located
in the directory** <qp>\qpc\ports\80x86\dos\tcpp101\1\

```
      #ifndef qep_port_h
      #define qep_port_h

                  /* special keyword used for ROM objects (none for 80x86) */
(1)   #define Q_ROM

               /* mechanism of accessing const objects in ROM (far pointers) */
(2)   #define Q_ROM_VAR       far
                                    /* 1-byte signal space (255 signals) */
(3)   #define Q_SIGNAL_SIZE

         /* exact-width types. WG14/N843 C99 Standard, Section 7.18.1.1 */
(4)   typedef signed   char int8_t;
      typedef signed   int  int16_t;
      typedef signed   long int32_t;
      typedef unsigned char uint8_t;
      typedef unsigned int  uint16_t;
      typedef unsigned long uint32_t;
(5)   #include "qep.h"     /* QEP platform-independent public interface */

      #endif                                              /* qep_port_h */
```

(1)   The Q_ROM macro allows enforcing placing the constant objects, such as lookup tables, constant strings, and the like, in ROM rather than in the precious RAM. On CPUs with the Harvard architecture (such as 8051 or the Atmel AVR), the code and data spaces are separate and are accessed through different CPU instructions. The compilers often provide specific extended keywords to designate code or data space, such as the "__code" extended keyword in the IAR 8051 compiler. Here, for the 80x86 CPU, the definition of the Q_ROM macro is empty.

(2)   The macro Q_ROM_VAR specifies the kind of the pointer to be used to access the ROM objects because many compilers provide different-sized pointers for accessing objects in various memories. Constant objects allocated in ROM often mandate the use of specific-size pointers (e.g., far pointers) to get access to ROM objects. An example of valid Q_ROM_VAR macro definition is __far (Freescale HC(S)08 compiler).

NOTE

Macros Q_ROM and Q_ROM_VAR refer to the different parts of the object declaration. The macro Q_ROM specifies the ROM memory type to allocate an object. This allows compilers generating different instructions for accessing such ROM objects for CPUs with the Harvard architecture. On the other hand, the macro Q_ROM_VAR specifies the size of the pointer (e.g., the "far" pointer) to access the ROM data, so it refers just to the size of the object's address, not to the object itself. The Q_ROM_VAR macro is useful for the von Neumann machines.

If you don't define macros Q_ROM or Q_ROM_VAR, the qep.h header file will provide default empty definitions, which means that no special extended keywords are necessary to correctly allocate and access the constant objects.

(3)   The macro Q_SIGNAL_SIZE configures the QSignal type (see Section 4.3.1). If the macro is not defined, the default of 1 byte will be chosen in qep.h. The valid Q_SIGNAL_SIZE values 1, 2, or 4 correspond to QSignal of uint8_t, uint16_t, and uint32_t, respectively. The QSignal data type determines the dynamic range of numerical values of signals in your application.

(4)   Porting QEP requires providing the C99-standard exact-width integer types that are consistent with the CPU and compiler options used. For newer C and C++ compilers, you simply need to include the standard header file <stdint.h> provided by the compiler vendor. For prestandard compilers, you need to provide the typedefs for the six basic exact-width integer types.

(5)   The `qep_port.h` platform-specific header file must include the `qep.h` platform-independent header file.

4.9 Summary

Almost all real-life state machines can vastly benefit from the reuse of behavior enabled by hierarchical state nesting. Traditionally, state hierarchy has been considered an advanced feature that mandates automatic code synthesis by CASE tools. However, the use of a generic event processor enables very straightforward manual coding of HSMs.

This chapter described the inner workings of a small, generic, hierarchical event processor called QEP. The event processor consists of just two classes: `QHsm` for derivation of state machines and `QEvent` for derivation of events with parameters. The event-dispatching algorithm implemented in the `QHsm` class has been carefully optimized over the years for both speed and space. The most recent QEP version requires only a single pointer to function per state machine in RAM and minimizes the stack usage by very judiciously sizing automatic variables and by avoiding any recursive calls to state handler functions. State-handler functions are an inexpensive commodity, and there are no limits (except for code space) of how many you can use.

Implementing HSMs with QEP is straightforward because the hierarchical event processor does most of the heavy lifting for you. In fact, coding of even the most complex HSM turns out to be a rather simple exercise in applying just a few straightforward rules. As your design evolves, QEP allows easily changing the state machine topology. In particular, no transition chains must be coded manually. To change the target of a transition, you modify the argument of the `Q_TRAN()` macro. Similarly, to change the superstate of a given state, you modify the argument of the `Q_SUPER()` macro. All these changes are confined to one line of code.

The most important perspective to keep in mind while coding state machines with QEP is that the source code is as much the implementation as it is the *executable specification* of your state machine. Instead of thinking in terms of individual C or C++ statements, you should think in terms of state machine elements, such as states, transitions, entry/exit actions, initial transitions, and guards. When you make this quantum leap, you will no longer struggle with convoluted `if-else` "spaghetti" code. You will start thinking at a *higher level of abstraction* about the best ways to partition behavior into states, about the events available at any given time, and about the best state hierarchy for your state machine.

CHAPTER 5

State Patterns

Science is a collection of successful recipes.
— Paul Valéry

In the previous chapter, you learned how to implement hierarchical state machines (HSMs) in C and C++ with the generic hierarchical event processor called QEP. In fact, QEP enabled a rather mechanical one-to-one mapping between state models and the code. With just a bit of practice, you will forget that you are laboriously translating state models into code; rather, you will directly build state machines in C or C++.

At this point, you will no longer struggle with 15 levels of if-else statements and gazillions of flags. You will start thinking at a higher level of abstraction about the best ways to partition behavior into states, about the structure of your state machine, and about the event exchange mechanisms.

However, coming up with a good structure for a nontrivial state machine isn't easy. Experienced developers know that a reusable and flexible state machine design is difficult to get right the first time. Yet experienced designers repeatedly realize good state machines, whereas newcomer are overwhelmed by the options available and tend to fall back on convoluted if-else constructs and the multitude of flags they have used before.

One thing that distinguishes an expert from a novice is the ability to recognize the similarities among problems encountered in the past and to reuse proven solutions that work. To share their expertise, OO designers began to catalog proven solutions to recurring problems as object-oriented design patterns [GoF 95]. Similarly, state patterns began to appear [Douglass 99]. In contrast to the OO patterns, which are concerned with optimal ways of structuring classes and objects, the state patterns focus on effective ways of structuring states, events, and transitions.

A state pattern has the following five essential elements, just as an OO pattern does:

- *The pattern name.* A word or two denoting the problem, the solution, and the consequences of a pattern. A good name is vital because it will become part of your vocabulary.

- *The problem.* An explanation of the problem the pattern addresses. A problem is often motivated by an example.

- *The solution.* A description of the elements (states, transitions, events, actions, and extended-state variables) that compose the solution and their relationships, responsibilities, and collaborations.

- *The sample code.* A presentation of a concrete implementation of an instance of the pattern. Usually the sample code implements the motivating example.

- *The consequences.* The results and trade-offs of applying the pattern.

In this chapter, I provide a mini-catalog of five basic state patterns (Table 5.1). The first two are relatively simple state machine solutions to common problems. The other three are just more advanced or expensive features that are found in the UML state machine package [OMG 07] but are not supported directly in the QEP event processor. The leading theme of all these patterns is reusing behavior through hierarchical state nesting, in contrast to the previously documented state patterns that all revolve primarily around orthogonal regions [Douglass 99]. The additional distinguishing aspect of the state patterns presented here is that all are illustrated with *executable code*. A state diagram alone is not enough to understand a state pattern, because the devil is always in the detail. To be genuinely useful, a pattern must be accompanied by a specific working example that will help you truly comprehend and evaluate the pattern and give you a good starting point for your own implementations.

Many examples in this chapter are implemented with the QF real-time framework that I will formally introduce in Chapter 6. The QEP component by itself is not sufficient, because it provides only the passive event processor that lacks such essential elements as the event loop, event queuing, and timing services. The QF framework provides these missing ingredients. However, all patterns can also be used in conjunction with any other event-driven infrastructure such as GUI systems (Windows, Mac, X11, etc.).

Table 5.1: Summary of state patterns covered in this chapter

Pattern Name	Intent
Ultimate Hook (Section 5.1)	Provide a common look and feel but let clients specialize every aspect of a system's behavior.
Reminder (Section 5.2)	Invent an event and post it to self.
Deferred Event (Section 5.3)	Control the sequence of events.
Orthogonal Component (Section 5.4)	Use state machines as components.
Transition to History (Section 5.5)	Transition to the most recent state configuration of a given composite state.

None of the state patterns described in this chapter captures new or unproven state machine designs. In fact, by definition, a state pattern is a proven solution to a recurring problem that is actually used in successful, real-life event-driven systems. However, most of the basic state patterns have never been documented before (at least not with such a level of detail and illustrated with *executable* code). They are either part of the folklore of various programming communities (e.g., the GUI community or the embedded systems community) or are elements of some successful systems, neither of which is easy for novice designers to learn from. So although these state machine designs are not new, they are offered here in a new and more accessible way.

5.1 Ultimate Hook

5.1.1 Intent

Provide common facilities and policies for handling events but let clients override and specialize every aspect of the system's behavior.

5.1.2 Problem

Many event-driven systems require consistent policies for handling events. In a GUI design, this consistency is part of the characteristic look and feel of the user interface. The challenge is to provide such a common look and feel in system-level software that client applications can use easily as the default. At the same time, the clients must be able to override every aspect of the default behavior easily if they so choose.

5.1.3   Solution

The solution is to apply *programming by difference* or, specifically in this case, the concept of *hierarchical state nesting*. A composite state can define the default behavior (the common look and feel) and supply an "outer shell" for nesting client substates. The semantics of state nesting provide the desired mechanism of handling all events, first in the context of the client code (the nested state) and of automatically forwarding of all unhandled events to the superstate (the default behavior). In that way, the client code intercepts every stimulus and can override every aspect of the behavior. To reuse the default behavior, the client simply ignores the event and lets the superstate handle it (the substate inherits behavior from the superstate).

Figure 5.1 shows the Ultimate Hook state pattern using the collaboration notation adapted for states [OMG 07]. The dashed oval labeled «state pattern» indicates collaboration among states. Dashed arrows emanating from the oval indicate state roles within the pattern. States playing these roles are shown with heavy borders. For example, the state "generic" plays the role of the generic superstate of the pattern, whereas the state "specific" plays the role of the specific substate.

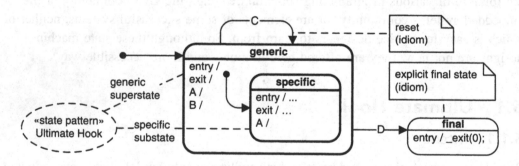

Figure 5.1: The Ultimate Hook state pattern.

A diagram like this attempts to convey an abstract pattern but can only show a concrete example (instance) of the pattern. In this instance, the concrete "generic" state in Figure 5.1 handles events A and B as internal transitions, event C as a transition to self, and event D as the termination of the state machine. The concrete "specific" state overrides event A and provides its own initialization and cleanup (in entry and exit actions, respectively). Of course, another instance of the pattern can implement completely different events and actions.

A few idioms worth noting are illustrated in this state diagram. First is the overall canonical structure of the state machine that, at the highest level, consists of only one composite state (the pattern role of the generic superstate). Virtually every application can benefit from having such a highest-level state because it is an ideal place for defining common policies subsequently inherited by the whole (arbitrary complex) submachine.

> **NOTE**
>
> As described in Section 2.3.2 in Chapter 2, every UML state machine is a submachine of an implicit top state and so has the canonical structure proposed here. However, because you cannot override the top state, you need another highest-level state that you can customize.

Within such a canonical structure, a useful idiom for resetting the state machine is an empty (actionless) transition to self in the "generic" superstate (transition C in Figure 5.1). Such a transition causes a recursive exit from all nested states (including the "generic" superstate), followed by initialization starting from the initial transition of the highest-level state. This way of resetting a state machine is perhaps the safest because it guarantees proper cleanup through the execution of exit actions and clean initialization by entry actions and nested initial transitions. Similarly, the safest way to terminate a state machine is through an explicit transition out of the generic superstate to a "final" state (transition D in Figure 5.1) because all pertinent exit actions are executed. The QEP event processor does not provide a generic final state (denoted as the bull's eye in the UML). Instead, the statechart in Figure 5.1 proposes an idiom, which consists of an explicit state named "final" with an application-specific termination coded in its entry action.[1]

5.1.4 Sample Code

The sample code for the Ultimate Hook state pattern is found in the directory `<qp>\qpc\examples\80x86\dos\tcpp101\1\hook\`. You can execute the application by double-clicking the file `HOOK.EXE` file in the `dbg\` subdirectory. Figure 5.2 shows the output generated by the `HOOK.EXE` application. Listing 5.1 shows the example implementation of the Ultimate Hook pattern from Figure 5.1.

---

[1] The calculator HSM designed in Chapter 2 and coded in Chapter 4 provides an example of the canonical state machine structure that uses the idioms to reset and terminate.

```
Command Prompt                                                    _ □ ×
QEP version: 4.0.00
Press 'a'..'c' to inject signals A..C
Press 'd' or ESC to inject signal D and quit
top-INIT;generic-ENTRY;generic-INIT;specific-ENTRY;
a: specific-A;
b: generic-B;
c: generic-C(reset);specific-EXIT;generic-EXIT;generic-ENTRY;generic-INIT;specif
ic-ENTRY;
d: specific-EXIT;generic-EXIT;final-ENTRY;
Bye!Bye!
```

Figure 5.2: Output generated by HOOK.EXE.

Listing 5.1 The Ultimate Hook sample code (file hook.c**).**

```
(1)  #include "qep_port.h"

     typedef struct UltimateHookTag {      /* UltimateHook state machine */
(2)      QHsm super;                                 /* derive from QHsm */
     } UltimateHook;

     void   UltimateHook_ctor      (UltimateHook *me);              /* ctor */
(3)  QState UltimateHook_initial  (UltimateHook *me, QEvent const *e);
     QState UltimateHook_generic  (UltimateHook *me, QEvent const *e);
     QState UltimateHook_specific (UltimateHook *me, QEvent const *e);
     QState UltimateHook_final    (UltimateHook *me, QEvent const *e);
(4)  enum UltimateHookSignals {                      /* enumeration of signals */
         A_SIG = Q_USER_SIG,
         B_SIG,
         C_SIG,
         D_SIG
     };
     /*..............................................................*/
     void UltimateHook_ctor(UltimateHook *me) {
         QHsm_ctor(&me->super, (QStateHandler)&UltimateHook_initial);
     }
     /*..............................................................*/
     QState UltimateHook_initial(UltimateHook *me, QEvent const *e) {
         printf("top-INIT;");
         return Q_TRAN(&UltimateHook_generic);
     }
     /*..............................................................*/
     QState UltimateHook_final(UltimateHook *me, QEvent const *e) {
         switch (e->sig) {
```

```
                case Q_ENTRY_SIG: {
                    printf("final-ENTRY(terminate);\nBye!Bye!\n");
                    exit(0);
                    return Q_HANDLED();
                }
            }
            return Q_SUPER(&QHsm_top);
        }
        /*..............................................................*/
        QState UltimateHook_generic(UltimateHook *me, QEvent const *e) {
            switch (e->sig) {
                . . .
                case Q_INIT_SIG: {
                    printf("generic-INIT;");
                    return Q_TRAN(&UltimateHook_specific);
                }
                case A_SIG: {
                    printf("generic-A;");
                    return Q_HANDLED();
                }
                case B_SIG: {
                    printf("generic-B;");
                    return Q_HANDLED();
                }
                case C_SIG: {
                    printf("generic-C(reset);");
(5)                 return Q_TRAN(&UltimateHook_generic);
                }
                case D_SIG: {
(6)                 return Q_TRAN(&UltimateHook_final);
                }
            }
            return Q_SUPER(&QHsm_top);
        }
        /*..............................................................*/
        QState UltimateHook_specific(UltimateHook *me, QEvent const *e) {
            switch (e->sig) {
(7)             case Q_ENTRY_SIG: {
                    printf("specific-ENTRY;");
                    return Q_HANDLED();
                }
(8)             case Q_EXIT_SIG: {
                    printf("specific-EXIT;");
                    return Q_HANDLED();
                }
```

Continued onto next page

```
(9)              case A_SIG: {
                     printf("specific-A;");
                     return Q_HANDLED();
                 }
             }
             return Q_SUPER(&UltimateHook_generic);     /* the superstate */
         }
```

(1)   Every QEP application needs to include qep_porth.h (see Section 4.8 in Chapter 4).

(2)   The structure UltimateHook derives from QHsm.

(3)   The UltimateHook declares the initial() pseudostate and three state-handler functions: generic(), specific(), and final().

(4)   The signals A through D are enumerated.

(5)   The transition-to-self in the "generic" state represents the reset idiom.

(6)   The transition to the explicit "final" state represents the terminate idiom.

(7,8) The entry and exit actions in the "specific" state provide initialization and cleanup.

(9)   The internal transition A in the "specific" state overrides the same transition in the "generic" superstate.

One option of deploying the Ultimate Hook pattern is to organize the code into a library that intentionally does not contain the implementation of the UltimateHook_specific() state-handler function. Clients would then have to provide their own implementation and link to the library to obtain the generic behavior. An example of a design using this technique is Microsoft Windows, which requires the client code to define the WinMain() function for the Windows application to link.

Another option for the C++ version is to declare the UltimateHook::specific() state handler as follows:

```
QState UltimateHook::specific(UltimateHook *me, QEvent const *e) {
    return me->v_specific(e);                    /* virtual call */
}
```

Where the member function `UltimateHook::v_specific(QEvent const *e)` is declared as a pure virtual member function in C++. This will force clients to provide implementation for the pure virtual state-handler function `v_specific()` by subclassing the `UltimateHook` class. This approach combines behavioral inheritance with traditional class inheritance. More precisely, Ultimate Hook represents, in this case, a special instance of the Template Method design pattern [GoF 95].

5.1.5 Consequences

The Ultimate Hook state pattern is presented here in its most limited version — exactly as it is used in GUI systems (e.g., Microsoft Windows). In particular, neither the generic superstate nor the specific substate exhibits any interesting state machine topology. The only significant feature is hierarchical state nesting, which can be applied recursively within the "specific" substate. For example, at any level, a GUI window can have nested child windows, which handle events before the parent.

Even in this most limited version, however, the Ultimate Hook state pattern is a fundamental technique for reusing behavior. In fact, every state model using the canonical structure implicitly applies this pattern.

The Ultimate Hook state pattern has the following consequences:

- The "specific" substate needs to know only those events it overrides.

- New events can be added easily to the high-level "generic" superstate without affecting the "specific" substate.

- Removing or changing the semantics of events that clients already use is difficult.

- Propagating every event through many levels of nesting (if the "specific" substate has recursively nested substates) can be expensive.

The Ultimate Hook state pattern is closely related to the Template Method OO design pattern and can be generalized by applying inheritance of entire state machines.

5.2 Reminder

5.2.1 Intent

Make the statechart topology more flexible by inventing an event and posting it to self.

5.2.2   Problem

Often in state modeling, loosely related functions of a system are strongly coupled by a common event. Consider, for example, periodic data acquisition, in which a sensor producing the data needs to be polled at a predetermined rate. Assume that a periodic TIMEOUT event is dispatched to the system at the desired rate to provide the stimulus for polling the sensor. Because the system has only one external event (the TIMEOUT event), it seems that this event needs to trigger both the polling of the sensor and the processing of the data. A straightforward but suboptimal solution is to organize the state machine into two distinct orthogonal regions (for polling and processing).[2] However, orthogonal regions increase the cost of dispatching events (see the "Orthogonal Component" pattern) and require complex synchronization between the regions because polling and processing are not quite independent.

5.2.3   Solution

A simpler and more efficient solution is to invent a stimulus (DATA_READY) and to propagate it to self as a reminder that the data is ready for processing (Figure 5.3). This new stimulus provides a way to decouple polling from processing without using orthogonal regions. Moreover, you can use state nesting to arrange these two functions in a hierarchical relation,[3] which gives you even more control over the behavior.

In the most basic arrangement, the "processing" state can be a substate of "polling" and can simply inherit the "polling" behavior so that polling occurs in the background to processing. However, the "processing" state might also choose to override polling. For instance, to prevent flooding the CPU with sensor data, processing might inhibit polling occasionally. The statechart in Figure 5.3 illustrates this option. The "busy" substate of "processing" overrides the TIMEOUT event and thus prevents this event from being handled in the higher-level "polling" superstate.

Further flexibility of this solution entails fine control over the generation of the invented DATA_READY event, which does not have to be posted at every occurrence of the original TIMEOUT event. For example, to improve performance, the "polling" state could buffer the raw sensor data and generate the DATA_READY event only when the buffer

---

[2] This example illustrates an alternative design for the Polling state pattern described in [Douglass 99].
[3] Using state hierarchy in this fashion is typically more efficient than using orthogonal regions.

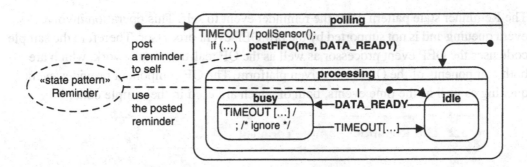

Figure 5.3: The Reminder state pattern.

fills up. Figure 5.3 illustrates this option with the `if (...)` condition, which precedes the `postFIFO(me, DATA_READY)` action in the "polling" state.

5.2.4 Sample Code

The sample code for the Reminder state pattern is found in the directory `<qp>\qpc \examples\80x86\dos\tcpp101\l\reminder\`. You can execute the application by double-clicking on the file `REMINDER.EXE` file in the dbg\ subdirectory. Figure 5.4 shows the output generated by the `REMINDER.EXE` application. The application prints every state entry (to "busy" or "idle") as well as the number of times the `TIMEOUT` event has been handled in "polling" and "processing," respectively. Listing 5.2 shows the example implementation of the Reminder pattern from Figure 5.3.

```
Command Prompt
Reminder state pattern
QEP version: 4.0.00
QF  version: 4.0.00
Press ESC to quit...
idle-ENTRY;
polling   1
polling   2
polling   3
polling   4
busy-ENTRY;
processing   1
processing   2
idle-ENTRY;
polling   5
polling   6
polling   7
polling   8
busy-ENTRY;
processing   3
final-ENTRY;
Bye!Bye!
```

Figure 5.4: Output generated by `REMINDER.EXE`.

The Reminder state pattern posts the reminder event to self. This operation involves event queuing and is not supported by the raw QEP event processor. Therefore the sample code uses the QEP event processor as well as the QF real-time framework, which are both components of the QP event-driven platform. The QF component provides event queuing as well as the time events, both of which are used in the sample code.

Listing 5.2 The Reminder sample code (file `reminder.c`**)**

```
(1)   #include "qp_port.h"                              /* QP interface */
      #include "bsp.h"                          /* board support package */

      enum SensorSignals {
          TIMEOUT_SIG = Q_USER_SIG,       /* the periodic timeout signal */
(2)       DATA_READY_SIG,                 /* the invented reminder signal */
          TERMINATE_SIG                   /* terminate the application */
      };
      /* . . . . . . . . . . . . . . . . . . . . . . . . . . . . . . . */
      typedef struct SensorTag {            /* the Sensor active object */
(3)       QActive super;                        /* derive from QActive */

(4)       QTimeEvt timeEvt;               /* private time event generator */
          uint16_t pollCtr;
          uint16_t procCtr;
      } Sensor;

      void Sensor_ctor(Sensor *me);
                                      /* hierarchical state machine ... */
      QState Sensor_initial   (Sensor *me, QEvent const *e);
      QState Sensor_polling    (Sensor *me, QEvent const *e);
      QState Sensor_processing(Sensor *me, QEvent const *e);
      QState Sensor_idle      (Sensor *me, QEvent const *e);
      QState Sensor_busy      (Sensor *me, QEvent const *e);
      QState Sensor_final     (Sensor *me, QEvent const *e);


      /* . . . . . . . . . . . . . . . . . . . . . . . . . . . . . . . */
      void Sensor_ctor(Sensor *me) {
          QActive_ctor_(&me->super, (QStateHandler)&Sensor_initial);
(5)       QTimeEvt_ctor(&me->timeEvt, TIMEOUT_SIG); /* time event ctor */
      }
      /* HSM definition----------------------------------------------*/
      QState Sensor_initial(Sensor *me, QEvent const *e) {
          me->pollCtr = 0;
          me->procCtr = 0;
(6)       return Q_TRAN(&Sensor_polling);
      }
```

```
    /*................................................................*/
    QState Sensor_final(Sensor *me, QEvent const *e) {
        switch (e->sig) {
            case Q_ENTRY_SIG: {
                printf("final-ENTRY;\nBye!Bye!\n");
                BSP_exit();                    /* terminate the application */
                return Q_HANDLED();
            }
        }
        return Q_SUPER(&QHsm_top);
    }
    /*................................................................*/
    QState Sensor_polling(Sensor *me, QEvent const *e) {
        switch (e->sig) {
            case Q_ENTRY_SIG: {
                                    /* periodic timeout every 1/2 second */
(7)             QTimeEvt_postEvery(&me->timeEvt, (QActive *)me,
                                   BSP_TICKS_PER_SEC/2);
                return Q_HANDLED();
            }
            case Q_EXIT_SIG: {
                QTimeEvt_disarm(&me->timeEvt);
                return Q_HANDLED();
            }
            case Q_INIT_SIG: {
                return Q_TRAN(&Sensor_processing);
            }
(8)         case TIMEOUT_SIG: {
                static const QEvent reminderEvt = { DATA_READY_SIG, 0 };
                ++me->pollCtr;
                printf("polling %3d\n", me->pollCtr);
                if ((me->pollCtr & 0x3) == 0) {            /* modulo 4 */
(9)                 QActive_postFIFO((QActive *)me, &reminderEvt);
                }
                return Q_HANDLED();
            }
            case TERMINATE_SIG: {
                return Q_TRAN(&Sensor_final);
            }
        }
        return Q_SUPER(&QHsm_top);
    }
    /*................................................................*/
    QState Sensor_processing(Sensor *me, QEvent const *e) {
        switch (e->sig) {
```

Continued onto next page

```
                case Q_INIT_SIG: {
                    return Q_TRAN(&Sensor_idle);
                }
            }
            return Q_SUPER(&Sensor_polling);
        }
        /*..............................................................*/
        QState Sensor_idle(Sensor *me, QEvent const *e) {
            switch (e->sig) {
                case Q_ENTRY_SIG: {
                    printf("idle-ENTRY;\n");
                    return Q_HANDLED();
                }
                case DATA_READY_SIG: {
(10)                return Q_TRAN(&Sensor_busy);
                }
            }
            return Q_SUPER(&Sensor_processing);
        }
        /*..............................................................*/
        QState Sensor_busy(Sensor *me, QEvent const *e) {
            switch (e->sig) {
                case Q_ENTRY_SIG: {
                    printf("busy-ENTRY;\n");
                    return Q_HANDLED();
                }
(11)            case TIMEOUT_SIG: {
                    ++me->procCtr;
                    printf("processing %3d\n", me->procCtr);
                    if ((me->procCtr & 0x1) == 0) {            /* modulo 2 */
                        return Q_TRAN(&Sensor_idle);
                    }
                    return Q_HANDLED();
                }
            }
            return Q_SUPER(&Sensor_processing);
        }
```

(1) The Reminder state pattern posts the reminder event to self. This operation involves event queuing and is not supported by the raw QEP event processor. The sample code uses the whole QP, which includes the QEP event processor and the QF real-time framework. QF provides event queuing as well as the time events, both of which are used in the sample code.

> **NOTE**
>
> Event queuing and event-driven timing services are available in virtually every event-driven infrastructure. For instance, Windows GUI applications can call the `PostMessage()` Win32 API to queue messages and the `WM_TIMER` message to receive timer updates.

(2)   The invented reminder event signal (`DATA_READY` in this case) is enumerated just like all other signals in the system.

(3)   The `Sensor` state machine derives from the QF class `QActive` that combines an HSM, an event queue, and a thread of execution. The `QActive` class actually derives from `QHsm`, which means that `Sensor` also indirectly derives from `QHsm` (see Chapter 6 for more details).

(4)   The `Sensor` state machine declares its own private time event. Time events are managed by the QF real-time framework. Section 7.7 in Chapter 7 covers the `QTimeEvt` facility in detail.

(5)   The time event must be instantiated, at which time it gets permanently associated with the given signal (`TIMEOUT_SIG` in this case).

(6)   The topmost initial transition enters the "polling" state, which in turn enters the "idle" substate.

(7)   Upon entry to the "polling" state, the time event is armed for generating periodic `TIMEOUT_SIG` events twice per second.

> **NOTE**
>
> In QF, as in every other RTOS, the time unit is the "time tick." The board support package (BSP) defines the constant `BSP_TICKS_PER_SEC` that ties the ticking rate to the second.

(8)   After being armed, the time event produces the `TIMEOUT_SIG` events at the programmed rate. Because neither the "idle" state nor the "processing" state handle the `TIMEOUT_SIG` signal, the signal is handled initially in the "polling" superstate.

(9)   At a lower rate (every fourth time, in this example), the "polling" state generates the reminder event (`DATA_READY`), which it posts to self. Event posting occurs

by calling the `QActive_postFIFO()` function provided in the QF real-time framework.

(10)   The reminder event causes a transition from "idle" to "busy."

(11)   The "busy" state overrides the `TIMEOUT_SIG` signal and after a few `TIMEOUT` events transitions back to "idle." The cycle then repeats.

5.2.5 Consequences

Although conceptually very simple, the Reminder state pattern has profound consequences. It can address many more problems than illustrated in the example. You could use it as a "Swiss Army knife" to fix almost any problem in the state machine topology.

For example, you also can apply the Reminder idiom to eliminate troublesome completion transitions, which in the UML specification are transitions without an explicit trigger (they are triggered implicitly by completion events, a.k.a. anonymous events). The QEP event processor requires that all transitions have explicit triggers; therefore, the QEP does not support completion transitions. However, the Reminder pattern offers a workaround. You can invent an explicit trigger for every transition and post it to self. This approach actually gives you much better control over the behavior because you can explicitly specify the completion criteria.

Yet another important application of the Reminder pattern is to break up longer RTC steps into shorter ones. As explained in more detail in Chapter 6, long RTC steps exacerbate the responsiveness of a state machine and put more stress on event queues. The Reminder pattern can help you break up CPU-intensive processing (e.g., iteration) by inventing a stimulus for continuation in the same way that you stick a Post-It note to your computer monitor to remind you where you left off on some lengthy task when someone interrupts you. You can also invent event parameters to convey the context, which will allow the next step to pick up where the previous step left off (e.g., the index of the next iteration). The advantage of fragmenting lengthy processing in such a way is that other (perhaps more urgent) events can "sneak in," allowing the state machine to handle them in a more timely way.

You have essentially two alternatives when implementing event posting: the first-in, first-out (FIFO) or the last-in, first-out (LIFO) policy, both of which are supported in the QF real-time framework (see Chapter 6). The FIFO policy is appropriate for breaking up longer RTC steps. You want to queue the Reminder event after other events that have potentially accumulated while the state machine was busy, to give the other events a chance to sneak in ahead of the Reminder. However, in other circumstances, you might want to process an uninterruptible sequence of posted events (such a sequence effectively forms an extended RTC step[4]). In this case, you need the LIFO policy because a reminder posted with that policy is guaranteed to be the next event to process and no other event can overtake it.

> **NOTE**
>
> You should always use the LIFO policy with great caution because it changes the order of events. In particular, if multiple events are posted with the LIFO policy to an event queue and no events are removed from the queue in the meantime, the order of these events in the queue will get reversed.

5.3 Deferred Event

5.3.1 Intent

Simplify state machines by modifying the sequencing of events.

5.3.2 Problem

One of the biggest challenges in designing reactive systems is that such systems must be prepared to handle every event at any time. However, sometimes an event arrives at a particularly inconvenient moment when the system is in the midst of some complex event sequence. In many cases, the nature of the event is such that it can be postponed (within limits) until the system is finished with the current sequence, at which time the event can be recalled and conveniently processed.

Consider, for example, the case of a server application that processes transactions (e.g., from ATM[5] terminals). Once a transaction starts, it typically goes through a sequence of processing, which commences with receiving the data from a remote terminal followed by the authorization of the transaction. Unfortunately, new transaction requests to the server arrive at random times, so it is possible to get a

---

[4] For example, state-based exception handling (see Section 6.7.4 in Chapter 6) typically requires immediate handling of exceptional situations, so you don't want other events to overtake the EXCEPTION event.

[5] *ATM* stands for *automated teller machine*, a.k.a. *cash machine*.

request while the server is still busy processing the previous transaction. One option is to ignore the request, but this might not be acceptable. Another option is to start processing the new transaction immediately, which can complicate things immensely because multiple outstanding transactions would need to be handled simultaneously.

5.3.3 Solution

The solution is to *defer* the new request and handle it at a more convenient time, which effectively leads to altering the sequence of events presented to the state machine.

UML statecharts support such a mechanism directly (see Section 2.3.11 in Chapter 2) by allowing every state to specify a list of deferred events. As long as an event is on the combined deferred list of the currently active state configuration, it is not presented to the state machine but instead is queued for later processing. Upon a state transition, events that are no longer deferred are automatically *recalled* and dispatched to the state machine.

Figure 5.5 illustrates a solution based on this mechanism. The transaction server state machine starts in the "idle" state. The NEW_REQUEST event causes a transition to a substate of the "busy" state. The "busy" state defers the NEW_REQUEST event (note the special "deferred" keyword in the internal transition compartment of the "busy" state). Any NEW_REQUEST arriving when the server is still in one of the "busy" substates gets automatically deferred. Upon the transition AUTHORIZED back to the "idle" state, the NEW_REQUEST is automatically recalled. The request is then processed in the "idle" state, just as any other event.

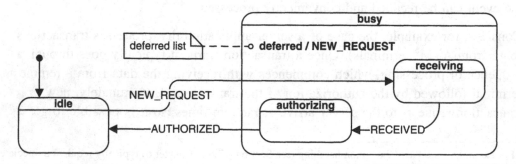

Figure 5.5: Event deferral using the built-in UML mechanism.

The lightweight QEP event processor does not support the powerful, but heavyweight, event deferral mechanism of the UML specification. However, you can achieve identical functionality by deferring and recalling events explicitly. In fact, the QF real-time framework supports event deferral by providing defer() and recall() operations.

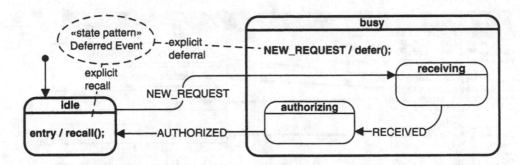

Figure 5.6: The Deferred Event state pattern.

Figure 5.6 shows how to integrate the explicit defer() and recall() operations into a HSM to achieve the desired effect. The internal transition NEW_REQUEST in the "busy" state traps any NEW_REQUEST received in any of the substates. This internal transition calls the defer() operation to postpone the event. The "idle" state explicitly recalls any deferred events by calling recall() in the entry action. The recall() operation posts the first of the deferred events (if available) to self. The state machine then processes the recalled event just as any other event.

5.3.4   Sample Code

The sample code for the Deferred Event state pattern is found in the directory
<qp>\qpc\examples\80x86\dos\tcpp101\l\defer\. You can execute the
application by double-clicking the file DEFER.EXE file in the dbg\ subdirectory.
Figure 5.7 shows the output generated by the DEFER.EXE application. The application
prints every state entry (to "idle," "receiving," and "authorizing"). Additionally, you
get notification of every NEW_REQUEST event and whether it has been deferred or
processed directly. You generate new requests by pressing the *n* key. Note that
request #7 is *not* deferred because the deferred event queue gets full. See the explanation
section following Listing 5.3 for an overview of options to handle this situation.

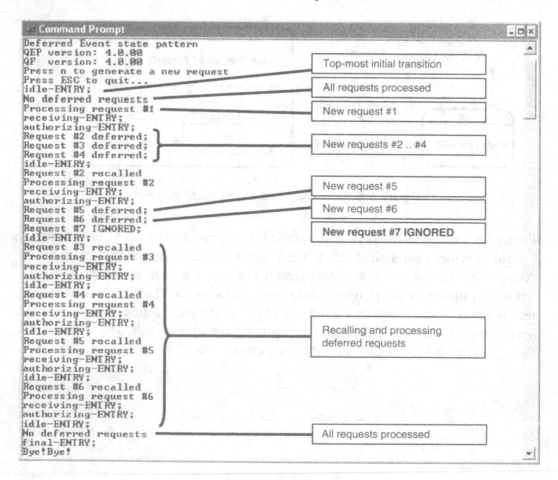

Figure 5.7: Annotated output generated by DEFER.EXE.

Listing 5.3 The Deferred Event sample code (file `defer.c`**)**

```
(1)   #include "qp_port.h"
      #include "bsp.h"

      /*..........................................................*/
      enum TServerSignals {
          NEW_REQUEST_SIG = Q_USER_SIG,            /* the new request signal */
          RECEIVED_SIG,                        /* the request has been received */
          AUTHORIZED_SIG,                     /* the request has been authorized */
          TERMINATE_SIG                          /* terminate the application */
      };
      /*..........................................................*/
(2)   typedef struct RequestEvtTag {
          QEvent super;                              /* derive from QEvent */
          uint8_t ref_num;                  /* reference number of the request */
      } RequestEvt;

      /*..........................................................*/
(3)   typedef struct TServerTag {            /* Transaction Server active object */
(4)       QActive super;                             /* derive from QActive */

(5)       QEQueue requestQueue;     /* native QF queue for deferred request events */
(6)       QEvent const *requestQSto[3]; /* storage for the deferred queue buffer */

(7)       QTimeEvt receivedEvt;                    /* private time event generator */
(8)       QTimeEvt authorizedEvt;                  /* private time event generator */
      } TServer;

      void TServer_ctor(TServer *me);                       /* the default ctor */
                                              /* hierarchical state machine ... */
      QState TServer_initial     (TServer *me, QEvent const *e);
      QState TServer_idle        (TServer *me, QEvent const *e);
      QState TServer_busy        (TServer *me, QEvent const *e);
      QState TServer_receiving   (TServer *me, QEvent const *e);
      QState TServer_authorizing(TServer *me, QEvent const *e);
      QState TServer_final       (TServer *me, QEvent const *e);

      /*..........................................................*/
      void TServer_ctor(TServer *me) {                      /* the default ctor */
          QActive_ctor(&me->super, (QStateHandler)&TServer_initial);
(9)       QEQueue_init(&me->requestQueue,
                  me->requestQSto, Q_DIM(me->requestQSto));
          QTimeEvt_ctor(&me->receivedEvt, RECEIVED_SIG);
          QTimeEvt_ctor(&me->authorizedEvt, AUTHORIZED_SIG);
      }
      /*HSM definition----------------------------------------------*/
      QState TServer_initial(TServer *me, QEvent const *e) {
          (void)e;          /* avoid the compiler warning about unused parameter */
          return Q_TRAN(&TServer_idle);
      }
      /*..........................................................*/
```

Continued onto next page

```
QState TServer_final(TServer *me, QEvent const *e) {
    (void)me;                    /* avoid the compiler warning about unused parameter */
    switch (e->sig) {
        case Q_ENTRY_SIG: {
            printf("final-ENTRY;\nBye!Bye!\n");
            BSP_exit();                               /* terminate the application */
            return Q_HANDLED();
        }
    }
    return Q_SUPER(&QHsm_top);
}
/*.........................................................................*/
QState TServer_idle(TServer *me, QEvent const *e) {
    switch (e->sig) {
        case Q_ENTRY_SIG: {
            RequestEvt const *rq;
            printf("idle-ENTRY;\n");

                    /* recall the request from the private requestQueue */
(10)                rq = (RequestEvt const *)QActive_recall((QActive *)me,
                                                    &me->requestQueue);
            if (rq != (RequestEvt *)0) {          /* recall posted an event? */
(11)                printf("Request #%d recalled\n", (int)rq->refNum);
            }
            else {
(12)                printf("No deferred requests\n");
            }
            return Q_HANDLED();
        }
        case NEW_REQUEST_SIG: {
            printf("Processing request #%d\n",
                    (int)((RequestEvt const *)e)->refNum);
            return Q_TRAN(&TServer_receiving);
        }
        case TERMINATE_SIG: {
            return Q_TRAN(&TServer_final);
        }
    }
    return Q_SUPER(&QHsm_top);
}
/*.........................................................................*/
QState TServer_busy(TServer *me, QEvent const *e) {
    switch (e->sig) {
        case NEW_REQUEST_SIG: {
(13)            if (QEQueue_getNFree(&me->requestQueue) > 0) {   /* can defer? */
                                                    /* defer the request */
(14)                QActive_defer((QActive *)me, &me->requestQueue, e);
                printf("Request #%d deferred;\n",
                        (int)((RequestEvt const *)e)->ref_num);
            }
```

```
                         else {
                             /* notify the request sender that the request was ignored.. */
(15)                         printf("Request #%d IGNORED;\n",
                                 (int)((RequestEvt const *)e)->ref_num);
                         }
                         return Q_HANDLED();
                     case TERMINATE_SIG: {
                         return Q_TRAN(&TServer_final);
                     }
                 }
                 return Q_SUPER(&QHsm_top);
             }
             /*..............................................................*/
             QState TServer_receiving(TServer *me, QEvent const *e) {
                 switch (e->sig) {
                     case Q_ENTRY_SIG: {
                         printf("receiving-ENTRY;\n");
                                                            /* one-shot timeout in 1 second */
                         QTimeEvt_fireIn(&me->receivedEvt, (QActive *)me,
                                         BSP_TICKS_PER_SEC);
                         return Q_HANDLED();
                     }
                     case Q_EXIT_SIG: {
                         QTimeEvt_disarm(&me->receivedEvt);
                         return Q_HANDLED();
                     }
                     case RECEIVED_SIG: {
                         return Q_TRAN(&TServer_authorizing);
                     }
                 }
                 return Q_SUPER(&TServer_busy);
             }
             /*..............................................................*/
             QState TServer_authorizing(TServer *me, QEvent const *e) {
                 switch (e->sig) {
                     case Q_ENTRY_SIG: {
                         printf("authorizing-ENTRY;\n");
                                                            /* one-shot timeout in 2 seconds */
                         QTimeEvt_fireIn(&me->authorizedEvt, (QActive *)me,
                                         2*BSP_TICKS_PER_SEC);
                         return Q_HANDLED();
                     }
                     case Q_EXIT_SIG: {
                         QTimeEvt_disarm(&me->authorizedEvt);
                         return Q_HANDLED();
                     }
                     case AUTHORIZED_SIG: {
                         return Q_TRAN(&TServer_idle);
                     }
```

Continued onto next page

```
        }
        return Q_SUPER(&TServer_busy);
    }
```

(1) The Deferred Event state pattern relies heavily on event queuing, which is not supported by the raw QEP event processor. The sample code uses the whole QP, which includes the QEP event processor and the QF real-time framework. QF provides specific direct support for deferring and recalling events.

(2) The RequestEvt event has a parameter ref_num (reference number) that uniquely identifies the request.

(3,4) The transaction server (TServer) state machine derives from the QF class QActive that combines an HSM, an event queue, and a thread of execution. The QActive class actually derives from QHsm, which means that TServer also indirectly derives from QHsm.

(5) The QF real-time framework provides a "raw" thread-safe event queue class QEQueue that is needed to implement event deferral. Here the TServer state machine declares the private requestQueue event queue to store the deferred request events. The QEQueue facility is discussed in Section 7.8.3 of Chapter 7.

(6) The QEQueue requires storage for the ring buffer, which the user must provide, because only the application designer knows how to size this buffer. Note that event queues in QF store just pointers to QEvent, not the whole event objects.

(7,8) The delays of receiving the whole transaction request (RECEIVED) and receiving the authorization notification (AUTHORIZED) are modeled in this example with the time events provided in QF.

(9) The private requestQueue event queue is initialized and given its buffer storage.

(10) Per the HSM design, the entry action to the "idle" state recalls the request events. The function QActive_recall() returns the pointer to the recalled event, or NULL if no event is currently deferred.

NOTE

Even though you can "peek" inside the recalled event, you should not process it at this point. By the time QActive_recall() function returns, the event is already posted to the active object's event queue using the LIFO policy, which guarantees that the recalled event will be the very next to process. (If other events were allowed to overtake the recalled event, the state machine might transition to a state where the recalled event would no longer be convenient.) The state machine will then handle the event like any other request coming at the convenient time. This is the central point of the Deferred Event design pattern.

(11,12)   The recalled event is inspected only to notify the user but not to handle it.

(13)   Before the "busy" superstate defers the request, it checks to see whether the private event queue can accept a new deferred event.

(14)   If so, the event is deferred by calling the QActive_defer() QF function.

(15)   Otherwise, the request is ignored and the user is notified about this fact.

NOTE

Losing events like this is often unacceptable. In fact, the default policy of QF is to fail an internal assertion whenever an event could be lost. In particular, the QActve_defer() function would fire an internal assertion if the event queue could not accept the deferred event. You can try this option by commenting out the if statement in Listing 5.3(13).

Figure 5.8 shows a variation of the Deferred Event state pattern, in which the state machine has the "canonical" structure recommended by the Ultimate Hook pattern. The "busy" state becomes the superstate of all states, including "idle." The "idle" substate overrides the NEW_REQUEST event. All other substates of "busy" rely on the default event handling inside the "busy" superstate, which defers the NEW_REQUEST event. You

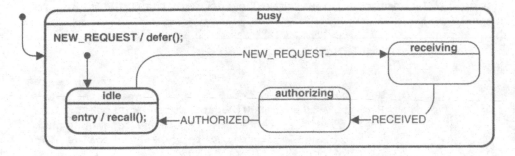

Figure 5.8: A variation of the Deferred Event state pattern.

can very easily try this option by reparenting the "idle" state. You simply change "return Q_SUPER(&QHsm_top)" to "return Q_SUPER(&TServer_busy)" in the TServer_idle() state-handler function.

Finally, I'd like to point out the true convenience of the QActive_defer() and QActive_recall() functions. The main difficulty in implementing the event deferral mechanism is actually not the explicit deferring and recalling but rather the *memory management* for the event objects. Consider, for example, that each request event must occupy some unique memory location, yet you don't know how long the event will be used. Some request events could be recycled just after the RTC step of the TServer state machine, but some will be deferred and thus will be used much longer. Recall that for memory efficiency and best performance the deferred event queue, as well as the queues of active objects in QF, store only *pointers* to events, not the whole event objects. How do you organize and manage memory for events?

This is where the QF real-time framework comes in. QF takes care of all the nitty-gritty details of managing event memory and does it very efficiently with "zero-copy" policy and in a *thread-safe* manner. As I will explain in Chapter 7, QF uses efficient event pools combined with a standard reference-counting algorithm to know when to recycle events back to the pools. The functions QActive_defer() and QActive_recall() participate in the reference-counting process so that QF does not recycle deferred events prematurely.

The whole event management mechanism is remarkably easy to use. You dynamically allocate an event, fill in the event parameters, and post it. QF takes care of the rest. In particular, you never explicitly recycle the event. Listing 5.4 shows how the request events are generated in the sample code for the Deferred Event pattern.

Listing 5.4 Generation of new request events with the Q_NEW() macro (file defer.c)

```
    void BSP_onConsoleInput(uint8_t key) {
        switch (key) {
            case 'n': {                                      /* new request */
                static uint8_t reqCtr = 0;        /* count the requests */
(1)             RequestEvt *e = Q_NEW(RequestEvt, NEW_REQUEST_SIG);
(2)             e->ref_num = (++reqCtr);    /* set the reference number */
                             /* post directly to TServer active object */
(3)             QActive_postFIFO((QActive *)&l_tserver, (QEvent *)e);
                break;
```

```
            }
        case 0x1B: {                                    /* ESC key */
(4)             static QEvent const terminateEvt = { TERMINATE_SIG, 0 };
(5)             QActive_postFIFO((QActive *)&l_tserver, &terminateEvt);
                break;
        }
    }
}
```

(1) When you press the *n* key, the QF macro Q_NEW() creates a new RequestEvt event and assigns it the signal NEW_REQUEST_SIG. The new event is allocated from an "event pool" that the application allocates at startup.

(2) You fill in the event parameters. Here the ref_num parameter is set from the incremented static counter.

(3) You post the event to an active object, such as the local l_tserver object.

(4) Constant, never-changing events can be allocated statically. Such events should have always the dynamic_ attribute set to zero (see Listing 4.1 and Section 4.3 in Chapter 4).

(5) You post such static event just like any other event. The QF real-time framework knows not to manage the static events.

5.3.5 Consequences

Event deferral is a valuable technique for simplifying state models. Instead of constructing an unduly complex state machine to handle every event at any time, you can defer an event when it comes at an inappropriate or awkward time. The event is recalled when the state machine is better able to handle it. The Deferred Event state pattern is a lightweight alternative to the powerful but heavyweight event deferral of UML statecharts. The Deferred Event state pattern has the following consequences.

- It requires explicit deferring and recalling of the deferred events.

- The QF real-time framework provides generic defer() and recall() operations.

- If a state machine defers more than one event type, it might use the same event queue (QEQueue) or different event queues for each event type. The generic QF defer() and recall() operations support both options.

- Events are deferred in a high-level state, often inside an internal transition in this state.

- Events are recalled in the entry action to the state that can conveniently handle the deferred event type.

- The event should not be processed at the time it is explicitly recalled. Rather, the recall() operation posts it using the LIFO policy so that the state machine cannot change state before processing the event.

- Recalling an event involves posting it to self; however, unlike the Reminder pattern, deferred events are usually external rather than invented.

5.3.6   Known Uses

The Real-Time Object-Oriented Modeling (ROOM) method [Selic+ 94] supports a variation of the Deferred Event pattern presented here. Just like the QF real-time framework, the ROOM virtual machine (infrastructure for executing ROOM models) provides the generic methods defer() and recall(), which clients need to call explicitly. The ROOM virtual machine also takes care of event queuing. Operations defer() and recall() in ROOM are specific to the interface component through which an event was received.

5.4   Orthogonal Component

5.4.1   Intent

Use state machines as components.

5.4.2   Problem

Many objects consist of relatively independent parts that have state behavior. As an example, consider a simple digital alarm clock. The device performs two largely independent functions: a basic timekeeping function and an alarm function. Each of these functions has its own modes of operation. For example, timekeeping can be in two modes: 12-hour or 24-hour. Similarly, the alarm can be either on or off.

The standard way of modeling such behavior in UML statecharts is to place each of the loosely related functions in a separate *orthogonal region*, as shown in Figure 5.9.

Orthogonal regions are a relatively expensive mechanism[6] that the current implementation of the QEP event processor does not support. In addition, orthogonal regions aren't often the desired solution because they offer little opportunity for reuse. You cannot reuse the "alarm" orthogonal region easily outside the context of the AlarmClock state machine.

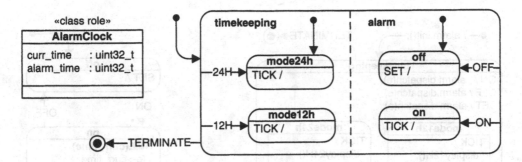

Figure 5.9: AlarmClock **class and its UML state machine with orthogonal regions.**

5.4.3 Solution

You can use object composition instead of orthogonal regions. As shown in Figure 5.10, the alarm function very naturally maps to the Alarm class that has both data (alarm_time) and behavior (the state machine). Indeed, Rumbaugh and colleagues [Rumbaugh+ 91] observe that this is a general rule. Concurrency virtually always arises within objects *by aggregation*; that is, multiple states of the components can contribute to a single state of the composite object.

The use of aggregation in conjunction with state machines raises three questions:

- How does the container state machine communicate with the component state machines?

- How do the component state machines communicate with the container state machine?

- What kind of concurrency model should be used?

---

[6] Each orthogonal region requires a separate state variable (RAM) and some extra effort in dispatching events (CPU cycles).

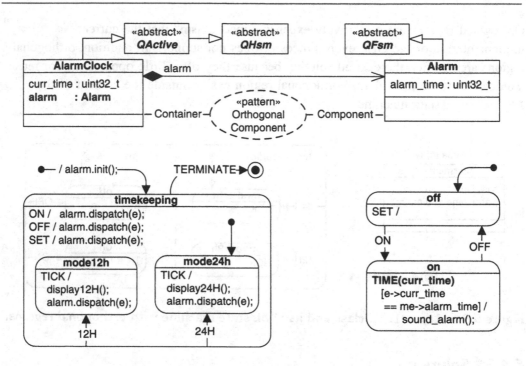

Figure 5.10: The Orthogonal Component state pattern.

The composite object interacts with its aggregate parts by synchronously dispatching events to them (by invoking dispatch() on behalf of the components). GUI systems, for instance, frequently use this model because it is how parent windows communicate with their child windows (e.g., dialog controls). Although, in principle, the container could invoke various functions of its components or access their data directly, dispatching events to the components should be the preferred way of communication. The components are state machines, and their behavior depends on their internal state.

You can view the event-dispatching responsibility as a liability given that errors will result if the container "forgets" to dispatch events in some states, but you can also view it as an opportunity to improve performance. Explicit event dispatching also offers more flexibility than the event dispatching of orthogonal regions because the container can choose the events it wants to dispatch to its components and even change the event type on the fly. I demonstrate this aspect, when the AlarmClock container generates the TimeEvt on the fly before dispatching it to the Alarm components (see Listing 5.8(9)).

To communicate in the opposite direction (from a component to the container), a component needs to post events to the container. Note that a component cannot call dispatch() on behalf of the container because this would violate RTC semantics. As a rule, the container is always in the middle of its RTC step when a component executes. Therefore, components need to asynchronously post (queue) events to the container.

This way of communication corresponds to a concurrency model in which a container shares its execution thread with the state machine components.[7] The container dispatches an event to a component by synchronously calling dispatch() state machine operation on behalf of the component. Because this function executes in the container's thread, the container cannot proceed until dispatch() returns, that is, until the component finishes its RTC step. In this way, the container and components can safely share data without any concurrency hazards (data sharing is also another method of communication among them). However, sharing the container's data makes the components dependent on the container and thus makes them less reusable.

As you can see on the right side of Figure 5.10, I decided to derive the Alarm component from the simpler QFsm base class to demonstrate that you have a choice of the base class for the components. You can decide to implement some components as HSMs and others as FSMs. The QEP event processor supports both options.

By implementing half of the problem (the AlarmClock container) as a hierarchical state machine and the other half as a classical "flat" FSM (the Alarm component), I can contrast the hierarchical and nonhierarchical solutions to essentially identical state machine topologies. Figure 5.10 illustrates the different approaches to representing mode switches in the HSM and in the FSM. The hierarchical solution demonstrates the "Device Mode" idiom [Douglass 99], in which the signals 12H and 24H trigger high-level transitions from the "timekeeping" superstate to the substates "mode12h" and "mode24h," respectively. The Alarm FSM is confined to only one level and must use direct transitions ON and OFF between its two modes. Although it is not clearly apparent with only two modes, the number of mode-switch transitions in the hierarchical technique scales up proportionally to the number of modes, *n*.

---

[7] Most commonly, all orthogonal regions in a UML statechart also share a common execution thread [Douglass 99].

The nonhierarchical solution requires many more transitions—$n \times (n-1)$, in general—to interconnect all states. There is also a difference in behavior. In the hierarchical solution, if a system is already in "mode12h," for example, and the 12H signal arrives, the system leaves this mode and re-enters it again. (Naturally, you could prevent that by overriding the high-level 12H transition in the "mode12h" state.) In contrast, if the flat state machine of the Alarm class is in the "off" state, for example, then nothing happens when the OFF signal appears. This solution might or might not be what you want, but the hierarchical solution (the Device Mode idiom) offers you both options and scales much better with a growing number of modes.

5.4.4 Sample Code

The sample code for the Orthogonal Component state pattern is found in the directory `<qp>\qpc\examples\80x86\dos\tcpp101\l\comp\`. You can execute the application by double-clicking the file `COMP.EXE` file in the `dbg\` subdirectory. Figure 5.11 shows the output generated by the `COMP.EXE` application. The application

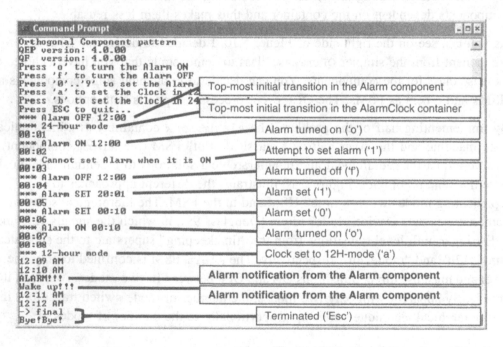

Figure 5.11: Annotated output generated by COMP.EXE.

prints the status of every mode change, both in the AlarmClock container and in the Alarm component. Additionally, you get feedback about the currently set alarm time and a notification when the alarm time is reached. The legend of the keystrokes at the top of the screen describes how to generate events for the application. Also, note that to make things happen a little faster, I made this alarm clock advance by one accelerated minute per one real second.

The sample code demonstrates the typical code organization for the Orthogonal Component state pattern, in which the component (Alarm) is implemented in a separate module from the container (AlarmClock). The modules are coupled through shared signals, events, and variables (Listing 5.5). In particular, the pointer to the container active object APP_alarmClock is made available to all components so that they can post events to the AlarmClock container.

Listing 5.5 Common signals and events (file `clock.h`)

```
#ifndef clock_h
#define clock_h
enum AlarmClockSignals {
    TICK_SIG = Q_USER_SIG,                          /* time tick event */
    ALARM_SET_SIG,                                  /* set the alarm */
    ALARM_ON_SIG,                                   /* turn the alarm on */
    ALARM_OFF_SIG,                                  /* turn the alarm off */
    ALARM_SIG, /* alarm event from Alarm component to AlarmClock container */
    CLOCK_12H_SIG,                              /* set the clock in 12H mode */
    CLOCK_24H_SIG,                              /* set the clock in 24H mode */
    TERMINATE_SIG                           /* terminate the application */
};
/*..............................................................*/
typedef struct SetEvtTag {
    QEvent super;                                   /* derive from QEvent */
    uint8_t digit;
} SetEvt;
typedef struct TimeEvtTag {
    QEvent super;                                   /* derive from QEvent */
    uint32_t current_time;
} TimeEvt;
extern QActive *APP_alarmClock;   /* AlarmClock container active object */
#endif                                              /* clock_h */
```

> **NOTE**
>
> Note that the `APP_alarmClock` pointer has the generic type `QActive*`. The components only "know" the container as a generic active object; they don't know its specific data structure or state-handler functions. This technique is called *opaque pointer* and is worth remembering for reducing dependencies among modules.

Listing 5.6 shows the declaration of the `Alarm` component (see Figure 5.10). Note that I don't actually need to expose the state-handler functions in the `alarm.h` header file. Instead, I provide only the generic interface to the `Alarm` component as the macros `Alarm_init()` and `Alarm_dispatch()` to let the container initialize and dispatch events to the component, respectively. This approach insulates the container from the choice of the base class for the component. If later on I decide to derive `Alarm` from `QHsm`, for example, I need to change only the definitions of the `Alarm_init()` and `Alarm_dispatch()` macros; I don't need to change the container code at all. Note that the macros are unnecessary in the C++ implementation because due to the compatibility between the `QHsm` and `QFsm` interfaces, the container state-handler functions always dispatch events to the `Alarm` component in the same way by calling `me->alarm.dispatch()`.

Listing 5.6 Alarm component declaration (file `alarm.h`)

```
#ifndef alarm_h
#define alarm_h

typedef struct AlarmTag {        /* the HSM version of the Alarm component */
    QFsm super;                                       /* derive from QFsm */
    uint32_t alarm_time;
} Alarm;

void Alarm_ctor(Alarm *me);
#define Alarm_init(me_)          QFsm_init     ((QFsm *)(me_), (QEvent *)0)
#define Alarm_dispatch(me_, e_) QFsm_dispatch((QFsm *)(me_), e_)

#endif                                                       /* alarm_h */
```

Listing 5.7 shows the implementation of the `Alarm` component state machine.

Listing 5.7 Alarm **state machine definition (file** `alarm.c`**)**

```
(1)  #include "alarm.h"
(2)  #include "clock.h"

     /* FSM state-handler functions */
(3)  QState Alarm_initial(Alarm *me, QEvent const *e);
     QState Alarm_off    (Alarm *me, QEvent const *e);
     QState Alarm_on     (Alarm *me, QEvent const *e);

     /*.................................................................*/
     void Alarm_ctor(Alarm *me) {
(4)      QFsm_ctor(&me->super, (QStateHandler)&Alarm_initial);
     }

     /* HSM definition ------------------------------------------------*/
     QState Alarm_initial(Alarm *me, QEvent const *e) {
         (void)e;                  /* avoid compiler warning about unused parameter */
         me->alarm_time = 12*60;
         return Q_TRAN(&Alarm_off);
     }
     /*.................................................................*/
     QState Alarm_off(Alarm *me, QEvent const *e) {
         switch (e->sig) {
             case Q_ENTRY_SIG: {
                     /* while in the off state, the alarm is kept in decimal format */
(5)              me->alarm_time = (me->alarm_time/60)*100 + me->alarm_time%60;
                 printf("*** Alarm OFF %02ld:%02ld\n",
                         me->alarm_time/100, me->alarm_time%100);
                 return Q_HANDLED();
             }
             case Q_EXIT_SIG: {
                         /* upon exit, the alarm is converted to binary format */
(6)              me->alarm_time = (me->alarm_time/100)*60 + me->alarm_time%100;
                 return Q_HANDLED();
             }
             case ALARM_ON_SIG: {
                 return Q_TRAN(&Alarm_on);
             }
             case ALARM_SET_SIG: {
                         /* while setting, the alarm is kept in decimal format */
                 uint32_t alarm = (10 * me->alarm_time
                                 +((SetEvt const *)e)->digit) % 10000;
                 if ((alarm / 100 < 24) && (alarm % 100 < 60)) {    /*alarm in range?*/
                     me->alarm_time = alarm;
                 }
                 else {                              /* alarm out of range -- start over */
                     me->alarm_time = 0;
                 }
                 printf("*** Alarm SET %02ld:%02ld\n",
```

Continued onto next page

```
                            me->alarm_time/100, me->alarm_time%100);
                    return Q_HANDLED();
                }
            }
            return Q_IGNORED();
        }
        /*..................................................................*/
        QState Alarm_on(Alarm *me, QEvent const *e) {
            switch (e->sig) {
                case Q_ENTRY_SIG: {
                    printf("*** Alarm ON %021d:%021d\n",
                            me->alarm_time/60, me->alarm_time%60);
                    return Q_HANDLED();
                }
                case ALARM_SET_SIG: {
                    printf("*** Cannot set Alarm when it is ON\n");
                    return Q_HANDLED();
                }
                case ALARM_OFF_SIG: {
                    return Q_TRAN(&Alarm_off);
                }
                case TIME_SIG: {
(7)                 if (((TimeEvt *)e)->current_time == me->alarm_time) {
                        printf("ALARM!!!\n");
                                /* asynchronously post the event to the container AO */
(8)                     QActive_postFIFO(APP_alarmClock, Q_NEW(QEvent, ALARM_SIG));
                    }
                    return Q_HANDLED();
                }
            }
            return Q_IGNORED();
        }
```

(1,2) The Alarm component needs both the alarm.h interface and the clock.h container interface.

(3)   The nonhierarchical state-handler functions have the same signature as the hierarchical state handlers (see Section 3.6 in Chapter 3).

(4)   The Alarm constructor must invoke the constructor of its base class.

(5)   Upon the entry to the "off" state, the alarm time is converted to the decimal format, in which 12:05 corresponds to decimal 1205.

(6)   Upon the exit from the "off" state, the alarm time is converted back to the binary format, in which 12:05 corresponds to $12 * 60 + 5 = 725$.

NOTE

The guaranteed initialization and cleanup provided by the entry and exit actions ensure that the time conversion will always happen, regardless of the way the state "off" is entered or exited. In particular, the alarm time will be always represented in decimal format while in the "off" state and in binary format outside the "off" state.

---

(7) The `Alarm` component keeps receiving the `TIME` event from the `AlarmClock` container. `AlarmClock` conveniently provides the `current_time` event parameter, which the `Alarm` component can directly compare to its `me->alarm_time` extended-state variable.

(8) When the `Alarm` component detects the alarm time, it notifies the container by posting an event to it. Here I am using a global pointer `APP_alarmClock` to the container active objects. An often used alternative is to store the pointer to the container inside each component.

Listing 5.8 `AlarmClock` **state machine definition (file** `clock.c`**)**

```
      #include "qp_port.h"
      #include "bsp.h"
(1)   #include "alarm.h"
(2)   #include "clock.h"
      /*..............................................................*/
      typedef struct AlarmClockTag {              /* the AlarmClock active object */

(3)       QActive super;                          /* derive from QActive */

          uint32_t current_time;                  /* the current time in seconds */
          QTimeEvt timeEvt;        /* time event generator (generates time ticks) */
(4)       Alarm alarm;                            /* Alarm orthogonal component */
      } AlarmClock;

      void AlarmClock_ctor(AlarmClock *me);                    /* default ctor */
                                              /* hierarchical state machine ... */
      QState AlarmClock_initial    (AlarmClock *me, QEvent const *e);
      QState AlarmClock_timekeeping(AlarmClock *me, QEvent const *e);
      QState AlarmClock_mode12hr   (AlarmClock *me, QEvent const *e);
      QState AlarmClock_mode24hr   (AlarmClock *me, QEvent const *e);
      QState AlarmClock_final      (AlarmClock *me, QEvent const *e);
      /*..............................................................*/
```

Continued onto next page

```
      void AlarmClock_ctor(AlarmClock *me) {                        /* default ctor */
          QActive_ctor(&me->super, (QStateHandler)&AlarmClock_initial);
(5)       Alarm_ctor(&me->alarm);                    /* orthogonal component ctor */
          QTimeEvt_ctor(&me->timeEvt, TICK_SIG);      /* private time event ctor */
      }


      /* HSM definition -------------------------------------------------------*/
      QState AlarmClock_initial(AlarmClock *me, QEvent const *e) {
          (void)e;              /* avoid compiler warning about unused parameter */
          me->current_time = 0;
(6)       Alarm_init(&me->alarm);        /* the initial transition in the component */
          return Q_TRAN(&AlarmClock_timekeeping);
      }
      /*.....................................................................*/
      QState AlarmClock_final(AlarmClock *me, QEvent const *e) {
          (void)me;           /* avoid the compiler warning about unused parameter */
          switch (e->sig) {
              case Q_ENTRY_SIG: {
                  printf("-> final\nBye!Bye!\n");
                  BSP_exit();                           /* terminate the application */
                  return Q_HANDLED();
              }
          }
          return Q_SUPER(&QHsm_top);
      }
      /*.....................................................................*/
      QState AlarmClock_timekeeping(AlarmClock *me, QEvent const *e) {
          switch (e->sig) {
              case Q_ENTRY_SIG: {
                                               /* periodic timeout every second */
                  QTimeEvt_fireEvery(&me->timeEvt,
                                  (QActive *)me, BSP_TICKS_PER_SEC);
                  return Q_HANDLED();
              }
              case Q_EXIT_SIG: {
                  QTimeEvt_disarm(&me->timeEvt);
                  return Q_HANDLED();
              }
              case Q_INIT_SIG: {
                  return Q_TRAN(&AlarmClock_mode24hr);
              }
              case CLOCK_12H_SIG: {
                  return Q_TRAN(&AlarmClock_mode12hr);
              }
              case CLOCK_24H_SIG: {
                  return Q_TRAN(&AlarmClock_mode24hr);
              }
              case ALARM_SIG: {
                  printf("Wake up!!!\n");
```

```
                        return Q_HANDLED();
                    }
                case ALARM_SET_SIG:
                case ALARM_ON_SIG:
                case ALARM_OFF_SIG: {
                             /* synchronously dispatch to the orthogonal component */
(7)                 Alarm_dispatch(&me->alarm, e);
                    return Q_HANDLED();
                }
                case TERMINATE_SIG: {
                    return Q_TRAN(&AlarmClock_final);
                }
        }
        return Q_SUPER(&QHsm_top);
    }
    /*.............................................................*/
    QState AlarmClock_mode24hr(AlarmClock *me, QEvent const *e) {
        switch (e->sig) {
            case Q_ENTRY_SIG: {
                printf("*** 24-hour mode\n");
                return Q_HANDLED();
            }
            case TICK_SIG: {
(8)             TimeEvt pe;    /* temporary synchronous event for the component */

                if (++me->current_time == 24*60) {  /* roll over in 24-hr mode? */
                    me->current_time = 0;
                }
                printf("%02ld:%02ld\n",
                        me->current_time/60, me->current_time%60);
(9)             ((QEvent *)&pe)->sig = TICK_SIG;
(10)            pe.current_time = me->current_time;
                            /* synchronously dispatch to the orthogonal component */
(11)            Alarm_dispatch(&me->alarm, (QEvent *)&pe);
                return Q_HANDLED();
            }
        }
        return Q_SUPER(&AlarmClock_timekeeping);
    }
    /*.............................................................*/
    QState AlarmClock_mode12hr(AlarmClock *me, QEvent const *e) {
        switch (e->sig) {
            case Q_ENTRY_SIG: {
                printf("*** 12-hour mode\n");
                return Q_HANDLED();
            }
            case TICK_SIG: {
                TimeEvt pe;    /* temporary synchronous event for the component */
                uint32_t h;                    /* temporary variable to hold hour */
```

Continued onto next page

```
                if (++me->current_time == 12*60) {  /* roll over in 12-hr mode? */
                    me->current_time = 0;
                }
                h = me->current_time/60;
                printf("%02ld:%02ld %s\n", (h % 12) ? (h % 12) : 12,
                        me->current_time % 60, (h / 12) ? "PM" : "AM");
                ((QEvent *)&pe)->sig = TICK_SIG;
                pe.current_time = me->current_time;

                        /* synchronously dispatch to the orthogonal component */
                Alarm_dispatch(&me->alarm, (QEvent *)&pe);
                return Q_HANDLED();
            }
        }
        return Q_SUPER(&AlarmClock_timekeeping);
    }
```

(1,2)   The `AlarmClock` container includes its own interface `clock.h` as well as all interfaces to the component(s) it uses.

(3)   The `AlarmClock` state machine derives from the QF class `QActive` that combines an HSM, an event queue, and a thread of execution. The `QActive` class actually derives from `QHsm`, which means that `AlarmClock` also indirectly derives from `QHsm`.

(4)   The container physically aggregates all the components.

(5)   The container must explicitly instantiate the components (in C).

(6)   The container is responsible for initializing the components in its topmost initial transition.

(7)   The container is responsible for dispatching the events of interest to the components. In this line, the container simply dispatches the current event e.

NOTE

The container's thread does not progress until the `dispatch()` function returns. In other words, the component state machine executes its RTC step in the container's thread. This type of event processing is called *synchronous*.

---

(8)   The temporary `TimeEvt` object to be synchronously dispatched to the component can be allocated on the stack. Note that the 'pe' variable represents the whole `TimeEvt` instance, not just a pointer.

(9,10) The container synthesizes the `TimeEvt` object on the fly and provides the current time.

(11) The temporary event is directly dispatched to the component.

5.4.5 Consequences

The Orthogonal Component state pattern has the following consequences.

- It partitions independent islands of behavior into separate state machine objects. This separation is deeper than with orthogonal regions because the objects have both distinct behavior and distinct data.

- Partitioning introduces a container–component (also known as parent–child or master–slave) relationship. The container implements the primary functionality and delegates other (secondary) features to the components. Both the container and the components are state machines.

- The components are often reusable with different containers or even within the same container (the container can instantiate more than one component of a given type).

- The container shares its execution thread with the components.

- The container communicates with the components by directly dispatching events to them. The components notify the container by posting events to it, never through direct event dispatching.

- The components typically use the Reminder state pattern to notify the container (i.e., the notification events are invented specifically for internal communication and are not relevant externally). If there are more components of a given type, the notification events must identify the originating component (the component passes its ID number in a parameter of the notification event).

- The container and components can share data. Typically, the data is a data member of the container (to allow multiple instances of different containers). The container typically grants friendship to the selected components.

- The container is entirely responsible for its components. In particular, it must explicitly trigger initial transitions in all components[8] as well as explicitly dispatch events to the components. Errors may arise if the container "forgets" to dispatch events to some components in some of its states.

- The container has full control over the dispatching of events to the components. It can choose not to dispatch events that are irrelevant to the components. It can also change event types on the fly and provide some additional information to the components.

- The container can dynamically start and stop components (e.g., in certain states of the container state machine).

- The composition of state machines is not limited to just one level. Components can have state machine subcomponents; that is, the components can be containers for lower-level subcomponents. Such a recursion of components can proceed arbitrarily deep.

5.4.6 Known Uses

The Orthogonal Component state pattern is popular in GUI systems. For example, dialog boxes are the containers that aggregate components in the form of dialog controls (buttons, check boxes, sliders, etc.). Both dialog boxes and dialog controls are event-driven objects with state behavior (e.g., a button has "depressed" and "released" states). GUIs also use the pattern recursively. For instance, a custom dialog box can be a container for the standard File-Select or Color-Select dialog boxes, which in turn contain buttons, check boxes, and so on.

The last example points to the main advantage of the Orthogonal Component state pattern over orthogonal regions. Unlike an orthogonal region, you can reuse a reactive component many times within one application and across many applications.

---

[8] In C, the container also must explicitly instantiate all components explicitly by calling their "constructors."

5.5   Transition to History

5.5.1   Intent

Transition out of a composite state, but remember the most recent active substate so you can return to that substate later.

5.5.2   Problem

State transitions defined in high-level composite states often deal with events that require immediate attention; however, after handling them, the system should return to the most recent substate of the given composite state.

For example, consider a simple toaster oven. Normally the oven operates with its door closed. However, at any time, the user can open the door to check the food or to clean the oven. Opening the door is an interruption; for safety reasons, it requires shutting the heater off and lighting an internal lamp. However, after closing the door, the toaster oven should resume whatever it was doing before the door was opened. Here is the problem: What was the toaster doing just before the door was opened? The state machine must remember the most recent state configuration that was active before opening the door in order to restore it after the door is closed again.

UML statecharts address this situation with two kinds of history pseudostates: shallow history and deep history (see Section 2.3.12 in Chapter 2). This toaster oven example requires the deep history mechanism (denoted as the circled H* icon in Figure 5.12). The QEP event processor does not support the history mechanism automatically for all states because it would incur extra memory and performance overheads. However, it is easy to add such support for selected states.

5.5.3   Solution

Figure 5.12 illustrates the solution, which is to store the most recently active *leaf* substate of the "doorClosed" state in the dedicated data member `doorClosed_history` (abbreviated to `history` in Figure 5.12). Subsequently, the Transition to History of the "doorOpen" state (transition to the circled H*) uses the attribute as the target of the transition.

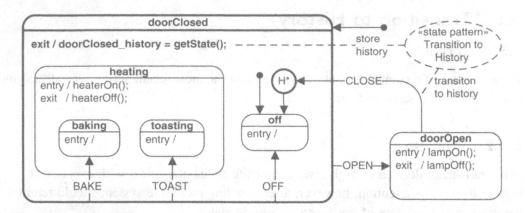

Figure 5.12: The Transition to History state pattern.

5.5.4 Sample Code

The sample code for the Transition to History state pattern is found in the directory `<qp>\qpc\examples\80x86\dos\tcpp101\l\history\`. You can execute the application by double-clicking the file `HISTORY.EXE` file in the `dbg\` subdirectory. Figure 5.13 shows the output generated by the `HISTORY.EXE` application. The application prints the actions as they occur. The legend of the keystrokes at the top of the screen describes how to generate events for the application. For example, you open the door by typing *o* and close the door by typing *c*.

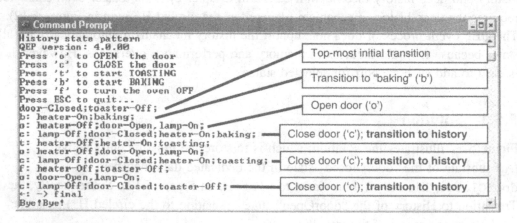

Figure 5.13: Annotated output generated by `HISTORY.EXE`.

Listing 5.9 shows the implementation of the Transition to History pattern.

Listing 5.9 The Transition to History sample code (file `history.c`)

```c
(1)   #include "qep_port.h"

      /*..........................................................*/
      enum ToasterOvenSignals {
          OPEN_SIG = Q_USER_SIG,
          CLOSE_SIG,
          TOAST_SIG,
          BAKE_SIG,
          OFF_SIG,
          TERMINATE_SIG                          /* terminate the application */
      };
      /*..........................................................*/
      typedef struct ToasterOvenTag {
          QHsm super;                                      /* derive from QHsm */
(2)       QStateHandler doorClosed_history;    /* history of the doorClosed state */
      } ToasterOven;

      void ToasterOven_ctor(ToasterOven *me);                   /* default ctor */

      QState ToasterOven_initial    (ToasterOven *me, QEvent const *e);
      QState ToasterOven_doorOpen   (ToasterOven *me, QEvent const *e);
      QState ToasterOven_off        (ToasterOven *me, QEvent const *e);
      QState ToasterOven_heating    (ToasterOven *me, QEvent const *e);
      QState ToasterOven_toasting   (ToasterOven *me, QEvent const *e);
      QState ToasterOven_baking     (ToasterOven *me, QEvent const *e);
      QState ToasterOven_doorClosed(ToasterOven *me, QEvent const *e);
      QState ToasterOven_final      (ToasterOven *me, QEvent const *e);

      /*..........................................................*/
      void ToasterOven_ctor(ToasterOven *me) {                  /* default ctor */
          QHsm_ctor(&me->super, (QStateHandler)&ToasterOven_initial);
      }

      /* HSM definitio ---------------------------------------------------*/
      QState ToasterOven_initial(ToasterOven *me, QEvent const *e) {
          (void)e;              /* avoid compiler warning about unused parameter */
(3)       me->doorClosed_history = (QStateHandler)&ToasterOven_off;
          return Q_TRAN(&ToasterOven_doorClosed);
      }
      /*..........................................................*/
      QState ToasterOven_final(ToasterOven *me, QEvent const *e) {
          (void)me;             /* avoid compiler warning about unused parameter */
          switch (e->sig) {
              case Q_ENTRY_SIG: {
                  printf("-> final\nBye!Bye!\n");
                  _exit(0);
                  return Q_HANDLED();
```

Continued onto next page

```
                    }
                }
            return Q_SUPER(&QHsm_top);
        }
/*.............................................................*/
QState ToasterOven_doorClosed(ToasterOven *me, QEvent const *e) {
    switch (e->sig) {
        case Q_ENTRY_SIG: {
            printf("door-Closed;");
            return Q_HANDLED();
        }
        case Q_INIT_SIG: {
            return Q_TRAN(&ToasterOven_off);
        }
        case OPEN_SIG: {
            return Q_TRAN(&ToasterOven_doorOpen);
        }
        case TOAST_SIG: {
            return Q_TRAN(&ToasterOven_toasting);
        }
        case BAKE_SIG: {
            return Q_TRAN(&ToasterOven_baking);
        }
        case OFF_SIG: {
            return Q_TRAN(&ToasterOven_off);
        }
        case TERMINATE_SIG: {
            return Q_TRAN(&ToasterOven_final);
        }
    }
    return Q_SUPER(&QHsm_top);
}
/*.............................................................*/
QState ToasterOven_off(ToasterOven *me, QEvent const *e) {
    (void)me;              /* avoid compiler warning about unused parameter */
    switch (e->sig) {
        case Q_ENTRY_SIG: {
            printf("toaster-Off;");
(4)         me->doorClosed_history = (QStateHandler)&ToasterOven_off;
            return Q_HANDLED();
        }
    }
    return Q_SUPER(&ToasterOven_doorClosed);
}
/*.............................................................*/
QState ToasterOven_heating(ToasterOven *me, QEvent const *e) {
    (void)me;              /* avoid compiler warning about unused parameter */
    switch (e->sig) {
        case Q_ENTRY_SIG: {
            printf("heater-On;");
```

```
                      return Q_HANDLED();
                  }
              case Q_EXIT_SIG: {
                  printf("heater-Off;");
                  return Q_HANDLED();
              }
          }
          return Q_SUPER(&ToasterOven_doorClosed);
      }
      /*..................................................................*/
      QState ToasterOven_toasting(ToasterOven *me, QEvent const *e) {
          (void)me;                  /* avoid compiler warning about unused parameter */
          switch (e->sig) {
              case Q_ENTRY_SIG: {
                  printf("toasting;");
(5)               me->doorClosed_history = (QStateHandler)&ToasterOven_toasting;
                  return Q_HANDLED();
              }
          }
          return Q_SUPER(&ToasterOven_heating);
      }
      /*..................................................................*/
      QState ToasterOven_baking(ToasterOven *me, QEvent const *e) {
          (void)me;                  /* avoid compiler warning about unused parameter */
          switch (e->sig) {
              case Q_ENTRY_SIG: {
                  printf("baking;");
(6)               me->doorClosed_history = (QStateHandler)&ToasterOven_baking;
                  return Q_HANDLED();
              }
          }
          return Q_SUPER(&ToasterOven_heating);
      }
      /*..................................................................*/
      QState ToasterOven_doorOpen(ToasterOven *me, QEvent const *e) {
          switch (e->sig) {
              case Q_ENTRY_SIG: {
                  printf("door-Open,lamp-On;");
                  return Q_HANDLED();
              }
              case Q_EXIT_SIG: {
                  printf("lamp-Off;");
                  return Q_HANDLED();
              }
              case CLOSE_SIG: {
(7)               return Q_TRAN(me->doorClosed_history); /* transition to HISTORY */
              }
          }
          return Q_SUPER(&QHsm_top);
      }
```

(1)   Every QEP application needs to include `qep_port.h` (see Section 4.8 in Chapter 4).

(2)   The `ToasterOven` state machine declares the history of the "doorClosed" state as a data member.

(3)   The `doorClosed_history` variable is initialized in the topmost initial transition according to the diagram in Figure 5.12.

(4-6)   The entry actions to all *leaf* substates of the "doorClosed" state record the history of entering those substates in the `doorClosed_history` variable. A leaf substate is a substate that has no further substates (see Section 2.3.8 in Chapter 2).

(7)   The transition to history is implemented with the standard macro `Q_TRAN()`, where the target of the transition is the `doorClosed_history` variable.

5.5.5   Consequences

The transition to history state pattern has the following consequences:

- It requires that a separate `QHsmState` pointer to the state-handler function (history variable) is provided for each composite state to store the history of this state.

- The Transition to History pseudostate (both deep and shallow history) is coded with the regular `Q_TRAN()` macro, where the target is specified as the history variable.

- Implementing the deep history pseudostate (see Section 2.3.12 in Chapter 2) requires explicitly setting the history variable in the entry action of each *leaf* substate of the corresponding composite state.

- Implementing the shallow history pseudostate (see Section 2.3.12 in Chapter 2) requires explicitly setting the history variable in each exit action from the desired level. For example, shallow history of the "doorClosed" state in Figure 5.12 requires setting `doorClosed_history` to `&ToasterOven_toasting` in the exit action from "toasting" to `&ToasterOven_baking` in the exit action from "baking," and so on for all direct substates of "doorClosed."

- You can explicitly clear the history of any composite state by resetting the corresponding history variable.

5.5.6 Known Uses

As a part of the UML specification, the history mechanism qualifies as a widely used pattern. The ROOM method [Selic+ 94] describes a few examples of transitions to history in real-time systems, whereas Horrocks [Horrocks 99] describes how to apply the history mechanism in the design of GUIs.

5.6 Summary

As Gamma and colleagues [GoF 95] observe: "One thing expert designers know not to do is solve every problem from first principles." Collecting and documenting design patterns is one of the best ways of capturing and disseminating expertise in any domain, not just in software design.

State patterns are specific design patterns that are concerned with optimal (according to some criteria) ways of structuring states, events, and transitions to build effective state machines. This chapter described just five such patterns and a few useful idioms for structuring state machines. The first two patterns, Ultimate Hook and Reminder, are at a significantly lower level than the rest, but they are so fundamental and useful that they belong in every state machine designer's bag of tricks.

The other three patterns (Deferred Event, Orthogonal Component, and Transition to History) are alternative, lightweight realizations of features supported natively in the UML state machine package [OMG 07]. Each one of these state patterns offers significant performance and memory savings compared to the full UML-compliant realization.

"Nice invention – how do you boot it up?"

www.CartoonStock.com

PART II REAL-TIME FRAMEWORK

The concept of a modern hierarchical state machine introduced in Part I is to event-driven programming as the invention of a wheel is to transportation. But just as wheels are useless without the infrastructure of roads, state machines are useless without an event-driven infrastructure that provides, at a minimum, a run-to-completion execution context for each state machine, queuing of events, and event-based timing services.

In Part II of this book, I describe such a reusable infrastructure for executing concurrent state machines in the form of a *real-time framework* called QF. QF is tailored specifically for developing real-time embedded (RTE) applications and in many respects resembles a real-time operating system (RTOS). Part II begins with Chapter 6, which introduces the real-time framework concepts. Chapter 7 describes the QF structure and implementation. Chapter 8 is devoted to porting and configuring QF, providing examples of using QF in a bare-metal system, with a traditional RTOS, and with a conventional OS (Linux). Chapter 9 describes how to develop QP applications that utilize both the QF framework and the QEP event processor described in Part I of this book. Chapter 10 presents a tiny preemptive, run-to-completion, real-time kernel called QK that beautifully complements QF. Chapter 11 describes a testing and debugging strategy based on software-tracing instrumentation built into all QP components. Chapter 12 concludes the book by presenting an ultralight version of the framework and the hierarchical event processor called QP-nano.

CHAPTER 6

Real-Time Framework Concepts

"Don't call us, we'll call you" (Hollywood Principle)
—*Richard E. Sweet,* The Mesa Programming Environment, *1985*

When you start combining multiple UML state machines into systems, you'll quickly learn that the problem is not so much in coding the state machines—Part I of this book showed that this is actually a nonissue. The next main challenge is to generate events, queue the events, and write all the code around state machines to make them execute and communicate with one another in a timely fashion and without creating concurrency hazards.

Obviously, you can develop all this "housekeeping"[1] code from scratch for each event-driven system at hand. But you could also reuse an event queue, an event dispatcher, or a time event generator across many projects. Ultimately, however, you can do much better than merely reusing specific elements as building blocks—you can achieve even greater leverage by reusing the whole infrastructure surrounding state machines. Such a reusable infrastructure is called a *framework*.

In this chapter I introduce the concepts associated with event-driven, real-time application frameworks. Most of the discussion is general and applicable to a wide range of event-driven frameworks. However, at times when I need to give more specific examples, I refer to the QF real-time framework, which is part of the QP platform and has been specifically designed for *real-time embedded* (RTE) systems. I begin with explaining why most event-driven infrastructures naturally take the form of a

---

[1] Published estimates claim that anywhere from 60 to 90 percent of an application is common "housekeeping" code that can be reused if properly structured [Douglass 99].

framework rather than a toolkit. Next, I present an overview of various CPU management policies and their relationship to the real-time framework design. In particular, I describe the modern active object computing model. Next, I discuss event management, memory management, and time management policies. I conclude with error- and exception-handling policies for a real-time framework.

6.1   Inversion of Control

Event-driven systems require a distinctly different way of thinking than traditional sequential programs. When a sequential program needs some incoming event, it *waits* in-line until the event arrives. The program remains in control all the time and because of this, while waiting for one kind of event, the sequential program cannot respond (at least for the time being) to other events.

In contrast, most event-driven applications are structured according to the *Hollywood principle*, which essentially means "Don't call us, we'll call you." So, an event-driven application is not in control while waiting for an event; in fact, it's not even active. Only once the event arrives, the event-driven application is called to process the event and then it quickly relinquishes the control again. This arrangement allows an event-driven program to wait for many events in parallel, so the system remains responsive to all events it needs to handle.

This scheme implies that in an event-driven system the control resides within the event-driven infrastructure, rather than in the application code. In other words, the control is inverted compared to a traditional sequential program. Indeed, as Ralph Johnson and Brian Foote observe [Johnson+ 88], this *inversion of control* gives the event-driven infrastructure all the defining characteristics of a *framework*.

> "One important characteristic of a framework is that the methods defined by the user to tailor the framework will often be called from within the framework itself, rather than from the user's application code. The framework often plays the role of the main program in coordinating and sequencing application activity. This inversion of control gives frameworks the power to serve as extensible skeletons. The methods supplied by the user tailor the generic algorithms defined in the framework for a particular application."
>
> —Ralph Johnson and Brian Foote

Inversion of control is key part of what makes a framework different from a toolkit. A toolkit, such as a traditional real-time operating system (RTOS), is essentially a set of predefined functions that you can call. When you use a toolkit, you write the main body of the application and call the various functions from the toolkit. When you use a framework, you reuse the main body and provide the application code that *it* calls, so the control resides in the framework rather than in your code.

Inversion of control is a common phenomenon in virtually all event-driven architectures because it recognizes the plain fact that the events are controlling the application, not the other way around. That's why most event-driven infrastructures naturally take the form of a framework rather than a toolkit.

6.2 CPU Management

An event-driven framework can work with a number of execution models, that is, particular policies of managing the central processor unit (CPU). In this section, I briefly examine the basic traditional CPU management policies and point out how they relate to the real-time framework design.

6.2.1 Traditional Sequential Systems

A traditional sequential program controls the CPU at all times.[2] Functions called directly or indirectly from the main program issue requests for external input and then wait for it; when input arrives, control resumes within the function that made the call. The location of the program counter, the tree of function calls on the stack, and local stack variables define the program state at any given time.

In the embedded space, the traditional sequential system corresponds to the background loop in the simple *foreground/background* architecture (a.k.a. super-loop or main +ISRs). As the name suggests, the foreground/background architecture consists of two main parts: the interrupt service routines (ISRs) that handle external interrupts in a timely fashion (foreground) and an infinite main loop that calls various functions (background). Figure 6.1 shows a typical flow of control within a background loop. This particular example depicts the control flow in the Quickstart application described in Section 1.9 in Chapter 1. The dashed boxes represent function calls. The heavy lines indicate the most frequently executed paths through the code.

---

[2] Except when the CPU processes asynchronous interrupts, but the interrupts always return control to the point of preemption.

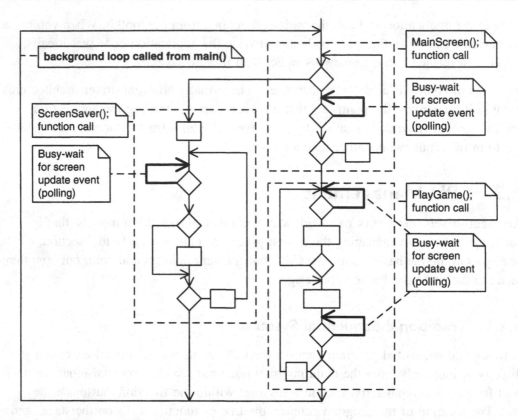

background loop called from main()

ScreenSaver(); function call

Busy-wait for screen update event (polling)

MainScreen(); function call

Busy-wait for screen update event (polling)

PlayGame(); function call

Busy-wait for screen update event (polling)

Figure 6.1: Flow of control in a typical background loop. The heavy lines indicate the most frequently executed paths through the code.

The major advantage of the traditional sequential control is that it closely matches the way the conventional procedural languages work. C and C++, for example, are exactly designed to represent the flow of control in the rich set of control statements, function call tree, and local stack variables. The main disadvantage is that a sequential system is unresponsive while waiting, which is actually most of the time. Asynchronous events cannot be easily handled within the background loop because the loop must explicitly poll for inputs. Flexible control systems, communication software, or user interfaces are hard to build using this style [Rumbaugh+ 91].

Due to the explicit polling for events scattered throughout the background code, the traditional sequential architecture is not compatible with the event-driven paradigm. However, it can be adapted to implement a single event-loop, as described in

Section 6.2.3. The simple sequential control flow is also an important stepping stone for understanding other, more advanced CPU management policies.

6.2.2 Traditional Multitasking Systems

Multitasking is the process of scheduling and switching the CPU among several sequential programs called *tasks* or *threads*. Multitasking is like foreground/background with multiple backgrounds [Labrosse 02]. Tasks share the same address space[3] and, just like the backgrounds, are typically structured as endless loops. In a multitasking system, control resides concurrently in all the tasks comprising the application.

A specific software component of a multitasking system, called the *kernel*, is responsible for managing the tasks in such a way as to create an illusion that each task has a dedicated CPU all to itself, even though the computer has typically only one CPU. The kernel achieves this by frequently switching the CPU from one task to the next in the process called *context switching*. As shown in Figure 6.2, each task is assigned its own stack area in memory and its own data structure, called a *task control block* (TCB). Context switching consists of saving the CPU registers into the current task's stack and restoring the registers from the next task's stack. Some additional

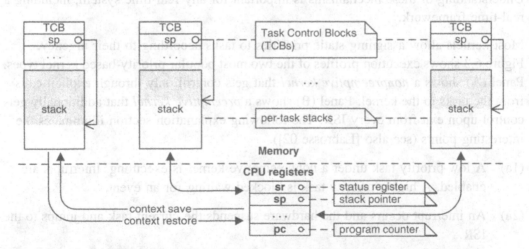

Figure 6.2: Multiple tasks with per-task stacks and task control blocks (TCBs).

---

[3] By sharing a common address space, tasks (threads) are much lighter than heavyweight processes, which execute in separate address spaces and contain one or more lightweight threads.

bookkeeping information is also updated in the TCBs. Context switches are generally transparent to the tasks and are activated from asynchronous interrupts (in case of a *preemptive kernel*) as well as synchronously from explicit calls to the kernel.

The multitasking kernel works hard behind the scenes to preserve the same state for each task as the state maintained automatically by a simple sequential program. As you can see in Figure 6.2, for each task the context-switching mechanism preserves the CPU registers, including the program counter as well as the whole private stack with the tree of nested function calls and local stack variables.

A big advantage of multitasking is better CPU utilization because when some tasks are waiting for events, other tasks can continue execution, so fewer CPU cycles are wasted on polling for events. The kernel enables the efficient waiting for events by providing special mechanisms for *blocking* tasks, such as semaphores, event flags, message mailboxes, message queues, timed blocking, and many others. A blocked task is simply switched away to memory and does not consume any CPU cycles.

Multiple tasks can wait on multiple events in parallel, so a multitasking system as a whole appears to be more *responsive* than a single background loop. The responsiveness of the system depends on how a kernel determines which task to run next. Understanding of these mechanisms is important for any real-time system, including a real-time framework.

Most kernels allow assigning static priorities to tasks according to their urgency. Figure 6.3 shows execution profiles of the two most popular priority-based kernel types. Panel (A) shows a *nonpreemptive kernel* that gets control only through explicit calls from the tasks to the kernel. Panel (B) shows a *preemptive kernel* that additionally gets control upon exit from every ISR. The following explanation section illuminates the interesting points (see also [Labrosse 02]).

(1a)   A low-priority task under a nonpreemptive kernel is executing. Interrupts are enabled. A higher-priority task is blocked waiting for an event.

(2a)   An interrupt occurs and the hardware suspends the current task and jumps to the ISR.

(3a)   ISR executes and, among other things, makes the high-priority task ready to run.

(4a)   The interrupt returns by executing a special interrupt-return instruction, which resumes the originally preempted task (the low-priority task) at the machine instruction following the interrupted instruction.

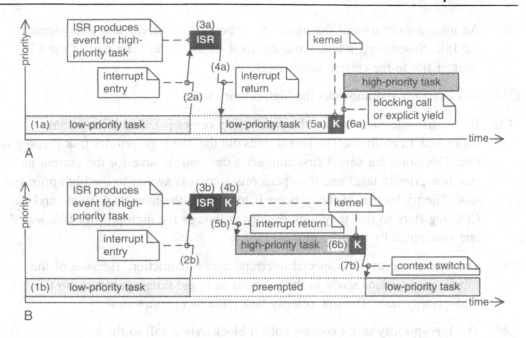

Figure 6.3: Execution profiles of a nonpreemptive kernel (A) and a preemptive kernel (B).

(5a)   The low-priority task executes until it makes an explicit blocking call to the kernel or an explicit yield, just to give the kernel a chance to run.

(6a)   The kernel runs and determines that the high-priority task is ready to run, so it performs a context switch to that task. The time delay between making the high-priority task ready to run in step (2a) and actually starting the task is called the *task-level response*.

The task-level response of a nonpreemptive kernel is nondeterministic because it depends on when other tasks voluntarily call the kernel. In other words, tasks must collaborate to share the CPU. Therefore, this form of multitasking is called *cooperative multitasking*. The upside is a much easer sharing of resources among the tasks. The kernel has an opportunity to perform a context switch only in explicitly known calls to the kernel, so tasks can safely access shared resources between any two kernel calls. In contrast, the execution profile of a preemptive kernel is as follows:

(1b)   A low-priority task under a preemptive kernel is executing. Interrupts are enabled. A higher-priority task is blocked waiting for an event.

(2b)   An interrupt occurs and the hardware suspends the current task and jumps to the ISR. Suspending a task typically involves saving at least part of the CPU register file to the current task's stack.

(3b)   ISR executes and unblocks the high-priority task.

(4b)   Before the interrupt returns, the preemptive kernel is called to determine which task to return to. The kernel finds out that the high-priority task is ready to run. Therefore, the kernel first completes the context save for the current task (the low-priority task) and then performs a context switch to the high-priority task. The tricky part of this process is to arrange the new stack frame and the CPU registers so that they look exactly as though the high-priority task was the one preempted by the interrupt.

(5b)   The kernel executes the special interrupt-return instruction. Because of the careful preparations made in the previous step, the interrupt returns to the high-priority task. The low-priority task remains preempted.

(6b)   The high-priority task executes until it blocks via a call to the kernel.

(7b)   The kernel determines that the low-priority task is still preempted and needs to run. The tricky part of resuming the low-priority task is to fake an interrupt stack frame and an interrupt CPU context to resume the low-priority task that has been preempted by an interrupt, even though the kernel is invoked via a regular function call.

> **NOTE**
>
> A preemptive kernel must actually make every context switch look like an interrupt return, even though some context switches occur from regular function calls to the kernel and don't involve asynchronous interrupts.

A preemptive kernel can guarantee a deterministic task-level response of the highest-priority tasks because the lower-priority tasks can always be preempted[4] and so it does not matter that they even exist. But this determinism comes at a huge price of increased complexity in sharing resources. A preemptive kernel can perform a context

---

[4] Using some kernel blocking mechanisms can lead to the situation in which a ready-to-run higher-priority task cannot preempt a lower-priority task. This condition is called *priority inversion*.

switch at any point of the task's code as long as the scheduler is not locked and interrupts are enabled. Any unexpected context switch might lead to corruption of shared memory or other shared resources, and kernels provide a special mechanism (such as mutexes or monitors) to guarantee a *mutually exclusive access* to shared resources. Unfortunately, programmers typically vastly underestimate the risks and skills needed to use these mechanisms safely and therefore underestimate the true costs of their use.

In summary, perhaps the most important benefit of multitasking is partitioning of the original problem into smaller, more manageable pieces (the tasks). In this respect, multitasking is a very powerful divide-and-conquer strategy. Multitasking kernels carefully preserve the private stack contents of each task so that tasks are as close as possible to simple sequential programs and thus map well to the traditional languages like C or C++.

Ultimately, however, when it comes to handling events, tasks have the same fundamental limitations as the simple sequential programs. A blocked task waiting for an event is unresponsive to all other events. Also, the whole intervening code around a blocking call is typically designed to handle only the one event that it explicitly waits for. To get a picture of what a task control flow might look like, you can simply replace the heavy polling loops in Figure 6.1 with blocking calls. Adding new events to such code is hard and typically requires deep changes to the whole task structure.

Due to the explicit blocking calls scattered throughout the task code, which the kernel encourages by providing a rich assortment of blocking mechanisms, the traditional multitasking architecture is not compatible with the event-driven paradigm. However, it can be adapted (actually simplified) for executing concurrent active objects, as I describe in the upcoming Section 6.3. Especially valuable in this respect is the thread-safe intertask communication mechanism based on message queues that most kernels or RTOSs provide. Message queues can typically be easily customized for sending events to active objects.

6.2.3   Traditional Event-Driven Systems

A traditional event-driven system is clearly divided into the event-driven infrastructure and the application (see Figure 6.4). The event-driven infrastructure consists of an event loop, an event dispatcher, and an event queue. The application consists of *event-handler functions* that all share common data.

All events in the system originating from asynchronous interrupts or from the event-handler functions are always inserted first into the event queue. The control

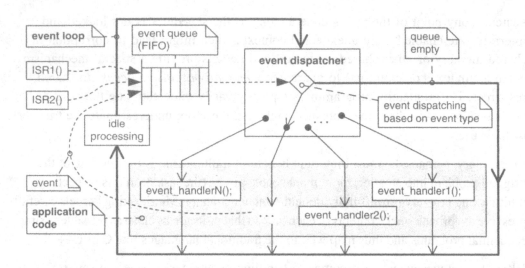

Figure 6.4: Traditional event-driven system with event loop and a single event queue.

resides within the event dispatcher that polls the event queue in the infinite event loop. For every event extracted from the queue the dispatcher calls the event-handler function associated with this event type. All event-handler functions contain essentially linear code that returns to the main event loop as quickly as possible. In particular, the event-handler functions don't poll for events and don't access the event queue. All the polling and event queuing is centralized and encapsulated within the dispatcher.

To keep the system responsive, the dispatcher must be able to check for events continuously and frequently. This implies that the event-handler functions must execute quickly. An errant event-handler function can halt the entire application, so care must be taken to avoid any "blocking" code or simply long-running code. The application cannot preserve its state using the program counter and stack because the stack contents disappear when the event-handler functions return control to the event loop. The application must rely on static variables to maintain state.

The event-loop architecture automatically guarantees that every event-handler function *runs to completion*, because the dispatcher can dispatch a new event only after the last event-handler function returns to the event loop. The need for an event queue is the direct consequence of the run-to-completion event processing. Queuing prevents losing events that arrive while the event-handler functions are running to completion and the dispatcher is unable to accept new events. The event queue is an essential part of the design.

The traditional event-driven architecture is immensely popular in event-driven graphical user interface (GUI) frameworks such as MFC, OWL, Delphi, Tcl/Tk, X-Windows, SunView, or, more recently, Java/AWT/Swing, ActionScript, Qt, .NET, and many, many others. The countless variations of the technique have mostly to do with the creative ways of associating events with event-handler functions, but all of them are ultimately based on the prevalent *event-action paradigm*, in which event types are mapped to the code that is supposed to be executed in response.

However, as explained in Chapter 2, the system response to an event depends as much on the event type as on the application context (state) in which the event arrives. The prevalent event-action paradigm recognizes only the dependency on the event type and leaves the handling of the context to largely ad hoc techniques. State machines provide very strong support for handling the context (state), but unfortunately, the event-action paradigm is incompatible with state machines because a single event-handler function contains pieces of many states. (That's exactly why event-handler functions become convoluted and brittle as they grow and evolve.) The complementary relation between the event-action paradigm and state machines is best visible in the state-table representation of a state machine (see Table 3.1 in Chapter 3), in which an event-handler function corresponds to the vertical cut through all the states in the state table along a given event column.

To summarize, the traditional event-driven architecture permits more flexible patterns of control than any sequential system [Rumbaugh+ 91]. Also, compared to any traditional sequential technique, an event-driven scheme uses the CPU more efficiently and tends to consume less stack space, which are all very desirable characteristics for embedded systems. However, the traditional event-driven architecture is not quite suitable for real-time frameworks. The remaining problems are at least threefold:

1. *Responsiveness.* The single event queue does not permit any reasonable prioritization of work. Every event, regardless of priority, must wait for processing until all events that precede it in the queue are handled.

2. *No support for managing the context of the application.* The prevalent event-action paradigm neglects the application context in responding to events, so application programmers improvise and end up creating "spaghetti" code. Unfortunately, the event-action paradigm is incompatible with state machines.

3. *Global data.* In the traditional event architecture all event-handler functions access the same global data. This hinders partitioning of the problem and can create concurrency hazards for any form of multitasking.

6.3   Active Object Computing Model

The *active object computing model* addresses most problems of the traditional event-driven architecture, retaining its good characteristics. As described in the sidebar "From Actors to Active Objects," the term *active object* comes from the UML and denotes "an object having its own thread of control" [OMG 07]. The essential idea of this model is to use multiple event-driven systems in a multitasking environment.

FROM ACTORS TO ACTIVE OBJECTS

The concept of autonomous software objects communicating by message passing dates back to the late 1970s, when Carl Hewitt and colleagues [Hewitt 73] developed a notion of an *actor*. In the 1980s, actors were all the rage within the distributed artificial intelligence community, much as agents are today. In the 1990s, methodologies like ROOM [Selic+ 94] adapted actors for real-time computing. More recently, the UML specification has introduced the concept of an *active object* that is essentially synonymous with the notion of a ROOM actor [OMG 07].

In the UML specification, an active object is "an object having its own thread of control" [OMG 07] that processes events in a run-to-completion fashion and that communicates with other active objects by asynchronously exchanging events. The UML specification further proposes the UML variant of state machines with which to model the behavior of event-driven active objects.

Active objects are most commonly implemented with *real-time frameworks*. Such frameworks have been in extensive use for many years and have proven themselves in a very wide range of real-time embedded (RTE) applications. Today, virtually every design automation tool that supports code synthesis for RTE systems incorporates a variant of a real-time framework. For instance, Real-time Object-Oriented Modeling (ROOM) calls its framework the "ROOM virtual machine" [Selic+ 94]. The VisualSTATE tool from IAR Systems calls it a "VisualSTATE engine" [IAR 00]. The UML-compliant design automation tool Rhapsody from Telelogic calls it "Object Execution Framework (OXF)" [Douglass 99].

Figure 6.5 shows a minimal active object system. The application consists of multiple active objects, each encapsulating a thread of control (event loop), a private event queue, and a state machine.

Active object = (thread of control + event queue + state machine).

The active object's event loop, shown in Figure 6.5(B), is a simplified version of the event loop from Figure 6.4. The simplified loop gets rid of the dispatcher and directly extracts events from the event queue, which efficiently blocks the loop as long as the queue is empty. For every event obtained from the queue, the event loop calls the `dispatch()` function associated with the active object. The `dispatch()` function performs both the dispatching and processing of the event, similarly to the event-handler functions in the traditional event-driven architecture.

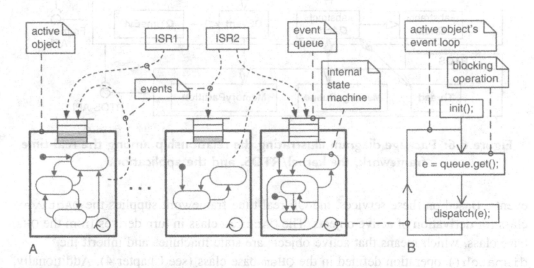

Figure 6.5: Active-object system (A) and active object's event loop (B).

6.3.1 System Structure

The event queues, event loops, and the event processor for state machines are all generic and as such are part of a generic *real-time framework*. The application consists of the specific state machine implementations, which the framework invokes indirectly through the `dispatch()`[5] state machine operation.

Figure 6.6 shows the relationship between the application, the real-time framework, and the real-time kernel or RTOS. I use the QF real-time framework as an example, but the general structure is typical for any other framework of this type. The design is layered, with an RTOS at the bottom providing the foundation for multitasking and basic services like message queues and deterministic memory partitions for storing

---

[5] The `dispatch()` operation is understood here generically and denotes any state machine implementation method, such as any of the techniques described in Chapters 3 or 4.

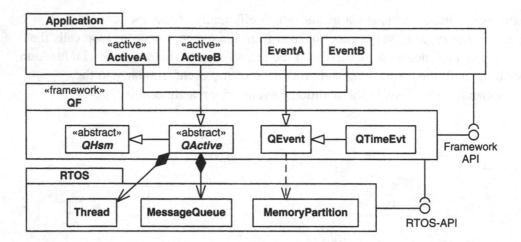

Figure 6.6: Package diagram illustrating the relationship among the real-time framework, the kernel/RTOS, and the application.

events. Based on these services, the QF real-time framework supplies the `QActive` class for derivation of active objects. The `QActive` class in turn derives from the `QHsm` base class, which means that active objects are state machines and inherit the `dispatch()` operation defined in the `QHsm` base class (see Chapter 4). Additionally, `QActive` contains a thread of execution and an event queue, typically based on the message queue of the underlying RTOS. An application extends the real-time framework by deriving active objects from the `QActive` base class and deriving events with parameters from the `QEvent` class.

NOTE

Most frameworks rely heavily on the object-oriented concepts of classes and inheritance as the key technique for extending and customizing the framework. If you program in C and the concepts are new to you, refer to the sidebar "Single Inheritance in C" in Chapter 1. In Chapter 7 you'll see that the QF real-time framework and the applications derived from it can be quite naturally implemented in standard, portable C.

The application uses the QF communication and timing services through the framework API (indicated by the ball-and-socket UML notation); however, the application typically should not need to directly access the RTOS API. Thus, a real-time framework can serve as an *RTOS abstraction layer*. The framework effectively insulates applications from the underlying RTOS. Changing the RTOS on which the framework is built requires porting the framework but does not affect applications. I'll demonstrate this aspect in Chapter 8, where I discuss porting QF.

6.3.2 Asynchronous Communication

As shown in Figure 6.5(A), active objects receive events exclusively through their event queues. All events are delivered *asynchronously*, meaning that an event producer merely posts an event to the event queue of the recipient active object but doesn't wait in line for the actual processing of the event.

The system makes no distinction between external events generated from interrupts and internal events originating from active objects. As shown in Figure 6.5(A), an active object can post events to any other active object, including to self. All events are treated uniformly, regardless of their origin.

6.3.3 Run-to-Completion

Each active object processes events in run-to-completion (RTC) fashion, which is guaranteed by the structure of the active object's event loop. As shown in Figure 6.5(B), the dispatch() operation must necessarily complete and return to the event loop before the next event from the queue can be extracted. RTC event processing is the essential requirement for proper execution of state machines.

In the case of active objects, where each object has its own thread of execution, it is very important to clearly distinguish the notion of RTC from the concept of thread preemption [OMG 07]. In particular, RTC does not mean that the active object thread has to monopolize the CPU until the RTC step is complete. Under a preemptive multitasking kernel, an RTC step can be preempted by another thread executing on the same CPU. This is determined by the scheduling policy of the underlying multitasking kernel, not by the active object computing model. When the suspended thread is assigned CPU time again, it resumes its event processing from the point of preemption and, eventually, completes its event processing. As long as the preempting and the preempted threads don't share any resources, there are no concurrency hazards.

6.3.4 Encapsulation

Perhaps the most important characteristic of active objects, from which active *objects* actually derive their name, is their strict *encapsulation*. Encapsulation means that active objects don't share data or any other resources. Figure 6.5(A) illustrates this aspect by a thick, opaque encapsulation shell around each active object and by showing the internal state machines in gray, since they are really not supposed to be visible from the outside.

As described in the previous section, no sharing of any resources (encapsulation) allows active objects to freely preempt each other without the risk of corrupting memory or other resources. The only allowed means of communication with the external world and among active objects is asynchronous event exchange. The event exchange and queuing are controlled entirely by the real-time framework, perhaps with the help of the underlying multitasking kernel, and are guaranteed to be *thread-safe*.

Even though encapsulation has been traditionally associated with object-oriented programming (OOP), it actually predates OOP and does not require object-oriented languages or any fancy tools. Encapsulation is not an abstract, theoretical concept but simply a disciplined way of designing systems based on the concept of *information hiding*. Experienced software developers have learned to be extremely wary of shared (global) data and various mutual exclusion mechanisms (such as semaphores). Instead, they bind the data to the tasks and allow the tasks to communicate only via message passing. For example, the embedded systems veteran, Jack Ganssle, offers the following advice [Ganssle 98].

> "Novice users all too often miss the importance of the sophisticated messaging mechanisms that are a standard part of all commercial operating systems. Queues and mailboxes let tasks communicate safely... the operating system's communications resources let you cleanly pass a message without fear of its corruption by other tasks. Properly implemented code lets you generate the real-time analogy of object-oriented programming's (OOP) first tenet: encapsulation. Keep all of the task's data local, bound to the code itself and hidden from the rest of the system."
>
> —Jack Ganssle

Although it is certainly true that the operating system mechanisms, such as message queues, critical sections, semaphores, or condition variables, can serve in the construction of a real-time framework, application programmers do not need to directly use these often troublesome mechanisms. Encapsulation lets programmers implement the internal structure of active objects without concern for multitasking. For example, application programmers don't need to know how to correctly use a semaphore or even know what it is. Still, as long as active objects are encapsulated, an active object system can execute safely, taking full advantage of all the benefits of multitasking, such as optimal responsiveness to events and good CPU utilization.

In Chapter 9 I will show you how to organize the application source code so that the internal structure of active objects is hidden and inaccessible to the rest of the application.

6.3.5   Support for State Machines

Event-driven systems are in general more difficult to implement with standard languages, such as C or C++, than procedure-driven systems [Rumbaugh+ 91]. The main difficulty comes from the fact that an event-driven application must return control after handling each event, so the code is fragmented and expected sequences of events aren't readily visible.

For example, Figure 6.7(A) shows a snippet of a sequential pseudocode, whereas panel (B) shows the corresponding flowchart. The boldface statements in the code and

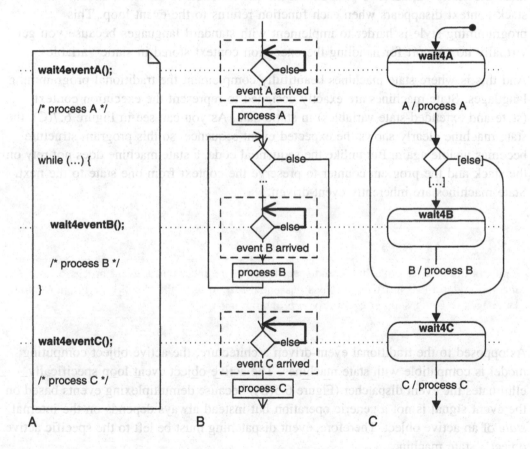

Figure 6.7: Sequential pseudocode (A), flowchart (B), and state machine (C).

heavy lines in the flowchart represent waiting for events (either polling or efficient blocking). Both the sequential code and the flowchart show the expected sequence of events (A, B...B, C) very clearly. Moreover, the sequential processing allows passing data from one processing stage to the next in temporary stack variables. Traditional programming languages and traditional multitasking kernels strongly support this style of programming that relies heavily on stack-intensive nested function calls and sophisticated flow of control (e.g., loops).

In contrast, the traditional event-driven code representing essentially the same behavior consists of three event-handler functions onA(), onB(), and onC(), and it is not at all clear that the expected sequence of calls should be onA(), onB()...onB(), onC(). This information is hidden inside the event-hander functions. Moreover, the functions must use static variables to pass data from one function to the next, because the stack context disappears when each function returns to the event loop. This programming style is harder to implement with standard languages because you get virtually no support for handling the execution context stored in static variables.

And this is where state machines beautifully complement the traditional programming languages. State machines are exactly designed to represent the execution context (state and extended-state variables) in static data. As you can see in Figure 6.7(C), the state machine clearly shows the expected event sequence, so this program structure becomes visible again. But unlike the sequential code, a state machine does not rely on the stack and the program counter to preserve the context from one state to the next. State machines are inherently event-driven.

NOTE

You can think of state machines, and specifically of the hierarchical event processor implementation described in Chapter 4, as an essential *extension* of the C and C++ programming languages to better support event-driven programming.

As opposed to the traditional event-driven architecture, the active object computing model is compatible with state machines. The active object event loop specifically eliminates the event dispatcher (Figure 6.5(B)) because demultiplexing events based on the event signal is not a generic operation but instead always depends on the internal *state* of an active object. Therefore, event dispatching must be left to the specific active object's state machine.

6.3.6 Traditional Preemptive Kernel/RTOS

In the most common implementations of the active object computing model, active objects map to threads of a traditional preemptive RTOS or OS. For example, the real-time framework inside the Telelogic Rhapsody design automation tool provides standard bindings to VxWorks, QNX, and Linux, to name a few [Telelogic 07]. In this standard configuration the active object computing model can take full advantage of the underlying RTOS capabilities. In particular, if the kernel is preemptive, the active object system achieves exactly the same optimal task-level response as traditional tasks.

Consider how the preemptive kernel scenario depicted in Figure 6.3(B) plays out in an active object system. The scenario begins with a low-priority active object executing its RTC step and a high-priority active object efficiently blocked on its empty event queue.

> **NOTE**
>
> The priority of an active object is the priority of its execution thread.

At point (2b) in Figure 6.3(B), an interrupt preempts the low-priority active object. The ISR executes and, among other things, posts an event to the high-priority active object (3b). The preemptive kernel called upon the exit from the ISR (4b) detects that the high-priority active object is ready to run, so it switches context to that active object (5b). The interrupt returns to the high-priority active object that extracts the just-posted event from its queue and processes the event to completion (6b). When the high-priority active object blocks again on its event queue, the kernel notices that the low-priority active object is still preempted. The kernel switches context to the low-priority active object (7b) and lets it run to completion.

Note that even though the high-priority active object preempted the low-priority one in the middle of the event processing, the RTC principle hasn't been violated. The low-priority active object resumed its RTC step exactly at the point of preemption and completed it eventually, before engaging in processing another event.

> **NOTE**
>
> In Chapter 8, I show how to adapt the QF real-time framework to work with a typical preemptive kernel (μC/OS-II) as well as a standard POSIX operating system (e.g., Linux, QNX, Solaris).

6.3.7   Cooperative Vanilla Kernel

The active object computing model can also work with nonpreemptive kernels. In fact, one particular cooperative kernel matches the active object computing model exceptionally well and can be implemented in an absolutely portable manner. For lack of a better name, I will call this kernel *plain vanilla* or just *vanilla*. I explain first how the vanilla kernel works and later I compare its execution profile with the profile of a traditional nonpreemptive kernel from Figure 6.3(A). Chapter 7 describes the QF implementation of the vanilla kernel.

> **NOTE**
>
> The vanilla kernel is so simple that many commercial real-time frameworks don't even call it a kernel. Instead this configuration is simply referred to as *without an RTOS*.[6] However, if you want to understand what it means to execute active objects "without an RTOS" and what execution profile you can expect in this case, you need to realize that a simple cooperative vanilla kernel is indeed involved.

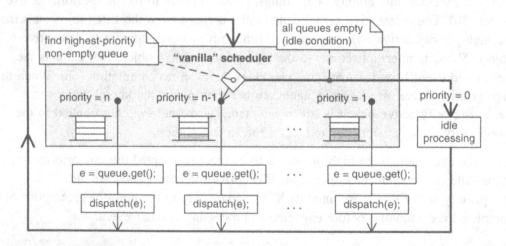

Figure 6.8: Active object system executing under the cooperative vanilla kernel.

---

[6] For example, the Interrupt Driven Framework (IDF) inside the Telelogic Rhapsody design automation tool executes "without an RTOS."

Figure 6.8 shows the architecture of the simple cooperative vanilla kernel. The most important element of the design is the *scheduler*, which is the part of a kernel responsible for determining which task to run next. The vanilla scheduler operates in a single loop. The scheduler constantly monitors all event queues of the active objects. Each event queue is assigned a unique priority, which is the priority of the active object that owns this queue. The scheduler always picks the highest-priority not-empty queue.

> **NOTE**
>
> The vanilla scheduler uses the event queues of active objects as *priority queues* and thus embodies the standard priority queue algorithm [Cormen+ 01]. Chapter 7 shows how the QF real-time framework implements the vanilla scheduler with a bitmask and a lookup table.

After finding the queue, the vanilla kernel extracts the event from the queue and dispatches it to the active object that owns this queue. Note that the queue get() operation cannot block because at this point the queue is guaranteed to be not empty. Of course, the vanilla kernel applies all the necessary safeguards to protect the internal state of the scheduler and the event queues from corruption by asynchronous interrupts, which can post events to the queues at any time.

The dispatch() operation always runs to completion and returns to the main loop. The scheduler takes over and the cycle repeats. As usual in event-driven systems, the main event loop and the event queues are all part of the vanilla kernel or the framework. The application code is not supposed to poll or block.

The vanilla scheduler very easily detects the condition when all event queues are empty. This situation is called the *idle condition* of the system. In this case, the scheduler performs idle processing, which can be customized by the application.

> **NOTE**
>
> In an embedded system, the idle processing is the ideal place to put the CPU into a low-power sleep mode. The power-saving hardware wakes up the CPU upon an interrupt, which is exactly right because at this point only an interrupt can provide new event(s) to the system.

Now consider how the scenario depicted in Figure 6.3(A) plays out under the vanilla kernel. The scenario begins with a low-priority active object executing its RTC step (dispatch() function) and a high-priority active object having its event queue empty. At point (2a) an interrupt preempts the low-priority active object. The ISR executes and, among other things, posts an event to the high-priority active object (3a). The interrupt returns and resumes the originally preempted low-priority active object (4a). The low-priority object runs to completion and returns to the main loop. At this point, the vanilla scheduler has a chance to run and picks the highest-priority nonempty queue, which is the queue of the high-priority active object (6a). The vanilla kernel calls the `dispatch()` function of the high-priority active object, which runs to completion.

As you can see, the task-level response of the vanilla kernel is exactly the same as any other nonpreemptive kernel. Even so, the vanilla kernel achieves this responsiveness without per-task stacks or complex context switching. The active objects naturally collaborate to share the CPU and implicitly yield to each other at the end of every RTC step. The implementation is completely portable and suitable for low-end embedded systems.

Because typically the RTC steps are quite short, the kernel can often achieve adequate task-level response even on a low-end CPU. Due to the simplicity, portability, and minimal overhead, I highly recommend the vanilla kernel as your first choice. Only if this type of kernel cannot meet your timing requirements should you move up to a preemptive kernel.

> **NOTE**
>
> The vanilla kernel also permits executing multiple active objects inside a single thread of a bigger multitasking system. In this case, the vanilla scheduler should efficiently block when all event queues are empty instead of wasting CPU cycles for polling the event queues. Posting an event to any of the active object queues should unblock the kernel. Of course, this requires integrating the vanilla kernel with the underlying multitasking system.

6.3.8   Preemptive RTC Kernel

Finally, if your task-level response requirements mandate a preemptive kernel, you can consider a super-simple, *run-to-completion preemptive kernel* that matches perfectly the active object computing model [Samek+ 06]. A preemptive RTC kernel implements in software exactly the same deterministic scheduling policy for tasks as most prioritized interrupt controllers implement in hardware for interrupts.

Prioritized interrupt controllers, such as the venerable Intel 8259A, the Motorola 68K and derivatives, the interrupt controllers in ARM-based MCUs by various vendors, the NVIC in the ARMv7 architecture (e.g., Cortex-M3), the M16C from Renesas, and many others allow prioritized nesting of interrupts on a *single stack*.

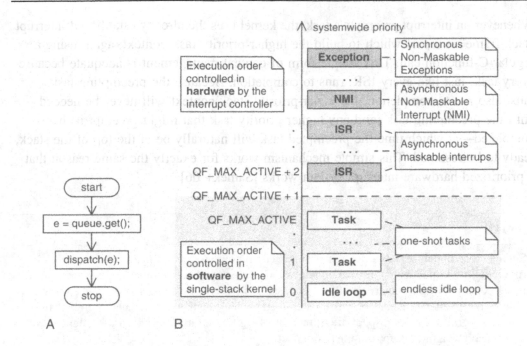

Figure 6.9: Systemwide priority of a single-stack kernel (A) and task structure (B).

In an RTC kernel tasks and interrupts are nearly symmetrical: both tasks and ISRs are one-shot, RTC functions (Figure 6.9(A)). In fact, an RTC kernel views interrupts very much like tasks of "super-high" priority, except that interrupts are prioritized in hardware by the interrupt controller, whereas tasks are prioritized in software by the kernel (Figure 6.9(B)).

> **NOTE**
>
> In all traditional kernels, tasks are generally structured as endless loops. An RTC kernel breaks with this arrangement entirely. Under an RTC kernel, tasks are one-shot functions that run to completion and return, very much like ISRs managed by a prioritized interrupt controller.

By requiring that all tasks run to completion and enforcing fixed-priority scheduling, an RTC kernel can use the machine's natural stack protocol. Whenever a task is preempted by a higher-priority task (perhaps as a result of the currently running task posting an event to a higher-priority task), the RTC kernel uses a regular C-function call to build the higher-priority task context on top of the preempted-task stack context.

Whenever an interrupt preempts a task, the kernel uses the already established interrupt stack frame on top of which to build the higher-priority task context, again using a regular C-function call. This simple form of context management is adequate because every task, just like every ISR, runs to completion. Because the preempting task must also run to completion, the lower-priority stack context will never be needed until the preempting task (and any higher-priority task that might preempt it) has completed—at which time the preempted task will naturally be at the top of the stack, ready to be resumed. This simple mechanism works for exactly the same reason that a prioritized hardware-interrupt system works [Samek+ 06].

NOTE

Such a close match between the active object computing model and prioritized, nested interrupt handling implemented directly in hardware suggests that active objects are in fact quite a basic concept. In particular, the RTC processing style and no need for blocking in active objects map better to actual processor architectures and incur less overhead than traditional blocking kernels. In this respect, traditional blocking tasks must be viewed as a higher-level, more heavyweight concept than active objects.

One obvious consequence of the stack-use policy, and the most severe limitation of an RTC kernel, is that *tasks cannot block*. The kernel cannot leave a high-priority task context on the stack and at the same time resume a lower-priority task. The lower-priority task context simply won't be accessible on top of the stack unless the higher-priority task completes. But as I keep repeating *ad nauseam* throughout this book, event-driven programming is all about writing nonblocking code. Event-driven active objects don't have a need for blocking.

In exchange for not being able to block, an RTC kernel offers many advantages over traditional blocking kernels. By nesting all task contexts in a single stack, the RTC kernel can be super-simple because it doesn't need to manage multiple stacks and all their associated bookkeeping. The result is not just significantly less RAM required for the stacks and task control blocks but a faster context switch and, overall, less CPU overhead. At the same time, an RTC kernel is as deterministic and responsive as any other fully preemptive priority-based kernel. In Chapter 10, I describe an RTC kernel called QK, which is part of the QP platform. QK is a tiny preemptive, priority-based RTC kernel specifically designed to provide preemptive multitasking support to the QF real-time framework.

If you are using a traditional preemptive kernel or RTOS for executing event-driven systems, chances are that you're overpaying in terms of CPU and memory overhead. You can achieve the same execution profile and determinism with a much simpler RTC kernel. The only real reason for using a traditional RTOS is compatibility with existing software. For example, traditional device drivers, communication stacks (such as TCP/IP, USB, CAN, etc.), and other legacy subsystems are often written with the blocking paradigm. A traditional blocking RTOS can support both active object and traditional blocking code, which the RTOS executes outside the real-time framework.

NOTE

Creating entirely event-driven, nonblocking device drivers and communication stacks is certainly possible but requires standardizing on specific event-queuing and event-passing mechanisms rather than blocking calls. Such widespread standardization simply hasn't occurred yet in the industry.

6.4 Event Delivery Mechanisms

One of the main responsibilities of every real-time framework is to efficiently deliver events from producers to consumers. The event delivery is generally asynchronous, meaning that the producers of events only insert them into event queues but do not wait for the actual processing of the events.

In addition, any part of the system can usually produce events, not necessarily only the active objects. For example, ISRs, device drivers, or legacy code running outside the framework can produce events. On the other hand, only active objects can consume events, because only active objects have event queues.

NOTE

A framework can also provide "raw" thread-safe event queues without active objects behind them. Such "raw" thread-safe queues can consume events as well, but they never block and are intended to deliver events to ISRs, that is, provide a communication mechanism from the task level to the ISR level.

Real-time frameworks typically support two types of event delivery mechanism (see Figure 6.10):

1. The simple mechanism of *direct event posting*, when the producer of an event directly posts the event to the event queue of the consumer active object.

2. A more sophisticated *publish-subscribe* event delivery mechanism, where a producer "publishes" an event to the framework, and the framework then delivers the event to all active objects that had "subscribed" to this event. The publish-subscribe mechanism provides lower coupling between event producers and consumers.

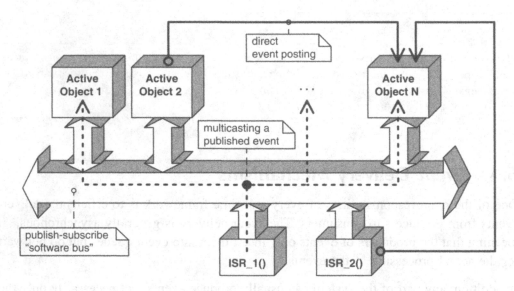

Figure 6.10: Direct event posting and publish-subscribe event delivery coexisting in a single application.

6.4.1   Direct Event Posting

The simplest mechanism lets producers post events directly to the event queue of the recipient active object. This method requires minimal participation from the framework. The framework merely provides a public operation (a function in the framework API), which allows any producer to post an event directly to the given active object. For example, the QF real-time framework provides the operation QActive_postFIFO(), which is the operation of the QActive class (see Figure 6.6). Of course the framework is responsible for implementing this function in a thread-safe manner. Figure 6.10 illustrates this form of communication as thick, solid arrows connecting event producers and the consumer active objects.

Direct event posting is a "push-style" communication mechanism, in which recipients receive unsolicited events whether they want them or not. Direct event posting is ideal in situations where a group of active objects, or an active object and an ISR, form a subsystem delivering a particular service, such as a communication stack, GPS capability, digital camera subsystem in a mobile phone, or the like. This style of event passing requires that the event producers intimately "know" the recipients. The "knowledge" that a sender needs is more than merely having a pointer to the recipient active object; the sender must also know the kind of events the particular object might be interested in. This intimate knowledge, distributed among the participating application components, makes the coupling among the components quite strong and inflexible at runtime. For example, it might be difficult to add new active objects to the subsystem, because existing event producers won't know about the newcomers and won't send them events.

6.4.2 Publish-Subscribe

The *publish–subscribe model* is a popular way of decoupling the event producers from the event consumers. Publish-subscribe is a "pull-style" communication mechanism in which recipients receive only solicited events. The properties of the publish-subscribe model are:

- Producers and consumers of events don't need to know each other (loose coupling).

- The events exchanged via this mechanism must be publicly known and must have the same semantics to all parties.

- A mediator[7] is required to accept published events and to deliver them to interested subscribers.

- Many-to-many interactions (object-to-object) are replaced with one-to-many (object-to-mediator) interactions.

The publish-subscribe event delivery is shown in Figure 6.10 as a "software bus" into which active objects "plug in" through the specified interface. Active objects interested in certain events subscribe to one or more event signals by the framework.

---

[7] The publish-subscribe event delivery is closely related to the Observer and Mediator design patterns [GoF 95].

Event producers make event publication requests to the framework. Such requests can originate asynchronously from many sources, not necessarily just active objects—for example, from interrupts or device drivers. The framework manages all these interactions by supplying the following services:

- Provide an API for active objects to subscribe and unsubscribe to particular event signals. For example, the QF real-time framework provides functions `QActive_subscribe()`, `QActive_unsubscribe()`, and `QActive_unsubscribeAll()`.

- Provide a generally accessible interface for publishing events. For example, QF provides `QF_publish()` function.

- Define and implement a thread-safe event delivery policy (including *multicasting events* when an event is subscribed by multiple active objects).

One obvious implication of publish-subscribe is that the framework must store the subscriber information, whereas it must allow associating more than one subscriber active object with an event signal. The framework must also allow modifying the subscriber information at runtime (dynamic subscribe and unsubscribe). The QF real-time framework supports dynamic subscriptions and cancellations of subscriptions.

6.5   Event Memory Management

In any event-driven system, events are frequently produced and consumed, so by nature they are highly dynamic. One of the most critical aspects of every real-time framework is managing the memory used by events, because obviously this memory must be frequently reused as new events are constantly produced. The main challenge for the framework is to guarantee that the event memory is not reused until all active objects have finished their RTC processing of the event. In fact, as described in Section 4.7.10 in Chapter 4, corrupting the current event while it is still in use constitutes a violation of the RTC semantics and is one of the hardest bugs to resolve.

6.5.1   Copying Entire Events

Section 4.7.10 offered the general solution, which is to use event queues. Indeed, as shown in Figure 6.11, entire events can be copied into an event queue and then copied out of the queue again before they can be processed. Many RTOSs support this style of event exchange through *message queues*. For example, the VxWorks RTOS provides functions `msgQSend()` to copy a chunk of memory (message) into a message queue, and `msgQReceive()` to copy the entire message out of the queue to the provided memory buffer.

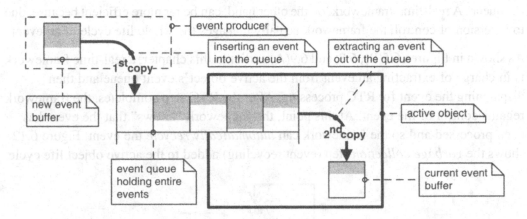

Figure 6.11: Copying entire events into the event queue and out of the event queue.

Copying entire events addresses all the potential problems with corrupting event memory prematurely, but the approach is terribly expensive in both space and time. In terms of space requirements, a message queue must typically be oversized so that all locations in the queue are able to accept the largest expected event. Additionally, every event producer needs an oversized memory buffer and every event consumer needs another oversized buffer to hold copies of the events. In terms of CPU overhead, each event passed through the queue requires making at least two copies of the data (see Figure 6.11). Moreover, the queue is inaccessible while the lengthy copy operations take place, which can negatively impact responsiveness of the system. Of course, the high overheads of copying events only multiply when multicasting events is required.

To mitigate the costs of message queues, some authors advise sending just pointers to larger chunks of data over a message queue and then let the recipient directly access the data via the provided pointer [Li+ 03]. You should be very careful with this approach. Due to the asynchronous nature of a message queue, the sender typically cannot know when the event actually gets processed, and the sender all too easily can prematurely corrupt the memory buffer by trying to reuse it for the next event. This is, of course, the classic concurrency problem caused by a shared memory buffer. Introducing such direct sharing of memory defeats the purpose of the message queue as a *safe* mechanism for passing messages (events) from producers to consumers.

6.5.2   Zero-Copy Event Delivery

The brute-force approach of copying entire events into message queues is the best a traditional RTOS can do, because an RTOS does not control the events after they leave the queue. A real-time framework, on the other hand, can be far more efficient because, due to inversion of control, the framework actually manages the whole life cycle of an event.

As shown in Figures 6.5(B), 6.8, and 6.9(A) earlier in this chapter, a real-time framework is in charge of extracting an event from the active object's event queue and then dispatching the event for RTC processing. After the RTC step completes, the framework regains control of the event. At this point, the framework "knows" that the event has been processed and so the framework can *automatically recycle* the event. Figure 6.12 shows the *garbage collection* step (event recycling) added to the active object life cycle.

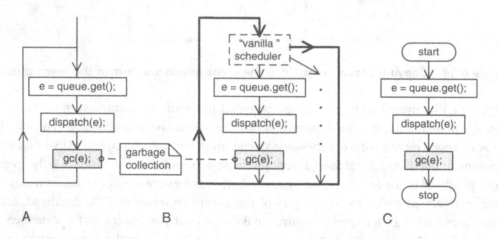

Figure 6.12: Adding garbage collection step to active object thread (A), to the cooperative vanilla kernel (B), and to the one-shot task of an RTC kernel (C).

A real-time framework can also easily control the allocation of events. The framework can simply provide an API function that application code must call to allocate new events. The QF framework, for example, provides the macro Q_NEW() for this purpose.

With the addition of the event creation and automatic garbage collection steps, the framework controls the life cycle of an event from cradle to grave. This in turn permits the framework to implement controlled, *thread-safe* sharing of event memory, which from the application standpoint is undistinguishable from true event copying. Such memory management is called *zero-copy* event delivery.

Figure 6.13 illustrates the zero-copy event delivery mechanism. The life cycle of an event begins when the framework allocates the event from an **event pool** and returns a pointer to this memory to the event producer, such as the ISR in Figure 6.13(1). The producer then fills the event parameters, writing directly to the provided event pointer. Next, the event producer posts just the pointer to the event to the queue of the recipient active object (Figure 6.13(2)).

> **NOTE**
>
> In the "zero-copy" event delivery scheme, event queues hold only pointers or references to events, not the entire events.

At some later time, the active object comes around to process the event. The active object reads the event data via the pointer extracted from the queue. Eventually, the framework automatically recycles the event in the garbage collection step. Note that the event is never copied. At the same time the framework makes sure that the event is not recycled prematurely. Of course, the framework must also guarantee that all these operations are performed in a thread-safe manner.

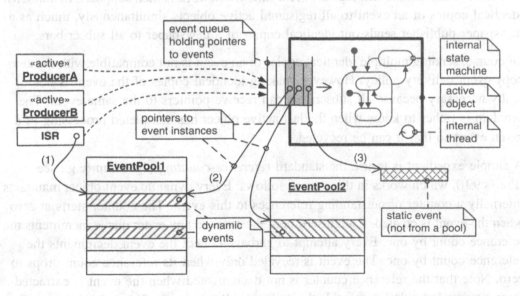

Figure 6.13: Passing events without copying them (zero-copy event delivery).

6.5.3 Static and Dynamic Events

Not all events in the system have parameters or changing parameters. For example, the TIME_TICK event or the PLAYER_TRIGGER button-press event in the "Fly 'n' Shoot" game from Chapter 1 don't really change. Such immutable event objects can be shared safely and can be allocated statically once, rather than being created and recycled every time. Figure 6.13(3) shows an example of a static event that does not come from an event pool.

The "zero-copy" event delivery mechanism can very easily accommodate such *static events* by simply not managing them at all. All static events must have a unique signature that indicates to the garbage collector to ignore such events. Conversely, events allocated dynamically must have a unique signature identifying them as *dynamic events* that the framework needs to manage. The applications use static and dynamic events in exactly the same way, except that static events are not allocated dynamically.

6.5.4 Multicasting Events and the Reference-Counting Algorithm

In the publish-subscribe mechanism, it is common for multiple active objects to subscribe to the same event signal. A real-time framework is then supposed to *multicast* identical copies of an event to all registered active objects simultaneously, much as a newspaper publisher sends out identical copies of a newspaper to all subscribers.

Of course, sending multiple identical copies of an event is not compatible with the zero-copy event delivery policy. However, making identical copies of the event is not really necessary because all subscribers can receive pointers to the same event. The problem is rather to know when the last active object has completed processing of a given event so that it can be recycled.

A simple expedient is to use the standard *reference-counting* algorithm (e.g., see [Preiss 99]), which works in this case as follows: Every dynamic event object maintains internally a counter of outstanding references to this event. The counter starts at zero when the event is created. Each insertion of the event to any event queue increments the reference count by one. Every attempt to garbage-collect the event decrements the reference count by one. The event is recycled only when its reference count drops to zero. Note that the reference counter is not decremented when the event is extracted from a queue but only later, inside the garbage collection step. This is because an event must be considered referenced as long as it is being processed, not just as long as it sits in a queue. Of course, the reference counting should only affect dynamic events and must be performed in a thread-safe manner.

> **NOTE**
>
> The garbage collection step is not equivalent to event recycling. The garbage collector function always decrements the reference counter of a dynamic event but actually recycles the event only when the counter reaches zero.

Reference counting allows more complex event exchange patterns than just multicasting. For example, a recipient of an event might choose to post the received event again, perhaps more than once. In any case, the reference-counting algorithm will correctly spare the event from recycling at the end of the first RTC step and will eventually recycle the event only when the last active object has finished processing the event.

6.5.5 Automatic Garbage Collection

The garbage collection step is part of the active object life cycle controlled by the real-time framework (Figure 6.12). The application has typically no need to recycle events explicitly. In fact, some automatic garbage collection systems, most notably Java, don't even expose a public API for recycling individual objects.

However, a real-time framework might decide to provide a way to explicitly garbage-collect an event object, but this is always intended for special purposes. For example, an event producer might start to build a dynamic event but eventually decide to bail out without posting or publishing the event. In this case the event producer must call the garbage collector explicitly to avoid leaking of the event.

> **NOTE**
>
> The garbage collection step must be performed explicitly when receiving events from "raw" thread-safe queues inside ISRs. The framework does not control ISRs and therefore ISRs are entirely responsible for implementing the whole event life cycle, including the garbage collection step.

6.5.6   Event Ownership

The zero-copy event delivery mechanisms are designed to be transparent to the application-level code. Even so, applications must obey certain ownership rules with respect to dynamic events, similar to the rules of working with objects allocated dynamically with `malloc()` or the C++ operator `new`.

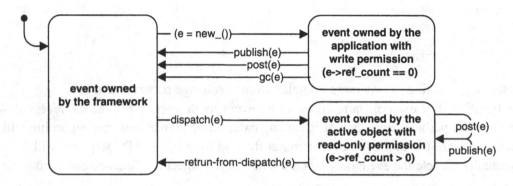

Figure 6.14: Transferring event ownership during the life cycle of a dynamic event.

Figure 6.14 illustrates the concept of event ownership and possible transfers of ownership rights. All dynamic events are initially owned by the framework. An event producer might gain ownership of a new event only by calling the `new_()` operation. At this point, the producer gains the ownership rights with the permission to write to the event. The event producer might keep the event as long as it needs, but eventually the producer must transfer the ownership back to the framework. Typically the producer posts or publishes the event. As a special case, the producer might decide that the event is not good, in which case the producer must call the garbage collector explicitly. After any of these three operations, the producer loses ownership of the event and can no longer access it.

The consumer active object gains ownership of the current event 'e' when the framework calls the `dispatch(e)` operation. This time, the active object gains merely the *read-only permission* to the current event. The consumer active object is also allowed to post or publish the event any number of times. The ownership persists over the entire RTC step. The ownership ends, however, when the `dispatch()` operation returns to the framework. The active object cannot use the event in any way past the RTC step.

6.5.7 Memory Pools

The dynamic, reference-counted events could, in principle, be allocated and freed with the standard `malloc()` and `free()` functions, respectively. However, as described in the sidebar "A Heap of Problems," using the standard heap for frequent allocation and recycling of events causes simply too many problems for any high-performance system.

A HEAP OF PROBLEMS

If you have been in the embedded real-time software business for a while, you must have learned to be wary of `malloc()` and `free()` (or their C++ counterparts `new` and `delete`) because embedded real-time systems are particularly intolerant of heap problems, which include the following pitfalls:

- Dynamically allocating and freeing memory can fragment the heap over time to the point that the program crashes because of an inability to allocate more RAM. The total remaining heap storage might be more than adequate, but no single piece satisfies a specific `malloc()` request.

- Heap-based memory management is wasteful. All heap management algorithms must maintain some form of header information for each block allocated. At the very least, this information includes the size of the block. For example, if the header causes a 4-byte overhead, a 4-byte allocation requires at least 8 bytes, so only 50 percent of the allocated memory is usable to the application. Because of these overheads and the aforementioned fragmentation, determining the minimum size of the heap is difficult. Even if you were to know the worst-case mix of objects simultaneously allocated on the heap (which you typically don't), the required heap storage is much more than a simple sum of the object sizes. As a result, the only practical way to make the heap more reliable is to massively oversize it.

- Both `malloc()` and `free()` can be (and often are) nondeterministic, meaning that they potentially can take a long (hard to quantify) time to execute, which conflicts squarely with real-time constraints. Although many RTOSs have heap management algorithms with bounded or even deterministic performance, they don't necessarily handle multiple small allocations efficiently.

Unfortunately, the list of heap problems doesn't stop there. A new class of problems appears when you use heap in a multithreaded environment. The heap becomes a shared resource and consequently causes all the headaches associated with resource sharing, so the list goes on:

- Both `malloc()` and `free()` can be (and often are) nonreentrant; that is, they cannot be safely called simultaneously from multiple threads of execution.

Continued onto next page

A HEAP OF PROBLEMS—CONT'D

- The reentrancy problem can be remedied by protecting `malloc()`, `free()`, `realloc()`, and so on internally with a mutex, which lets only one thread at a time access the shared heap. However, this scheme could cause excessive blocking of threads (especially if memory management is nondeterministic) and can significantly reduce parallelism. Mutexes can also be subject to priority inversion. Naturally, the heap management functions protected by a mutex are not available to ISRs because ISRs cannot block.

Finally, all the problems listed previously come on top of the usual pitfalls associated with dynamic memory allocation. For completeness, I'll mention them here as well.

- If you destroy all pointers to an object and fail to free it or you simply leave objects lying about well past their useful lifetimes, you create a memory leak. If you leak enough memory, your storage allocation eventually fails.

- Conversely, if you free a heap object but the rest of the program still believes that pointers to the object remain valid, you have created dangling pointers. If you dereference such a dangling pointer to access the recycled object (which by that time might be already allocated to somebody else), your application can crash.

- Most of the heap-related problems are notoriously difficult to test. For example, a brief bout of testing often fails to uncover a storage leak that kills a program after a few hours or weeks of operation. Similarly, exceeding a real-time deadline because of nondeterminism can show up only when the heap reaches a certain fragmentation pattern. These types of problems are extremely difficult to reproduce.

However, simpler, higher-performance, and safer options exist to the general-purpose, variable-block-size heap. A well-known alternative, commonly supported by RTOSs, is a fixed-block-size heap, also known as a memory partition or *memory pool*. Memory pools are a much better choice for a real-time framework to manage dynamic event allocation than the general-purpose heap.

Unlike the conventional (variable-block-size) heap, a memory pool has guaranteed capacity. It is not subject to fragmentation, because all blocks are exactly the same size. Because all blocks have identical size, no header is associated with each block allocated, thus reducing the system overhead per block. Furthermore, allocation through a memory pool can be very fast and completely deterministic. This aspect allows the kernel to protect a memory pool with a critical section of code (briefly disabling interrupts) rather than a mutex. In the case of a memory pool, the access is so fast that interrupts need to be disabled only briefly (no longer than other critical sections in the system), which does not increase interrupt latency and allows access to a memory pool, even from ISRs.

A memory pool is no different from any other multitasking kernel object. For example, accessing a semaphore also requires briefly turning off interrupts (after all, a semaphore is also a shared resource). The QF real-time framework provides a native implementation of a thread-safe memory pool.

The most obvious drawback of a memory pool is that it does not support variable-sized blocks. Consequently, the blocks have to be oversized to handle the biggest possible allocation. Such a policy is often too wasteful if the actual sizes of allocated objects (events, in this case) vary a lot. A good compromise is often to use not one but a few memory pools with blocks of different sizes. The QF real-time framework, for example, can manage up to three event pools with different block sizes (e.g., small, medium, and large, like shirt sizes).

When multiple memory pools are used, each dynamic event object must remember which pool it came from, so that the framework can recycle the event to the same pool. The QF real-time framework combines the pool ID and the reference count into one data member "dynamic_" of the QEvent structure (see Listings 4.2 and 4.3 in Chapter 4).

6.6 Time Management

Time management available in traditional RTOSs includes delaying a calling task (sleep()) or timed blocking on various kernel objects (e.g., semaphores or event flags). These blocking mechanisms are not very useful in active object-based systems where blocking is not allowed. Instead, to be compatible with the active object computing model, time management must be based on the event-driven paradigm in which every interesting occurrence manifests itself as an event instance.

6.6.1 Time Events

A real-time framework manages time through *time events*, often called timers. *Time event* is a UML term and denotes a point in time. At the specified time, the event occurs

[OMG 07]. The basic usage model of these time events is as follows: An active object allocates one or more time event objects (provides the storage for them). When the active object needs to arrange for a timeout, it arms one of its time events to post itself at some time in the future.

Figure 6.15 shows the time event facility in the QF real-time framework. The QTimeEvt class derives from QEvent, which means that time events can be used in all the same contexts as regular events. Time events can be further specialized to add more information (event parameters).

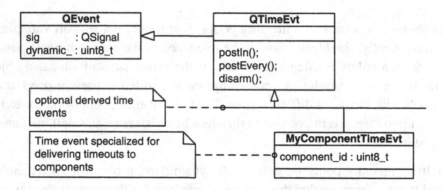

Figure 6.15: **QF class** QTimeEvt **derived from** QEvent.

The time event provides public operations for that purpose: postIn() for a one-shot timeout and postEvery() for a periodic timeout event. Each timeout request has a different time event associated with it, so the application can make multiple parallel requests (from the same or different active objects). When the framework detects that the appropriate moment has arrived, the framework posts the requested time event directly into the recipient's event queue (direct event posting). The recipient then processes the time event just like any other event.

The application can explicitly disarm any time event (periodic or one-shot) at any time using the disarm() operation. After disarming (explicitly or implicitly, as in the case of the one-shot time event), the time event can be reused for one-shot or periodic timeouts. In addition, as long as the time event remains armed it can be rearmed with a different number of ticks through the rearm() operation. For one-shot time events, rearming is useful, for example, to implement watchdog timers that need to be periodically "tickled" to prevent them from ever timing out. Rearming might also be useful to adjust the phasing of periodic time events (often you need to extend or shorten one period).

6.6.2 System Clock Tick

Every real-time system, including traditional blocking kernels, requires a periodic time source called the *system clock tick*. The system clock tick is typically a periodic interrupt that occurs at a predetermined rate, typically between 10Hz and 100Hz. You can think of the system clock tick as the heartbeat of the system. The actual frequency of the system clock tick depends on the desired tick resolution of your application. The faster the tick rate, the more overhead the time management implies.

The system clock tick must call a special framework function to give the framework a chance to periodically update the armed time events. The QF real-time framework, for example, updates time events in the function `QF_tick()`.

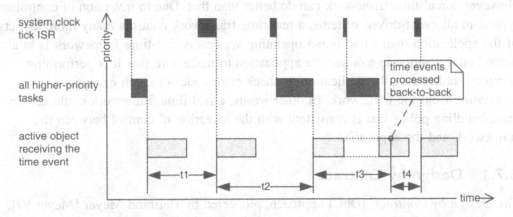

Figure 6.16: Jitter of a periodic time event firing every tick.

The delivery of time events in a real-time framework is subject to various delays, as is also the case with all real-time kernels or RTOSs [Labrosse 02]. Figure 6.16 shows in a somewhat exaggerated manner the various delays of a periodic time event programmed with one tick interval. As indicated by the varying time intervals in Figure 6.16, the time event delivery is always subject to jitter. The jitter gets worse as the priority of the recipient active object gets lower. In heavily loaded systems, the jitter might even exceed one clock tick period.[8] In particular, a time event armed for just one tick might expire immediately because the system clock tick is asynchronous with respect to active object execution. To guarantee at least one tick timeout, you need to arm a time event for two clock ticks. Note too that time events are generally

---

[8] This might be indicative of incorrect system design.

not lost due to event queuing. This is in contrast to clock ticks of a traditional RTOS, which can be lost during periods of heavy loading.

6.7 Error and Exception Handling

A real-time framework, just like any piece of system-level software, must implement a policy of handling erroneous conditions within the framework and—more important—within the application based on the framework. Of course, a framework could use the usual techniques, such as return error codes from the framework API calls, set error codes in the standard `errno` facility, or throw C++ exceptions. In fact, most operating systems and commercial RTOSs use these methods.

However, a real-time framework can do better than that. Due to inversion of control so typical in all event-driven systems, a real-time framework controls many more aspects of the application than a traditional operating system. A real-time framework is in a much better position to monitor the application to make sure that it is performing correctly, rather than the application to check error codes or catch exceptions originating from the framework. In other words, a real-time framework could use an error-handling policy that is consistent with the inversion of control between the framework and the application.

6.7.1 Design by Contract

The Design by Contract[9] (DbC) approach, pioneered by Bertrand Meyer [Meyer 97], provides an excellent methodology for implementing a very robust error-handling policy within a real-time framework that makes the most of the control inversion. The DbC philosophy views a software system as a set of components whose collaboration is based on precisely defined specifications of mutual obligations—the *contracts*. The central idea of this method is to inherently embed the contracts in the code and validate them automatically at runtime.

In C or C++, the most important aspects of DbC (the contracts) can be implemented with *assertions*. The standard C-library macro `assert()` takes a Boolean argument and terminates the application if the argument evaluates to FALSE. A real-time framework can of course use a customized version of the macro, which would invoke an application-specific handler function when the assertion fails (see upcoming Section 6.7.3), but the general idea of asserting certain conditions at runtime is the same.

---

[9] Design by Contract is a registered trademark of Interactive Software Engineering.

Assertions built into a real-time framework are consistent with inversion of control because through the assertions a real-time framework can enforce software contracts without relying on the application to check error codes or catch thrown exceptions.

The most important point to realize about software contracts (assertions in C/C++) is that they neither prevent errors nor really handle them, in the same way as contracts between people do not prevent fraud. For example, the QF real-time framework asserts that a published event signal is in the preconfigured range. Naturally, such an assertion neither handles nor prevents the application from publishing an event out of range. However, the assertion establishes a contract, which spells out that an attempt to publish an event out of range is an error. And sure enough, the framework will quite brutally abort the application that violates this contract. At first you might think that this must be backward. Contracts not only do nothing to prevent (let alone handle) errors, but they actually make things worse by turning every asserted condition, however benign, into a fatal error! However, when you really think about it, you must admit that publishing an event out of range is not really all right. It indicates that the application somehow lost consistency of event signals, which is a sure sign of a larger problem (a broken build, perhaps).

The DbC philosophy is the exact opposite of the popular *defensive programming* strategy, which denotes a programming style that aims at making operations more robust to errors, by accepting a wider range of inputs or allowing order of operations not necessarily consistent with the object's state. Defensive programming is often advertised as a better coding style, but unfortunately, it often hides bugs. To use the same example again, the QF framework could very easily make the `publish()` operation more "robust" simply by ignoring an event that is out of range. Defensive programming is not necessarily harder to implement than DbC. Rather, the problem with defensive programming is that it allows the code to "wonder around," silently taking care of various invalid conditions. The DbC approach, in contrast, represents the point of view that either a program is in full control of what's going on or it isn't, whereas assertions define what it means that the program is in control. When any of these safeguards fails, the method prescribes that it is better to face up to the problem as soon as possible and put the system in a fail-safe state (whatever this might mean for a particular system) than to let a runaway program continue. This practice is especially advisable for safety-critical applications such as medical devices.

Due to their simplicity, assertions are sometimes viewed as too primitive error-checking mechanisms—something that's perhaps good enough for smaller programs but must be replaced with a "real" error handling in the industry-strength software. This view is inconsistent with the DbC philosophy, which regards contracts as the integral part of the software design. Software contracts embody important design decisions, namely declaring certain conditions as errors rather than exceptional conditions, and therefore embedding them in large-scale, mission-critical software is even more important than in quick-and-dirty solutions. As Bertrand Meyer [Meyer 97b] observes:

> "It is not an exaggeration to say that applying assertion-based development will completely change your view of software construction ... It puts the whole issue of errors, the unsung part of the software developer's saga, in a completely different light."
>
> — Bertrand Meyer

6.7.2 Errors versus Exceptional Conditions

Another critical point to understand about the DbC philosophy is that the purpose of software contracts is to detect errors but *not* to handle exceptional conditions.

An *error* (known otherwise as a *bug*) means a persistent defect due to a design or implementation mistake (e.g., overrunning an array index or dereferencing a NULL pointer). Software contracts (assertions in C/C++) should be used to document such logically impossible situations to discover programming errors. If the "impossible" occurs, something fundamental is clearly wrong and you cannot trust the program anymore.

In contrast to an error, an *exceptional condition* is a specific circumstance that can legitimately arise during the system lifetime but is relatively rare and lies off the main execution path of your software. You need to design and implement a recovery strategy that handles the exceptional condition.

The distinction between errors and *exceptional conditions* is important because errors require the exact opposite programming strategy from dealing with exceptional conditions. The first priority in dealing with errors is to detect them as early as possible. Any attempt to handle a bug as an exceptional condition only increases the risks of damage that a runaway program can cause. It also tends to introduce immense complications to the code only camouflaging the bug. In the worst case, the attempts to "handle" a bug introduce new bugs.

A big often overlooked advantage of assertions is that they lead to considerable simplification of the software by flagging many situations as errors (that you don't need to handle) rather than exceptional conditions (that you do need to handle). Often, the application code is much simpler when it does not need to check and handle error codes returned by the framework but instead can rely on the framework policies enforced by assertions (see Section 6.1.6).

> **NOTE**
>
> Assertions can be an important source of information for modern static source code analyzing tools to test the correctness of your code.

6.7.3 Customizable Assertions in C and C++

Listing 6.1 shows the simple, customizable, embedded systems-friendly assertions that I've found adequate for a wide range of projects, embedded or otherwise. These simple assertions are consistently used in all QP components, such as the QF real-time framework and the QEP event processor.

The qassert.h header file shown in Listing 6.1 is similar to the standard <assert.h> header file, except (1) qassert.h allows customizing the error response, (2) it conserves memory by avoiding proliferation of multiple copies of the filename string, and (3) it provides additional macros for testing and documenting preconditions (Q_REQUIRE), postconditions (Q_ENSURE), and invariants (Q_INVARIANT). The names of these three latter macros are a direct loan from Eiffel, the programming language that natively supports DbC.

Listing 6.1 The qassert.h header file

```
(1)    #ifdef Q_NASSERT              /* Q_NASSERT defined-assertion checking disabled */

       #define Q_DEFINE_THIS_FILE
       #define Q_DEFINE_THIS_MODULE(name_)
       #define Q_ASSERT(test_)        ((void)0)
       #define Q_ALLEGE(test_)        ((void)(test_))
       #define Q_ERROR()              ((void)0)

    #else                            /* Q_NASSERT not defined-assertion checking enabled */

            /* callback invoked in case the condition passed to assertion fails */
       #ifdef __cplusplus
            extern "C"
       #endif
(2)    void Q_onAssert(char const Q_ROM * const Q_ROM_VAR file, int line);
```

Continued onto next page

```
(3)        #define Q_DEFINE_THIS_FILE \
              static char const Q_ROM Q_ROM_VAR l_this_file[] = __FILE__;

(4)        #define Q_DEFINE_THIS_MODULE(name_) \
              static char const Q_ROM Q_ROM_VAR l_this_file[] = #name_;

                                                /* general purpose assertion */
(5)        #define Q_ASSERT(test_) \
              if (test_) { \
              } \
              else (Q_onAssert(l_this_file, __LINE__))

              /* general purpose assertion that ALWAYS evaluates the test_ argument */
(6)        #define Q_ALLEGE(test_)    Q_ASSERT(test_)

                                                /* Assertion that always fails */
(7)        #define Q_ERROR() \
              (Q_onAssert(l_this_file, __LINE__))

       #endif                                          /* Q_NASSERT */

                              /* assertion that checks for a precondition */
(8)    #define Q_REQUIRE(test_)    Q_ASSERT(test_)

                              /* assertion that checks for a postcondition */
(9)    #define Q_ENSURE(test_)    Q_ASSERT(test_)

                              /* assertion that checks for an invariant */
(10)   #define Q_INVARIANT(test_) Q_ASSERT(test_)

                                                /* compile-time assertion */
(11)   #define Q_ASSERT_COMPILE(test_) \
              extern char Q_assert_compile[ (test_)]

       #endif/                                          /* qassert_h */
```

(1)   Defining the macro Q_NASSERT disables assertions. When disabled, all assertion macros expand to empty statements that don't generate any code.

(2)   The function Q_onAssert() prototyped in this line is invoked whenever an assertion fails. This function is application-specific and you need to define it somewhere in your program. In embedded systems, Q_onAssert() typically first disables interrupts to monopolize the CPU, then possibly attempts to put the system in a fail-safe state and eventually triggers a system reset. If possible, the

function should also leave a "trail of bread crumbs" from the cause, perhaps by storing the filename and line number in a nonvolatile memory. In addition, Q_onAssert() is an ideal place to set a breakpoint during development and debugging.

NOTE

The macros Q_ROM and Q_ROM_VAR used in the signature of Q_onAssert() are explained in Listing 4.14 in Chapter 4.

---

(3)   The macro Q_DEFINE_THIS_FILE defines a static and constant string l_this_file[] as the name of the source file provided in the standard macro __FILE__. You need to place the Q_DEFINE_THIS_FILE macro at the top of every .C or .CPP file.

Compared to the standard assert(), the assertion macros defined in Listing 6.1 conserve memory (typically ROM) by using l_this_file[] string as the first argument to Q_onAssert() rather than the standard preprocessor macro __FILE__. This avoids proliferation of the multiple copies of the __FILE__ string for each use of the assert() macro (see [Maguire 93]).

(4)   This macro Q_DEFINE_THIS_MODULE() defines a static and constant string l_this_file[] as the string provided in the argument. This macro provides an alternative to Q_DEFINE_THIS_FILE (so you use one or the other). The __FILE__ macro often expands to the full path name of the translation unit, which might be too long to log.

(5)   The macro Q_ASSERT() defines a general-purpose assertion. The empty block in the if statement might seem strange, but you need both the if and the else statements to prevent unexpected dangling if problems.

(6)   When assertions are disabled by defining Q_NASSERT, the assertion macros don't generate any code; in particular, they don't test the expressions passed as arguments, so you should be careful to avoid any side effects required for normal program operation inside the expressions tested in assertions. The macro Q_ALLEGE() is a notable exception. This assertion macro always tests the condition, although when assertions are disabled it does not invoke the Q_onAssert() callback function. Q_ALLEGE() is useful in situations where avoiding side effects of the test would require introducing temporary

variables on the stack—something you often want to minimize in embedded systems.

(7)   The macro Q_ERROR() always fails. The use of this macro is equivalent to Q_ASSERT(0) but is more descriptive.

(8-10)   The macros Q_REQUIRE(), Q_ENSURE(), and Q_INVARIANT() are intended for validating preconditions, postconditions, and invariants, respectively. They all map to Q_ASSERT(). Their different names serve only to better document the specific intent of the contract.

(11)   The macro Q_ASSERT_COMPILE() validates a contract at compile time. The macro exploits the fact that the dimension of an array in C cannot be zero. Note that the macro does not actually allocate any storage, so there is no penalty in using it (see [Murphy 01]).

6.7.4 State-Based Handling of Exceptional Conditions

An exceptional condition is a specific situation in the lifetime of a system that calls for a special behavior. In a state-driven active object, a change in behavior corresponds to a change in state (state transition). Hence, in active object systems, the associated state machines are the most natural way to handle all conditions, including exceptional conditions.

Using state hierarchy can be very helpful to separate the "exceptional" behavior from the "normal" behavior. Such state-based exception handling is typically a combination of the Ultimate Hook and the Reminder state patterns (Chapter 5) and works as follows: A common superstate defines a high-level transition to the "exceptional" parts of the state machine. The submachine of this superstate implements the "normal" behavior. Whenever an action within the submachine encounters an exceptional condition, it posts a Reminder event to self to trigger the high-level transition to handle the exception. The exit actions executed upon the high-level transition perform the cleanup of the current context and cleanly enter the "exceptional" context.

State-based exception handling offers a safe and *language-independent* alternative to the built-in exception-handling mechanism of the underlying programming language. As described in Section 3.7.2 in Chapter 3, throwing and catching exceptions in C++ is risky in any state machine implementation because it conflicts with the fundamental RTC semantics of state machines. Stack unwinding inherent in propagating of thrown exceptions is also less relevant in event-driven systems than traditional sequential code because event-driven systems rely less on the stack and more on storing the state information in the static data. As Tom Cargill noticed in the seminal paper *Exception handling: A false sense of security* [Cargill 94]:

> "Counter-intuitively, the hard part of coding exceptions is not the explicit throws and catches. The really hard part of using exceptions is to write all the intervening code in such a way that an arbitrary exception can propagate from its throw site to its handler, arriving safely and without damaging other parts of the program along the way."
>
> —Tom Cargill

If you only can, consider leaving out the C++ throw-and-catch exception handling from your event-driven software. If you cannot avoid it, make sure to catch all exceptions before they can cause any damage.

6.7.5 Shipping with Assertions

The standard practice is to use assertions during development and testing but to disable them in the final product. The often-quoted opinion in this matter comes from C.A.R. Hoare, who considered disabling assertions in the final product like using a lifebelt during practice but then not bothering with it for the real thing.

The question of shipping with assertions really boils down to two issues. First is the overhead that assertions add to your code. Obviously, if the overhead is too big, you have no choice. But then you must ask yourself how you will build and test your application. It's much better to consider assertions as an integral part of the software and size the hardware adequately to accommodate them. As the prices of computer hardware rapidly drop while the capabilities increase, it simply makes sense to trade a small fraction of the raw CPU horsepower and some extra code memory for better system integrity. In practice, assertions often pay for themselves by eliminating reams of "error-handling" code that tries to camouflage bugs.

The other issue is the correct system response when an assertion fires in the field. This response must be obviously designed carefully and safety-critical systems might require some redundancy and recovery strategy. For many less critical applications a simple system reset turns out to be the least inconvenient action from the user's perspective—certainly less inconvenient than locking up the application and denying service. You should also try to leave some "bread crumbs" of information as to what went wrong. To this end, assertions provide a great starting point for debugging and ultimately fixing the root cause of the problem.

6.7.6   Asserting Guaranteed Event Delivery

Traditional sequential systems communicate predominately by means of synchronous function calls. When module A wants to communicate with module B, module A calls a function in B. The communication is implicitly assumed to be reliable; that is, the programmer takes for granted that the function call mechanism will work, that the parameters will be passed to the callee, and that the return value will be delivered to the caller. The programmer does not conceive of any recovery strategy to handle a failure in the function call mechanism itself. But in fact, any function call can fail due to insufficient stack space. Consequently, the reliability of synchronous communication is in fact predicated on the implicit assumption of adequate stack resource.

Event-driven systems communicate predominately by asynchronous event exchange. When active object A wants to communicate with active object B, object A allocates an event and posts it to the event queue of object B. The programmer should be able to take for granted that the event delivery mechanism will work, that the event will be available, and that the event queue will accept the event. However, in fact, asynchronous communication can fail due to event pool depletion or insufficient queue depth. Consequently, the reliability of asynchronous communication is in fact predicated on the assumption of adequate event pool size and event queue depth. If those two resources are sized properly, the asynchronous event posting in the same address space should be as reliable as a synchronous function call.

> **NOTE**
>
> At this point, I limit the discussion to nondistributed systems executing in a single address space. Distributed systems connected with unreliable communication media pose quite different challenges. In this case neither synchronous communications such as remote procedure call (RPC) nor asynchronous communications via message passing can make strong guarantees.

I hope that this argument helps you realize that event pools and event queues should be treated on equal footing as the execution stacks in that depletion of any of these resources represents an error. In fact, event pools and queues fulfill in event-driven systems many responsibilities of the execution stacks in traditional multitasking systems. For example, parameters passed inside events play the same role as the parameters of function calls passed on the call stacks. Consequently, event-driven systems use less stack space than sequential systems but instead require event queues and event pools.

A real-time framework can use assertions to detect event pool and event queue overruns, in the same way that many commercial RTOSs detect stack overflows. For example, the QF real-time framework asserts internally that a requested dynamic event can always be allocated from one of the event pools. Similarly, the QF framework asserts that an event queue can always accept a posted event. It's up to the application implementer to adequately size all event pools and event queues in the system in the same exact way as it is the implementer's responsibility to adequately size the execution stacks for a multitasking kernel.

> ### NOTE
>
> Standard message queues available in traditional RTOSs allow many creative ways of circumventing the event delivery guarantee. For example, message queues allow losing events when the queue is full or blocking until the queue can accept the event. The QF real-time framework does not use these mechanisms. QF simply asserts that an event queue accepts every event without blocking.

6.8 Framework-Based Software Tracing

A running application built of active objects is a highly structured affair where all important system interactions funnel through the real-time framework and the event processor executing the state machines. This arrangement offers a unique opportunity for applying *software-tracing* techniques. In a nutshell, software tracing is similar to peppering the code with `printf()` statements, which is called *instrumenting the code,* to log interesting discrete events for subsequent retrieval from the target system and analysis. Of course, a good software-tracing instrumentation can be much less intrusive and more powerful than the primitive `printf()`.

By instrumenting just the real-time framework code you can gain an unprecedented wealth of information about the running system, far more detailed and comprehensive than any traditional RTOS can provide. (This is, of course, yet another benefit of control inversion.) The software trace data from the framework alone allows you to produce complete, time-stamped sequence diagrams and detailed state machine activity for all active objects in the system. This ability can form the foundation of the whole testing strategy for your application. In addition, individual active objects are natural entities for unit testing, which you can perform simply by injecting events into the active objects and collecting the trace data. Software tracing at the framework level makes all this comprehensive information available to you, even with no instrumentation added to the application-level code.

Most commercial real-time frameworks routinely use software tracing to provide visualization and animation of the state machines in the system. The QF real-time framework is also instrumented, and the software trace data can be extracted by means of the QS (Q-SPY) component, described in Chapter 11.

6.9   Summary

Event-driven programming requires a paradigm shift compared to traditional sequential programming. This paradigm shift leads to inversion of control between the event-driven application and the infrastructure on which it is based. The event-driven infrastructure can be generic and typically takes the form of a real-time framework. Using such a framework to implement your event-driven applications can spare you reinventing the wheel for each system you implement.

A real-time framework can employ a number of various CPU management policies, so it is important to understand the basic real-time concepts, starting from simple foreground/background systems through cooperative multitasking to fully preemptive multitasking. Traditionally, these execution models have been used with the *blocking* paradigm. *Blocking* means that the program frequently waits for events, either hanging in tight polling loops or getting efficiently blocked by a multitasking kernel. The blocking model is not compatible with the event-driven paradigm, but it can be adapted. The general strategy is to centralize and encapsulate all the blocking code inside the event-driven infrastructure (the framework) so that the application code never blocks.

The active object computing model combines multiple traditional event loops with a multitasking environment. In this model, applications are divided into multiple autonomous active objects, each encapsulating an execution thread (event loop), an event queue, and a state machine. Active objects communicate with one another asynchronously by posting events to each other's event queues. Within an active object, events are always processed in run-to-completion (RTC) fashion while a real-time framework handles all the details of thread-safe event exchange and queuing.

The active object computing model can work with a traditional preemptive RTOS, with just a basic cooperative vanilla kernel, or with the super-simple, preemptive, RTC kernel. In all these configurations, the active object model can take full advantage of the underlying CPU management policy, achieving optimal responsiveness and CPU utilization. As long as active objects are strictly encapsulated (i.e., they don't share resources), they can be programmed internally without concern for multitasking. In particular, the application programmer does not need to use or understand semaphores, mutexes, monitors, or other such troublesome mechanisms.

Most real-time frameworks support the simple direct event posting, and some frameworks also support the more sophisticated publish-subscribe event delivery mechanism. Direct event posting is a "push-style" communication in which the recipient active object gets unsolicited events, whether it wants them or not. Publish-subscribe is a "pull-style" communication in which active objects subscribe to event signals by the framework and then the framework delivers only the solicited events to the active objects. Publish-subscribe promotes loose coupling between event producers and event consumers.

Perhaps the biggest responsibility of a real-time framework is to guarantee thread-safe RTC event processing within active objects. This includes event management policies so that the current event is not corrupted over the entire RTC step. A simple but horrendously expensive method to protect events from corruption is to copy entire events into and out of event queues. A far more efficient way is for a framework to implement a *zero-copy* event delivery. *Zero-copy* really means that the framework controls *thread-safe* sharing of event memory, which at the application level is indistinguishable from true event copying. A real-time framework can do it because the framework actually controls the whole life cycle of events as well as active objects.

A real-time framework manages time through time events, also known as *timers*. Time events are time-delayed requests for posting events. The framework can handle many such requests in parallel. The framework uses the system clock tick to periodically gain control to manage the time events. The resolution of time events is one system clock tick, but it does not mean that the accuracy is also one clock tick due to various delays that cause jitter.

A real-time framework can use the traditional error-handling policies such as returning error-codes from framework API calls. However, a real-time framework can also use error management that takes advantage of the inversion of control between the framework and the application. Such techniques are based on assertions, or more generally on the Design by Contract (DbC) methodology. A framework can use assertions to constantly monitor the application. Among others, a framework can enforce guaranteed event delivery, which immensely simplifies event-driven application design.

Finally, a real-time framework can use software-tracing techniques to provide more detailed and comprehensive information about the running application than any traditional RTOS. The software-tracing instrumentation of the framework can form the backbone of a testing and debugging strategy for active-object systems.

CHAPTER 7

Real-Time Framework Implementation

Let us change our traditional attitude to the construction of programs. Instead of imagining that our main task is to instruct a computer what to do, let us concentrate rather on explaining to human beings what we want a computer to do.
— Donald E. Knuth

In this chapter I describe the implementation of a lightweight real-time framework called QF. As shown in Figure 7.1, QF is the central component of the QP event-driven platform, which also includes the QEP hierarchical event processor (described in Part I

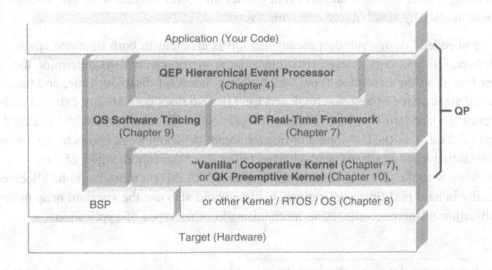

Figure 7.1: QP Components (in gray) and their relationship to the target hardware, board support package (BSP), and the application.

of this book) as well as the preemptive run-to-completion (RTC) kernel (QK) and the software tracing instrumentation (QS).

The focus of this chapter is on the generic, platform-independent QF source code. I devote Chapter 8 entirely to describing the platform-specific code that depends on the specific processor, compiler, and operating system/RTOS (including the case where QF is used without an RTOS).

I describe QF in the top-down fashion beginning with an overview of the QF features, then presenting the framework structure, both logical (partitioning into classes) and physical (partitioning into files). In the remaining bulk of the chapter I explain the implementation of the QF services. As usual, I mostly refer to the C source code (located in the `<qp>\qpc\` directory in the accompanying code). I mention the C++ version (located in the `<qp>\qpcpp\` directory) only when the differences from C become important.

7.1   Key Features of the QF Real-Time Framework

QF is a generic, portable, scalable, lightweight, real-time framework designed specifically for the domain of *real-time embedded systems* (RTES). QF can manage up to 63 concurrently executing active objects,[1] which are encapsulated tasks (each embedding a state machine and an event queue) that communicate with one another asynchronously by sending and receiving events.

The "embedded" design mindset means that QF is efficient in both time and space. Moreover, QF uses only deterministic algorithms, so you can always determine the upper bound on the execution time, the maximum interrupt disabling time, and the required stack space of any given QF service. QF also does not call any external code, not even the standard C or C++ libraries. In particular, QF does not use the standard heap (`malloc()` or the C++ operator `new`). Instead, the framework leaves to the clients the instantiation of any framework-derived objects and the initialization of the framework with the memory it needs for operation. All this memory could be allocated statically in hard real-time applications, but you could also use the standard heap or any combination of memory allocation mechanisms in other types of applications.

---

[1] This does not mean that your application is limited to 63 state machines. Each active object can manage an unlimited number of stateful components, as described in the "Orthogonal Component" state pattern in Chapter 5.

7.1.1 Source Code

The companion Website to this book at www.quantum-leaps.com/psicc2/ contains the complete source code for all QP components, including QF. I hope that you will find the source code very clean and consistent. The code has been written in strict adherence to the coding standard documented at www.quantum-leaps.com/doc/ AN_QL_Coding_Standard.pdf.

All QP source code is "lint-free." The compliance was checked with PC-lint/FlexLint static analysis tool from Gimpel Software (www.gimpel.com). The QP distribution includes the <qp>\qpc\ports\lint\ subdirectory, which contains the batch script make.bat for compiling all the QP components with PC-lint.

The QP source code is also 98 percent compliant with the Motor Industry Software Reliability Association (MISRA) *Guidelines for the Use of the C Language in Vehicle-Based Software* [MISRA 98]. MISRA created these standards to improve the reliability and predictability of C programs in critical automotive systems. Full details of this standard can be obtained directly from the MISRA Website at www.misra.org.uk. The PC-lint configuration used to analyze QP code includes the MISRA rule checker.

Finally and most important, I believe that simply giving you the source code is not enough. To gain real confidence in event-driven programming, you need to understand how a real-time framework is ultimately implemented and how the different pieces fit together. This book, and especially this chapter, provides this kind of information.

7.1.2 Portability

All QF source code is written in portable ANSI-C, or in the Embedded C++ subset[2] in case of QF/C++, with all processor-specific, compiler-specific, or operating system-specific code abstracted into a clearly defined *platform abstraction layer* (PAL).

In the simplest standalone configurations, QF runs on "bare-metal" target CPU completely replacing the traditional RTOS. As shown in Figure 7.1, the QP event-driven platform includes the simple nonpreemptive "vanilla" scheduler as well as the fully preemptive kernel QK. To date, the standalone QF configurations have been ported to over 10 different CPU architectures, ranging from 8-bit (e.g., 8051, PIC,

---

[2] Embedded C++ subset is defined online at www.caravan.net/ec2plus/.

AVR, 68H(S)08), through 16-bit (e.g., MSP430, M16C, x86-real mode) to 32-bit architectures (e.g., traditional ARM, ARM Cortex-M3, Cold Fire Altera Nios II, x86).

The QF framework can also work with a traditional OS/RTOS to take advantage of the existing device drivers, communication stacks, middleware, or any legacy code that requires a conventional "blocking" kernel. To date, QF has been ported to six major operating systems and RTOSs, including Linux (POSIX) and Win32.

As you'll see in the course of this chapter, high portability is the main challenge in writing a widely useable real-time framework like QF. Obviously, coming up with an efficient PAL that would correctly capture all possible platform variances required many iterations and actually porting the framework to several CPUs, operating systems, and compilers. (I describe porting the QF framework in Chapter 8.) The www.quantum-leaps.com Website contains the steadily growing number of QF ports, examples, and documentation.

7.1.3 Scalability

All components of the QP event-driven platform, especially the QF real-time framework, are designed for scalability so that your final application image contains only the services that you actually use. QF is designed for deployment as a fine-granularity object library that you statically link with your applications. This strategy puts the onus on the linker to eliminate any unused code automatically at link time instead of burdening the application programmer with configuring the QF code for each application at compile time.

As shown in Figure 7.2, a minimal QP/C or QP/C++ system requires some 8KB of code space (ROM) and about 1KB of data space (RAM) to leave enough room for a meaningful application code and data. This code size corresponds to the footprint of a typical, small, bare-bones RTOS application except that the RTOS approach typically requires more RAM for the stacks.

NOTE

A typical, standalone QP configuration with QEP, QF, and the "vanilla" scheduler or the QK preemptive kernel, with all major features enabled, requires around 2-4KB of code. Obviously you need to budget additional ROM and RAM for your own application code and data. Figure 7.2 shows the application footprint.

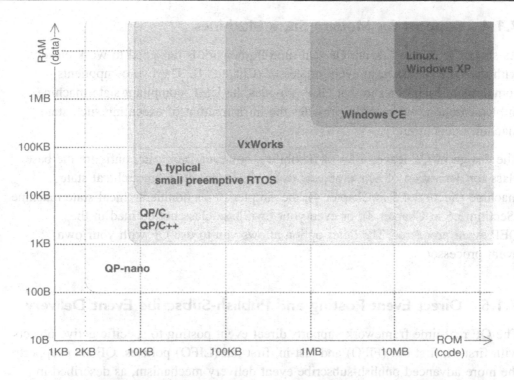

**Figure 7.2: RAM/ROM footprints of QP, QP-nano, and other RTOS/OS.
The chart shows approximate total system size as opposed to just the RTOS/OS
footprints. Note the logarithmic axes.**

However, the event-driven approach scales down even further, beyond the reach of
any conventional RTOS. To address still smaller systems, a reduced QP version called
QP-nano implements a subset of features provided in QP/C or QP/C++. QP-nano has
been specifically designed to enable active object computing with hierarchical state
machines on low-end 8- and 16-bit embedded MCUs. As shown in Figure 7.2, a
meaningful QP-nano application starts from about 100 bytes of RAM and 2KB of
ROM. I describe QP-nano in Chapter 11.

On the opposite end of the complexity spectrum, QP applications can also scale up
to very big systems with gigabytes of RAM and multiple or multicore CPUs. The
large-scale applications, such as various servers, have often large numbers of
stateful components to manage, so the efficiency per component becomes critical. It
turns out that the lightweight, event-driven, state machine-based approach easily scales
up and offers many benefits over the traditional thread-per-component paradigm.

7.1.4   Support for Modern State Machines

As shown in Figure 7.1, the QF real-time framework is designed to work closely with the QEP hierarchical event processor (Chapter 4). The two components complement each other in that QEP provides the UML-compliant state machine implementation, whereas QF provides the infrastructure of executing such state machines concurrently.

The design of QF leaves a lot of flexibility, however. You can configure the base class for derivation of active objects to be either the QHsm hierarchical state machine (Section 4.5 in Chapter 4), the simpler QFsm nonhierarchical state machine (Section 3.6 in Chapter 3), or even your own base class not defined in the QEP event processor. The latter option allows you to use QF with your own event processor.

7.1.5   Direct Event Posting and Publish-Subscribe Event Delivery

The QF real-time framework supports direct event posting to specific active objects with first-in, first-out (FIFO) and last-in, first-out (LIFO) policies. QF also supports the more advanced publish-subscribe event delivery mechanism, as described in Section 6.4 in Chapter 6. Both mechanisms can coexist in a single application.

7.1.6   Zero-Copy Event Memory Management

Perhaps that most valuable feature provided by the QF real-time framework is the efficient "zero-copy" event memory management, as described in Section 6.5 in Chapter 6. QF supports event multicasting based on the reference-counting algorithm, automatic garbage collection for events, efficient static events, "zero-copy" event deferral, and up to three event pools with different block sizes for optimal memory utilization.

7.1.7   Open-Ended Number of Time Events

QF can manage an open-ended number of time events (timers). QF time events are extensible via structure derivation (inheritance in C++). Each time event can be armed as a one-shot or a periodic timeout generator. Only armed (active) time events consume CPU cycles.

7.1.8 Native Event Queues

QF provides two versions of native event queues. The first version is optimized for active objects and contains a portability layer to adapt it for either blocking kernels, the simple cooperative "vanilla" kernel (Section 6.3.7), or the QK preemptive kernel (Section 6.3.8 in Chapter 6). The second native queue version is a simple "thread-safe" queue not capable of blocking and designed for sending events to interrupts as well as storing deferred events. Both native QF event queue types are lightweight, efficient, deterministic, and thread-safe. They are optimized for passing just the pointers to events and are probably smaller and faster than full-blown message queues available in a typical RTOS.

7.1.9 Native Memory Pool

QF provides a fast, deterministic, and thread-safe memory pool. Internally, QF uses memory pools as event pools for managing dynamic events, but you can also use memory pools for allocating any other objects in your application.

7.1.10 Built-in "Vanilla" Scheduler

The QF real-time framework contains a portable, cooperative "vanilla" kernel, as described in Section 6.3.7 of Chapter 6. Chapter 8 presents the QF port to the "vanilla" kernel.

7.1.11 Tight Integration with the QK Preemptive Kernel

The QF real-time framework can also work with a deterministic, preemptive, nonblocking QK kernel. As described in Section 6.3.8 in Chapter 6, run-to-completion kernels, like QK, provide preemptive multitasking to event-driven systems at a fraction of the cost in CPU and stack usage compared to traditional blocking kernels/RTOSs. I describe QK implementation in Chapter 10.

7.1.12 Low-Power Architecture

Most modern embedded microcontrollers (MCUs) provide an assortment of low-power sleep modes designed to conserve power by gating the clock to the CPU and various peripherals. The sleep modes are entered under the software control and are exited upon an external interrupt.

The event-driven paradigm is particularly suitable for taking advantage of these power-savings features because every event-driven system can easily detect situations in which the system has no more events to process, called the *idle condition* (Section 6.3.7). In both standalone QF configurations, either with the cooperative "vanilla" kernel or with the QK preemptive kernel, the QF framework provides callback functions for handling the idle condition. These callbacks are carefully designed to place the MCU into a low-power sleep mode safely and without creating race conditions with active interrupts.

7.1.13   Assertion-Based Error Handling

The QF real-time framework consistently uses the Design by Contract (DbC) philosophy described in Section 6.7 in Chapter 6. QF constantly monitors the application by means of assertions built into the framework. Among others, QF uses assertions to enforce the event delivery guarantee, which immensely simplifies event-driven application design.

7.1.14   Built-in Software Tracing Instrumentation

As described in Section 6.8 in Chapter 6, a real-time framework can use software-tracing techniques to provide more comprehensive and detailed information about the running application than any traditional RTOS. The QF code contains the software-tracing instrumentation so it can provide unprecedented visibility into the system. Nominally the instrumentation is inactive, meaning that it does not add any code size or runtime overhead. But by defining the macro Q_SPY, you can activate the instrumentation. I devote all of Chapter 11 to software tracing.

NOTE

The QF code is instrumented with QS (Q-Spy) macros to generate software trace output from active object execution. However, the instrumentation is disabled by default and for better clarity will not be shown in the listings discussed in this chapter. Refer to Chapter 11 for more information about the QS software-tracing implementation.

7.2 QF Structure

Figure 7.3 shows the main QF classes and their relation to the application-level code, such as the "Fly 'n' Shoot" game example from Chapter 1. As all real-time frameworks, QF provides the central base class QActive for derivation of active object classes. The QActive class is abstract, which means that it is not intended for direct instantiation but rather only for derivation of concrete[3] active object classes, such as Ship, Missile, and Tunnel shown in Figure 7.3.

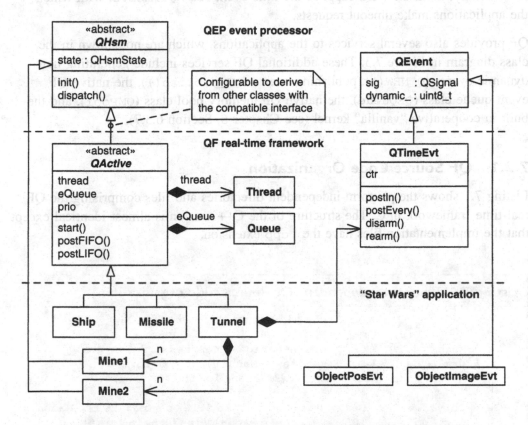

Figure 7.3: QEP event processor, QF real-time framework, and the "Fly 'n' Shoot" application.

---

[3] *Concrete class* is the OOP term and denotes a class that has no abstract operations or protected constructors. Concrete class can be instantiated, as opposed to *abstract class,* which cannot be instantiated.

By default, the QActive class derives from the QHsm hierarchical state machine class defined in the QEP event processor (Chapter 4). This means that by virtue of inheritance active objects are HSMs and inherit the init() and dispatch() state machine interface. QActive also contains a thread of execution and an event queue, which can be native QF classes, or might be coming from the underlying RTOS.

QF uses the same QEvent class for representing events as the QEP event processor. Additionally, the framework supplies the time event class QTimeEvt, with which the applications make timeout requests.

QF provides also several services to the applications, which are not shown in the class diagram in Figure 7.3. These additional QF services include generating new dynamic events (Q_NEW()), publishing events (QF_publish()), the native QF event queue class (QEQueue), the native QF memory pool class (QMPool), and the built-in cooperative "vanilla" kernel (see Chapter 6, Section 6.3.7).

7.2.1   QF Source Code Organization

Listing 7.1 shows the platform-independent directories and files comprising the QF real-time framework in C. The structure of the C++ version is almost identical except that the implementation files have the .cpp extension.

```
Listing 7.1  Platform-independent QF source code organization

qpc\                          - QP/C root directory (qpcpp for QP/C++)
  |
  +-doxygen\                  - QP/C documentation generated with Doxygen
  | +-html\                   - "QP/C Reference Manual" in HTML format
  | | +-index.html - The starting HTML page for the "QP/C Reference Manual"
  | | +- . . .
  | +-Doxyfile                - Doxygen configuration file to generate the Manual
  | +-qpc.chm                 - "QP/C Reference Manual" in CHM Help format
  | +-qpc_rev.h               - QP/C revision history
  |
  +-include\                  - QP platform-independent header files
  | +-qf.h                    - QF platform-independent interface
  | +-qequeue.h               - QF native event queue facility
  | +-qmpool.h                - QF native memory pool facility
```

```
| +-qpset.h       - QF native priority set facility
| +-qvanilla.h    - QF native "vanilla" cooperative kernel interface
|
+-qf\             - QF real-time framework
| +-source\       - QF platform-independent source code (*.C files)
| | +-qf_pkg.h   - internal, interface for the QF implementation
| | +-qa_defer.c - definition of QActive_defer()/QActive_recall()
| | +-qa_ctor.c  - definition of QActive_ctor()
| | +-qa_fifo.c  - definition of QActive_postFIFO()
| | +-qa_fifo_.c - definition of QActive_postFIFO_()
| | +-qa_get_.c  - definition of QActive_get_()
| | +-qa_lifo.c  - definition of QActive_postLIFO()
| | +-qa_lifo_.c - definition of QActive_postLIFO_()
| | +-qa_sub.c   - definition of QActive_subscribe()
| | +-qa_usub.c  - definition of QActive_unsubscribe()
| | +-qa_usuba.c - definition of QActive_unsubscribeAll()
| | +-qeq_fifo.c - definition of QEQueue_postFIFO()
| | +-qeq_get.c  - definition of QEQueue_get()
| | +-qeq_init.c - definition of QEQueue_init()
| | +-qeq_lifo.c - definition of QEQueue_postLIFO()
| | +-qf_act.c   - definition of QF_active_[]
| | +-qf_gc.c    - definition of QF_gc_()
| | +-qf_log2.c  - definition of QF_log2Lkup[]
| | +-qf_new.c   - definition of QF_new_()
| | +-qf_pool.c  - definition of QF_poolInit()
| | +-qf_psini.c - definition of QF_psInit()
| | +-qf_pspub.c - definition of QF_publish()
| | +-qf_pwr2.c  - definition of QF_pwr2Lkup_[]
| | +-qf_tick.c  - definition of QF_tick()
| | +-qmp_get.c  - definition of QMPool_get()
| | +-qmp_init.c - definition of QMPool_init()
| | +-qmp_put.c  - definition of QMPool_put()
| | +-qte_arm.c  - definition of QTimeEvt_arm_()
| | +-qte_ctor.c - definition of QTimeEvt_ctor()
| | +-qte_darm.c - definition of QTimeEvt_darm()
| | +-qte_rarm.c - definition of QTimeEvt_rearm()
| | +-qvanilla.c - "vanilla" cooperative kernel implementation
| |
| +-lint\         - QF options for lint
| | +-opt_qf.lnt - PC-lint options for linting QF
|
+-ports\          - Platform-specific QP ports
| +- . . .
+-examples\       - Platform-specific QP examples
| +- . . .
```

The QF source files contain typically just one function or a data structure definition per file. This design aims at deploying QF as a fine-granularity library that you statically link with your applications. *Fine granularity* means that the QF library consists of several small, loosely coupled modules (object files) rather than a single module that contains all functionality. For example, a separate module `qa_lifo.c` implements the `QActive_postLIFO()` function; therefore, if your application never calls this function, the linker will not pull in the `qa_lifo.obj` module. This strategy puts the burden on the linker to do the "heavy lifting" of automatically eliminating any unused code at link time, rather than on the application programmer to configure the QF code for each application at compile time.

7.3   Critical Sections in QF

QF, just like any other system-level software, must protect certain sequences of instructions against preemptions to guarantee thread-safe operation. The sections of code that must be executed indivisibly are called *critical sections*.

In an embedded system environment, QF uses the simplest and most efficient way to protect a section of code from disruptions, which is to lock interrupts on entry to the critical section and unlock interrupts at the exit from the critical section. In systems where locking interrupts is not allowed, QF can employ other mechanisms supported by the underlying operating system, such as a mutex.

NOTE

The maximum time spent in a critical section directly affects the system's responsiveness to external events (interrupt latency). All QF critical sections are carefully designed to be as short as possible and are of the same order as critical sections in any commercial RTOS. Of course, the length of critical sections depends on the processor architecture and the quality of the code generated by the compiler.

To hide the actual critical section implementation method available for a particular processor, compiler, and operating system, the QF platform abstraction layer includes two macros, `QF_INT_LOCK()` and `QF_INT_UNLOCK()`, to lock and unlock interrupts, respectively.

7.3.1 Saving and Restoring the Interrupt Status

The most general critical section implementation involves saving the
interrupt status before entering the critical section and restoring the status upon
exit from the critical section. Listing 7.2 illustrates the use of this critical section
type.

Listing 7.2 Example of the "saving and restoring interrupt status" policy

```
      {
(1)      unsigned int lock_key;

(2)      lock_key = get_int_status();
(3)      int_lock();
         . . .
(4)      /* critical section of code */
         . . .
(5)      set_int_status(lock_key);
      }
```

---

(1) The temporary variable lock_key holds the interrupt status across the critical
     section.

(2) Right before entering the critical section, the current interrupt status is
     obtained from the CPU and saved in the lock_key variable. Of course,
     the name of the actual function to obtain the interrupt status can be different
     in your system. This function could actually be a macro or inline assembly
     statement.

(3) Interrupts are locked using the mechanism provided by the compiler.

(4) This section of code executes indivisibly because it cannot be interrupted.

(5) The original interrupt status is restored from the lock_key variable. This step
     unlocks interrupts only if they were unlocked at step 2. Otherwise, interrupts
     remain locked.

Listing 7.3 shows an example of the "saving and restoring interrupt status" policy.

Listing 7.3 QF macro definitions for the "saving and restoring interrupt status" policy

```
(1)   #define QF_INT_KEY_TYPE        unsigned int

(2)   #define QF_INT_LOCK(key_)      do { \
          (key_) = get_int_status();     \
          int_lock();                    \
      } while (0)

(3)   #define QF_INT_UNLOCK(key_) set_int_status(key_)
```

---

(1)   The macro `QF_INT_KEY_TYPE` denotes a data type of the "interrupt key" variable, which holds the interrupt status. Defining this macro in the `qf_port.h` header file indicates to the QF framework that the policy of "saving and restoring interrupt status" is used, as opposed to the policy of "unconditional locking and unlocking interrupts" described in the next section.

(2)   The macro `QF_INT_LOCK()` encapsulates the mechanism of interrupt locking. The macro takes the parameter `key_`, into which it saves the interrupt lock status.

NOTE

The `do {...} while (0)` loop around the `QF_INT_LOCK()` macro is the standard practice for syntactically correct grouping of instructions. You should convince yourself that the macro can be used safely inside the `if-else` statement (with the semicolon after the macro) without causing the "dangling-else" problem. I use this technique extensively in many QF macros.

---

(3)   The macro `QF_INT_UNLOCK()` encapsulates the mechanism of restoring the interrupt status. The macro restores the interrupt status from the argument `key_`.

The main advantage of the "saving and restoring interrupt status" policy is the ability to *nest critical sections*. The QF real-time framework is carefully designed to never nest critical sections internally. However, nesting of critical sections can easily occur when QF functions are invoked from within an already established critical section, such as an interrupt service routine (ISR). Most processors lock interrupts in hardware upon the interrupt entry and unlock upon the interrupt exit, so the whole ISR is a critical section. Sometimes you can unlock interrupts inside ISRs, but often you cannot. In the latter case, you have no choice but to invoke QF services, such as event posting or publishing, with interrupts locked. This is exactly when you must use this type of critical section.

7.3.2 Unconditional Locking and Unlocking Interrupts

The simpler and faster critical section policy is to always unconditionally unlock interrupts in QF_INT_UNLOCK(). Listing 7.4 provides an example of the QF macro definitions to specify this type of critical section.

Listing 7.4 QF macro definitions for the "unconditional interrupt locking and unlocking" policy

```
(1)   /* QF_INT_LOCK_KEY not defined */
(2)   #define QF_INT_LOCK(key_)        int_lock()
(3)   #define QF_INT_UNLOCK(key_)      int_unlock()
```

---

(1)   The macro QF_INT_KEY_TYPE is not defined in this case. The absence of the QF_INT_KEY_TYPE macro indicates to the QF framework that the interrupt status is not saved across the critical section.

(2)   The macro QF_INT_LOCK() encapsulates the mechanism of interrupt locking. The macro takes the parameter key_, but this parameter is not used in this case.

(3)   The macro QF_INT_UNLOCK() encapsulates the mechanism of unlocking interrupts. The macro always unconditionally unlocks interrupts. The parameter key_ is ignored in this case.

The policy of "unconditional locking and unlocking interrupts" is simple and fast, but it does *not* allow nesting of critical sections, because interrupts are always unlocked upon exit from a critical section, regardless of whether interrupts were already locked on entry.

The inability to nest critical sections does not necessarily mean that you cannot nest interrupts. Many processors are equipped with a prioritized interrupt controller, such as the Intel 8259A Programmable Interrupt Controller (PIC) in the PC or the Nested Vectored Interrupt Controller (NVIC) integrated inside the ARM Cortex-M3. Such interrupt controllers handle interrupt prioritization and nesting before the interrupts reach the processor core. Therefore, you can safely unlock interrupts at the processor level, thus avoiding nesting of critical sections inside ISRs. Listing 7.5 shows the general structure of an ISR in the presence of an interrupt controller.

Listing 7.5 General structure of an ISR in the presence of a prioritized interrupt controller

```
(1) void interrupt ISR(void) {   /* entered with interrupts locked in hardware */
(2)     Acknowledge the interrupt to the interrupt controller (optional)
(3)     Clear the interrupt source, if level triggered
(4)     QF_INT_UNLOCK(dummy); /* unlock the interrupts at the processor level */

(5)     Handle the interrupt, use QF calls, e.g., QF_tick(), Q_NEW or QF_publish()

(6)     QF_INT_LOCK(dummy);     /* lock the interrupts at the processor level */
(7)     Write End-Of-Interrupt (EOI) instruction to the Interrupt Controller
(8) }
```

---

(1) Most processors enter the ISR with interrupts locked in hardware.

(2) The interrupt controller must be notified about entering the interrupt. Often this notification happens automatically in hardware before vectoring (jumping) to the ISR. However, sometimes the interrupt controller requires a specific notification from the software. Check your processor's datasheet.

(3) You need to explicitly clear the interrupt source, if it is level triggered. Typically you do it before unlocking interrupts at the CPU level, but a prioritized interrupt controller will prevent the same interrupt from preempting itself, so it really does not matter if you clear the source before or after unlocking interrupts.

(4) Interrupts are explicitly unlocked at the CPU level, which is the key step of this ISR. Enabling interrupts allows the interrupt controller to do its job, that is, to prioritize interrupts. At the same time, enabling interrupts terminates the critical section established upon the interrupt entry. Note that this step is only necessary when the hardware actually locks interrupts upon the interrupt entry (e.g., the ARM Cortex-M3 leaves interrupts unlocked).

(5) The main ISR body executes outside the critical section, so QF services can be safely invoked without nesting critical sections.

> **NOTE**
>
> The prioritized interrupt controller remembers the priority of the currently serviced interrupt and allows only interrupts of higher priority than the current priority to preempt the ISR. Lower- and same-priority interrupts are locked at the interrupt controller level, even though the interrupts are unlocked at the processor level. The interrupt prioritization happens in the interrupt controller hardware until the interrupt controller receives the end-of-interrupt (EOI) instruction.

(6)   Interrupts are locked to establish critical sections for the interrupt exit.

(7)   The end-of-interrupt (EOI) instruction is sent to the interrupt controller to stop prioritizing this interrupt level.

(8)   The interrupt exit synthesized by the compiler restores the CPU registers from the stack, which includes restoring the CPU status register. This step typically unlocks interrupts.

7.3.3   Internal QF Macros for Interrupt Locking/Unlocking

The QF platform abstraction layer (PAL) uses the interrupt locking/unlocking macros QF_INT_LOCK(), QF_INT_UNLOCK(), and QF_INT_KEY_TYPE in a slightly modified form. The PAL defines internally the parameterless macros, shown in Listing 7.6. Please note the trailing underscores in the internal macros' names.

**Listing 7.6 Internal macros for interrupt locking/unlocking
(file** `<qp>\qpc\qf\source\qf_pkg.h`**)**

```
#ifndef QF_INT_KEY_TYPE      /* simple unconditional interrupt locking/unlocking */
    #define QF_INT_LOCK_KEY_
    #define QF_INT_LOCK_()        QF_INT_LOCK    (ignore)
    #define QF_INT_UNLOCK_()      QF_INT_UNLOCK (ignore)
#else                        /* policy of saving and restoring interrupt status */
    #define QF_INT_LOCK_KEY_      QF_INT_KEY_TYPE intLockKey_;
    #define QF_INT_LOCK_()        QF_INT_LOCK    (intLockKey_)
    #define QF_INT_UNLOCK_()      QF_INT_UNLOCK (intLockKey_)
#endif
```

The internal macros `QF_INT_LOCK_KEY_`, `QF_INT_LOCK_()`, and `QF_INT_UNLOCK_()` enable writing the same code for the case when the interrupt key is defined and when it is not. The following code snippet shows the usage of the internal QF macros. Convince yourself that this code works correctly for both interrupt-locking policies.

```
void QF_service_xyz(arguments) {
    QF_INT_LOCK_KEY_
    . . .
    QF_INT_LOCK_();
    . . .
    /* critical section of code */
    . . .
    QF_INT_UNLOCK_();
}
```

7.4   Active Objects

As shown in Figure 7.3, the QF real-time framework provides the base structure `QActive` for deriving application-specific active objects. `QActive` combines the following three essential elements:

- It is a state machine (derives from `QHsm` or some other class with a compatible interface).

- It has an event queue.

- It has an execution thread with a unique priority.

Listing 7.7 shows the declaration of the `QActive` base structure and related functions.

Listing 7.7 The `QActive` base class for derivation of active objects (file `<qp>\qpc\include\qf.h`)

```
(1)   #ifndef QF_ACTIVE_SUPER_
(2)       #define QF_ACTIVE_SUPER_                QHsm
(3)       #define QF_ACTIVE_CTOR_(me_, initial_) QHsm_ctor((me_), (initial_))
(4)       #define QF_ACTIVE_INIT_(me_, e_)       QHsm_init((me_), (e_))
(5)       #define QF_ACTIVE_DISPATCH_(me_, e_) QHsm_dispatch((me_), (e_))
(6)       #define QF_ACTIVE_STATE_               QState
      #endif

      typedef struct QActiveTag {
```

```
(7)       QF_ACTIVE_SUPER_ super;       /* derives from QF_ACTIVE_SUPER_ */

(8)       QF_EQUEUE_TYPE  eQueue;       /* event queue of active object */
     #ifdef QF_OS_OBJECT_TYPE
(9)       QF_OS_OBJECT_TYPE osObject; /* OS-object for blocking the queue */
     #endif

     #ifdef QF_THREAD_TYPE
(10)      QF_THREAD_TYPE  thread; /* execution thread of the active object */
     #endif

(11)      uint8_t       prio;        /* QF priority of the active object */
(12)      uint8_t       running;    /* flag indicating if the AO is running */
     } QActive;

(13)  void QActive_start(QActive *me, uint8_t prio,
                         QEvent const *qSto[], uint32_t qLen,
                         void *stkSto, uint32_t stkSize,
                         QEvent const *ie);
(14)  void QActive_postFIFO(QActive *me, QEvent const *e);
(15)  void QActive_postLIFO(QActive *me, QEvent const *e);

(16)  void QActive_ctor(QActive *me, QState initial);
(17)  void QActive_stop(QActive *me);

(18)  void QActive_subscribe(QActive const *me, QSignal sig);
(19)  void QActive_unsubscribe(QActive const *me, QSignal sig);
(20)  void QActive_unsubscribeAll(QActive const *me);

(21)  void QActive_defer(QActive *me, QEQueue *eq, QEvent const *e);
(22)  QEvent const *QActive_recall(QActive *me, QEQueue *eq);

(23)  QEvent const *QActive_get_(QActive *me);
```

(1) The macro QF_ACTIVE_SUPER_ specifies the ultimate base class for deriving active objects. This macro lets you define (in the QF port) any base class for QActive as long as the base class supports the state machine interface. (See Chapter 3, "Generic State Machine Interface.")

(2) When the macro QF_ACTIVE_SUPER_ is not defined in the QF port, the default is the QHsm class provided in the QEP hierarchical event processor.

(3) The macro QF_ACTIVE_CTOR_() specifies the name of the base class constructor.

(4)   The macro `QF_ACTIVE_INIT_()` specifies the name of the base class `init()` function.

(5)   The macro `QF_ACTIVE_DISPATCH_()` specifies the name of the base class `dispatch()` function.

(6)   The macro `QF_ACTIVE_STATE_` specifies the type of the parameter for the base class constructor.

By defining the macros `QF_ACTIVE_XXX_` to your own class, you can eliminate the dependencies between the QF framework and the QEP event processor. In other words, you can replace QEP with your own event processor, perhaps based on one of the techniques discussed in Chapter 3, or not based on state machines at all (e.g., you might want to try *protothreads* [Dunkels+ 06]). Consider the following definitions:

```
#define QF_ACTIVE_SUPER_              MyClass
#define QF_ACTIVE_CTOR_(me_, ini_)    MyClass_ctor((me_), (ini_))
#define QF_ACTIVE_INIT_(me_, e_)      MyClass_init((me_), (e_))
#define QF_ACTIVE_DISPATCH_(me_, e_)  MyClass_dispatch((me_), (e_))
#define QF_ACTIVE_STATE_              void*
```

(7)   The first member `super` specifies the base class for `QActive` (see the sidebar "Single Inheritance in C" in Chapter 1).

(8)   The type of the event queue member `eQueue` is platform-specific. For example, in the standalone QF configurations, the macro `QF_EQUEUE_TYPE` is defined as the native QF event queue `QEqueue` (see Section 7.8). However, when QF is based on an external RTOS, the event queue might be implemented with a message queue of the underlying RTOS.

(9)   The data member `osObject` is used in some QF ports to block the native QF event queue. The `osObject` data member is necessary when the underlying OS does not provide an adequate queue facility, so the native QF queue must be used. In that case the `osObject` data member holds an OS-specific primitive to efficiently block the native QF event queue when the queue is empty. See Chapter 8, "POSIX QF Port," for an example of using the osObject data member.

(10) The data member `thread` is used in some QF ports to hold the thread handle associated with the active object.

(11) The data member `prio` holds the priority of the active object. In QF, each active object has a unique priority. The lowest possible task priority is 1 and higher-priority values correspond to higher-urgency active objects. The maximum allowed active object priority is determined by the macro `QF_MAX_ACTIVE`, which currently cannot exceed 63.

(12) The data member `running` is used in some QF ports to represent whether the active object is `running`. In these ports, writing zero to the running member causes exit from the active object's event loop and cleanly terminates the active object thread.

(13) The function `QActive_start()` starts the active object thread. This function is platform-specific and is explained in Section 7.4.3.

(14) The function `QActive_postFIFO()` is used for direct event posting to the active object's event queue using the FIFO policy.

(15) The function `QActive_postLIFO()` is used for direct event posting to the active object's event queue using the LIFO policy.

(16) The function `QActive_ctor()` is the "constructor" of the `QActive` class. This constructor has the same signature as the constructor of `QHsm` or `QFsm` (see Section 4.5.1 in Chapter 4). In fact, the main job of the `QActive` constructor is to initialize the state machine base class (the member `super`).

NOTE

In the C++ version, the `QActive` constructor is protected. This prevents direct instantiation of the `QActive` class, since it is intended only for derivation (the abstract class).

---

(17) The function `QActive_stop()` stops the execution thread of the active object. This function is platform-specific and is discussed in Chapter 8. Not all QF ports need to define this function.

> **NOTE**
>
> In the C++ version, `QActive::stop()` is not equivalent to the active object destructor. The function merely causes the active object thread to eventually terminate, which might not happen immediately.

(18-20)   The functions `QActive_subscribe()`, `QActive_usubscribe()`, and `QActive_unsubscribeAll()` are used for subscribing and unsubscribing to events. I discuss these functions in the upcoming Section 7.6.2.

(21,22)   The functions `QActive_defer()` and `QActive_recall()` are used for efficient ("zero copy") deferring and recalling of events, respectively. I describe these functions in the upcoming Section 7.5.4.

(23)   The function `QActive_get_()` is used to remove one event at a time from the active object's event queue. This function is used only inside QF and never at the application level. In some QF ports the function `QActive_get_()` can block. I describe this function in the upcoming Section 7.4.2.

7.4.1   Internal State Machine of an Active Object

As shown in Figure 7.3, every concrete active object, such as `Ship`, `Missile`, or `Tunnel` in the "Fly 'n' Shoot" game example from Chapter 1, *is a* state machine because it derives indirectly from the `QHsm` base class or a class that supports a generic state machine interface (see the data member `super` in Listing 7.7(7)). *Derivation* means simply that every pointer to `QActive` or a structure derived from `QActive` can always be safely used as a pointer to the base structure `QHsm`. Such a pointer can therefore always be passed to any function designed to work with the state machine structure. At the application level, you can mostly ignore the other aspects of your active objects and view them predominantly as state machines. In fact, your main job in developing a QF application consists of elaborating the state machines of your active objects.

7.4.2   Event Queue of an Active Object

Event queues are essential components of any event-driven system because they reconcile the asynchronous production of events with the RTC semantics of their consumption. An event queue makes the corresponding active object appear to always be responsive to events, even though the internal state machine can accept events only between RTC steps. Additionally, the event queue provides buffer space that protects the internal state machine from bursts in event production that can, at times, exceed the available processing capacity.

You can view the active object's event queue as an outer rind that provides an external interface for injecting events into the active object and at the same time protects the internal state machine during RTC processing. To perform these functions, the event queue must allow any thread of execution (as well as an ISR) to asynchronously post events, but only one thread—the local thread of the active object—needs to be able to remove events from the queue. In other words, the event queue in QF needs multiple-write but only single-read access.

From the description so far, it should be clear that the event queue is quite a sophisticated mechanism. One end of the queue—the end where producers insert events—is obviously shared among many tasks and interrupts and must provide an adequate mutual exclusion mechanism to protect the internal consistency of the queue. The other end—the end from which the local active object thread extracts events—must provide a mechanism for blocking the active object when the queue is empty. In addition, an event queue must manage a buffer of events, typically organized as a ring buffer.

As shown in Figure 6.13 in Chapter 6, the "zero copy" event queues do not store actual events, only pointers to event instances. Typically these pointers point to event instances allocated dynamically from event pools (see Section 7.5.2), but they can also point to statically allocated events. You need to specify the maximum number of event pointers that a queue can hold at any one time when you start the active object with the QActive_start() function (see the next section). The correct sizing of event queues depends on many factors and generally is not a trivial task. I discuss sizing event queues in Chapter 9.

Many commercial RTOSs natively support queuing mechanisms in the form of message queues. Standard message queues are far more complex than required by active objects because they typically allow multiple-write as well as multiple-read access (the QF requires only single-read access) and often support variable-length data (not only pointer-sized data). Usually message queues also allow blocking when the queue is empty and when the queue is full, and both types of blocking can be timed out. Naturally, all this extra functionality, which you don't really need in QF, comes at an extra cost in CPU and memory usage. The QF port to the µC/OS-II RTOS described in Chapter 8 provides an example of an event queue implemented with a message queue of an RTOS. The standalone QF ports to x86/DOS and ARM Cortex-M3 (used in the "Fly 'n' Shoot" game from Chapter 1) provide examples of using the native QF event queue. I discuss the native QF active object queue implementation in Section 7.8.3.

7.4.3   Thread of Execution and Active Object Priority

Every QF active object executes in its own thread. The actual control flow within the active object thread depends on the multitasking model actually used, but the event processing always consists of the three essential steps shown in Listing 7.8.

Listing 7.8 The three steps of an active object thread

```
(1)   QEvent const *e = QActive_get_(a);    /* get the next event for AO 'a' */
(2)   QF_ACTIVE_DISPATCH_(&a->super, e);          /* dispatch to the AO 'a' */
(3)   QF_gc(e);     /* determine if event 'e' is garbage and collect it if so */
```

---

(1) The event is extracted from the active object's event queue by means of the function `QActive_get_()`. This function might block in blocking kernels. In Section 7.8.3 I describe the implementation of `QActive_get_()` for the native QF active object queue. In Chapter 8 I describe the `QActive_get_()` implementation when a message queue of an RTOS is used instead of the native QF event queue.

(2) The event is dispatched to the active object's state machine for processing (see Listing 7.7(5) for the definition of the `QF_ACTIVE_DISPATCH_()` macro).

NOTE

Step 2 constitutes the RTC processing of the active object's state machine. The active object's thread continues only after step 2 completes.

---

(3) The event is passed to the QF garbage collector for recycling. As described in Section 6.5.5 in Chapter 6, the garbage collector actually recycles the event only when it determines that the event is no longer referenced.

In the presence of a traditional RTOS (e.g., VxWorks) or a multitasking operating system (e.g., Linux), the three event processing steps just explained are enclosed by the usual endless loop, as shown in Figure 6.12(A) in Chapter 6. Under a cooperative "vanilla" kernel (Figure 6.12(B)) or an RTC kernel (Figure 6.12(C)), the three steps are executed in one-shot fashion for every event.

The `QActive_start()` function creates the active object's thread and notifies QF to start managing the active object. A QF application needs to call the `QActive_start()` function on behalf of every active object in the system. In principle, active objects can be started and stopped (with `QActive_stop()`) multiple times during the lifetime of the application. However, in most cases, all active objects are started just once during the system initialization.

The `QActive_start()` function is one of the central elements of the framework, but obviously it strongly depends on the underlying multitasking kernel. Listing 7.9 shows the pseudocode of `QActive_start()`.

Listing 7.9 `QActive_start()` **function pseudocode**

```
(1)    void QActive_start(QActive *me,
(2)                       uint8_t prio,                    /* the unique priority */
(3)                       QEvent const *qSto[], uint32_t qLen, /* event queue */
(4)                       void *stkSto, uint32_t stkSize,  /* per-task stack */
(5)                       QEvent const *ie)       /* the initialization event */
       {
(6)        me->prio = prio;                               /* set the QF priority */
(7)        QF_add_(me);                      /* make QF aware of this active object */
(8)        QF_ACTIVE_INIT_(me, ie);          /* execute the initial transition */

(9)        Initialize the event queue object 'me->eQueue' using qSto and qLen
(10)       Create and start the thread 'me->thread' of the underlying kernel
       }
```

(1)   The argument 'me' is the pointer to the active object being started.

(2)   The argument 'prio' is the priority you assign to the active object. In QF, every active object must have a *unique priority,* which you assign at startup and cannot change later. QF uses a priority numbering system in which priority 1 is the lowest and higher numbers correspond to higher priorities.

> **NOTE**
>
> You can think of QF priority 0 as corresponding to the idle task, which has the absolute lowest priority not accessible to the application-level tasks.

---

(3)   The arguments 'qSto' and 'qLen' are a pointer to the storage for the event queue buffer and the length of that buffer (in units of QEvent*), respectively. If the underlying RTOS cannot accept externally allocated storage for the queue, the 'qSto' pointer should be set to NULL.

(4)   The argument 'stkSto' is the pointer to the storage for the private stack, and the argument 'stkSize' is the size of that stack (in bytes), respectively. If the underlying kernel/RTOS does not need per-task stacks or cannot accept externally allocated storage for the stack, the 'stkSto' pointer should be set to NULL.

(5)   The argument 'ie' is a pointer to initialization event for the topmost initial transition in the active object state machine. This argument is very specific to the active object being initialized and can be NULL.

> **NOTE**
>
> The "initialization event" 'ie' gives you an opportunity to provide some information to the active object, which is only known later in the initialization sequence (e.g., a window handle in a GUI system). Note that the active object constructor runs even before main() (in C++), at which time you typically don't have all the information to initialize all aspects of an active object.

---

(6)   The QF priority of the active object is set.

(7)   The active object is registered with the QF framework. The QF_add_() function asserts that the priority of the active object is in range and is not already used (unique priority).

(8)   The topmost initial transition in the active object's state machine is taken (see Listing 7.7(4) for the definition of the QF_ACTIVE_INIT_() macro). Note that the initial transition is executed in the same thread that called QActive_start(), which often is the main() thread.

> **NOTE**
>
> This design allows the initialization event (passed to QActive_start() as the 'ie' pointer) to be allocated on the stack of the caller. Note that the initialization event is *not* recycled.

(9)   The QActive_start() function initializes the event queue attribute me->eQueue, typically using the storage for the queue buffer (qSto[]), and the length of this buffer (qLen).

(10)   Finally, the thread (task) of the active object is created and the further execution of the active object occurs in that newly created task context. The priority of the thread should correspond to the relative QF priority passed as the argument 'prio' to QActive_start(). If the underlying scheduler uses a different priority numbering scheme, the concrete implementation of QActive_start() must remap the QF priority to the priority required by the scheduler before invoking the platform-specific thread creation routine.

7.5   Event Management in QF

QF implements the efficient "zero-copy" event delivery scheme, as described in Section 6.5 in Chapter 6. QF supports two kinds of events: (1) dynamic events managed by the framework, and (2) other events (typically statically allocated) not managed by QF. For each dynamic event, QF keeps track of the reference counter of the event (to know when to recycle the event) as well as the event pool from which the dynamic event was allocated (to recycle the event back to the same pool).

7.5.1   Event Structure

QF uses the same event representation as the QEP event processor described in Part I. Events in QF are represented as instances of the QEvent structure (shown in Listing 7.10), which contains the event signal sig and a byte dynamic_ to represent the internal "bookkeeping" information about the event.

Listing 7.10 QEvent **structure defined in** `<qp>\qpc\include\qevent.h`

```
typedef struct QEventTag {                        /* QEvent base structure */
    QSignal sig;                        /* public signal of the event instance */
    uint8_t dynamic_;    /* attributes of a dynamic event (0 for static event) */
} QEvent;
```

As shown in Figure 7.4, the QF framework uses the `QEvent.dynamic_` data byte in the following way.[4] The six least-significant bits [0..5] represent the reference counter of the event, which has the dynamic range of 0..63. The two most significant bits [6..7] represent the event pool ID of the event, which has the dynamic range of 1..3. The pool ID of zero is reserved for static events, that is, events that do not come from any event pool. With this representation, a static event has a unique, easy-to-check signature (`QEvent.dynamic_ == 0`). Conversely, the signature (`QEvent.dynamic_ != 0`) unambiguously identifies a dynamic event.

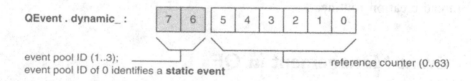

Figure 7.4: Allocation of bits in the `QEvent.dynamic_` **byte.**

NOTE

The `QEvent` data member `dynamic_` is used only by the QF framework for managing dynamic events (see the following section). For every static event, you must initialize this member to zero. Otherwise, the `QEvent.dynamic_` data member should never be of interest to the application code.

---

[4] I avoid using bit fields because they are not quite portable. Also, the use of bit fields would be against the required MISRA rule 111.

7.5.2 Dynamic Event Allocation

Dynamic events allow reusing the same memory over and over again for passing different events. QF allocates such events dynamically from one of the event pools managed by the framework. An event pool in QF is a fixed-block-size heap, also known as a *memory partition* or *memory pool*.

The most obvious drawback of a fixed-block-size heap is that it does not support variable-sized blocks. Consequently, the blocks have to be oversized to handle the biggest possible allocation. A good compromise to avoid wasting memory is to use not one but a few heaps with blocks of different sizes. QF can manage up to three event pools (e.g., small, medium, and large events, like shirt sizes).

Event pools require initialization through QF_poolInit() function shown in Listing 7.11. An application may call this function up to three times to initialize up to three event pools in QF.

**Listing 7.11 Initializing an event pool to be managed by QF
(file** <qp>\qpc\qf\source\qf_pool.c**)**

```
      /* Package-scope objects --------------------------------------------*/
(1)   QF_EPOOL_TYPE_ QF_pool_[3];                        /* allocate 3 event pools */
(2)   uint8_t QF_maxPool_;                        /* number of initialized event pools */


      /*.................................................................*/
(3)   void QF_poolInit(void *poolSto, uint32_t poolSize, QEventSize evtSize) {
                        /* cannot exceed the number of available memory pools */
(4)       Q_REQUIRE(QF_maxPool_ < (uint8_t)Q_DIM(QF_pool_));
                    /* please initialize event pools in ascending order of evtSize: */
(5)       Q_REQUIRE((QF_maxPool_ == (uint8_t)0)
                    || (QF_EPOOL_EVENT_SIZE_(QF_pool_[QF_maxPool_ - 1]) < evtSize));
                    /* perfom the platform-dependent initialization of the pool */
(6)       QF_EPOOL_INIT_(QF_pool_[QF_maxPool_], poolSto, poolSize, evtSize);
(7)       ++QF_maxPool_;                                    /* one more pool */
      }
```

---

(1) The macro QF_EPOOL_TYPE_ represents the QF event pool type. This macro lets the QF port define a particular memory pool (fixed-size heap) implementation that might be already provided with the underlying kernel or RTOS. If QF is used standalone or if the underlying RTOS does not provide an adequate memory pool, the QF framework provides the efficient native QMPool class.

Note that an event pool object is quite small because it does not contain the actual memory managed by the pool (see Section 7.9).

(2) The variable `QF_maxPool_` holds the number of pools actually used, which can be 0 through 3.

> **NOTE**
>
> All QP components, including the QF framework, consistently assume that variables without an explicit initialization value are initialized to zero upon system startup, which is a requirement of the ANSI-C standard. In embedded systems, this initialization step corresponds to clearing the `.BSS` section. You should make sure that in your system the `.BSS` section is indeed cleared before `main()` is called.

(3) According to the general policy of QF, all memory needed for the framework operation is provided to the framework by the application. Therefore, the first parameter 'poolSto' of `QF_poolInit()` is a pointer to the contiguous chunk of storage for the pool. The second parameter 'poolSize' is the size of the pool storage in bytes, and finally, the last parameter 'evtSize' is the maximum event size that can be allocated from this pool.

> **NOTE**
>
> The number of events in the pool might be smaller than the ratio `poolSize/evtSize` because the pool might choose to internally align the memory blocks. However, the pool is guaranteed to hold events of at least the specified size `evtSize`.

(4) This precondition (see Chapter 6, "Customized Assertions in C/C++") asserts that the application does not attempt to initialize more than the supported number of event pools (currently three).

(5) For possibly quick event allocation, the event pool array `QF_pool_[]` must be sorted in ascending order of block sizes. This precondition asserts that the application initializes event pools in the increasing order of the event sizes. This assertion significantly simplifies the `QF_poolInit()` function without causing any true inconvenience for the application implementer.

NOTE

The subsequent calls to `QF_poolInit()` function must be made with progressively increasing values of the `evtSize` parameter.

(6) The macro `QF_EPOOL_INIT_()` specifies the initialization function for the event pool object. In case of the native QF memory pool, the macro is defined as the `QMPool_init()` function.

(7) Finally, the variable `QF_maxPool_` is incremented to indicate that one more pool has been initialized.

Listing 7.12 shows the implementation of the `QF_new_()` function, which allocates a dynamic event from one of the event pools managed by QF. The basic policy is to allocate the event from the first pool that has a block size big enough to fit the requested event size.

Listing 7.12 Simple policy of allocating an event from the smallest event-size pool (file `<qp>\qpc\qf\source\qf_new.c`)

```
(1)   QEvent *QF_new_(QEventSize evtSize, QSignal sig) {
          QEvent *e;
                           /* find the pool id that fits the requested event size ... */
          uint8_t idx = (uint8_t)0;
(2)       while (evtSize > QF_EPOOL_EVENT_SIZE_(QF_pool_[idx])) {
              ++idx;
(3)           Q_ASSERT(idx < QF_maxPool_);   /* cannot run out of registered pools */
          }
(4)       QF_EPOOL_GET_(QF_pool_[idx], e);              /* get e -- platform-dependent */
(5)       Q_ASSERT(e != (QEvent *)0);               /* pool must not run out of events */

(6)       e->sig = sig;                             /* set signal for this event */
                                        /* store the dynamic attributes of the event:
                                         * the pool ID and the reference counter == 0
                                         */
(7)       e->dynamic_ = (uint8_t)((idx + 1) << 6);
(8)       return e;
      }
```

(1) The function `QF_new_()` allocates a dynamic event of the requested size 'evtSize' and sets the signal 'sig' in the newly allocated event. The function returns a pointer to the event.

(2)   This `while` loop scans through the `QF_pool_[]` array starting from pool `id = 0` in search of a pool that would fit the requested event size. Obtaining the event size of a pool is a platform-specific operation because various RTOSs that support fixed-size heaps might report the event size in a different way. This platform dependency is hidden in the QF code by the indirection layer of the macro `QF_EPOOL_EVENT_SIZE_()`.

(3)   This assertion fires when the `while` loop runs out of the event pools, which means that the requested event is too big for all initialized event pools.

(4)   The macro `QF_EPOOL_GET_()` obtains a memory block from the pool found in the previous step.

(5)   The assertion fires when the pool returns the `NULL` pointer, which indicates depletion of this pool.

NOTE

The QF framework treats the inability to allocate an event as an error. The assertions in lines 3 and 5 are part of the event delivery guarantee policy. It is the application designer's responsibility to size the event pools adequately so that they never run out of events.

---

(6)   The signal of the event is initialized.

(7)   The two most significant bits of the `e->dynamic_` byte are set to the pool ID, whereas the pool ID is the index into the `QF_pool_[]` array incremented by one to fall in the range 1..3. At the same time, the reference counter of the event in the six least significant bits of the `e->dynamic_` byte is set to zero.

(8)   The event is returned to the caller.

Typically, you will not use `QF_new_()` directly but through the `Q_NEW()` macro defined as follows:

```
#define Q_NEW(evtT_, sig_) ((evtT_ *)QF_new_(sizeof(evtT_), (sig_)))
```

The `Q_NEW()` macro dynamically creates a new event of type `evT_` with the signal `sig_`. It returns a pointer to the event already cast to the event type `(evT_*)`. Here is an example of dynamic event allocation with the macro `Q_NEW()`:

```
MyEventXYZ *e_xyz = Q_NEW(MyEventXYZ, XYZ_SIG);  /* dynamically allocate */
/* NOTE: no need to check for validity of the event pointer */
e_xyz->foo = ...;                                /* fill the event parameters... */
QF_publish((QEvent *)e_xyz);                     /* publish the event */
```

The assertions inside QF_new_() guarantee that the pointer is valid, so you don't
need to check the pointer returned from Q_NEW(), unlike the value returned
from malloc(), which you should check.

NOTE

In C++, the Q_NEW() macro does not invoke the constructor of the event. This is not a problem for the QEvent base struct and simple structs derived from it. However, you need
to keep in mind that subclasses of QEvent should not introduce virtual functions because the
virtual pointer won't be set up during the dynamic allocation through Q_NEW().[5]

7.5.3 Automatic Garbage Collection

Most of the time, you don't need to worry about recycling dynamic events, because
QF does it automatically when it detects that an event is no longer referenced.

NOTE

The explicit garbage collection step is necessary only in the code that is out of the framework's control, such as ISRs receiving events from "raw" thread-safe queues (see upcoming
Section 7.8.4).

QF uses the standard reference-counting algorithm to keep track of the outstanding
references to each dynamic event managed by the framework. The reference
counter for each event is stored in the six least significant bits of the event attribute
dynamic_. Note that the data member dynamic_ of a dynamic event cannot be
zero because the two most significant bits of the byte hold the pool ID, with valid
values of 1, 2, or 3.

---

[5] A simple solution would be to use the placement new() operator inside the Q_NEW() macro to enforce
full instantiation of an event object, but it is currently not used, for better efficiency and compatibility
with older C++ compilers, which might not support placement new().

The reference counter of each event is always updated and tested in a critical section of code to prevent data corruption. The counter is incremented whenever a dynamic event is inserted into an event queue. The counter is decremented by the QF garbage collector, which is called after every RTC step (see Listing 7.8(3)). When the reference counter of a dynamic event drops to zero, the QF garbage collector recycles the event back to the event pool number stored in the two most significant bits of the `dynamic_` attribute.

```
(1)   void QF_gc(QEvent const *e) {
(2)       if (e->dynamic_ != (uint8_t)0) {                /* is it a dynamic event? */
(3)           QF_INT_LOCK_KEY_

(4)           QF_INT_LOCK_();
(5)           if ((e->dynamic_ & 0x3F) > 1) {      /* isn't this the last reference? */
(6)               --((QEvent *)e)->dynamic_;   /* decrement the reference counter */
(7)               QF_INT_UNLOCK_();
              }
(8)           else {          /* this is the last reference to this event, recycle it */
(9)               uint8_t idx = (uint8_t)((e->dynamic_ >> 6) - 1);
(10)              QF_INT_UNLOCK_();

(11)              Q_ASSERT(idx < QF_maxPool_);               /* index must be in range */
(12)              QF_EPOOL_PUT_(QF_pool_[idx], (QEvent *)e);
              }
          }
      }
```

(1)   The function `QF_gc()` garbage-collects one event at a time.

(2)   The function checks the unique signature of a dynamic event. The garbage collector handles only dynamic events.

(3)   The critical section status is allocated on the stack (see Section 7.3.3).

(4)   Interrupts are locked to examine and decrement the reference count.

(5)   If the reference count (lowest 6 bits of the `e->dynamic_` byte) is greater than 1, the event should not be recycled.

(6)   The reference count is decremented. Note that the `const` attribute of the event pointer is "cast away," but this is safe after checking that this must be a dynamic event (and not a static event possibly placed in ROM).

(7)   Interrupts are unlocked for the `if`-branch.

(8)   Otherwise, reference count is becoming zero and the event must be recycled.

(9)   The pool ID is extracted from the two most significant bits of the `e->dynamic_` byte and decremented by one to form the index into the `QF_pool_[]` array.

(10)   Interrupts are unlocked for the `else` branch. It is safe at this point because you know for sure that the event is not referenced by anybody else, so it is exclusively owned by the garbage collector thread.

(11)   The index must be in the expected range of initialized event pools.

(12)   The macro `QF_EPOOL_PU_()` recycles the event to the pool `QF_pool_[idx]`. The explicit cast removes the const attribute.

7.5.4 Deferring and Recalling Events

Event deferral comes in very handy when an event arrives in a particularly inconvenient moment but can be deferred for some later time, when the system is in a much better position to handle the event (see "Deferred Event" state pattern in Chapter 5). QF supports very efficient event deferring and recalling mechanisms consistent with the "zero-copy" policy.

QF implements explicit event deferring and recalling through `QActive` class functions `QActive_defer()` and `QActive_recall()`, respectively. These functions work in conjunction with the native "raw" event queue provided in QF (see upcoming Section 7.8.4). Listing 7.13 shows the implementation.

Listing 7.13 QF event deferring and recalling
(file `<qp>\qpc\qf\source\qa_defer.c`**)**

```
        void QActive_defer(QActive *me, QEQueue *eq, QEvent const *e) {
            (void)me;                    /* avoid compiler warning about 'me' not used */
(1)         QEQueue_postFIFO(eq, e);     /* increments ref-count of a dynamic event */
        }
        /*................................................................*/
(2)     QEvent const *QActive_recall(QActive *me, QEQueue *eq) {
(3)         QEvent const *e = QEQueue_get(eq);   /* get an event from deferred queue */
            if (e != (QEvent *)0) {                        /* event available? */
                QF_INT_LOCK_KEY_
(4)             QActive_postLIFO(me, e);     /* post it to the front of the AO's queue */

(5)             QF_INT_LOCK_();
(6)             if (e->dynamic_ != (uint8_t)0) {          /* is it a dynamic event? */
```

Continued onto next page

```
(7)                Q_ASSERT((e->dynamic_ & 0x3F) > 1);
(8)                --((QEvent *)e)->dynamic_;  /* decrement the reference counter */
                }
(9)            QF_INT_UNLOCK_();
            }
(10)       return e; /*pass the recalled event to the caller (NULL if not recalled) */
        }
```

(1)   The function QActive_defer() takes posts the deferred event into the given "raw" queue 'eq.' The event posting increments the reference counter of a dynamic event, so the event is not recycled at the end of the current RTC step (because it is referenced by the "raw" queue).

(2)   The function QActive_recall() attempts recalling an event from the provided "raw" thread-safe queue 'eq.' The function returns the pointer to the recalled event or NULL if the provided queue is empty.

(3)   The event is extracted from the queue. The "raw" queue never blocks and returns NULL if it is empty.

(4)   If an event is available, it is posted using the last-in, first-out (LIFO) policy into the event queue of the active object. The LIFO policy is employed to guarantee that the recalled event will be the very next to process. If other already queued events were allowed to precede the recalled event, the state machine might transition to a state where the recalled event would no longer be convenient.

(5)   Interrupts are locked to decrement the reference counter of the event, to account for removing the event from the "raw" thread-safe queue.

(6)   The unique signature of a dynamic event is checked.

(7)   The reference counter must be at this point at least 2 because the event is referenced by at least two event queues (the deferred queue and the active object's queue).

(8)   The reference counter is decremented by one to account for removing the event from the deferred queue.

(9)   Interrupts are unlocked.

(10) The recalled event pointer is returned to the caller.

> **NOTE**
>
> Even though you can "peek" inside the event right at the point it is recalled, you should typically handle the event only after it arrives through the active object's queue. See the "Deferred Event" state pattern in Chapter 5.

7.6   Event Delivery Mechanisms in QF

QF supports only asynchronous event exchange within the application, meaning that the producers post events into event queues, but do not wait for the actual processing of the events. QF supports two types of asynchronous event delivery:

1. The simple mechanism of direct event posting, when the producer of an event directly posts the event to the event queue of the consumer active object.

2. A more sophisticated publish-subscribe event delivery mechanism, where the producers of events publish them to the framework and the framework then delivers the events to all active objects that had subscribed to this event.

7.6.1   Direct Event Posting

QF supports direct event posting through the `QActive_postFIFO()` and `QActive_postLIFO()` functions. These functions depend on the active object's employed queue mechanism. In the upcoming Section 7.8.3, I show how these functions are implemented when the native QF active object queue is used. In Chapter 8, I demonstrate how to implement these functions to use a message queue of a traditional RTOS.

> **NOTE**
>
> Direct event posting should not be confused with event dispatching. In contrast to asynchronous event posting through event queues, direct event dispatching is a simple synchronous function call. Event dispatching occurs when you call the `QHsm_dispatch()` function, as in Listing 7.8(2), for example.

Direct event posting is illustrated in the "Fly 'n' Shoot" example from Chapter 1, when an ISR posts a PLAYER_SHIP_MOVE event directly to the Ship active object:

```
QActive_postFIFO(AO_ship, (QEvent *)e);   /* post event 'e' to the Ship AO */
```

Note that the producer of the event (ISR) in this case must only "know" the recipient (Ship) by an "opaque pointer" QActive*, and the specific definition of the Ship active object structure is not required. The AO_ship pointer is declared in the game.h header file as:

```
extern QActive * const AO_Ship;           /* opaque pointer to the Ship AO */
```

The Ship structure definition is in fact entirely encapsulated in the ship.c module and is inaccessible to the rest of the application. I recommend using this variation of the "opaque pointer" technique in your applications.

7.6.2   Publish-Subscribe Event Delivery

QF implements publish-subscribe event delivery through the following services:

- The function QF_psInit() to initialize the publish-subscribe mechanism

- Functions QActive_subscribe(), QActive_unsubscribe(), and QActive_unsubscribeAll() for active objects to subscribe and unsubscribe to particular event signals

- The function QF_publish() for publishing events

Delivering events is the most frequently performed function of the framework; therefore, it is important to implement it efficiently. As shown in Figure 7.5, QF uses a lookup table indexed by the event signal to efficiently find all subscribers to a given signal. For each event signal index (e->sig), the lookup table stores a *subscriber list*. A subscriber list (typedef'd to QSubscrList) is just a densely packed bitmask where each bit corresponds to the unique priority of the active object. If the bit is set, the corresponding active object is the subscriber to the signal, otherwise the active object is not the subscriber.

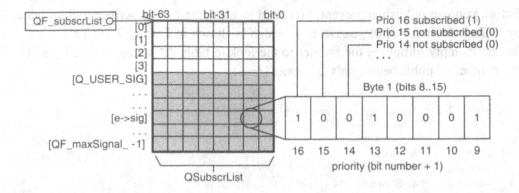

Figure 7.5: Signal to Subscriber-List lookup table `QF_subscrList_[]`.

The actual size of the `QSubscrList` bitmask is determined by the macro
`QF_MAX_ACTIVE`, which specifies the maximum active objects in the system (the
current range of `QF_MAX_ACTIVE` is 1..63). Subscriber list type `QSubscrList`
is `typedef`'ed in Listing 7.14.

Listing 7.14 `QF_psInit()` **(file** `<qp>\qpc\init\qf.h`**)**

```
typedef struct QSubscrListTag {
    uint8_t bits[((QF_MAX_ACTIVE - 1) / 8) + 1];
} QSubscrList;
```

To reduce the memory taken by the subscriber lookup table, you have options to reduce
the number of published signals and reduce the number of potential subscribers
`QF_MAX_ACTIVE`. Typically, however, the table is quite small. For example, the table
for a complete real-life GPS receiver application with 50 different signals and up to
eight active objects costs 50 bytes of RAM.

NOTE

Not all signals in the system are published. To conserve memory, you can enumerate the
published signals before other nonpublished signals and thus arrive at a lower limit for the
number of published signals.

Before you can publish any event, you need to initialize the subscriber lookup table by calling the function `QF_psInit()`, which is shown in Listing 7.15. This function simply initializes the pointer to the lookup table `QF_subsrcrList_` and the number of published signals `QF_maxSignal`.

Listing 7.15 `QF_psInit()` **(file** `<qp>\qpc\qf\source\qf_psini.c`**)**

```
QSubscrList *QF_subscrList_;      /* initialized to zero per C-standard */
QSignal QF_maxSignal_;            /* initialized to zero per C-standard */

void QF_psInit(QSubscrList *subscrSto, QSignal maxSignal) {
    QF_subscrList_ = subscrSto;
    QF_maxSignal_  = maxSignal;
}
```

Active objects subscribe to signals through `QActive_subscribe()`, shown in Listing 7.16.

Listing 7.16 `QActive_subscribe()` **function**
(file `<qp>\qpc\qf\source\qa_sub.c`**)**

```
(1)   void QActive_subscribe(QActive const *me, QSignal sig) {
          uint8_t p = me->prio;
(2)       uint8_t i = Q_ROM_BYTE(QF_div8Lkup[p]);
          QF_INT_LOCK_KEY_

(3)       Q_REQUIRE(((QSignal)Q_USER_SIG <= sig)
                    && (sig < QF_maxSignal_)
                    && ((uint8_t)0 < p) && (p <= (uint8_t)QF_MAX_ACTIVE)
                    && (QF_active_[p] == me));

          QF_INT_LOCK_();
(4)       QF_subscrList_[sig].bits[i] |= Q_ROM_BYTE(QF_pwr2Lkup[p]);
          QF_INT_UNLOCK_();
      }
```

---

(1)   The function `QActive_subscribe()` subscribes a given active object 'me' to the event signal 'sig.'.

(2)   The index 'i' represents the byte index into the multibyte `QSubscrList` bitmask (see Listing 7.14). The array `QF_div8Lkup[]` is a lookup table that stores the

precomputed values of the following expression: `QF_div8Lkup[p] = (p - 1)/8`, where $0 < p < 64$. The `QF_div8Lkup[]` lookup table is defined in the file `<qp>\qpc\qf\source\qf_pwr2.c` and occupies 64 bytes of ROM.

NOTE

Obviously, you don't want to use precious RAM for storing constant lookup tables. However, some compilers for Harvard architecture MCUs (e.g., GCC for AVR) cannot generate code for accessing data allocated in the program space (ROM), even though the compiler can allocate constants in ROM. The workaround for such compilers is to explicitly add assembly code to access data allocated in the program space. The macro `Q_ROM_BYTE()` retrieves a byte from the given ROM address. This macro is transparent (i.e., copies its argument) for compilers that can correctly access data in ROM.

---

(3) This precondition asserts that the signal is in range and that the priority of the active object is in range as well. In addition, the assertion makes sure that the active object is known to the framework under the priority it claims (the active object becomes known to the framework through `QActive_start()`, which invokes `QF_add()`).

(4) The bit corresponding to the active object's priority is set in the subscriber list within a critical section. The array `QF_pwr2Lkup[]` is a lookup table that stores the precomputed values of the following expression: `QF_pwr2Lkup[p] = 1 << ((p - 1) % 8)`, where $0 < p < 64$. The `QF_pwr2Lkup[]` lookup table is defined in the file `<qp>\qpc\qf\source\qf_pwr2.c` and occupies 64 bytes of ROM.

I don't explicitly discuss the mirror function `QActive_unsubscribe()`, but it is virtually identical to `QActive_subscribe()` except that it clears the appropriate bit in the subscriber bitmask. Note that both `QActive_subscribe()` and `QActive_unsubscribe()` require an active object as the first parameter " me, " which means that only active objects are capable of subscribing or unsubscribing to events.

The QF real-time framework implements event publishing with the function `QF_publish()` shown in Listing 7.17. This function performs efficient "zero-copy" event multicasting. `QF_publish()` is designed to be callable from both the task level and the interrupt level.

Listing 7.17 `QF_publish()` **function
(file** `<qp>\qpc\qf\source\qa_pspub.c`**)**

```
(1)    void QF_publish(QEvent const *e) {
           QF_INT_LOCK_KEY_

           /* make sure that the published signal is within the configured range */
(2)        Q_REQUIRE(e->sig < QF_maxSignal_);

           QF_INT_LOCK_();
           if (e->dynamic_ != (uint8_t)0) {              /* is it a dynamic event? */
               /*lint -e1773                      Attempt to cast away const */
(3)            ++((QEvent *)e)->dynamic_;  /* increment reference counter, NOTE01 */
           }
           QF_INT_UNLOCK_();

(4)    #if (QF_MAX_ACTIVE <= 8)
           {
(5)            uint8_t tmp = QF_subscrList_[e->sig].bits[0];
(6)            while (tmp != (uint8_t)0) {
(7)                uint8_t p = Q_ROM_BYTE(QF_log2Lkup[tmp]);
(8)                tmp &= Q_ROM_BYTE(QF_invPwr2Lkup[p]);  /* clear subscriber bit */
(9)                Q_ASSERT(QF_active_[p] != (QActive *)0);  /* must be registered */

                                       /* internally asserts if the queue overflows */
(10)               QActive_postFIFO(QF_active_[p], e);
           }
           }
(11)   #else
           {
(12)           uint8_t i = Q_DIM(QF_subscrList_[0].bits);
               do {                     /* go through all bytes in the subscription list */
               uint8_t tmp;
               --i;
(13)           tmp = QF_subscrList_[e->sig].bits[i];
               while (tmp != (uint8_t)0) {
                   uint8_t p = Q_ROM_BYTE(QF_log2Lkup[tmp]);
                   tmp &= Q_ROM_BYTE(QF_invPwr2Lkup[p]);  /*clear subscriber bit */
(14)               p = (uint8_t)(p + (i << 3));              /* adjust the priority */
                   Q_ASSERT(QF_active_[p] != (QActive *)0);  /*must be registered*/

                                       /* internally asserts if the queue overflows */
                   QActive_postFIFO(QF_active_[p], e);
               }
           } while (i != (uint8_t)0);
           }
       #endif

(15)       QF_gc(e);                        /* run the garbage collector, see NOTE01 */
       }
```

(1)   The function QF_publish() publishes a given event 'e' to all subscribers.

(2)   The precondition checks that the published signal is in initialized range (see Listing 7.15).

(3)   The reference counter of a dynamic event is incremented in a critical section. This protects the event from being prematurely recycled before it reaches all subscribers.

The QF_publish() function must ensure that the event is not recycled by a subscriber before all the subscribers receive the event. For example, consider the following scenario: A low-priority active object dynamically allocates an event with Q_NEW() and publishes it by calling QF_publish() in its own thread of execution. In the course of multicasting the event, QF_publish() posts the event to a high-priority active object, which immediately preempts the current thread and starts processing the event. After the RTC step, the high-priority active object calls the garbage collector (see Listing 7.8(3)). If QF_publish() did not increment the event counter in step 3, the counter would be only 1 because the event has only been posted once. The high-priority active object would recycle the event. After resuming the low-priority thread, the QF_publish() might want to keep posting the event to some other subscribers, but the event would be already recycled.

(4)   The conditional compilation is used to distinguish the simpler and faster case of single-byte QSubscrList (see Listing 7.14).

(5)   The entire subscriber bitmask is placed in a temporary byte.

(6)   The while loop runs over all 1 bits set in the subscriber bitmask until the bitmask becomes empty.

(7)   The log-base-2 lookup quickly determines the most significant 1 bit in the bitmask, which corresponds to the highest-priority subscriber. The structure of the lookup table QF_log2Lkup[tmp], where 0 < tmp <= 255, is shown in Figure 7.6. The QF_log2Lkup[] lookup table is defined in the file <qp>\qpc\qf\source\qf_log2.c and occupies 256 bytes of ROM.

NOTE

To avoid priority inversions, the event is multicast starting from the highest-priority subscriber.

(8)   The highest-priority subscriber bit is cleared in the temporary bitmask.

(9)   The assertion makes sure that the active object with the given priority has been registered in the QF framework.

(10)   The event is posted to the subscriber, which always increments the reference counter of a dynamic event.

(11)   This conditional compilation branch is taken when the subscriber list contains more than 1 byte. The algorithm in this case requires an additional loop to run over all the bytes in the subscriber list.

(12)   The counter of the loop over the bytes is initialized. The loop starts with the highest-order bytes, which correspond to highest-priority subscribers.

(13)   The algorithm in this case is essentially the same as for the single-byte bitmask except that additional loop is added to run over all the bytes in the subscriber list.

(14)   The active object priority is adjusted by the byte number times 8, equivalent to (i << 3).

(15)   The garbage collection step balances the incrementing of the reference counter in step 3. The call to garbage collector also covers the case when the event is not subscribed by any active object, in which case the event needs to be recycled right away.

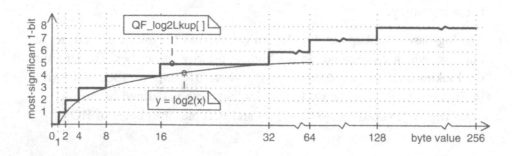

Figure 7.6: The binary logarithm lookup table QF_log2Lkup[] **maps byte value to the most significant 1-bit number (bits are numbered starting with 1 for the LSB).**

7.7 Time Management

QF manages time through time events, as described in Section 6.6.1 of Chapter 6. In the current QF version, time events cannot be dynamic and must be allocated statically. Also, a time event must be assigned a signal upon instantiation (in the constructor) and the signal cannot be changed later. This latter restriction prevents unexpected changes of the time event while it still might be held inside an event queue.

7.7.1 Time Event Structure and Interface

QF represents time events as instances of the QTimeEvt class (see Figure 7.3). QTimeEvt, as all events in QF, derives from the QEvent base structure. Typically, you will instantiate the QTimeEvt structure directly, but you might also further derive more specialized time events from it to add some more data members and/or specialized functions that operate on the derived time events. Listing 7.18 shows the QTimeEvt class, that is, the QTimeEvt structure declaration and the functions to manipulate it.

> **Listing 7.18** QTimeEvt **structure and interface**
> (**file** <qp>\qpc\include\qf.h)

```
        typedef struct QTimeEvtTag {
(1)         QEvent super;                              /* derives from QEvent */
(2)         struct QTimeEvtTag *prev; /* link to the previous time event in the list */
(3)         struct QTimeEvtTag *next;    /* link to the next time event in the list */
(4)         QActive *act;            /* the active object that receives the time event */
(5)         QTimeEvtCtr ctr;              /* the internal down-counter of the time event */
(6)         QTimeEvtCtr interval;         /* the interval for the periodic time event */
        } QTimeEvt;

(7)    void QTimeEvt_ctor(QTimeEvt *me, QSignal sig);

(8)    #define QTimeEvt_postIn(me_, act_, nTicks_) do { \
           (me_)->interval = (QTimeEvtCtr)0; \
           QTimeEvt_arm_((me_), (act_), (nTicks_)); \
       } while (0)

(9)    #define QTimeEvt_postEvery(me_, act_, nTicks_) do { \
           (me_)->interval = (nTicks_); \
           QTimeEvt_arm_((me_), (act_), (nTicks_)); \
       } while (0)
```

Continued onto next page

```
(10)   uint8_t QTimeEvt_disarm(QTimeEvt *me);

(11)   uint8_t QTimeEvt_rearm(QTimeEvt *me, QTimeEvtCtr nTicks);

       /* private helper function */
(12)   void QTimeEvt_arm_(QTimeEvt *me, QActive *act, QTimeEvtCtr nTicks);
```

(1)   The QTimeEvt structure derives from QEvent.

(2,3) The two pointers 'prev' and 'next' are used as links to chain the time events into a bidirectional list (see Figure 7.7).

(4)   The active object pointer 'act' stores the recipient of the time event.

(5)   The member 'ctr' is the internal down-counter decremented in every QF_tick() invocation (see the next section). The time event is posted when the down-counter reaches zero.

(6)   The member 'interval' is used for the periodic time event (it is set to zero for the one-shot time event). The value of the interval is reloaded to the 'ctr' down-counter when the time event expires, so the time event keeps timing out periodically.

(7)   Every time event must be initialized with the constructor QTimeEvt_ctor(). You should call the constructor exactly once for every time event object before arming the time event. The most important action performed in this function is assigning a signal to the time event. You can reuse the time event any number of times, but you should not change the signal. This is because a pointer to the time event might still be held in an event queue and changing the signal could lead to subtle and hard-to-find errors.

(8)   The macro QTimeEvt_postIn() arms a time event 'me_' to fire once in 'nTicks_' clock ticks (a one-shot time event). The time event gets directly posted (using the FIFO policy) into the event queue of the active object 'act_.' After posting, a one-shot time event gets automatically disarmed and can be reused for a one-shot or periodic timeout requests.

(9)   The macro QTimeEvt_postEvery() arms a time event 'me_' to fire periodically every 'nTicks_' clock ticks (periodic time event). The time event gets directly posted (using the FIFO policy) into the event queue of the active object 'act_'. After posting, the periodic time event gets automatically rearmed to fire again in the specified 'nTicks_' clock ticks.

(10) The function `QTimeEvt_disarm()` explicitly disarms any time event (one-shot or periodic). The time event can be reused immediately after the call to `QTimeEvt_disarm()`. The function returns the status of the disarming operation: 1 if the time event has been actually disarmed and 0 if the time event has already been disarmed.

(11) The function `QTimeEvt_rearm()` reloads the down-counter 'ctr' with the specified number of clock ticks. The function returns the status of the rearming operation: 1 if the time event has been actually armed and 0 if the time event has been disarmed. In the latter case, the `QTimeEvt_rearm()` function arms the time event.

(12) The helper function `QTimeEvt_arm_()` inserts the time event into the linked list of armed timers. This function is used in the `QTimeEvt_postIn()` and `QTimeEvt_postEvery()` macros.

NOTE

An attempt to arm an already armed time event (one-shot or periodic) raises an assertion. If you're not sure that the time event is disarmed, call the `QTimeEvt_disarm()` function before reusing the time event.

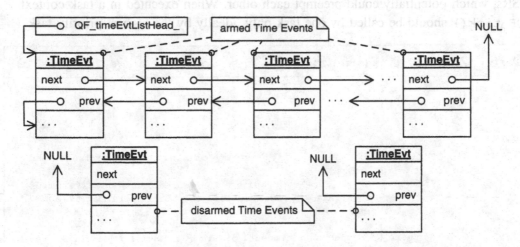

Figure 7.7: Armed QTimeEvt objects linked in a bidirectional linked list and disarmed time events outside the list.

Figure 7.7 shows the internal representation of armed and disarmed time events. QF chains all armed time events in a bidirectional linked list. The list is scanned from the head at every system clock tick. The list is not sorted in any way. Newly armed time events are always inserted at the head. When a time event gets disarmed, either automatically when a one-shot timer expires or explicitly when the application calls QTimeEvt_disarm(), the time event is simply removed from the list. Removing an object from a bidirectional list is a quick, deterministic operation. In particular, the list does not need to be rescanned from the head. Disarmed time events remain outside the list and don't consume any CPU cycles.

7.7.2 The System Clock Tick and the QF_tick() Function

To manage time events, QF requires that you invoke the QF_tick() function from a periodic time source called the *system clock tick* (see Chapter 6, "System Clock Tick"). The system clock tick typically runs at a rate between 10Hz and 100Hz.

Listing 7.19 shows the implementation of QF_tick(). This function is designed to be called from both the interrupt context and the task-level context, in case the underlying OS/RTOS does not allow accessing interrupts or you want to keep the ISRs very short. QF_tick() must always run to completion and never preempt itself. In particular, if QF_tick() runs in an ISR context, the ISR must not be allowed to preempt itself. In addition, QF_tick() should not be called from two different ISRs, which potentially could preempt each other. When executed in a task context, QF_tick() should be called by one task only, ideally by the highest-priority task.

Listing 7.19 QF_tick() **function (file** <qp>\qpc\qf\source\qf_tick.c**)**

```
      void QF_tick(void) {                                    /* see NOTE01 */
          QTimeEvt *t;
          QF_INT_LOCK_KEY_

(1)       QF_INT_LOCK_();
(2)       t = QF_timeEvtListHead_;           /* start scanning the list from the head */
(3)       while (t != (QTimeEvt *)0) {
(4)           if (--t->ctr == (QTimeEvtCtr)0) {     /* is time evt about to expire? */
(5)               if (t->interval != (QTimeEvtCtr)0) {  /* is it periodic timeout? */
(6)                   t->ctr = t->interval;                  /* rearm the time event */
              }
(7)           else {   /* one-shot timeout, disarm by removing it from the list */
(8)               if (t == QF_timeEvtListHead_) {
```

```
(9)                          QF_timeEvtListHead_ = t->next;
                    }
(10)              else {
(11)                  if (t->next != (QTimeEvt *)0) {     /* not the last event? */
(12)                      t->next->prev = t->prev;
                      }
(13)                  t->prev->next = t->next;
                  }
(14)              t->prev = (QTimeEvt *)0;              /* mark the event disarmed */
              }
(15)          QF_INT_UNLOCK_();/* unlock interrupts before calling QF service */
                      /* postFIFO() asserts internally that the event was accepted */
(16)          QActive_postFIFO(t->act, (QEvent *)t);
          }
(17)      else {
              static uint8_t volatile dummy;
(18)          QF_INT_UNLOCK_();
(19)          dummy = (uint8_t)0;    /* execute a few instructions, see NOTE02 */
          }

(20)      QF_INT_LOCK_();          /* lock interrupts again to advance the link */
(21)      t = t->next;
      }
(22)  QF_INT_UNLOCK_();
  }
```

(1) Interrupts are locked before accessing the linked list of time events.

(2) The internal QF variable `QF_timeEvtListHead_` holds the head of the linked list.

(3) The loop continues until the end of the linked list is reached (see Figure 7.7).

(4) The down-counter of each time event is decremented. When the counter reaches zero, the time event expires.

(5) The 'interval' member is nonzero only for a periodic time event.

(6) The down-counter of a periodic time event is simply reset to the interval value. The time event remains armed in the list.

(7) Otherwise the time event is a one-shot and must be disarmed by removing it from the list.

(8-13) These lines of code implement the standard algorithm of removing a link from a bidirectional list.

(14) A time event is internally marked as disarmed by writing NULL to the 'prev' link.

(15) Interrupts can be unlocked after the bookkeeping of the linked list is done.

(16) The time event posts itself to the event queue of the active object.

(17) The else branch is taken when the time event is not expiring on this tick.

(18) Interrupts can be unlocked.

(19) On many CPUs, the interrupt unlocking takes effect only on the next machine instruction, which happens here to be an interrupt lock instruction (line (20)). The assignment of the volatile 'dummy' variable requires a few machine instructions, which the compiler cannot optimize away. This ensures that the interrupts actually get unlocked so that the interrupt latency stays low.

NOTE

The critical section lasts for only one time event, not for the whole list.

---

(20) Interrupts are locked again for another pass through the loop.

(21) The link is advanced to the next timer in the list.

(22) Interrupts are unlocked before the function returns.

7.7.3 Arming and Disarming a Time Event

Listing 7.20 shows the helper function QTimeEvt_arm_() for arming a time event. This function is used inside the macros QTimeEvt_postIn() and QTimeEvt_postEvery() for arming a one-shot or periodic time event, respectively.

Listing 7.20 QTimeEvt_arm_() **(file** `<qp>\qpc\qf\source\qte_arm.c`**)**

```
void QTimeEvt_arm_(QTimeEvt *me, QActive *act, QTimeEvtCtr nTicks) {
    QF_INT_LOCK_KEY_
    Q_REQUIRE((nTicks > (QTimeEvtCtr)0)      /* cannot arm a timer with 0 ticks */
            && (((QEvent *)me)->sig >= (QSignal)Q_USER_SIG)/*valid signal */
```

```
(1)                  && (me->prev == (QTimeEvt *)0)     /* time evt must NOT be used */
                     && (act != (QActive *)0));   /* active object must be provided */
        me->ctr = nTicks;
(2)     me->prev = me;                                    /* mark the timer in use */
        me->act = act;

        QF_INT_LOCK_();
(3)     me->next = QF_timeEvtListHead_;
(4)     if (QF_timeEvtListHead_ != (QTimeEvt *)0) {
(5)         QF_timeEvtListHead_->prev = me;
        }
(6)     QF_timeEvtListHead_ = me;
        QF_INT_UNLOCK_();
    }
```

(1) The preconditions include checking that the time event is not already in use. A used time event has always the 'prev' pointer set to non-NULL value.

(2) The 'prev' pointer is initialized to point to self, to mark the time event in use (see also Figure 7.7).

(3) Interrupts are locked to insert the time event into the linked list. Note that until that point the time event is not armed, so it cannot unexpectedly change due to asynchronous tick processing.

(3-6) These lines of code implement the standard algorithm of inserting a link into a bidirectional list at the head position.

Listing 7.21 shows the function QTimeEvt_disarm() for explicitly disarming a time event.

Listing 7.21 QTimeEvt_disarm()
(file <qp>\qpc\qf\source\qte_darm.c**)**

```
        uint8_t QTimeEvt_disarm(QTimeEvt *me) {
            uint8_t wasArmed;
            QF_INT_LOCK_KEY_
(1)         QF_INT_LOCK_();
(2)         if (me->prev != (QTimeEvt *)0) {      /* is the time event actually armed? */
                wasArmed = (uint8_t)1;
(3)             if (me == QF_timeEvtListHead_) {
(4)                 QF_timeEvtListHead_ = me->next;
                }
```

Continued onto next page

```
(5)              else {
(6)                  if (me->next != (QTimeEvt *)0) {        /* not the last in the list? */
(7)                      me->next->prev = me->prev;
                     }
(8)                  me->prev->next = me->next;
                 }
(9)              me->prev = (QTimeEvt *)0;                 /* mark the time event as disarmed */
             }
             else {                                        /* the time event was not armed */
                 wasArmed = (uint8_t)0;
             }
             QF_INT_UNLOCK_();
(10)         return wasArmed;
         }
```

(1)   Critical section is established right away.

(2)   The time event is still armed if the 'prev' pointer is not NULL.

(3-8) These lines of code implement the standard algorithm of removing a link from a bidirectional list (compare also Listing 7.19(8-13)).

(9)   A time event is internally marked as disarmed by writing NULL to the 'prev' link.

(10)   The function returns the status: 1 if the time event was still armed at the time of the call and 0 if the time event was disarmed before the function QTimeEvt_disarm() was called. In other words, the return value of 1 ensures the caller that the time event has not been posted and never will be, because disarming takes effect immediately. Conversely, the return value of 0 informs the caller that the time event has been posted to the event queue of the recipient active object and was automatically disarmed.

The status information returned from QTimeEvt_disarm() could be useful in the state machine design. For example, consider a state machine fragment shown in Figure 7.8. The entry action to "stateA" arms a one-shot time event me->timer1. Upon expiration, the time event generates signal TIMER1, which causes some internal or regular transition. However, another event, say BUTTON_PRESS, triggers a transition to "stateB." The events BUTTON_PRESS and TIMER1 are inherently set up to race each other and so it is possible that they arrive very close in time. In particular, when the BUTTON_PRESS event arrives, the TIMER1 event could potentially follow very shortly thereafter and might get queued as well. If that happens, the state machine receives *both* events. This might be a problem if, for example, the next state tries to reuse the time event for a different purpose.

Figure 7.8 shows the solution. The exit action from "stateA" stores the return value of QTimeEvt_disarm() in the extended state variable me->g1. Subsequently, the variable is used as a guard condition on transition TIMER1 in "stateB." The guard allows the transition only if the me->g1 flag is set. However, when the flag is zero, it means that the TIMER1 event was already posted. In this case the TIMER1 event sets only the flag but otherwise is ignored. Only in the next TIMER1 instance is the true timeout event requested in "stateB."

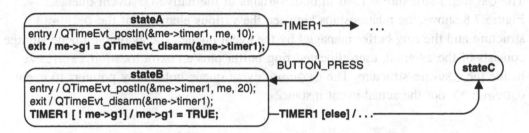

Figure 7.8: Reusing a one-shot time event.

7.8   Native QF Event Queue

Many RTOSs natively support message queues, which provide a superset of functionality needed for event queues of active objects. QF is designed up front for easy integration of such external message queues. However, in case no such support exists or the available implementation is inefficient or inadequate, QF provides a robust and efficient native event queue that you can easily adapt to virtually any underlying operating system or kernel.

The native QF event queues come in two flavors, which share the same data structure (QEQueue) and initialization but differ significantly in behavior. The first variant is the event queue specifically designed and optimized for active objects (see Section 7.8.3). The implementation omits several commonly supported features of traditional message queues, such as variable-size messages (native QF event queues store only pointers to events), blocking on a full queue (QF event queue cannot block on insertion), and timed blocking on empty queues (QF event queues block indefinitely), to name just a few. In exchange, the native QF event queue implementation is small and probably faster than any full-blown message queue of an RTOS.

The other, simpler variant of the native QF event queue is a generic "raw" thread-safe queue not capable of blocking but useful for thread-safe event delivery from active objects to other parts of the system that lie outside the framework, such as ISRs or device drivers. I explain the "raw" queue in Section 7.8.4.

7.8.1   The QEQueue Structure

The QEQueue structure is used in both variants of the native QF event queues. Figure 7.8 shows the relationships between the various elements of the QEQueue structure and the ring buffer managed by the event queue. The available queue storage consists of the external, user-allocated ring buffer plus an extra location frontEvt inside the QEvent structure. The QEQueue event queue holds only pointers to events (QEvent *), not the actual event instances.

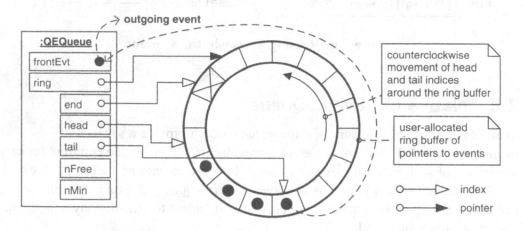

Figure 7.9: The relationship between the elements of the QEQueue **structure and the ring buffer.**

As indicated by the dashed lines in Figure 7.9, all outgoing events must pass through the frontEvt data member. This extra location outside the ring buffer optimizes queue operation by allowing it to frequently bypass, the buffering because very often queues alternate between empty and nonempty states with just one event present in the queue at a time. In addition, the `frontEvt` pointer serves as a queue status indicator, whereas the `NULL` value of `frontEvt` indicates that the queue is empty. The indices `head`, `tail`, and `end` are relative to the `ring` pointer. Events are always extracted from the buffer at the `tail` index. New events are typically inserted at the `head` index. Inserting events at the `head` and extracting from the `tail` corresponds to FIFO queuing (the `postFIFO()` operation). `QEQueue` also allows inserting new events at the `tail`, which corresponds to LIFO queuing (the `postLIFO()` operation). Either way, the `tail` always decrements when the event is extracted, as does the `head` index when an event is inserted. The index 0 limits the range of the `head` and `tail` indices that must "wrap around" to `end` once they reach 0. The effect is a counterclockwise movement of the indices around the ring buffer, as indicated by the arrow in Figure 7.9. Other data members of the `QEQueue` structure include the current number of free events in the buffer (`nFree`) and the minimum number of free events ever present in the buffer (`nMin`). The `nMin` member tracks the worst-case queue utilization (the low-watermark of free events in the queue), which provides a valuable data point for fine-tuning the ring buffer size.

Listing 7.22 shows the declaration of the `QEQueue` structure. The `QEQueueCtr` data type determines the dynamic range of the queue indices and counters. It is `typedef`'ed to `uint8_t`, `uint16_t`, or `uint32_t`, depending on the macro `QF_EQUEUE_CTR_SIZE`. You can define the macro `QF_EQUEUE_CTR_SIZE` in the QF port file `qf_port.h` in the correct port directory. If the macro is not defined, the default size of 1 byte is assumed, which results in `QEQueueCtr` data type being `typdef`'ed to `uint8_t` (up to 255 events in the ring buffer).

Listing 7.22 `QEQueue` **structure (file** `<qp>\qpc\include\qequeue.h`**)**

```
#ifndef QF_EQUEUE_CTR_SIZE
    #define QF_EQUEUE_CTR_SIZE 1
#endif
#if (QF_EQUEUE_CTR_SIZE == 1)
    typedef uint8_t QEQueueCtr;
#elif (QF_EQUEUE_CTR_SIZE == 2)
    typedef uint16_t QEQueueCtr;
#elif (QF_EQUEUE_CTR_SIZE == 4)
    typedef uint32_t QEQueueCtr;
#else
    #error "QF_EQUEUE_CTR_SIZE defined incorrectly, expected 1, 2, or 4"
#endif
```

Continued onto next page

```
typedef struct QEQueueTag {
    QEvent const *frontEvt;   /* pointer to event at the front of the queue */
    QEvent const **ring;        /* pointer to the start of the ring buffer */

    QEQueueCtr end;  /* offset of the end of the ring buffer from the start */
    QEQueueCtr head;        /* offset to where next event will be inserted */
    QEQueueCtr tail;      /* offset of where next event will be extracted */
    QEQueueCtr nFree;          /* number of free events in the ring buffer */
    QEQueueCtr nMin;   /* minimum number of free events ever in the buffer */
} QEQueue;
```

7.8.2 Initialization of QEQueue

Listing 7.23 shows the event queue initialization function QEQueue_init(). The function takes the preallocated contiguous storage for the ring buffer (an array of QEvent* pointers, qSto[]) and the length of the buffer qLen, which is the number of preallocated event pointers. The function sets the QEQueue data members to emulate an empty event queue. The body of the function is not protected with a critical section because the application should never access a queue before it is initialized.

Listing 7.23 QEQueue_init() (**file** <qp>\qpc\qf\source\qeq_init.c)

```
void QEQueue_init(QEQueue *me, QEvent const *qSto[], QEQueueCtr qLen) {
    me->frontEvt = (QEvent *)0;                      /* no events in the queue */
    me->ring     = &qSto[0];
    me->end      = qLen;
    me->head     = (QEQueueCtr)0;
    me->tail     = (QEQueueCtr)0;
    me->nFree    = qLen;                                /* all events are free */
    me->nMin     = qLen;                               /* the minimum so far */
}
```

Note that you can initialize an event queue with parameters qSto == NULL and qLen == 0. Such an event queue will still be able to hold one event because the frontEvt location also counts toward the queue capacity.

7.8.3 The Native QF Active Object Queue

The QEQueue structure is not quite complete to serve as the event queue of an active object because it does not provide any data member for implementing blocking of the active object thread when the queue is empty. Such a mechanism is always platform-specific and typically it is an operating-system primitive, such as a semaphore, a condition variable in POSIX, or a Win32 object that can be used with the WaitForSingleObject() and SetEvent() Win32 APIs.

The OS-specific blocking primitive is intentionally not included in the QEQueue structure to enable using this structure for both the active object event queue and the generic "raw" thread-safe queue discussed in the next section. Instead, the OS-specific thread-blocking primitive is included directly in the level QActive structure as the data member osObject of type QF_OS_OBJECT_TYPE (Listing 7.7(9)). The macro QF_OS_OBJECT_TYPE is obviously part of the platform abstraction layer and is defined differently for different QF ports. The osObject member is initialized in the platform-specific QActive_start() function (see Listing 7.9). You can think of the native QF active object queue as the aggregate of the QEQueue structure and the QActive.osObject data member.

The interface of the active object event queue consists of three functions: QActive_postFIFO(), QActive_postLIFO(), and QActive_get_(). The implementation of these functions is located in the files qa_fifo.c, qa_lifo.c, and qa_get_.c, respectively. These files should be included in the QF port only when the port uses the native event queue. Otherwise, the functions QActive_postFIFO(), QActive_postLIFO(), and QActive_get_() should be implemented differently, perhaps with the RTOS-specific message queue.

Listing 7.24 shows how the QActive_get_() function extracts events from the queue. This function is called only from the thread routine of the active object that owns this queue (see Listing 7.8(1)). You should never call QActive_get_() from the application-level code (hence the trailing underscore in the function name).

Listing 7.24 Extracting events from the event queue with QActive_get_() **(file** <qp>\qpc\qf\source\qa_get_.c**)**

```
(1)   QEvent const *QActive_get_(QActive *me) {
          QEvent const *e;
          QF_INT_LOCK_KEY_
          QF_INT_LOCK_();

(2)       QACTIVE_EQUEUE_WAIT_(me);          /* wait for event queue to get an event */

(3)       e = me->eQueue.frontEvt;
```

Continued onto next page

```
(4)        if (me->eQueue.nFree != me->eQueue.end) {  /* any events in the buffer? */
                                                       /* remove event from the tail */
(5)            me->eQueue.frontEvt = me->eQueue.ring[me->eQueue.tail];
(6)            if (me->eQueue.tail == (QEQueueCtr)0) {    /* need to wrap the tail? */
(7)                me->eQueue.tail = me->eQueue.end;                  /* wrap around */
               }
(8)            --me->eQueue.tail;

(9)            ++me->eQueue.nFree;         /* one more free event in the ring buffer */
           }
(10)       else {
(11)           me->eQueue.frontEvt = (QEvent *)0;           /* queue becomes empty */
(12)           QACTIVE_EQUEUE_ONEMPTY_(me);
           }
           QF_INT_UNLOCK_();
(13)       return e;
       }
```

(1) The function QActive_get_() returns a read-only (const) pointer to an event that has been previously posted to the active object 'me.' The function always returns a valid event pointer.

(2) In some QF ports, the function must block until an event arrives. Blocking is always a platform-specific operation and the function handles it through the platform-specific macro QACTIVE_EQUEUE_WAIT_(). Note that this macro is invoked from the critical section. The macro might unlock interrupts momentarily, but it must restore critical section before it returns. I describe the implementation of the QACTIVE_EQUEUE_WAIT_() macro for POSIX threads in Chapter 8.

(3) At this point the queue cannot be empty anymore—it either was not empty to begin with or it just received an event after blocking. The event at the front of the queue is copied for delivery to the caller from the front event.

(4) If not all events in the buffer are free, the buffer must contain some events.

(5) The event pointer is copied from the tail index in the buffer to the front event.

(6) The tail index is checked for a wraparound.

(7) If wraparound is required, the tail index is moved to the end of the buffer. This makes the buffer circular.

(8) The `tail` index is always decremented, including just after the wraparound. I've chosen to decrement the tail (and also the head) index because it leads to a more efficient implementation than incrementing the indices. The wraparound occurs in this case at zero rather than at the end. Comparing a variable to a constant zero is more efficient than any other comparison.

(9) The `nFree` counter is incremented to account for freeing one event in the buffer.

(10) Otherwise the queue is becoming empty.

(11) The front event is set to `NULL`.

(12,13) Additionally, a platform-specific macro `QACTIVE_EQUEUE_ONEMPTY_()` is called. The job of this macro is to inform the underlying kernel that the queue is becoming empty, which is required in some QF ports. I show this macro implementation for the QF port to the cooperative "vanilla" kernel discussed in Chapter 8 as well as in the QF port to the preemptive run-to-completion QK kernel that I cover in Chapter 10.

Listing 7.25 shows the implementation of the `QActive_postFIFO()` queue operation. This function is used for posting an event directly to the recipient active object. Direct event posting can be performed from any part of the application, including interrupts.

Listing 7.25 Inserting events into the event queue with `QActive_postFIFO()` **(file** `<qp>\qpc\qf\source\qa_fifo.c`**)**

```
    void QActive_postFIFO(QActive *me, QEvent const *e) {
        QF_INT_LOCK_KEY_
(1)     QF_INT_LOCK_();

        if (e->dynamic_ != (uint8_t)0) {          /* is it a dynamic event? */
(2)         ++((QEvent *)e)->dynamic_;        /* increment the reference counter */
        }

(3)     if (me->eQueue.frontEvt == (QEvent *)0) {          /* empty queue? */
(4)         me->eQueue.frontEvt = e;                  /* deliver event directly */
(5)         QACTIVE_EQUEUE_SIGNAL_(me);              /* signal the event queue */
        }
(6)     else {          /* queue is not empty, insert event into the ring-buffer */
                /* the queue must be able to accept the event (cannot overflow) */
(7)         Q_ASSERT(me->eQueue.nFree != (QEQueueCtr)0);
                              /* insert event into the ring buffer (FIFO) */
```

Continued onto next page

```
(8)          me->eQueue.ring[me->eQueue.head] = e;
(9)          if (me->eQueue.head == (QEQueueCtr)0) {    /* need to wrap the head? */
(10)             me->eQueue.head = me->eQueue.end;                    /* wrap around */
             }
(11)         --me->eQueue.head;

(12)         --me->eQueue.nFree;                          /* update number of free events */
(13)         if (me->eQueue.nMin > me->eQueue.nFree) {
(14)             me->eQueue.nMin = me->eQueue.nFree;           /* update min so far */
             }
         }
         QF_INT_UNLOCK_();
     }
```

(1)   The whole function body runs in a critical section.

(2)   The reference count of a dynamic event is incremented to account for another outstanding reference to the event.

NOTE

Incrementing the reference count is essential and must be performed in every `QActive_postFIFO()` implementation, including implementations not based on the native QF event queue but, for example, on a message queue of an RTOS.

---

(3)   If the front event is `NULL`, the queue is empty.

(4)   The event pointer is copied directly to the front event, bypassing the whole buffering mechanism.

(5)   Additionally, a platform-specific macro `QACTIVE_EQUEUE_SIGNAL_()` is called. The job of this macro is to signal the thread waiting on the event queue. Note that this macro is invoked from the critical section. I describe the implementation of the `QACTIVE_EQUEUE_SIGNAL_()` macro for POSIX-threads in Chapter 8.

(6)   Otherwise, the queue is not empty, so the event must be inserted into the ring buffer.

(7)   The assertion makes sure that the queue can accept this event.

> **NOTE**
>
> The QF framework treats the inability to post an event as an error. This assertion is part of the event delivery guarantee policy. It's the application designer's responsibility to size the event queues adequately for the job at hand.

(8)   The event is inserted at the `head` index. This corresponds to the FIFO queuing policy.

(9)   The `head` index is checked for a wraparound.

(10) If wraparound is required, the `head` index is moved to the end of the buffer. This makes the buffer circular.

(11) The `head` index is always decremented, including just after the wraparound. I've chosen to decrement the `head` (and also the `tail`) index because it leads to a more efficient implementation than incrementing the indices. The wraparound occurs in this case at zero rather than at the end. Comparing a variable to a constant zero is more efficient than any other comparison.

(12) The `nFree` counter is decremented to account for using one event in the buffer.

(13,14) Finally, the function updates the low-watermark `nMin` of the queue. This step is not necessary for the correct operation of the queue, but the low-watermark provides valuable empirical data for proper sizing of the event queue.

The QF framework provides also the `QActive_postLIFO()` queue operation. I don't discuss the code (located in the file `<qp>\qpc\qf\source\qa_lifo.c`) because it is very similar to `QActive_postFIFO()` except that the event is inserted at the `tail` index.

7.8.4   The "Raw" Thread-Safe Queue

The `QEQueue` structure can be used directly as the native QF "raw" thread-safe queue. The basic operations of the "raw" thread-safe queue are `QEQueue_postFIFO()`, `QEQueue_postLIFO()`, and `QEQueue_get()`. None of these functions can block. This type of queue is employed for deferring events (see Section 7.5.4) and also can be very useful for passing events from active objects to ISRs, as shown in Listing 7.26.

Listing 7.26 Using the "raw" thread-safe queue to send events to an ISR

```
     /* Application header file -------------------------------------------*/
     #include "qequeue.h"

(1)  extern QEQueue APP_isrQueue;                          /* global "raw" queue */

(2)  typedef struct IsrEvtTag {    /* event with parameters to be passed to the ISR */
         QEvent super;
         . . .
     } IsrEvt;

     /* ISR module -----------------------------------------------------*/
(3)  QEQueue APP_isrQueue;                       /* definition of the "raw" queue */

     void interrupt myISR() {
         QEvent const *e;
         . . .
(4)      e = QEQueue_get(&APP_isrQueue);        /* get an event from the "raw" queue */
(5)      if (e != (QEvent *)0) {                           /* event available? */
(6)          Process the event e (could be dispatching to a state machine)
             . . .
(7)          QF_gc(e);                              /* explicitly recycle the event */
         }
         . . .
     }

     /* Active object module ----------------------------------------------*/
     QState MyAO_stateB(MyAO *me, QEvent const *e) {
         switch (e->sig) {
         . . .
             case SOMETHING_INTERESTING_SIG: {
                 IsrEvt *pe = Q_NEW(IsrEvt, ISR_SIG);
                 pe->... = ...                        /* set the event attributes */
(8)              QEQueue_postFIFO(&APP_isrQueue, (QEvent *)pe);
                 return (QSTATE)0;
             }
         . . .
         }
         return (QState)&MyAO_stateA;
     }

     /* main module -------------------------------------------------------*/
     static QEvent *l_isrQueueSto[10];    /* allocate a buffer for the "raw" queue */

     main() {
         . . .
                                               /* initialize the "raw" queue */
(9)      QEQueue_init(&APP_isrQueue, l_isrQueueSto, Q_DIM(l_isrQueueSto));
         . . .
     }
```

(1) In the application header file, you declare the external "raw" event queue so that various parts of the code can access the queue.

(2) In the same header file, you'll typically also declare all event types that the raw queue can accept.

(3) In the ISR module, you define the "raw" queue object.

(4) Inside an ISR, you call QEQueue_get() to get an event.

NOTE

The function QEQueue_get() uses internally a critical section of code. If you are using the simple unconditional interrupt-locking policy (see Section 7.3.2), you must be careful not to call QEQueue_get() with interrupts locked, as might be the case inside an ISR.

---

(5) If the event is available, the returned pointer is not NULL.

(6) You process the event. Please note that you have read-only access to the event.

(7) After the processing, you must not forget to call the QF garbage collector, because now QF is no longer in charge of event processing and you are solely responsible for not leaking the event.

(8) In an active object state machine you call QEQueue_get() to post an event (dynamic or static).

(9) You must not forget to initialize the "raw" queue object, which is typically done upon system startup.

The actual implementation of the QEQueue functions QEQueue_postFIFO(), QEQueue_postLIFO(), and QEQueue_get() is very straightforward since no platform-specific macros are necessary. All these functions are reentrant because they preserve the integrity of the queue by using critical sections of code.

7.9 Native QF Memory Pool

In Section 7.5.2, I introduced the concept of an event pool—a fixed block–size heap specialized to hold event instances. Some RTOSs natively support such fixed block–size heaps (often called *memory partitions* or *memory pools*). However, many platforms don't. This section explains the native QF implementation of a memory pool based on the QMPool structure.

The native QF memory pool is generic and can be used in your application for storing any objects, not just events. All native QF memory pool functions are reentrant and deterministic. They can be used in any parts of the code, including the ISRs. Figure 7.10 explains the relationships between the various elements of the QMPool structure and the memory buffer managed by the memory pool.

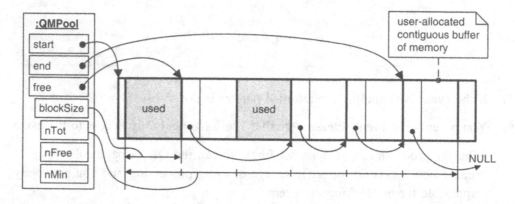

Figure 7.10: The relationship between QMPool structure elements and the memory buffer.

The native QF memory pool requires the actual pool storage to be allocated externally and provided to the pool upon initialization. Internally, the memory pool tracks memory by dividing it into blocks of the requested size and linking all unused memory blocks in a singly-linked list (the free list originating at the QMPool.free pointer). This technique is standard for organizing stack-like data structures, where the structure is accessed from one end only (LIFO). QMPool also uses a handy trick to link free blocks together in the free list without consuming extra storage for the pointers. Because the blocks in the free list are not used for anything else, QMPool can reuse the blocks as linked list pointers. This use implies that the block size must be big enough to hold a pointer [Lafreniere 98, Labrosse 02]. Listing 7.27 shows the declaration of the QMPool structure.

Listing 7.27 QMPool **structure (file** `<qp>\qpc\include\qmpool.h`**)**

```
       typedef struct QMPoolTag {
(1)        void *free;            /* the head of the linked list of free blocks */
(2)        void *start;                /* the start of the original pool buffer */
(3)        void *end;                           /* the last block in this pool */
(4)        QMPoolSize blockSize;         /* maximum block size (in bytes) */
(5)        QMPoolCtr nTot;                      /* total number of blocks */
(6)        QMPoolCtr nFree;               /* number of free blocks remaining */
(7)        QMPoolCtr nMin;  /* minimum number of free blocks ever in this pool */
       } QMPool;
```

---

(1)   The only data member strictly required for allocating and freeing blocks in the pool is the head of the free list 'free.' The other data members are for making the memory pool operations more robust.

(2,3)   The start and end pointers are used as delimiters of the valid range of memory blocks managed by this pool. I have specifically added them to enable writing an assertion to ensure that every memory block returned to the pool is in the range of memory managed by the pool.

(4)   The member blockSize holds the block size of this pool in bytes.

(5)   The member nTot holds the total number of blocks in the pool. This member allows me to assert the invariant that the number of free blocks in the pool at any given time cannot exceed nTot.

(6)   The member nFree holds the current number of free blocks in the pool.

(7)   The member nMin holds the lowest number of free blocks ever present in the pool.

The QMPoolSize data type is typedef'ed as uint8_t, uint16_t, or uint32_t, configurable by the macro QF_MPOOL_SIZ_SIZE. The dynamic range of the QMPoolSize data type determines the maximum size of blocks that can be managed by the native QF event pool. Similarly, the QMPoolCtr data type is typedef'ed

as `uint8_t`, `uint16_t`, or `uint32_t`, depending on the macro
`QF_MPOOL_CTR_SIZE`. The dynamic range of the `QMPoolCtr` data type determines
the maximum number of blocks that can be stored in the pool. The macros
`QF_MPOOL_SIZ_SIZE` and `QF_MPOOL_CTR_SIZE` should be defined in the QF port
file `qf_port.h` in the correct port directory. If the macros are not defined, the
default size of 2 is assumed for both of them, which results in `QMPoolSize` and
`QMPoolCtr` data types `typedef`'ed to `uint16_t`.

7.9.1   Initialization of the Native QF Memory Pool

You must initialize a memory pool before you can use it by calling the function
`QMPool_init()`, to which you provide the pool storage, the size of the
storage, and the block size managed by the pool. A general challenge in writing
this function is portability, because storage allocation is intrinsically
machine-dependent [Kernighan 88]. Perhaps the trickiest aspect here is the proper
and optimal alignment of the blocks within the contiguous memory buffer.
In particular, the alignment of blocks must be such that every new block can be
treated as a pointer to the next block. The code in listing 7.28 illustrates
how to control the machine dependencies, at the cost of extensive but careful
typecasting.

> **NOTE**
>
> Many CPU architectures put special requirements on the proper alignment of pointers.
> For example, the ARM processor requires a pointer to be allocated at an address divisible
> by 4. Other CPUs, such as the Pentium, can accept pointers allocated at odd addresses but
> perform substantially better when pointers are aligned at addresses divisible by 4.

To achieve better portability and optimal alignment of blocks, the QF memory
pool implementation uses internally a helper structure `QFreeBlock`, which represents a
node in the linked-list of free blocks. `QFreeBlock` is declared as follows in the file
`<qp>\qpc\qf\source\qf_pkg.h`:

```
typedef struct QFreeBlockTag {
    struct QFreeBlockTag *next;              /* link to the next free block */
} QFreeBlock;
```

Listing 7.28 Initialization of the QF memory pool with `QMPool_init()`
(file `<qp>\qpc\qf\source\qmp_init.c`**)**

```
        void QMPool_init(QMPool *me, void *poolSto,
                           uint32_t poolSize, QMPoolSize blockSize)
        {
            QFreeBlock *fb;
            uint32_t corr;
            uint32_t nblocks;

            /* The memory block must be valid
             * and the poolSize must fit at least one free block
             * and the blockSize must not be too close to the top of the dynamic range
             */
(1)         Q_REQUIRE((poolSto != (void *)0)
(2)                   && (poolSize >= (uint32_t)sizeof(QFreeBlock))
(3)                   && ((QMPoolSize)(blockSize + (QMPoolSize)sizeof(QFreeBlock))
                            > blockSize));

            /*lint -e923                        ignore MISRA Rule 45 in this expression */
(4)         corr = ((uint32_t)poolSto & ((uint32_t)sizeof(QFreeBlock) - (uint32_t)1));
(5)         if (corr != (uint32_t)0) {                         /* alignment needed? */
(6)            corr = (uint32_t)sizeof(QFreeBlock) - corr; /*amount to align poolSto*/
(7)            poolSize -= corr;                       /* reduce the available pool size */
            }
            /*lint -e826 align the head of free list at the free block-size boundary*/
(8)         me->free = (void *)((uint8_t *)poolSto + corr);
            /* round up the blockSize to fit an integer # free blocks, no division */
(9)         me->blockSize = (QMPoolSize)sizeof(QFreeBlock); /* start with just one */
(10)        nblocks = (uint32_t)1;       /* # free blocks that fit in one memory block */
(11)        while (me->blockSize < blockSize) {
(12)           me->blockSize += (QMPoolSize)sizeof(QFreeBlock);
(13)           ++nblocks;
            }
(14)        blockSize = me->blockSize;           /* use the rounded-up value from now on */

                                  /* the pool buffer must fit at least one rounded-up block */
(15)        Q_ASSERT(poolSize >= (uint32_t)blockSize);

                                     /* chain all blocks together in a free-list... */
(16)        poolSize -= (uint32_t)blockSize; /*don't link the last block to the next */
(17)        me->nTot  = (QMPoolCtr)1;                /* the last block already in the pool */
(18)        fb        = (QFreeBlock *)me->free; /*start at the head of the free list */
(19)        while (poolSize >= (uint32_t)blockSize) {      /* can fit another block? */
(20)           fb->next = &fb[nblocks];        /* point the next link to the next block */
               fb = fb->next;                           /* advance to the next block */
               poolSize -= (uint32_t)blockSize; /* reduce the available pool size */
               ++me->nTot;                      /* increment the number of blocks so far */
            }
```

Continued onto next page

```
(21)        fb->next  = (QFreeBlock *)0;          /* the last link points to NULL */
(22)        me->nFree = me->nTot;                      /* all blocks are free */
(23)        me->nMin  = me->nTot;           /* the minimum number of free blocks */
(24)        me->start = poolSto;          /* the original start this pool buffer */
(25)        me->end   = fb;                     /* the last block in this pool */
     }
```

(1)   The precondition requires a valid pointer to the pool storage.

(2)   The pool size must be able to fit at least one free block. Later, after aligning the pool storage and rounding up the block size, this assertion will be strengthened (see label (15)).

(3)   The argument blockSize must not be too close to the upper limit of its dynamic range, to avoid unexpected wraparound at rounding up the block size.

(4)   The expression assigned to corr computes the misalignment of the provided storage poolSto with respect to free block (pointer) size boundary.

(5)   Nonzero value of corr indicates misalignment of the pool storage.

(6)   Now corr holds the correction needed to align poolSto.

(7)   The available pool size is reduced by the amount of the correction.

(8)   The head of the free list is set at the start of the pool storage aligned at the nearest free block boundary.

(9-14)   To achieve alignment of all blocks in the pool, I round up the specified blockSize to the nearest integer multiple of QFreeBlock size. With the head of the free list already aligned at the QFreeBlock size and all blocks being integer multiples of the QFreeBlock size, I can be sure that every block is aligned as well. Note that instead of computing the number of free blocks going into the blockSize as (nblocks = (blockSize + sizeof (QFreeBlock) – 1)/sizeof(QFreeBlock) + 1), I compute the value iteratively in a loop. I decided to do this to avoid integer division, which would be the only division in the whole QP code base. On many CPUs division requires a sizable library function and I didn't want to pull this code.

(15)   For the correctness of the following algorithm, the pool must fit at least one rounded-up block.

(16)   The very last memory block is not linked to the next, so it is excluded.

(17)   The total count of blocks in the pool starts with one, to account for the last block.

(18)   The free block pointer starts at the head of the free list.

(19)   The loop chains all free blocks in a single-linked list until the end of the provided buffer storage.

(20)   The next link points to the free block, which is an integer multiple of the free block size `nblocks`, computed at step 13.

(21)   The last link is explicitly pointed to NULL.

(22,23)   Initially, all blocks in the pool are free.

(24)   The original pool buffer pointer is stored in me->start.

(25)   The pointer to the last block is stored in me->end.

7.9.2   Obtaining a Memory Block from the Pool

The implementation of the `QMPool_get()` function shown in Listing 7.29 is straightforward. The function returns a pointer to a new memory block or NULL if the pool runs out of blocks. This means that when you use QMPool directly as a general-purpose memory manager, you must validate the pointer returned from QMPool_get() before using it in your code. Note, however, that when QF uses QMPool internally as the event pool, the framework asserts that the pointer is valid (see Listing 7.12(3) in Section 7.5.2). QF considers running out of events in an event pool as an error.

Listing 7.29 Obtaining a block from a pool with QMPool_get()
(file <qp>\qpc\qf\source\qmp_get.c**)**

```
void *QMPool_get(QMPool *me) {
    QFreeBlock *fb;
    QF_INT_LOCK_KEY_

    QF_INT_LOCK_();
```

Continued onto next page

```
    fb = (QFreeBlock *)me->free;              /* get a free block or NULL */
    if (fb != (QFreeBlock *)0) {              /* free block available? */
        me->free = fb->next;    /* adjust list head to the next free block */
        --me->nFree;                          /* one less free block */
        if (me->nMin > me->nFree) {
            me->nMin = me->nFree;             /* remember the minimum so far */
        }
    }
    QF_INT_UNLOCK_();
    return fb;              /* return the block or NULL pointer to the caller */
}
```

7.9.3   Recycling a Memory Block Back to the Pool

Listing 7.30 shows the QMPool_put() function for recycling blocks back to the
pool. The most interesting aspects of this implementation are the preconditions.
Assertion at label (1) makes sure that the recycled block pointer lies in range of a
memory buffer managed by the pool (see Figure 7.10). Assertion 2 checks that the
number of free blocks is less than the total number of blocks (a new block is just
about to be inserted into the pool).

NOTE

The C standard guarantees meaningful pointer comparisons, such as the precondition (1) in
Listing 7.30, only if compared pointers point to the same array. Strictly speaking, this is only
the case when the pointer 'b' is indeed in range. When pointer 'b' is out of range, the
comparison might not be meaningful, and theoretically the precondition might not catch
the foreign block being recycled into the pool.

Listing 7.30 Recycling a block back to the pool with QMPool_put()
(file <qp>\qpc\qf\source\qmp_put.c)

```
    void QMPool_put(QMPool *me, void *b) {
        QF_INT_LOCK_KEY_

(1)     Q_REQUIRE((me->start <= b) && (b <= me->end)          /* must be in range */
(2)             && (me->nFree <= me->nTot));    /* # free blocks must be < total */

        QF_INT_LOCK_();
        ((QFreeBlock *)b)->next = (QFreeBlock *)me->free;   /* link into free list */
```

```
        me->free = b;                    /* set as new head of the free list */
        ++me->nFree;                     /* one more free block in this pool */
        QF_INT_UNLOCK_();
    }
```

7.10 Native QF Priority Set

The QF native priority set is generally useful for representing sets of up to 64
elements numbered 1..64. For example, you can use such a set to represent groups
of GPS satellites (numbered 1..32) or any other elements. The set provides
deterministic and efficient operations for inserting, removing, and testing elements
as well as determining the largest-number element in the set. The latter operation
is very helpful for quickly finding out the highest-priority active object ready to
run, and I use it inside the cooperative "vanilla" kernel (see the next Section 7.11)
and also inside the preemptive run-to-completion QK kernel (see Chapter 10).
The QF priority set implementation is adapted from the algorithm described in
[Bal Sathe 88 and Labrosse 02]. Listing 7.31 shows the declaration of the QPset64
data structure.

Listing 7.31 QPSet64 **structure**

```
typedef struct QPSet64Tag {
    uint8_t bytes;          /* condensed representation of the priority set */
    uint8_t bits[8];        /* bitmasks representing elements in the set */
} QPSet64;
```

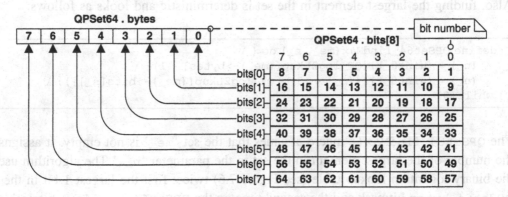

Figure 7.11: Dependency between QPSet64.bits[] **and** QPSet64.bytes.

Figure 7.11 graphically summarizes the semantics of QPSet64 data members. The bits of the array QPSet64.bits[8] correspond to the set elements as follows:

```
QPSet64.bits[0] represent elements 1..8
QPSet64.bits[1] represent elements 9..16
. . .
QPSet64.bits[7] represent elements 57..64
```

In addition, to speed up access to the bitmasks, the redundant summary of the bitmasks is stored in the member QPSet64.bytes with the following semantics of the bits:

```
bit 0 in QPSet64.bytes is 1 when any bit in QPSet64.bits[0] is 1
bit 1 in QPSet64.bytes is 1 when any bit in QPSet64.bits[1] is 1
. . .
bit 7 in QPSet64.bytes is 1 when any bit in QPSet64.bits[7] is 1
```

With this data representation, all operations on the set are fast and deterministic, meaning that the operations always take the same number of CPU cycles to execute, regardless of how many elements are in the set. All QPSet64 operations are implemented "inline" as macros to avoid the overhead of a function call.

For example, determining whether the set is not empty is remarkably simple:

```
#define QPSet64_notEmpty(me_) ((me_)->bytes != (uint8_t)0)
```

Also, finding the largest element in the set is deterministic and looks as follows:

```
#define QPSet64_findMax(me_, n_) do { \
    (n_) = (uint8_t)(QF_log2Lkup[(me_)->bytes] - 1); \
    (n_) = (uint8_t)(((n_) << 3) + QF_log2Lkup[(me_)->bits[n_]]); \
} while(0)
```

The QPSet64_findMax() macro assumes that the set 'me_' is not empty. It assigns the number of the largest element in the set to the parameter 'n_.' The algorithm uses the binary logarithm lookup table (see Figure 7.6) twice: first the largest 1 bit in the QPSet64.bytes bitmask and the second time on the QPSet64.bits[n_] bitmask to determine the largest 1 bit in the bitmask. The largest set element is the combination of the bit number returned by the lookup (a number in the range 1..8) plus the index multiplied by 8, to account for the byte position in the `bits[]` array.

Inserting an element 'n_' into the set 'me_' is implemented as follows:

```
#define QPSet64_insert(me_, n_) do { \
    (me_)->bits[QF_div8Lkup[n_]] |= QF_pwr2Lkup[n_]; \
    (me_)->bytes |= QF_pwr2Lkup[QF_div8Lkup[n_] + 1]; \
} while(0)
```

Finally, here is the macro for removing an element 'n_' from the set 'me_':

```
#define QPSet64_remove(me_, n_) do { \
    (me_)->bits[QF_div8Lkup[n_]] &= QF_invPwr2Lkup[n_]; \
    if ((me_)->bits[QF_div8Lkup[n_]] == (uint8_t)0) { \
        (me_)->bytes &= QF_invPwr2Lkup[QF_div8Lkup[n_] + 1]; \
    } \
} while(0)
```

7.11 Native Cooperative "Vanilla" Kernel

QF contains a simple cooperative "vanilla" kernel, which works as I described in Section 6.3.7 in Chapter 6. The "vanilla" kernel is implemented in two files: the `qvanilla.h` header file located in `<qp>\qpc\include\` directory and the `qvanilla.c` source file found in `<qp>\qpc\qf\source\` directory.

The "vanilla" kernel operates by constantly polling all event queues of active objects in an endless loop. The kernel always selects the highest-priority active object ready to run, which is the highest-priority active object with a nonempty event queue (see Figure 6.8 in Chapter 6). The scheduler maintains the global status of all event queues in the application in the priority set called the `QF_readySet_`. As shown in Figure 7.12, `QF_readySet_` represents a "ready-set" of all nonempty event queues in the system. For example, an element number 'p' is present in the ready-set if and only if the event queue of the active object with priority 'p' is nonempty. With this representation, posting an event to an empty queue with priority 'p' inserts the element number 'p' to the `QF_readySet_` set. Conversely, retrieving the last event from the queue with priority 'q' removes the element number 'q' from the ready-set `QF_readySet_`.

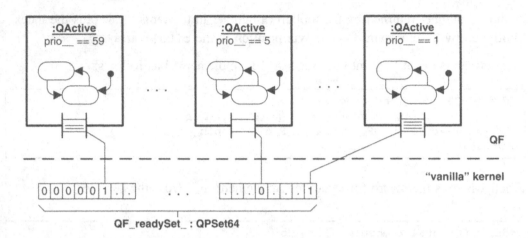

**Figure 7.12: Representing state of all event queues
in the** `QF_readySet_` **priority set.**

7.11.1   The `qvanilla.c` Source File

Listing 7.32 shows the complete implementation of the "vanilla" kernel.

Listing 7.32 The "vanilla" kernel implementation
(`<qp>\qpc\qf\source\qvanilla.c`)

```
(1)  #include "qf_pkg.h"
     #include "qassert.h"

     Q_DEFINE_THIS_MODULE(qvanilla)

     /* Package-scope objects -------------------------------------------------*/
(2)  QPSet64 volatile QF_readySet_;              /* QF-ready set of active objects */

     /*.......................................................................*/
     void QF_init(void) {
         /* nothing to do for the "vanilla" kernel */
     }
     /*.......................................................................*/
(3)  void QF_stop(void) {
         /* nothing to cleanup for the "vanilla" kernel */
         QF_onCleanup();                                        /* cleanup callback */
     }
     /*.......................................................................*/
(4)  void QF_run(void) {                                        /* see NOTE01 */
         uint8_t p;
```

```
            QActive *a;
            QEvent const *e;
            QF_INT_LOCK_KEY_

  (5)       QF_onStartup();                              /* invoke the QF startup callback */

  (6)       for (;;) {                                          /* the background loop */
  (7)           QF_INT_LOCK_();
  (8)           if (QPSet64_notEmpty(&QF_readySet_)) {
  (9)               QPSet64_findMax(&QF_readySet_, p);
 (10)              a = QF_active_[p];
 (11)              QF_INT_UNLOCK_();

 (12)              e = QActive_get_(a);              /* get the next event for this AO */
 (13)              QF_ACTIVE_DISPATCH_(&a->super, e);       /* dispatch to the AO */
 (14)              QF_gc(e);  /* determine if event is garbage and collect it if so */
              }
 (15)          else {                              /* all active object queues are empty */
       #ifndef QF_INT_KEY_TYPE
 (16)              QF_onIdle();                                    /* see NOTE02 */
       #else
 (17)              QF_onIdle(intLockKey_);                         /* see NOTE02 */
       #endif                                                 /* QF_INT_KEY_TYPE */
              }
          }
      }
      /*................................................................*/
 (18) void QActive_start(QActive *me, uint8_t prio,
                         QEvent const *qSto[], uint32_t qLen,
                         void *stkSto, uint32_t stkSize,
                         QEvent const *ie)
      {
 (19)     Q_REQUIRE(((uint8_t)0 < prio) && (prio <= (uint8_t)QF_MAX_ACTIVE)
                  && (stkSto == (void *)0));   /* does not need per-actor stack */

          (void)stkSize;           /* avoid the "unused parameter" compiler warning */
 (20)     QEQueue_init(&me->eQueue, qSto, (QEQueueCtr)qLen);/* initialize QEQueue */
 (21)     me->prio = prio;              /* set the QF priority of this active object */
 (22)     QF_add_(me);                          /* make QF aware of this active object */
 (23)     QF_ACTIVE_INIT_(&me->super, ie);         /* execute initial transition */
      }
      /*................................................................*/
      void QActive_stop(QActive *me) {
          QF_remove_(me);
      }
```

(1)   As every QF source file, the qvanilla.c file includes to the wider "package-scope"
      QF interface qf_pkg.h, located in <qp>\qpc\qf\source\. The qf_pkg.h
      header file includes the platform-specific QF port header file qf_port.h, but it
      additionally defines some internal macros and objects shared only internally within QF.

(2)   QF_readySet_ priority set represents the ready-set of the scheduler. I declared it volatile to inform the compiler to never cache this variable because it can change unexpectedly in interrupts (e.g., when ISRs post or publish events).

(3)   The function QF_stop() stops execution of the QF framework. In the case of the "vanilla" kernel, this function has nothing to do except invoke the QF_onCleanup() callback function to give the application a chance to clean up and exit to the underlying operating system (e.g., consider a "vanilla" kernel running on top of DOS). I summarize all QF callback functions in Section 8.1.8 in Chapter 8.

(4)   Applications call the function QF_run() from main() to transfer the control to the framework. This function implements the entire "vanilla" kernel.

(5)   The QF_onStartup() callback function configures and starts interrupts. This function is typically implemented at the application level (in the BSP). I summarize all QF callback functions in Section 8.1.8 in Chapter 8.

(6)   This is the *event loop* of the "vanilla" kernel.

(7)   Interrupts are locked to access the QF_readySet_ ready-set.

(8)   If the ready-set QF_readySet_ is not empty, the "vanilla" kernel has some events to process.

(9)   The QF priority set quickly discovers the highest-priority, not-empty event queue, as described in Section 7.10.

(10) The active object pointer 'a' is resolved through the QF_active_[] priority-to-active object lookup table maintained internally by QF.

(11) Interrupts can be unlocked.

(12-14) These are the three steps of the active object thread (see Listing 7.8).

(15) The else branch is taken when all active object event queues are empty, which is by definition the *idle condition* of the "vanilla" kernel.

(16,17) The "vanilla" kernel calls the QF_onIdle() callback function to give the application a chance to put the CPU to a low-power sleep mode or to perform other processing (e.g., software-tracing output; see Chapter 11). The

QF_onIdle() function is typically implemented at the application level (in the BSP). Note that the signature of QF_onIdle() depends on the critical section mechanism you choose. The function takes no parameters when the simple "unconditional interrupt unlocking" policy is used but needs the interrupt status parameter when the "saving and restoring interrupt status" policy is used (see Section 7.3).

NOTE

Most MCUs provide software-controlled low-power sleep modes, which are designed to reduce power dissipation by gating the clock to the CPU and various peripherals. To ensure a safe transition to a sleep mode, the "vanilla" kernel calls QF_onIdle() with interrupts locked. The QF_onIdle() function *must* always unlock interrupts internally, ideally atomically with the transition to a sleep mode.

---

(18)   The QActive_start() function initializes the event queue and starts the active object thread under the "vanilla" kernel.

(19)   The precondition asserts that the provided priority is within range and that the stack pointer is NULL because the "vanilla" kernel does not need the per-task stack.

(20)   The "vanilla" kernel uses the native QF event queue QEQueue, which needs to be initialized with the function QEQueue_init().

(21)   The QF priority of the active object is set inside the active object.

(22)   The active object is added to the QF framework.

(23)   The internal state machine of the active object is initialized.

Figure 7.13 shows a typical execution scenario in the "vanilla" kernel. As long as events are available, the event loop calls various active objects to process the events in run-to-completion fashion. When all event queues run out of events, the event loop calls the QF_onIdle() function to give the application a chance to switch the MCU to a low-power sleep mode. The "vanilla" kernel must invoke QF_onIdle() with interrupts *locked*. If the interrupts were enabled after the event loop determines that the ready-set is empty (Listing 7.32(20)), but before calling QF_onIdle() (where the switching to the low-power mode actually takes place), an interrupt could preempt the event loop at this exact point and an ISR could post new events to active objects, thus invalidating the idle condition.

By the simplistic nature of the "vanilla" kernel, the event loop always resumes exactly at the point it was interrupted, so the event loop would enter the low-power sleep mode while events would be waiting for processing! The MCU will be stopped for a nondeterministic period of time until the next interrupt wakes it up. Thus unlocking interrupts before transitioning to a low-power state opens a time window for a race condition between any enabled interrupt and the transition to the low-power mode.

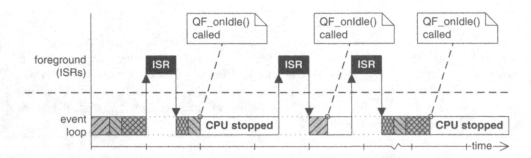

Figure 7.13: Entering low-power sleep modes in the "vanilla" kernel.

Entering a sleep mode while interrupts are disabled poses a chicken-and-egg problem for waking the system up, because only an interrupt can terminate the low-power sleep mode. To operate in the "vanilla" kernel, the MCU must allow entering the low-power sleep mode and enabling the interrupts at the same time, without creating the race condition described above.

Many MCUs indeed allow such an atomic transition to the sleep mode. Other MCUs support multiple levels of disabling interrupts and can accomplish low-power transitions with interrupts disabled at one level. Yet other class of MCUs doesn't provide any way of entering the low-power mode with interrupts disabled and requires some different approaches. Refer the ESD article "Use an MCU's low-power modes in foreground/ background systems" [Samek 07b] for an overview of safe sleep mode transitions in various popular MCUs.

7.11.2   The qvanilla.h Header File

The qvanilla.h header file, shown in Listing 7.33, integrates the "vanilla" kernel into the QF framework. The most important function of this header file is to codify the updates to the ready-set (QF_readySet_) as events are posted and removed from the active object event queues.

Listing 7.33 The "vanilla" kernel interface
(`<qp>\qpc\include\qvanilla.h`)

```
        #ifndef qvanilla_h
        #define qvanilla_h

(1)     #include "qequeue.h"      /* "Vanilla" kernel uses the native QF event queue */
(2)     #include "qmpool.h"       /* "Vanilla" kernel uses the native QF memory pool */
(3)     #include "qpset.h"        /* "Vanilla" kernel uses the native QF priority set */

                          /* the event queue and thread types for the "Vanilla" kernel */
(4)     #define QF_EQUEUE_TYPE              QEQueue

                                                /* native QF event queue operations */
(5)     #define QACTIVE_EQUEUE_WAIT_(me_) \
            Q_ASSERT((me_)->eQueue.frontEvt != (QEvent *)0)

(6)     #define QACTIVE_EQUEUE_SIGNAL_(me_) \
            QPSet64_insert(&QF_readySet_, (me_)->prio)
(7)     #define QACTIVE_EQUEUE_ONEMPTY_(me_) \
            QPSet64_remove(&QF_readySet_, (me_)->prio)

                                                /* native QF event pool operations */
(8)     #define QF_EPOOL_TYPE_              QMPool
(9)     #define QF_EPOOL_INIT_(p_, poolSto_, poolSize_, evtSize_) \
            QMPool_init(&(p_), poolSto_, poolSize_, evtSize_)
(10)    #define QF_EPOOL_EVENT_SIZE_(p_)    ((p_).blockSize)
(11)    #define QF_EPOOL_GET_(p_, e_)       ((e_) = (QEvent *)QMPool_get(&(p_)))
(12)    #define QF_EPOOL_PUT_(p_, e_)       (QMPool_put(&(p_), e_))

(13)    extern QPSet64 volatile QF_readySet_;    /** QF-ready set of active objects */

        #endif                                                      /* qvanilla_h */
```

(1) The "vanilla" kernel uses the native QF event queue, so it needs to include the `qequeue.h` header file.

(2) The "vanilla" kernel uses the native QF memory pool, so it needs to include the `qmpool.h` header file.

(3) The "vanilla" kernel uses the native QF priority set, so it needs to include the `qpset.h` header file.

(4) The "vanilla" kernel uses `QEQueue` as the event queue for active objects (see also Listing 7.7(8)).

(5)   The "vanilla" kernel never blocks. It calls `QActive_get_()` only when it knows for sure that the event queue contains at least one event (see Listing 7.32(12)). Since this is certainty in this type of kernel, the `QACTIVE_EQUEUE_WAIT_()` macro (see Listing 7.24(2)) asserts that the event queue is indeed not empty.

(6)   The macro `QACTIVE_EQUEUE_SIGNAL_()` is called from `QActive_postFIFO()` and `QActive_postLIFO()` when an event is posted to an empty queue (see Listing 7.25(5)). This is exactly when the priority of the active object needs to be inserted into the ready-set `QF_readySet_`. Note that `QF_readySet_` is modified within a critical section.

(7)   The macro `QACTIVE_EQUEUE_ONEMPTY_()` is called from `QActive_get_()` when the queue is becoming empty (see Listing 7.24(12)). This is exactly when the priority of the active object needs to be removed from the ready-set `QF_readySet_`. Note that `QF_readySet_` is modified within a critical section.

(8-12)   The "vanilla" kernel uses `QMPool` as the QF event pool. The platform abstraction layer (PAL) macros are set to access the `QMPool` operations (see Section 7.9).

(13)   The `QF_readySet_` is declared as volatile because it can change asynchronously in an ISR.

7.12   QP Reference Manual

The source code available from the companion Website to this book at `www.quantumleaps.com/psicc2/` contains the complete "QP Reference Manual" in HTML and CHM-Help formats (see Figure 7.14). The Reference Manual has been generated by Doxygen (`www.doxygen.org`), which is an open-source documentation-generation system for C, C++, Java, and other languages. The HTML documentation is found in `<qp>\qpc\doxygen\html\`, while the CHM Help format is located in `<qp>\qpc\qpc.chm`.

NOTE

The "QP/C++ Reference Manual" for the QP/C++ version is located in `<qp>\qpcpp\`.

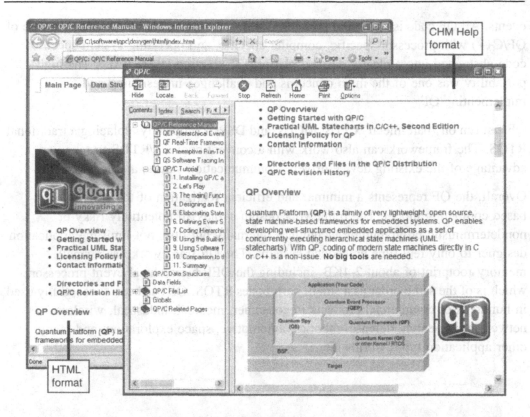

Figure 7.14: Screen shots of the "QP/C Reference Manual," which is available in HTML and CHM Help formats.

The "QP Reference Manual" is quite detailed. Every file, structure (class), function, macro, and `typedef` is documented. The Doxygen tool does a superb job in cross-referencing the manual, so you can find information quickly. The manual also contains plenty of examples and useful code snippets that you can conveniently cut and paste into your own code. Finally, if you choose to modify the QP source code, you can regenerate the Doxygen documentation by yourself because I provide the `Doxyfile`.

7.13 Summary

QF is a generic, portable, scalable, lightweight, deterministic, real-time framework for embedded systems. QF supports many advances features, such as "zero-copy" event management, publish-subscribe event delivery, and automatic garbage collection for events. All QF code is written in portable ANSI-C (or Embedded C++ subset in case of QF/C++) with processor-specific, compiler-specific, and operating system-specific code abstracted into a clearly defined platform abstraction layer (PAL). The high portability was one of the main concerns and challenges in designing and implementing QF.

QF can run on "bare-metal" CPUs, MCUs, and DSPs, completely replacing a traditional RTOS. The framework can also work with a conventional OS/RTOS to take advantage of the existing device drivers, communication stacks, and legacy code.

Overall, the QF represents a minimal and efficient realization of the active object–based computing model. The framework design avoids all potentially risky or nondeterministic programming techniques internally but does not limit the application designer to only real-time embedded systems. The QF framework has a small memory footprint of about 2-4KB, including the QEP hierarchical event processor, which is of the same order as a small, bare-bones RTOS. QF has been successfully used in hundreds of commercial products in consumer, medical, industrial, wireless, networking, research, defense, robotics, automotive, space exploration, and many other application types worldwide.

CHAPTER 8

Porting and Configuring QF

As a rule, software systems do not work well until they have been used, and have failed repeatedly,
in real applications.
—David Parnas

In this chapter I describe how to adapt the QF real-time framework to various
processors, compilers, and operating systems, which is a process called *porting*.
Porting QF is relatively easy because QF has been designed from the ground up to
be portable. In particular, QF contains a clearly defined *platform abstraction layer*
(PAL), which encapsulates all the platform-specific code and cleanly separates
it from the platform-neutral code. Depending on the chosen RTOS/OS, the CPU
architecture, and the compiler, porting QF might require writing or modifying
between 5 and 100 lines of code within the PAL.

This chapter starts with a summary of the files, macros, and functions comprising the
PAL. Next, I describe the following three QF ports:

- The QF port to the "vanilla" cooperative kernel built into QF

- The QF port to the μC/OS-II RTOS as an example of using QF with a traditional
  real-time, preemptive, priority-based RTOS

- The QF port to Linux as an example of using QF with a conventional POSIX-
  compliant operating system

Note that the QF ports I discuss in this chapter do not include the port to the QK
preemptive kernel. That's because I devote the whole of Chapter 10 to QK.

8.1   The QP Platform Abstraction Layer

All software components of the QP event-driven platform, such as the QEP event
processor and the QF real-time framework, contain a PAL. The PAL is an indirection
layer that hides the differences in hardware and software environments in which
QP operates so that the QP source code does not need to be changed to run in a
different environment. Instead, all the changes required to adapt QP are confined
to the PAL.

In the previous Chapter 7, you already saw quite a few PAL macros, such as macros for
locking and unlocking interrupts or macros for hiding the data types of operating
system-specific objects like threads or event queues. However, the abstraction layer
consists of more than just macros and typedef's. The PAL also includes the directory
structure to hold all the platform variations, platform-specific header files, platform-
specific source files, and build scripts or Makefiles.

8.1.1   Building QP Applications

The PAL actually serves a dual purpose. Obviously, its goal is to ease the porting effort.
But even more important, the other objective of the PAL is to simplify the use of
QP in the applications.

Figure 8.1 shows the process of building a QP application. Each QP component requires
inclusion of only one platform-specific header file and linking one platform-specific
library. For example, to use the QF real-time framework, you need to include the
qf_port.h header file[1] and you need to link the qf.lib library file from the specific
QP port directory. It really doesn't get any simpler than that.

> **NOTE**
>
> All QP components are designed to be deployed in fine-granularity object libraries. QP
> libraries allow the linker to eliminate any unreferenced QP code at link time, which results
> in *automatic scaling* of every QP component for a wide range of applications. This approach
> eliminates the need to manually configure and recompile the QP source code for each appli-
> cation at hand.

---

[1] You typically include qf_port.h indirectly via the qp_port.h, discussed in Section 8.1.7.

The QP port you are using is determined by the directory branch in which the `qf_port.h` header file and the QF library file are located. Listing 8.1 in Section 8.1.3 shows some examples of such port directories. Typically you need to instruct the C/C++ compiler to include header files from the specific QP port directory and also from the platform-independent include directory `<qp>\qpc\include\`. I strongly discourage hardcoding full pathnames of the include files in your source code. You should simply include the QP port header file (`#include "qf_port.h"`) without any path. Then you specify to the compiler to search the QP port directory for include files, typically through the `-I` option.

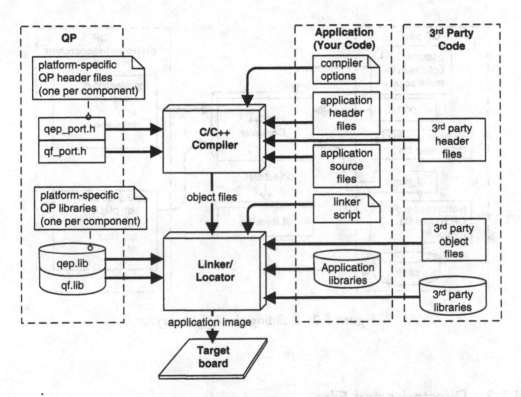

Figure 8.1: Building a QP-based application.

8.1.2 Building QP Libraries

Figure 8.2 illustrates the steps required to build the QF library. The process of building other QP components, such as QEP or QK, is essentially identical.

The key point of the design is that all platform-independent QF source files include *the same* qf_port.h header file as the application source files (see Figure 8.2). At this point you can clearly see that the PAL plays the dual role of facilitating the porting of QP as well as using it in the applications.

Figure 8.2 also shows that every QP component, such as QF, can contain a platform-specific source file (qf_port.c in this case). The platform-specific source file is optional and many ports don't require it.

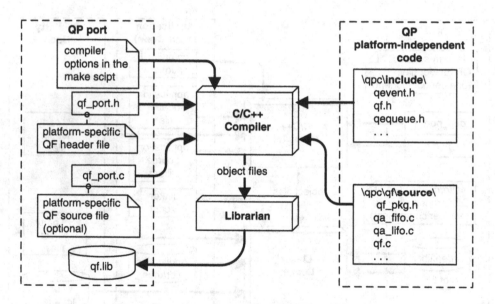

Figure 8.2: Building the QF library.

8.1.3 Directories and Files

The PAL uses a consistent directory structure that allows you to find the QP port to a given CPU, operating system, and compiler quite easily. Listing 8.1 shows the platform-specific QP code organization.

Listing 8.1 Platform-specific QP code organization

```
(1)   qpc\                      - QP/C root directory (qpcpp\ for QP/C++)
      |
(2)   +-ports\                  - Platform-specific QP ports
(3)   | +-80x86\                - Ports to the 80x86 processor
(4)   | | +-dos\                - Ports to DOS with the "vanilla" cooperative kernel
(5)   | | | +-tcpp101\          - Ports with the Turbo C++ 1.01 compiler
(6)   | | |   +-l\              - Ports using the Large memory model
(7)   | | |     +-dbg\    - Debug build
(8)   | | |     | +-qf.lib    - QF library
      | | |     | +-qep.lib   - QEP library
(9)   | | |     +-rel\    - Release build
(10)  | | |     +-spy\    - Spy build (with software instrumentation)
(11)  | | |     +-make.bat    - batch script for building the QP libraries
      | | |     +-qep_port.h  - QEP platform-dependent include file
(12)  | | |     +-qf_port.h   - QFplatform-dependent include file
      | | |     +-qs_port.h   - QSplatform-dependent include file
      | | |     +-qp_port.h   - QPplatform-dependent include file
      | | |
      | | +-qk\              - Ports to the QK preemptive kernel
      | | | +-. . .
      | | |
      | | +-ucos2\           - Ports to the µC/OS-II RTOS
      | | | +-tcpp101\       - Ports with the Turbo C++ 1.01 compiler
      | | | | +-l\           - Ports using the Large memory model
(13)  | | | | | +-ucos2.86\  - µC/OS-II v2.86 object code and header files
      | | | | | +-src\       - Port-specific source files
(14)  | | | | | | +-qf_port.c - QF port to µC/OS-II source file
      | | | | | +-. . .
      | | |
      | | +-linux\           - Ports to the Linux operating system (POSIX)
      | |   +-gnu\           - Ports with the GNU compiler
      | | | | +-src\         - Port-specific source files
(15)  | | | | | +-qf_port.c  - QF port to Linux source file
      | |     +-. . .
      | |
      | +-cortex-m3\         - Ports to the Cortex-M3 processor
      | | +-vanilla\         - Ports to the "vanilla" cooperative kernel
      | | | +-iar\           - Ports with the IAR compiler
      | | |   | +-dbg\       - Debug build
      | | |   | +-rel\       - Release build
      | | |   | +-spy\       - Spy build (with software instrumentation)
      | | |   | +-make.bat   - batch script for building QP libraries
      | | |   | +-qep_port.h - QEP platform-dependent include file
      | | |   | +-qf_port.h  - QF platform-dependent include file
      | | |   | +-qs_port.h  - QS platform-dependent include file
      | | |   | +-qp_port.h  - QP platform-dependent include file
      | | | . . .
```

Continued onto next page

```
        | | +-qk\              - Ports to the QK preemptive kernel
        | | +-iar\             - Ports with the IAR compiler
        | +-. . .              - Ports to other CPUs
        |
(16)    +-examples\            - Platform-specific QP examples
        | +-80x86\             - Examples for the 80x86 processor
        | | +-dos\             - Examples for DOS with the "vanilla" cooperative kernel
        | |   +-tcpp101\       - Examples with the Turbo C++ 1.01 compiler
        | |     +-l\           - Examples using the Large memory model
(17)    | |       +-dpp\       - DPP example
(18)    | |       | +-dbg\     - Debug build
        | |       | | +-dpp.exe - Debug executable
        | |       | +-rel\       - Release build
        | |       | | +-dpp.exe - Release executable
        | |       | +-spy\       - Spy build (with software instrumentation)
        | |       | | +-dpp.exe - Spy executable
        | |       | +-DPP-DBG.PRJ - Turbo C++ project to build the Debug version
        | |       +-game\       - "Fly 'n' Shoot" game example
        | |         +-. . .
        | +-cortex-m3\        - Examples for the Cortex-M3 processor
        | | +-vanilla\        - Examples for the "vanilla" cooperative kernel
        | | | +-iar\          - Examples with the IAR compiler
        | |   +-dpp\          - DPP example
        | |   +-game\         - "Fly 'n' Shoot" game example
        | |   +-. . .         - Other examples
        | +-. . .             - Examples for other CPUs
        |
(19)    +-include\            - Platform independent QP header files
        | +-qep.h             - QEP platform-independent interface
        | +-qf.h              - QF platform-independent interface
        | +-qk.h              - QK platform-independent interface
        | +-qs.h              - QS platform-independent interface
        | +-. . .             - Other platform-independent QP header files
        |
(20)    +-qep\               - QEP event processor
        | +-source\          - QEP platform-independent source code (*.C files)
        | | +-. . .
(21)    +-qf\                - QF real-time framework
        | +-source\          - QF platform-independent source code (*.C files)
        | | +-. . .
(22)    +-qk\                - QK preemptive kernel
        | +-source\          - QK platform-independent source code (*.C files)
        | | +-. . .
(23)    +-qs\                - QS software tracing
        | +-source\          - QS platform-independent source code (*.C files)
        | | +-. . .
```

(1)   Every QP version such as QP/C and QP/C++ resides in the separate directory branch, called henceforth the QP Root Directory. The essential element of the design is that the QP Root Directory can be "plugged into" any branch of a hierarchical file system and you can move the QP Root Directory around, or even have multiple versions of the QP Root Directories. You can also freely choose the name of the QP Root Directory, although I recommend the directory names <qp>\qpc\ for QP/C and <qp>\qpcpp\ for QP/C++. The ability to relocate the QP Root Directory means that only relative paths should be used in the Makefiles, build scripts, workspaces, or project files.

(2)   The directory ports\ contains platform-specific header files and libraries to be used by QP applications. This directory structure is the most complicated because of the large number of choices available, such as CPU architectures, compilers, operating systems, and compiler options. Each of those choices is represented as a separate level of nesting in a hierarchical directory tree, so each dimension in the multidimensional space of options can be extended independently from the others. In addition, the directory branch for each port is individually customizable, so each branch can represent only choices relevant for a given CPU, operating system, compiler, etc.

(3)   I've decided to put the CPU architecture as the first level of nesting within the ports\ directory. Examples of CPU architectures are 80x86, Cortex-M3, ARM, AVR, MSP430, and M16C. Note that a separate directory is needed whenever the CPU architecture is significantly different. For example, even though the traditional ARM and the new ARM Cortex-M3 are related, the differences are significant enough to require a separate directory branch for ARM and Cortex-M3.

(4)   The second level of nesting under the CPU architecture is the operating system used. For example, in the 80x86 architecture, QP can operate under DOS (with the "vanilla" cooperative kernel), under the QK preemptive kernel, under the µC/OS-II RTOS, or under Linux (and perhaps other OSs such as Win32).

NOTE

The ordering of directory levels reflects the embedded focus in the QF design. In most standalone QF applications the CPU architecture is typically more important than the RTOS/OS. For general-purpose operating systems such as Linux, the reversed order (operating system at a higher level than the CPU architecture) would perhaps feel more natural.

(5) The next level of nesting, under each operating system directory, is the directory for the compiler used. For example, the DOS port can be compiled with the Turbo C++ 1.01 or perhaps with Visual C++ 1.52. Similarly, the port to Cortex-M3 with QK kernel can be compiled with the IAR, RealView, or GNU compilers.

(6) In some ports, the compiler can emit code for various modes of the CPU. For example, a compiler for 80x86 under DOS can produce small, compact, large, or huge memory models. These different modes result in incompatible object code, and therefore each of them requires a separate branch. Note that the compiler options level is optional. For example, the Cortex-M3 CPU branch does not need the compiler options level.

(7) Finally, the QP libraries can be compiled with different compile-time switches and optimization options. For example, the dbg\ directory holds the Debug configuration, which contains the symbolic debug information.

(8) Each specific build directory contains the QP library files. The actual library names should comply with the conventions used on a particular platform. For example, on Linux the libraries are typically named lib???.a (e.g., libqep.a, libqf.a, etc.).

(9) The rel\ directory holds the Release configuration, which typically does not contain debug information but might use aggressive optimizations for best performance.

(10) The spy\ directory holds the Spy configuration, which uses the QS software-tracing instrumentation (see Chapter 11).

(11) The standard QP ports often contain a simple make.bat script or a Makefile for building all the QP libraries for the port. You typically can choose the build configuration by providing a target to the make.bat script or to the Makefile. The default target is "dbg." Other possible targets are "rel" and "spy." Table 8.1 summarizes the targets accepted by the make.bat scripts or the Makefiles.

(12) The qf_port.h header file is the most important part of the port. This header file contains the definitions of the PAL macros and typedef's as well as other elements. I discuss the qf_port.h header file in detail in Section 8.1.5.

Table 8.1: Build targets accepted by make.bat scripts or Make

Build Configuration	Build Command
Debug	make
Release	make rel
Spy	make spy

(13)   The subdirectory `ucos2.86\` contains the headers and object files of µC/OS-II v2.86, compiled for 80x86 with the Turbo C++ 1.01 compiler. This directory is provided only to allow you to rebuild the example applications based on the QF port to µC/OS-II. Typically, however, you will need to obtain the µC/OS-II source code to rebuild it for the actual processor you're using. Refer to Section 8.3 for more details about the µC/OS-II port.

(14,15) QP ports to an external RTOS or OS such as µC/OS-II or Linux require some "glue-code" to bolt the QF framework to the external RTOS/OS. This source code is placed in the file `qf_port.c` in the subdirectory `src\`.

(16)   The `examples\` directory contains the application examples that accompany the ports. The structure of the `examples\` branch closely mirrors the structure of the `ports\` directory, except that it adds one more level for various example applications.

(17)   For example, the `dpp\` directory contains the "Dining Philosopher Problem" application example for this particular port. I describe the DPP test application in Chapter 9.

(18)   The `dpp\dbg\` directory contains the object files and the executable for the Debug configuration of the DPP application build.

(19)   The `include\` directory contains the platform-independent header files. You always need to include this directory in the compiler's search path to build applications.

(20-23) The platform-independent source code of each QP component is located in the separate directory. The source files are only needed to rebuild QP libraries, but you don't need to include these directories in the compiler's search path to build applications.

8.1.4 The `qep_port.h` Header File

The header file `qep_porth.h` adapts and configures the QEP event processor component of QP. I already discussed this header file in Section 4.8 of Chapter 4. However, `qep_porth.h` also provides macros and `typedef`'s that affect all the other QP components. Therefore, for the completeness of the PAL description, I decided to include here again the explanation of this important header file.

Listing 8.2 The `qep_port.h` header file

```
     #ifndef qep_port_h
     #define qep_port_h
                                    /* special keyword used for ROM objects */
(1)  #define Q_ROM                     ????
        /* specific pointer variant for accessing const objects in ROM */
(2)  #define Q_ROM_VAR             ????
             /* platform-specific access to constant data bytes in ROM */
(3)  #define Q_ROM_BYTE(rom_var_)  ????


                                    /* size of the QSignal data type */
(4)  #define Q_SIGNAL_SIZE          ?


                                         /* exact-width integer types */
(5)  #include <stdint.h>    /* WG14/N843 C99 Standard, Section 7.18.1.1 */

(6)  typedef signed    char int8_t;                /* signed 8-bit integer */
(7)  typedef signed    short int16_t;             /* signed 16-bit integer */
(8)  typedef signed    long int32_t;              /* signed 32-bit integer */
(9)  typedef unsigned char uint8_t;             /* unsigned 8-bit integer */
(10) typedef unsigned short uint16_t;          /* unsigned 16-bit integer */
(11) typedef unsigned long uint32_t;           /* unsigned 32-bit integer */

(12) #include "qep.h"    /* QEP platform-independent public interface */

     #endif                                           /* qep_port_h */
```

---

(1) The `Q_ROM` macro allows enforcing placement of the constant objects, such as lookup tables, constant strings, and the like, in ROM rather than in the precious RAM. On CPUs with the Harvard architecture (such as 8051 or the Atmel AVR), the code and data spaces are separate and are accessed through different CPU instructions. Various compilers often provide specific extended keywords to designate code or data space, such as the "__code" extended keyword in the

IAR 8051 compiler. You don't need to provide this macro, in which case it will be defined to nothing in the `qep.h` platform-independent header file.

(2) The macro `Q_ROM_VAR` specifies the kind of the pointer to be used to access the ROM objects because many compilers provide different size pointers for accessing objects in various memories. Constant objects allocated in ROM often mandate the use of specific-size pointers (e.g., `far` pointers) to get access to ROM objects. An example of a valid `Q_ROM_VAR` macro definition is `__far` (Freescale HC(S)08 compiler). You don't need to provide this macro, in which case it will be defined to nothing in the `qep.h` platform-independent header file.

NOTE

Note that macros `Q_ROM` and `Q_ROM_VAR` refer to the different parts of the object declaration. The macro `Q_ROM` specifies the ROM memory type to allocate an object. This allows compilers to generate different instructions for accessing such ROM objects for CPUs with the Harvard architecture. On the other hand, the macro `Q_ROM_VAR` specifies the size of the pointer (e.g., "far" pointer) to access the ROM data, so it refers just to the size of the object's address, not to the object itself. The `Q_ROM_VAR` macro is useful for the von Neumann machines.

If you don't define macros `Q_ROM` or `Q_ROM_VAR`, the `qep.h` header file will provide default empty definitions, which means that no special extended keywords are necessary to correctly allocate and access the constant objects.

---

(3) The macro `Q_ROM_BYTE()` encapsulates a custom mechanism of retrieving a data byte from the given ROM address, which is useful for some compilers for Harvard architecture CPUs (e.g., GCC for AVR). Some compilers cannot generate code for accessing data allocated in the program space (ROM) when data and program spaces require different machine instructions, even though the compiler can allocate constants in ROM. The workaround for such compilers is to explicitly add custom assembly code to access data allocated in the program space. The `Q_ROM_BYTE()` macro is really for special occasions and you typically don't need to define it. The default that QF provides assumes a compiler capable of correctly accessing data objects in ROM.

(4) The macro `Q_SIGNAL_SIZE` configures the size of the event signal (the `QSignal` data type). If the macro is not defined, the default of 1 byte will be chosen in `qep.h`. The valid `Q_SIGNAL_SIZE` values 1, 2, or 4 correspond to

QSignal of uint8_t, uint16_t, and uint32_t, respectively. The QSignal data type determines the dynamic range of numerical values of signals in your application. The default for Q_SIGNAL_SIZE is 1 (256 signals).

(5)   Porting QEP and any other QP component requires providing the C99-standard exact-width integer types that are consistent with the CPU and compiler options used. For newer C and C++ compilers, you can simply include the standard header file <stdint.h> provided by the compiler vendor.

(6-11)   For prestandard compilers that don't provide the <stdint.h> header file, you need to typedef the six basic exact-width integer types used in QP. Consult your compiler documentation to find out which combination of the basic C data types map to the C99-standard integer types. Also note that the mapping might depend on the compiler options you're using (e.g., the memory model).

(12)   The qep_port.h platform-specific header file must include the qep.h platform-independent header file.

8.1.5   The qf_port.h Header File

The qf_port.h header file contains the definitions of the PAL macros, typedef's, include files, as well as constants for porting and configuring the QF real-time framework. This is by far the most complex and important file in the whole QP PAL.

Listing 8.3 shows the general layout of the qf_port.h header file. Note that I have placed all elements into this example file for completeness, even though some sections of the file are mutually exclusive. In the text immediately following the listing, I clarify the purpose of each qf_port.h section and explain when you should use it. The concrete QF ports described later in this chapter provide the examples of valid qf_port.h header files.

Listing 8.3 The qf_port.h header file

```
      #ifndef qf_port_h
      #define qf_port_h

      /* Types of platform-specific QActive data members ************************/
(1)   #define QF_EQUEUE_TYPE                ????
(2)   #define QF_OS_OBJECT_TYPE             ????
(3)   #define QF_THREAD_TYPE                ????
```

```
        /* Base class for derivation of QActive ***********************************/
(4)     #define QF_ACTIVE_SUPER_                    ????
(5)     #define QF_ACTIVE_CTOR_(me_, initial_)      ????
(6)     #define QF_ACTIVE_INIT_(me_, e_)            ????
(7)     #define QF_ACTIVE_DISPATCH_(me_, e_)        ????
(8)     #define QF_ACTIVE_STATE_                    ????


        /* The maximum number of active objects in the application ****************/
(9)     #define QF_MAX_ACTIVE                       ????


        /* Various object sizes within the QF framework **************************/
(10)    #define QF_EVENT_SIZ_SIZE          2
(11)    #define QF_EQUEUE_CTR_SIZE         1
(12)    #define QF_MPOOL_SIZ_SIZE          2
(13)    #define QF_MPOOL_CTR_SIZE          2
(14)    #define QF_TIMEEVT_CTR_SIZE        2


        /* QF critical section mechanism *****************************************/
(15)    #define QF_INT_KEY_TYPE                     ????
(16)    #define QF_INT_LOCK(key_)                   ????
(17)    #define QF_INT_UNLOCK(key_)                 ????


        /* Include files used by this QF port ***********************************/
(18)    #include <????.h>                 /* underlying OS/RTOS/Kernel interface */
(19)    #include "qep_port.h"                                      /* QEP port */
(20)    #include "qequeue.h"                              /* native QF event-queue */
(21)    #include "qmpool.h"                              /* native QF memory-pool */
(22)    #include "qvanilla.h"                         /* native QF "vanilla" kernel */
(23)    #include "qf.h"                        /* platform-independent QF interface */

        /************************************************************************
        * Interface used only inside QF, but not in applications
        */
        /* Active object event queue operations ********************************/
(24)    #define QACTIVE_EQUEUE_WAIT_(me_)           ????
(25)    #define QACTIVE_EQUEUE_SIGNAL_(me_)         ????
(26)    #define QACTIVE_EQUEUE_ONEMPTY_(me_)        ????


        /* QF event pool operations ********************************************/
(27)    #define QF_EPOOL_TYPE_                      ????
(28)    #define QF_EPOOL_INIT_(p_, poolSto_, poolSize_, evtSize_) ????
(29)    #define QF_EPOOL_EVENT_SIZE_(p_)            ????
(30)    #define QF_EPOOL_GET_(p_, e_)               ????
(31)    #define QF_EPOOL_PUT_(p_, e_)               ????


        /* Global objects required by the QF port ******************************/
        extern ????;
        . . .


        #endif                                                  /* qf_port_h */
```

Types of Platform-Specific `QActive` Data Members

This section is required only when QF is ported to an external OS/RTOS. All standalone QF ports, such as the port to the native "vanilla" cooperative kernel or the QK preemptive kernel, do *not* need this section.

(1)   You must tell QF the data type of the event queue for active objects. The event queue can be implemented with a message queue of the RTOS/OS. But it is also possible to use the native QF event queue `QEQueue` type if the underlying RTOS/OS does not provide an adequate queue.

(2)   The `QF_OS_OBJECT_TYPE` data member is necessary when the underlying OS does not provide an adequate queue facility, so the native QF queue `QEQueue` must be used. In this case the `QF_OS_OBJECT_TYPE` data member holds an operating system-specific primitive to efficiently block the native QF event queue when the queue is empty. Section 8.4, "Conventional POSIX-Compliant OS (Linux)," provides an example of specifying the `QF_OS_OBJECT_TYPE` data type.

(3)   The data type `QF_THREAD_TYPE` holds the thread handle associated with the active object.

Base Class for Derivation of `QActive`

This section is required only when you want to derive the `QActive` base class from a class different than `QHsm`, which is the default. The macros in this section allow you to replace the QEP component with your own event processor. The macros defined in this segment should generally not be used in the applications. I apply the naming convention of terminating such macro names with an underscore, which you should consider as a red flag to avoid using them in the application code.

(4)   The macro `QF_ACTIVE_SUPER_` specifies the ultimate base class for deriving active objects. This macro lets you define any base class for `QActive`, as long as the base class supports the state machine interface (see Chapter 3, "Generic State Machine Interface"). If you don't provide this macro, QF will default to the `QHsm` base class.

(5)   The macro `QF_ACTIVE_CTOR_()` specifies the name of the base class constructor.

(6)   The macro `QF_ACTIVE_INIT_()` specifies the name of the base class `init()` function.

(7) The macro QF_ACTIVE_DISPATCH_() specifies the name of the base class dispatch() function.

(8) The macro QF_ACTIVE_STATE_ specifies the type of the parameter for the base class constructor.

NOTE

The macros QF_ACTIVE_CTOR_(), QF_ACTIVE_INIT_(), and QF_ACTIVE_DISPATCH_() are not needed in the C++ version of QF. Instead of providing these macros, the base class specified with QF_ACTIVE_SUPER_ must provide the constructor, init(), and dispatch() member functions with signatures compatible with the QHsm class interface.

The Maximum Number of Active Objects in the Application

(9) The macro QF_MAX_ACTIVE determines the maximum number of active objects, which at the same time is the maximum active object priority in the system. Currently QF_MAX_ACTIVE cannot exceed 63. Defining QF_MAX_ACTIVE to 8 or less results in a slightly better performance of the native "vanilla" scheduler, the QK scheduler, and the QF_publish() function.

NOTE

You need to always define QF_MAX_ACTIVE at the QF port level because QF provides no default value.

Various Object Sizes Within the QF Framework

This section defines various object sizes within the QF framework. All macros in this section have default values as specified, so you don't need to define these macros if the defaults are adequate.

(10) The macro QF_EVENT_SIZ_SIZE determines the size (in bytes) of the event-size representation in the QF. Valid values: 1, 2, or 4; default 2.

(11) The macro QF_EQUEUE_CTR_SIZE determines the size (in bytes) of the ring-buffer counters used in the native QF event queue implementation. Valid values: 1, 2, or 4; default 1.

(12) The macro `QF_MPOOL_SIZ_SIZE` determines the size (in bytes) of the block-size representation in the native QF memory pool. Valid values: 1, 2, or 4; default 2.

(13) The macro `QF_MPOOL_CTR_SIZE` determines the size (in bytes) of the block-counter representation in the native QF memory pool. Valid values: 1, 2, or 4; default 2.

(14) The macro `QF_TIMEEVT_CTR_SIZE` determines the size (in bytes) of the time event-counter representation in the `QTimeEvt struct`. Valid values: 1, 2, or 4; default 2.

QF Critical Section Mechanism

This section defines the critical section mechanism used within the QF framework, which you always need to provide. Refer to Section 7.3, "Critical Sections in QF," in Chapter 7 for the detailed discussion of critical sections in QF.

(15) The macro `QF_INT_KEY_TYPE` defines the data type of the "interrupt key" variable, which holds the interrupt status. When you define this macro, you indicate to the QF framework that the policy of "saving and restoring interrupt status" is used. Conversely, when you don't define the macro, the QF framework assumes the policy of "unconditional locking and unlocking interrupts."

(16) The macro `QF_INT_LOCK()` encapsulates the mechanism of interrupt locking. The macro takes a parameter into which it saves the interrupt lock status. The parameter is not used if you use the simple policy of "unconditional locking and unlocking interrupts."

(17) The macro `QF_INT_UNLOCK()` encapsulates the mechanism of unlocking interrupts. The macro takes a parameter from which it restores the interrupt lock status. The parameter is not used if you use the simple policy of "unconditional locking and unlocking interrupts."

Include Files Used by this QF Port

In this section you include the header files actually used in this particular QF port. Note that you generally don't include all files listed in this section at the same time.

(18) You include the header file(s) of the underlying operating system, RTOS, or kernel on which you base this QF port. If you use the simple cooperative "vanilla" kernel, you need to include the `qvanilla.h` header file (see

Section 8.2). If you use the QK preemptive kernel, you need to include the `qk_port.h` header file (see Chapter 10).

(19) You include the `qep_port.h` header file (see Section 8.1.4) if you derive `QActive` from `QHsm` or `QFsm`. Since the `QHsm` base class is the default, most of the time you need to include `qep_port.h`.

NOTE

If you choose to replace QEP with your own event processor, you need to include the `qevent.h` header file instead of `qep_port.h`. In addition, you might need to include the `<stdint.h>` header file and define the macros described in Listing 8.2.

---

(20) You need to include the `qequeue.h` header file if you use the native QF event queue for active objects. You might also want to include `qequeue.h` if your applications use the "raw" thread-safe queues (see Section 7.8.4 in Chapter 7) or event deferral via `QActive_defer()`/`QActive_recall()` mechanism (see Section 7.5.4 in Chapter 7).

(21) You need to include the `qmpool.h` header file if you use the native QF memory pool for event pools. You might also want to include `qmpool.h` if your applications use the native QF memory pools for allocating memory (see Section 7.9 in Chapter 7).

(22) You need to include the `qvanilla.h` header file only if you are using the native QF "vanilla" kernel. Note that `qvanilla.h` already includes `qequeue.h`, `qmpool.h`, and `qpset.h`, so you don't need to repeat them again.

(23) You *always* need to include the platform-independent QF header file `qf.h`. Typically, you include `qf.h` as the last header file because it generally depends on the other header files. For example, the type definitions of the operating system-dependent data members of `QActive` must be defined before the `QActive` class declaration located in `qf.h`.

Interface Used Only Inside QF, But Not in Applications

Below this line, you specify the elements of the PAL that are used only inside QF to adapt the framework source code to the particular platform. These elements are generally not used in the applications. I apply the naming convention of terminating such internal elements with an underscore, which you should consider a red flag to avoid using them in the application code.

Active Object Event Queue Operations

This section is required only when you use the native QF active object event queue with a traditional blocking operating system or RTOS. Chapter 7, "The Native QF Active Object Queue," explains the context in which these macros are used in the QF source code.

(24)   The macro QACTIVE_EQUEUE_WAIT_() encapsulates the mechanism of blocking the native QF event queue. Note that this macro is invoked from the critical section. The macro might unlock interrupts momentarily, but it must restore critical section before it returns. I provide an example of this macro in the Linux port (Section 8.3).

(25)   The macro QACTIVE_EQUEUE_SIGNAL_() encapsulates the mechanism of signaling the thread waiting on the event queue. Note that this macro is invoked from the critical section. The macro must exit the critical section before it returns. I provide an example of this macro in the Linux port (Section 8.4).

(26)   The macro QACTIVE_EQUEUE_ONEMPTY_() informs the underlying kernel that the active object event queue is becoming empty. Such notification is required by the cooperative "vanilla" kernel discussed in Section 8.2 as well as in the QF port to the preemptive run-to-completion QK kernel that I cover in Chapter 10.

QF Event Pool Operations

This section is required only when QF is ported to an external OS/RTOS. All standalone QF ports, such as the port to the native "vanilla" cooperative kernel or the QK preemptive kernel, do *not* need this section. I provide an example of this section that uses the memory partitions of the µC/OS-II RTOS in Section 8.3.

(27)   The macro QF_EPOOL_TYPE_ specifies the data type of the event pool used in this port.

(28)   The macro QF_EPOOL_INIT_() specifies the initialization function for the event pool object.

(29)   The macro QF_EPOOL_EVENT_SIZE_() returns the block size of a given event pool.

(30) The macro `QF_EPOOL_GET_()` obtains a memory block from a given event pool.

(31) The macro `QF_EPOOL_PUT_()` recycles a memory block to a given event pool.

8.1.6 The `qf_port.c` Source File

The `qf_port.c` source file defines platform-specific code for the QF port. Not all QF ports require this file. In fact, only the ports to the external RTOS or OS usually need some "glue-code" to bolt the framework to the external OS/RTOS. In particular, you don't need to provide any such "glue-code" for the simple "vanilla" kernel because it is totally portable and is already fully integrated with the QF (see Section 7.11 in Chapter 7). The QK preemptive kernel might require porting, just like any other preemptive real-time kernel, but in this case you will port QK and not QF (see Chapter 10). Again, `qf_port.c` won't be necessary for QK. Listing 8.4 shows the general layout of the `qf_port.c` source file for an external RTOS/OS.

Listing 8.4 The `qf_port.c` source file

```
(1)   #include "qf_pkg.h"
(2)   #include "qassert.h"

(3)   Q_DEFINE_THIS_MODULE(qf_port)

      /* Global objects -----------------------------------------------------*/
      . . .

      /* Local objects------------------------------------------------------*/
      . . .

      /*...................................................................*/
(4)   char const Q_ROM * Q_ROM_VAR QF_getPortVersion(void) {
          static const char Q_ROM Q_ROM_VAR version[] = "4.0.00";
          return version;
      }
      /*...................................................................*/
(5)   void QF_init(void) {

      }
      /*...................................................................*/
(6)   void QF_run(void) {
          . . .
      }
```

Continued onto next page

```
          /*...........................................................*/
(7)   void QF_stop(void) {
          . . .
      }
          /*...........................................................*/
(8)   void QActive_start(QActive *me,
                         uint8_t prio,                     /* the unique priority */
                         QEvent const *qSto[], uint32_t qLen,    /* event queue */
                         void *stkSto, uint32_t stkSize,       /* per-task stack */
                         QEvent const *ie)             /* the initialization event */
      {
(9)       me->prio = prio;                                  /* set the QF priority */
(10)      QF_add_(me);                        /* make QF aware of this active object */
(11)      QF_ACTIVE_INIT_(me, ie);                  /* execute the initial transition */

(12)      /* Initialize the event queue object 'me->eQueue' using qSto and qLen */
(13)      /* Create and start the thread 'me->thread' of the underlying RTOS     */
      }
          /*...........................................................*/
(14)  void QActive_stop(QActive *me) {
(15)      /* Cleanup me->eQueue or me->osObject */
      }
          /*...........................................................*/
      /* You need to define QActive_postFIFO(), QActive_postLIFO(), and
       * QActive_get_() only if your QF port uses the queue facility from
       * the underlying OS/RTOS.
       */
      void QActive_postFIFO(QActive *me, QEvent const *e) {
(16)      QF_INT_LOCK_KEY_
(17)      QF_INT_LOCK_();
(18)      if (e->dynamic_ != (uint8_t)0) {                  /* is it a dynamic event? */
(19)          ++((QEvent *)e)->dynamic_;       /* increment the reference counter */
          }
(20)      QF_INT_UNLOCK_();
          /* Post event pointer 'e' to the message queue of the RTOS 'me->eQueue'
           * using the FIFO policy without blocking. Also assert that the queue
           * accepted the event pointer.
           */
      }
          /*...........................................................*/
      void QActive_postLIFO(QActive *me, QEvent const *e) {
          QF_INT_LOCK_KEY_
          QF_INT_LOCK_();
          if (e->dynamic_ != (uint8_t)0) {                  /* is it a dynamic event? */
              ++((QEvent *)e)->dynamic_;       /* increment the reference counter */
          }
          QF_INT_UNLOCK_();
          /* Post event pointer 'e' to the message queue of the RTOS 'me->eQueue'
           * using the LIFO policy without blocking. Also assert that the queue
```

```
        * accepted the event pointer.
        */
}
/*...........................................................................*/
QEvent const *QActive_get_(QActive *me) {
        /* Get the next event from the active object queue 'me->eQueue'.
        * Block indefinitely as long as the queue is empty. Assert no errors
        * in the queue operation. Return the event pointer to the caller.
        */
}
```

(1) The qf_port.c source file is considered a part of the QF source code, and as such it needs access to the wider "package-scope" QF interface qf_pkg.h, located in <qp>\qpc\qf\source\. The qf_pkg.h header file includes qf_port.h, but it additionally defines some internal macros and objects shared only internally within the QF component.

(2) Typically, the qf_port.c source files uses QP assertions.

(3) As described in Section 6.7.3, "Customizable Assertions in C and C++," in Chapter 6, the macro Q_DEFINE_THIS_MODULE() defines the name of the module, which is subsequently referenced in all assertions implemented in this module.

(4) If you want to track the version of this particular QF port, you can define the function QF_getPortVersion() that returns the version string number. The string number is typically placed in ROM.

(5) The function QF_init() handles the specific initialization of the underlying OS/RTOS.

(6) The function QF_run() transfers control to QF to run the application. QF_run() is typically called from main() after you initialize the QF and start at least one active object with QActive_start(). QF_run() does not return to the caller as long as QF is in control.

(7) The function QF_stop() stops execution of the QF framework. The effect of this function might not be immediate. For example, it might only set an internal flag to terminate a loop inside QF_run() if QF_run() is implemented that way. When you design QF_stop(), you should also make sure that the callback QF_onCleanup() is called before the control is transferred back to the

underlying OS/RTOS. The specific QF ports described later in this chapter give examples of `QF_stop()` implementations. I discuss QF callback functions in the upcoming Section 8.1.8.

(8)   The function `QActive_start()` creates the active object's thread and notifies QF to start managing the active object. The argument 'me' is the pointer to the active object being started. The argument 'prio' is the QF priority you assign to the active object. In QF, every active object must have a unique priority, which you assign at startup and cannot change later. QF uses a priority numbering in which priority 1 is the lowest and higher numbers correspond to higher priorities. The arguments 'qSto' and 'qLen' are a pointer to the storage for the event queue buffer and the length of that buffer (in units of QEvent*), respectively. If the underlying RTOS cannot accept externally allocated storage for the queue, the 'qSto' pointer should be set to NULL. The argument 'stkSto' is the pointer to the storage for the private stack, and the argument 'stkSize' is the size of that stack (in bytes), respectively. If the underlying kernel/RTOS does not need per-task stacks or cannot accept externally allocated storage for the stack, the 'stkSto' pointer should be set to NULL. Finally, the argument 'ie' is a pointer to the initialization event for the topmost initial transition in the active object state machine. This argument is very specific to the active object being initialized and can be NULL.

(9)   The function `QActive_start()` must always set the active object priority.

(10) The function `QActive_start()` must always notify QF to add this active object at the priority level set previously.

(11) The function `QActive_start()` must trigger the initial transition in the active object's state machine (see also Listing 8.3(6)).

(12) The function `QActive_start()` must initialize the OS-specific event queue object.

(13) The function `QActive_start()` must start the active object's thread. Note that the priority of thread should correspond to the relative QF priority passed as the argument 'prio' to `QActive_start()`. If the underlying scheduler uses a different priority numbering scheme than QF, the concrete implementation of `QActive_start()` must remap the QF priority to the priority required by the underlying scheduler before invoking the QS-specific thread creation routine.

(14)   The function `QActive_stop()` stops the active object's thread and performs cleanup after the active object.

(15)   The function `QActive_stop()` must perform the OS-specific cleanup of the event queue or the OS-specific object for blocking the native QF event queue.

NOTE

You need to define the functions `QActive_postFIFO()`, `QActive_postLIFO()`, and `QActive_get_()` only if your QF port uses the queue facility from the underlying OS/RTOS. As you define these functions in `qf_port.c`, you should exclude the following three QF source files from the QF library build: `qa_fifo.c`, `qa_lifo.c`, and `qa_get_.c` (see also Table 8.2). The upcoming QF port to μC/OS-II (Section 8.2) provides an example of such a QF port.

(16-20)   The function `QActive_postFIFO()` (as well as `QActive_postLIFO()`) must increment the reference counter of a dynamic event exactly as shown.

8.1.7   The `qp_port.h` Header File

Every application C-file needs to include platform-specific header files for all QP components used in this application, as illustrated in Figure 8.1. To simplify this even further, you can combine all QP components used in the port so that the applications need to include just one `qp_port.h` header file. Listing 8.5 provides an example of the `qp_port.h` file. Typically, if you use the standard QP components, you don't need to change it much, although you might want to add to the `qp_port.h` file some elements that you always use with your QP applications.

Note that the `qp_port.h` header file is intended exclusively for the applications and is not used at all in building the QP libraries for the port.

Listing 8.5 The `qp_port.h` **header file**

```
#ifndef qp_port_h
#define qp_port_h

#include "qf_port.h"        /* includes qep_port.h and qk_port.h, if used */
#include "qassert.h"                                /* QP assertions */

#endif                                              /* qp_port_h */
```

8.1.8   Platform-Specific QF Callback Functions

A QF port cannot and should not define all the functions that it calls, because this would render the port too inflexible. Some functionality is simply much better left to the application, or perhaps to the Board Support Package (BSP). The functions that QF calls but doesn't actually implement are referred to as *callback functions*. All these functions in QF (as well as all other QP components) are easily identifiable by the "on" preposition used in the function name (e.g., QF_**on**Startup()). This section summarizes all QF callback functions.

```
void QF_onStartup(void)
```

This callback function is called just before the QF takes over control of the application. The main intent of the QF_onStartup() callback is to initialize and start interrupts. The timeline for calling QF_onStartup() depends on the particular QF port. However, in most cases, QF_onStartup() is called from QF_run(), right before starting any multitasking kernel or the background loop.

```
void QF_onCleanup(void)
```

QF_onCleanup() is called in some QF ports before QF returns to the underlying operating system or RTOS. The intent of the QF_onCleanup() callback is to give the application a chance to perform cleanup before exiting. This function might be empty, if the particular application has nothing to clean up or if the application never returns.

```
void QF_onIdle(void) or void QF_onIdle(QF_INT_KEY_TYPE
lockKey)
```

QF_onIdle() is called by the cooperative "vanilla" kernel built into QF. The signature of this callback depends on the interrupt-locking policy used in the QF port. I discussed the QF_onIdle() callback in Section 7.11.1 of Chapter 7 as well as in Section 8.2.4 in this chapter.

```
void Q_onAssert(char const Q_ROM * const Q_ROM_VAR file,
int line)
```

The callback Q_onAssert() is used by all QP components, not just QF. This callback is invoked in case the condition passed to Q_ASSERT(), Q_REQUIRE(), Q_ENSURE(),

Q_ERROR(), or Q_ALLEGE() evaluates to FALSE. The parameter 'file' denotes the filename where the assertion failed. The parameter 'line' holds the line number at which the assertion failed. I discuss the Q_onAssert() callback in Section 6.7.3 in Chapter 6.

8.1.9   System Clock Tick (Calling QF_tick())

As you design you port, you must decide how you are going to provide the system clock tick to call the QF_tick() function (see Section 7.7.2 in Chapter 7). Ideally, QF_tick() should be called from a periodic interrupt running at a rate between 10Hz and 100Hz. In case you can use the system clock interrupt, you don't need to do anything special at the QF port level. Simply don't forget to call QF_tick() from the system clock tick interrupt service routine (ISR) in your application. Naturally, your application can use the system clock tick ISR for other purposes as well.

However, in QF ports to the general-purpose operating systems, such as Linux or Windows, you cannot easily access the system clock tick ISR. In this case, you can call QF_tick() from the task level.[2] Typically, you dedicate a separate thread (the "ticker thread"), which is structured as an endless loop that calls QF_tick() and goes to sleep for the rest of the time slice. If you use this technique, you should implement the "ticker thread" in the qf_port.c source file. The Linux port in Section 8.4 provides an example of this approach.

8.1.10   Building the QF Library

A QP port should include a make.bat script, a Makefile, or a workspace/project file to build the QP libraries (see Listing 8.1(11)). Whichever way you actually provide to build the QF library, you should remember that not all QF source files need to be incorporated in every port. Table 8.2 contains the list of source files that you might need to *exclude* from the final QF build to avoid multiply defined symbols while linking the applications:

---

[2] As described in Section 7.7.2 in Chapter 7, QF_tick() is designed to be called from an interrupt or from the task level.

Table 8.2: QF source files that you might need to leave out

Filename	Comments
qa_fifo.c	Include in the build only if the port uses the native QF active object event queue. Do not include in the build if you provide definitions of functions `QActive_postFIFO()`, `QActive_postLIFO()`, and `QActive_get_()` in the qf_port.c source file.
qa_lifo.c	
qa_get_.c	
qvanilla.c	Include in the build only when you use the "vanilla" cooperative kernel.

8.2 Porting the Cooperative "Vanilla" Kernel

In Section 7.11 of Chapter 7 I described the native QF cooperative "vanilla" kernel. When you use QF with the "vanilla" kernel, you don't need to port the framework to the kernel—I already did it for you in Chapter 7. But you still need to port the "vanilla" kernel itself to the target CPU and compiler that you are using. Fortunately, this is quite easy due to the simplistic nature of the "vanilla" kernel. All you need to provide is the compiler-specific exact-width integer types in qep_porth.h and the interrupt-locking policy in qf_port.h. You typically don't need to provide any platform-specific source files.

In this section I show two examples of "vanilla" kernel ports, both of which you already used in Chapter 1 to run the "Fly 'n' Shoot" game example. The first one is for the 80x86 CPU under DOS, with the legacy Turbo C++ 1.01 compiler configured to generate code for "large" memory model. This port is located in <qp>\qpc\ports\80x86\dos\tcpp101\l\. The second "vanilla" port is for the ARM Cortex-M3 CPU with the latest IAR compiler and is located in <qp>\qpc\ports\cortex-m3\vanilla\iar\ (see Listing 8.1).

8.2.1 The qep_port.h Header File

Listing 8.6 shows the qep_port.h header file for 80x86/DOS/Turbo C++ 1.01/Large memory model. The legacy Turbo C++ 1.01 is a prestandard compiler, so I typedef the six platform-specific exact-with integer types used in QP.

Listing 8.6 The `qep_port.h` header file for 80x86/DOS/Turbo C++ 1.01/Large memory model

```
#ifndef qep_port_h
#define qep_port_h

   /* Exact-width integer types for DOS/Turbo C++ 1.01/Large memory model */
typedef signed   char int8_t;
typedef signed   int  int16_t;
typedef signed   long int32_t;
typedef unsigned char uint8_t;
typedef unsigned int  uint16_t;
typedef unsigned long uint32_t;

#include "qep.h"            /* QEP platform-independent public interface */

#endif                                              /* qep_port_h */
```

Listing 8.7 shows the `qep_port.h` header file for Cortex-M3/IAR. The IAR compiler is a standard C99 compiler, so I simply include the `<stdint.h>` header file that defines the platform-specific exact-with integer types.

Listing 8.7 The `qep_port.h` header file for Cortex-M3/IAR

```
#ifndef qep_port_h
#define qep_port_h

#include <stdint.h>           /* C99-standard exact-width integer types */
#include "qep.h"         /* QEP platform-independent public interface */

#endif                                              /* qep_port_h */
```

8.2.2 The `qf_port.h` Header File

The most important porting decision you need to make in the `qf_port.h` header file is the policy for locking and unlocking interrupts. To make this decision correctly, you need to learn a bit about your target CPU and the compiler to find out the most efficient way of enabling and disabling interrupts from C or C++. Generally, your first safe choice should be the more advanced policy of "saving and restoring the interrupt status" (Section 7.3.1 in Chapter 7). However, if you find out that it is safe to unlock interrupts within ISRs because your target processor can prioritize interrupts in hardware, you can use the simple and fast policy of "unconditional interrupt unlocking" (Section 7.3.2 in Chapter 7). With the fast policy you must always make sure that no QF functions are invoked with interrupts already locked, or more generally, that critical sections don't nest. Note that interrupts could be implicitly locked in the ISRs.

Listing 8.8 shows the `qf_port.h` header file for 80x86/DOS/Turbo C++ 1.01/Large memory model. I decided to use the simple "unconditional interrupt unlocking" policy because the standard PC has an external 8259A Programmable Interrupt Controller (PIC) and the Turbo C++ 1.01 compiler provides the pair of functions `disable()` and `enable()` to unconditionally lock and unlock interrupts, respectively. With this simple interrupt-locking policy, I need to be careful in calling QF services from ISRs. Listing 8.10 later in this section shows an example of the system clock tick ISR in this case. The QF port to µC/OS-II discussed in the upcoming Section 8.2 demonstrates the "saving and restoring the interrupt status" for the same CPU/compiler combination.

Listing 8.8 The `qf_port.h` header file for 80x86/DOS/Turbo C++ 1.01/Large memory model

```
#ifndef qf_port_h
#define qf_port_h

                                    /* DOS critical section entry/exit */
/* QF_INT_KEY_TYPE not defined: "unconditional interrupt unlocking" policy */
#define QF_INT_LOCK(dummy)        disable()
#define QF_INT_UNLOCK(dummy)      enable()

#include <dos.h>   /* DOS API, including disable()/enable() prototypes */
#undef outportb /*don't use the macro because it has a bug in Turbo C++ 1.01*/

#include "qep_port.h"                               /* QEP port */
#include "qvanilla.h"                /* The "Vanilla" cooperative kernel */
#include "qf.h"           /* QF platform-independent public interface */

#endif                                             /* qf_port_h */
```

Listing 8.9 shows the `qf_port.h` header file for Cortex-M3/IAR. Again, I use the simple "unconditional interrupt unlocking" policy because Cortex-M3 is equipped with the standard nested vectored interrupt controller (NVIC) and generally runs ISRs with interrupts unlocked.

Listing 8.9 The `qf_port.h` **header file for Cortex-M3/IAR**

```
#ifndef qf_port_h
#define qf_port_h

                                           /* QF critical section entry/exit */
/* QF_INT_KEY_TYPE not defined: "unconditional interrupt unlocking" policy */
#define QF_INT_LOCK(dummy)       __disable_interrupt()
#define QF_INT_UNLOCK(dummy)     __enable_interrupt()

#include <intrinsics.h>                             /* IAR intrinsic functions */

#include "qep_port.h"                                            /* QEP port */
#include "qvanilla.h"                   /* The "Vanilla" cooperative kernel */
#include "qf.h"            /* QF platform-independent public interface */

#endif                                                       /* qf_port_h */
```

8.2.3 The System Clock Tick (`QF_tick()`)

Strictly speaking, the "vanilla" QF port usually does not contain the system clock tick ISR because it is more convenient to place this ISR in the application. However, when developing any QF port, you need to have a pretty good idea how you are going to handle interrupts in general and the system clock interrupt in particular.

Listing 8.10 shows the system clock tick ISR for DOS, which is triggered by channel-0 of the 8253/8254 timer-counter chip connected to IRQ0.

Listing 8.10 The system clock tick ISR in 80x86/DOS/Turbo C++ 1.01/Large memory model

```
(1)   void interrupt ISR_tmr0(void) {    /* entered with interrupts LOCKED */
(2)       QF_INT_UNLOCK(dummy);                        /* unlock interrupts */
(3)       QF_tick();
          /* do some application-specific work ... */
(4)       QF_INT_LOCK(dummy);                       /* lock interrupts again */
(5)       outportb(0x20, 0x20);      /* write EOI to the master 8259A PIC */
      }
```

(1)   The Turbo C++ 1.01 compiler provides an extended keyword "interrupt" that enables you to program ISRs in C/C++.

(2)   The 80x86 processor locks interrupts in hardware before vectoring to the ISR. The interrupts can be unlocked right away, though, because the 8259A programmable interrupt controller prioritizes interrupts before they reach the CPU.

(3)   The `QF_tick()` service is called outside of critical section.

(4)   Interrupts are locked before exiting from the ISR.

(5)   The end-of-interrupt (EOI) instruction is sent to the master 8259A PIC so that it ends prioritization of this interrupt level.

Listing 8.11 shows the system clock tick ISR for Cortex-M3, which is triggered by the periodic timer called SysTick specifically designed for that purpose. In Cortex-M3 ISRs don't require any special instructions on entry or exit, so it is exceptionally easy to program ISR directly in C (isn't that nice?). This is actually unusual for most processor architectures. Furthermore, because Cortex-M3 enters ISRs with interrupts unlocked, there is no need to unlock interrupts to avoid nesting critical sections. Finally, the NVIC interrupt controller receives the EOI instruction implicitly by the special interrupt return code deposited in the LR register before the ISR entry, which means that you don't need to code EOI explicitly. All this enables using `QF_tick()` directly as the system clock tick ISR by placing it directly into the Cortex-M3 vector table. Typically, however, you need to add some more application-specific functionality to the SysTick interrupt than just `QF_tick()`, so I keep it as a separate function.

Listing 8.11 The SysTick ISR for Cortex-M3/IAR

```
void ISR_SysTick(void) {              /* entered with interrupts UNLOCKED */
    QF_tick();
    /* do some application-specific work ... */
}
```

8.2.4   Idle Processing (`QF_onIdle()`)

The "vanilla" kernel calls the `QF_onIdle()` callback function whenever it detects that all active object event queues in the system are empty. As I explained at the end of Section 7.11.1, the `QF_onIdle()` callback is invoked with interrupts locked and *must* always unlock interrupts; otherwise the system locks up.

Similarly to the system clock tick, the `QF_onIdle()` callback function usually is not part of the QF port because it is more convenient to place the idle processing in the application. However, idle processing is such an important issue to the "vanilla" kernel that I must mention it here.

Listing 8.12 shows the `QF_onIdle()` function for 80x86/DOS. The function simply unlocks interrupts because there is no standard way of turning low-power sleep mode for 80x86.

Listing 8.12 The `QF_onIdle()` **callback for 80x86/DOS**

```
void QF_onIdle(void) {                    /* entered with interrupts LOCKED */
    QF_INT_UNLOCK(dummy);                         /* always unlock interrupts */
    /* do some more application-specific work ... */
}
```

In contrast, the Cortex-M3 processor can be put into a low-power mode, as shown in Listing 8.13.

Listing 8.13 The `QF_onIdle()` **callback for Cortex-M3/IAR**

```
        void QF_onIdle(void) {                /* entered with interrupts LOCKED */
(1)  #ifdef NDEBUG
            /* Put the CPU and peripherals to the low-power mode.
            * NOTE: You might need to customize the clock management for your
            * application, by gating the clock to the selected peripherals.
            * See the datasheet for your particular Cortex-M3 MCU.
            */
(2)         __asm("WFI");                          /* Wait-For-Interrupt */
        #endif
(3)         QF_INT_UNLOCK(dummy);                  /* always unlock interrupts */
            /* optionally do some application-specific work ... */
        }
```

---

(1) I use conditional compilation to enter low-power sleep mode only when not debugging (NDEBUG defined). The transition to low-power sleep mode typically stops the CPU clock, which often makes debugging impossible.

(2) The Thumb-2 instruction set used in Cortex-M3 provides a special instruction WFI (Wait-For-Interrupt) for stopping the CPU clock. As described in the

"ARMv7-M Reference Manual" [ARM 06a], the WFI instruction can be executed with the PRIMASK bit set (interrupts locked). The Cortex-M3 core stops executing code immediately after the WFI instruction. Even so, any asynchronous exception (e.g., an interrupt) can wake the CPU by restarting the clock. The CPU resumes code execution but can handle the interrupt only when interrupts are unlocked.

(3) Interrupts are always unlocked. If an interrupt woke up the CPU from low-power sleep mode, the interrupt would be serviced only after interrupts get unlocked.

NOTE

Cortex-M3 allows entering the WFI mode atomically (with interrupts disabled), which is exactly how it should be done in the "vanilla" kernel or, more generally, in any foreground/background architecture.

8.3 QF Port to μC/OS-II (Conventional RTOS)

The QF real-time framework can work with virtually any traditional real-time operating system (RTOS), such as VxWorks, Nucleus, μC/OS-II, eCos, RTOS-32, and others alike. Combined with a conventional RTOS, QF takes full advantage of the multitasking capabilities of the RTOS by executing each active object in a separate task (see Section 6.3.6 in Chapter 6). The QF PAL includes an abstract RTOS interface to enable tight integration between QF and the underlying RTOS. Specifically, the PAL allows adapting most message queue variants as event queues of active objects as well as most memory partitions as QF event pools.

The most important reason why you should consider using a traditional blocking RTOS for executing event-driven QF applications is compatibility with the existing software. For example, most communication stacks (TCP/IP, USB, CAN) are designed for a traditional blocking kernel. In addition, a lot of legacy code requires blocking mechanisms, such as semaphores. A conventional RTOS allows you to run the existing software components as regular "blocking" tasks in parallel to the event-driven QF application.

On the other hand, if your project does not include legacy software or if you can afford to rewrite it in a nonblocking way, a conventional blocking RTOS could be an unnecessary complication and overkill for an event-driven system. You don't need an RTOS to partition the application into tasks, because active objects already achieve this goal. If the responsiveness of the simple "vanilla" kernel is not sufficient, you can employ a nonblocking, run-to-completion (RTC) *preemptive* kernel. As described in Section 6.3.8 in Chapter 6, such an RTC kernel provides deterministic, preemptive priority-based execution of event-driven systems at a fraction of the cost, complexity, and porting effort associated with any conventional blocking RTOS. In Chapter 10, I describe a lightweight, preemptive RTC kernel called QK, which is part of the QP event-driven platform.

In this section I describe porting QF to the µC/OS-II RTOS. I have chosen µC/OS-II as an example because it provides an excellent case study of a traditional priority-based, preemptive RTOS that it is superbly documented in the book *Micro-C/OS-II: The Real-Time Kernel*, by Jean J. Labrosse [Labrosse 02].

To focus the discussion of the port, I employ identical configuration to that described in the *Micro-C/OS-II* book. I use a standard 80x86-based PC running any variant of Windows and the Borland Turbo C++ 1.01 compiler[3] configured to generate code for "large" memory models. This configuration allows you to easily try the working port on your desktop without any investment in embedded hardware or the specific compiler. The QF port to µC/OS-II is located in the `<qp>\qpc\ports\80x86\ucos2\tcpp101\l\` directory.

> **NOTE**
>
> µC/OS-II is a commercial product of Micrium Inc. (www.micrium.com). With the kind permission from Micrium, the code accompanying this book contains precompiled object-modules for µC/OS-II v2.86 (the latest version as of this writing) as well as the external header files (see Listing 8.1(13) for the location of these files). Note that the version 2.86 is several generations newer than the version 2.60 published in the *Micro-C/OS-II* book. You need to contact Micrium Inc. to obtain the latest µC/OS-II source code and documentation.

Even though I had to settle on a concrete CPU and compiler to actually let you execute the code, I have carefully designed the provided QF port to µC/OS-II to be generic and applicable to most CPUs and compilers to which µC/OS-II has been ported. In the case of porting QF to an external RTOS (µC/OS-II in this case), the RTOS forms an indirection layer that insulates QF from the CPU and the nonportable compiler extensions. What this means is that you still need to port the RTOS to the specific CPU and compiler, but you don't need to modify the QF port to the RTOS because the RTOS API does not change.

---

[3] The *Micro-C/OS-II* book uses the newer Borland C++ 4.5 compiler.

One specific note I need to make at this point is that μC/OS-II employs a diametrically different porting and configuring policy than QF. Whereas QF is deployed as a fine-granularity library that allows the linker to eliminate any unused object modules at link time, μC/OS-II is configured mostly at compile time via configuration macros located in the `os_cfg.h` header file [Labrosse 02]. Most of the μC/OS-II code is compiled into just one monolithic object module (`ucos_ii.obj`), which the linker cannot chop into pieces easily.[4] The point to remember is that you need to recompile the QF library whenever you change the μC/OS-II configuration (the `os_cfg.h` file), because QF relies on some of this configuration.

8.3.1   The `qep_port.h` Header File

The platform-specific `qep_port.h` header file, shown in Listing 8.14, reuses the μC/OS-II configuration defined in the `os_cpu.h` header file. You typically don't need to modify `qep_port.h` to work with any other μC/OS-II port.

Listing 8.14 The `qep_port.h` header file for μC/OS-II

```
        #ifndef qep_port_h
        #define qep_port_h

(1)     #include "ucos_ii.h"                        /* uC/OS-II include file */

                        /* Exact-width integer types, as defined in uC/OS-II */
(2)     typedef INT8S   int8_t;
        typedef INT16S int16_t;
        typedef INT32S int32_t;
        typedef INT8U   uint8_t;
        typedef INT16U uint16_t;
        typedef INT32U uint32_t;

(3)     #include "qep.h"     /* QEP platform-independent public interface */

        #endif                                           /* qep_port_h */
```

---

[4] Some linkers can still remove unused code, even from a single object module, but most linkers simply pull in the entire object module.

(1) The `ucos_ii.h` header file contains the platform-independent μC/OS-II API as well as platform-specific declarations, such as μC/OS-II portable integer data types defined in the `os_cpu.h` header file.

(2) I decided to define the C99-standard exact-width integers in terms of μC/OS-II portable integer data types. The idea is to consistently reuse the same data types that you need to define for μC/OS-II anyway.

(3) As always, the platform-specific `qep_port.h` header file must include the platform-independent QEP interface.

8.3.2   The `qf_port.h` Header File

The integration between QF and μC/OS-II occurs at a higher level than in the "vanilla" port. The QF port to μC/OS-II uses less of the native QF facilities and more μC/OS-II services. Listing 8.15 shows the `qf_port.h` header file for μC/OS-II. The only piece of this file that you need to potentially adapt from one μC/OS-II port to another is the QF critical section definition.

Listing 8.15 The `qf_port.h` header file for μC/OS-II. The μC/OS-II API calls are shown in boldface

```
     #ifndef qf_port_h
     #define qf_port_h
                                            /* uC/OS-II event queue and thread types */
(1)  #define QF_EQUEUE_TYPE          OS_EVENT *
(2)  #define QF_THREAD_TYPE          INT8U

                    /* The maximum number of active objects in the application */
(3)  #define QF_MAX_ACTIVE           OS_MAX_TASKS
             /* uC/OS-II critical section operations (critical section method 3) */
(4)  #define QF_INT_KEY_TYPE         OS_CPU_SR
(5)  #define QF_INT_LOCK(key_)       ((key_) = OSCPUSaveSR())
(6)  #define QF_INT_UNLOCK(key_)     OSCPURestoreSR(key_)

(7)  #include "qep_port.h"     /* QEP port, includes the master uC/OS-II include */
(8)  #include "qequeue.h"          /* native QF event queue for deferring events */
(9)  #include "qf.h"                  /* QF platform-independent public interface */

     /*******************************************************************************
      * interface used only inside QF, but not in applications
      */
(10) typedef struct UCosMemPartTag {  /* uC/OS-II memory pool and block-size */
```

Continued onto next page

```
            OS_MEM *pool;                              /* uC/OS-II memory pool */
            QEventSize block_size;                  /* the block size of the pool */
      } UCosMemPart;


                                                  /* uC/OS-II event pool operations */
(11)  #define QF_EPOOL_TYPE_              UCosMemPart
      #define QF_EPOOL_INIT_(p_, poolSto_, poolSize_, evtSize_) do { \
            INT8U err; \
(12)        (p_).block_size = (evtSize_); \
            (p_).pool = OSMemCreate(poolSto_, (INT32U)((poolSize_)/(evtSize_)), \
                              (INT32U)(evtSize_), &err); \
            Q_ASSERT(err == OS_NO_ERR); \
      } while (0)


(13)  #define QF_EPOOL_EVENT_SIZE_(p_)  ((p_).block_size)
      #define QF_EPOOL_GET_(p_, e_) do { \
            INT8U err; \
            ((e_) = (QEvent *)OSMemGet(((p_).pool, &err)); \
      } while (0)


      #define QF_EPOOL_PUT_(p_, e_)   OSMemPut((p_).pool, (void *)(e_))

      #endif                                                   /* qf_port_h */
```

(1) The active object event queue type is set to the µC/OS-II message queue.

(2) In µC/OS-II, the task (thread) is unambiguously identified by its priority typed as INT8U.

(3) The maximum number of active object QF_MAX_ACTIVE is set to the maximum number of µC/OS-II tasks, configured in os_cfg.h.

Most of µC/OS-II ports use the critical section method 3 [Labrosse 02]. You specify the critical section method in the os_cpu.h header file for that port. The µC/OS-II critical section method 3 corresponds exactly to the "saving and restoring interrupt status" policy in QF. (Incidentally, the µC/OS-II critical section method 1 corresponds precisely to the QF policy of "unconditional interrupt locking and unlocking".)

(4) The policy of "saving and restoring interrupt status" is established by defining QF_INT_KEY_TYPE to the µC/OS-II type OS_CPU_SR.

(5) The QF_INT_LOCK() macro is defined consistently with the µC/OS-II OS_ENTER_CRITICAL() macro from os_cpu.h. Note that I cannot use the macro OS_ENTER_CRITICAL() directly because it hardcodes the interrupt lock key as cpu_sr.

(6)   The QF_INT_UNLOCK() macro is defined consistently with the μC/OS-II OS_EXIT_CRITICAL() macro from os_cpu.h.

(7)   Since this port uses QEP, the qep_port.h header file is included (see Listing 8.3 (19)).

(8)   I also decided to include the qequeue.h header file to be able to include the QF facilities for deferring and recalling events. Note that the native QF event queue is not used for posting events to active objects, which is accomplished with the μC/OS-II message queue.

(9)   The qf_port.h header file must always include the platform-independent QF interface (see Listing 8.3(23)).

This QF port uses the μC/OS-II memory partitions as event pools. In principle, you could define the QF_EPPOL_TYPE directly to OS_MEM*. The μC/OS-II memory partition provides all services that QF needs. In particular, you could get access to block size of the partition directly through the OS_MEM.OSMemBlk data member. However, memory partitions in other RTOSs often do not provide a way to obtain the block size managed by the partition (e.g., eCos). Therefore, in this QF port I decided to demonstrate a general solution, which is to combine the memory partition with the block-size data member into a single structure.

(10)   The UCosMemPart structure combines the μC/OS-II memory partition handle with the block size data member.

(11)   The QF_EPPOL_TYPE data type is defined to the UCosMemPart structure.

(12)   The initialization of an event pool includes storing the block size in the UCosMemPart.block_size data member.

(13)   The block-size information is available in the UCosMemPart.block_size data member.

8.3.3   The qf_port.c Source File

The QF port to μC/OS-II, as most QF ports to external RTOSs, require some glue-code to bolt the framework to the external RTOS. You place such code in the qf_port.c source file, which is shown in Listing 8.16. You typically don't need to change this file for different μC/OS-II ports.

Listing 8.16 The `qf_port.c` **source file for μC/OS-II; the μC/OS-II API calls are shown in boldface**

```
        #include "qf_pkg.h"
        #include "qassert.h"

        Q_DEFINE_THIS_MODULE(qf_port)

        /*.............................................................*/
        void QF_init(void) {
(1)         OSInit();                              /* initialize uC/OS-II */
        }
        /*.............................................................*/
        void QF_run(void) {
(2)         OSStart();                       /* start uC/OS-II multitasking */
        }
        /*.............................................................*/
        void QF_stop(void) {
(3)         QF_onCleanup();                    /* call the QF cleanup callback */
        }
        /*.............................................................*/
(4)     static void task_function(void *pdata) {      /* the expected signature */
(5)         ((QActive *)pdata)->running = (uint8_t)1;  /* enable the thread-loop */
(6)         while (((QActive *)pdata)->running) {            /* event loop */
(7)             QEvent const *e = QActive_get_((QActive *)pdata);
(8)             QF_ACTIVE_DISPATCH_(&((QActive *)pdata)->super, e);
(9)             QF_gc(e);     /* check if the event is garbage, and collect it if so */
            }
(10)        QF_remove_((QActive *)pdata); /* remove this object from the framework */
(11)        OSTaskDel(OS_PRIO_SELF);        /* make uC/OS-II forget about this task */
        }
        /*.............................................................*/
        void QActive_start(QActive *me, uint8_t prio,
                           QEvent const *qSto[], uint32_t qLen,
                           void *stkSto, uint32_t stkSize,
                           QEvent const *ie)
        {
            INT8U err;
(12)        me->eQueue = OSQCreate((void **)qSto, qLen);
(13)        Q_ASSERT(me->eQueue != (OS_EVENT *)0);        /* uC/OS-II queue created */
(14)        me->prio = prio;                             /* save the QF priority */
(15)        QF_add_(me);                        /* make QF aware of this active object */
(16)        QF_ACTIVE_INIT_(&me->super, ie);         /* execute initial transition */

                                /* uC/OS task is represented by its unique priority */
(17)        me->thread = (uint8_t)(QF_MAX_ACTIVE - me->prio); /* map to uC/OS prio. */
(18)        err = OSTaskCreateExt(&task_function,          /* the task function */
                me,                                     /* the 'pdata' parameter */
                &(((OS_STK *)stkSto)[(stkSize / sizeof(OS_STK)) - 1]), /* ptos */
```

```
                    me->thread,                    /* uC/OS-II task priority */
                    me->thread,                                      /* id */
                    (OS_STK *)stkSto,                               /* pbos */
                    stkSize/sizeof(OS_STK),    /* size of the stack in OS_STK units */
                    (void *)0,                                      /* pext */
                    (INT16U)OS_TASK_OPT_STK_CLR);                    /* opt */
(19)     Q_ASSERT(err == OS_NO_ERR);                /* uC/OS-II task created */
     }
     /*..............................................................*/
     void QActive_stop(QActive *me) {
         INT8U err;
(20)     ((QActive *)me)->running = (uint8_t)0;        /* stop the thread loop */
(21)     OSQDel(((QActive *)pdata)->eQueue, OS_DEL_ALWAYS, &err);/*cleanup queue */
(22)     Q_ASSERT(err == OS_NO_ERR);                /* uC/OS-II queue deleted */
     }
     /*..............................................................*/
     void QActive_postFIFO(QActive *me, QEvent const *e) {
(23)     QF_INT_LOCK_KEY_
(24)     QF_INT_LOCK_();
(25)     if (e->dynamic_ != (uint8_t)0) {
(26)         ++((QEvent *)e)->dynamic_;
         }
(27)     QF_INT_UNLOCK_();
(28)     Q_ALLEGE(OSQPost((OS_EVENT *)me->eQueue, (void *)e)== OS_NO_ERR);
     }
     /*..............................................................*/
     void QActive_postLIFO(QActive *me, QEvent const *e) {
         QF_INT_LOCK_KEY_
         QF_INT_LOCK_();
         if (e->dynamic_ != (uint8_t)0) {
             ++((QEvent *)e)->dynamic_;
         }
         QF_INT_UNLOCK_();
(29)     Q_ALLEGE(OSQPostFront((OS_EVENT *)me->eQueue, (void *)e) == OS_NO_ERR);
     }
     /*..............................................................*/
     QEvent const *QActive_get_(QActive *me) {
         INT8U err;
(30)     QEvent const *e = (QEvent *)OSQPend((OS_EVENT *)me->eQueue, 0, &err);
(31)     Q_ASSERT(err == OS_NO_ERR);
         return e;
     }
```

(1) The function `QF_init()` initializes the framework. In the case of QF port to µC/OS-II, the function must initialize the underlying RTOS by the µC/OS-II call `OSInit()`.

(2) The function `QF_run()` transfers control to the framework to run the application. In the case of QF port to µC/OS-II, the function starts multitasking by the µC/OS-II call `OSStart()`.

NOTE

Note that neither QF_init() nor QF_run() call the QF_onStartup() callback to enable interrupts. This is intentional in the μC/OS-II port. As described in Section 3.11 of the *Micro-C/OS-II* book [Labrosse 02], μC/OS-II requires you to start interrupts, including the system clock tick, only after you start multitasking with OSStart(). The recommended method is to start interrupts from a μC/OS-II task. The example application located in `<qp>\qpc\examples\80x86\ucos2\tcpp101\l\dpp\` illustrates this aspect.

---

(3)   The function QF_stop() stops the framework. There is nothing you can do to stop μC/OS-II, you simply abort. Therefore the only action is to call the cleanup callback.

(4)   Under a traditional RTOS, all active object threads execute the same function task_function(), which has the structure shown in Figure 6.12(A) in Chapter 6. The task function has the exact signature expected by μC/OS-II. The parameter pdata is set to the active object owning the task.

(5)   The task function sets the QActive.running flag to continue the local event loop.

(6)   The event loop continues as long as the QActive.running flag is set.

(7-9)   These are the three steps of the active object thread (see Listing 7.8 in Chapter 7).

(10)   After the event loop terminates, the active object is removed from the framework.

(11)   The task is deleted by the μC/OS-II call OSTaskDel().

(12)   The first step in starting an active object is creating the event queue by the μC/OS-II call OSQCreate().

(13)   The queue creation must be successful; otherwise the application cannot continue.

(14)   The active object's priority is set.

(15)   The active object is registered with the QF framework.

(16)   The active object's state machine is initialized.

(17)   The QF priority is mapped to the μC/OS-II task priority.

μC/OS-II uses a priority numbering scheme in which 0 represents the highest possible priority and higher numerical values represent lower priority of the tasks. This happens to be exactly the opposite of the QF priority numbering scheme.

(18)   The active object thread is created by the μC/OS-II call `OSTaskCreateExt()`.

NOTE

Traditional RTOSs, such as μC/OS-II, require per-task stacks. The `QActive_start()` parameters 'stkSto' and 'stkSize' are designed specifically to support conventional RTOSs.

---

(19)   The task creation must be successful; otherwise the application cannot continue.

(20)   Clearing the `QActive.running` flag terminates the event loop and exits the active object thread (see line (5)).

(21)   The event queue is deleted by the μC/OS-II call `OSQDel()`.

(22)   The deletion of the queue must be successful.

The QF port to μC/OS-II does not use the native QF active object queues. Therefore, the QF implementation of `QActive_postFIFO()`, `QActive_postLIFO()`, and `QActive_get_()` must be replaced by the μC/OS-II-specific code. The rest of the `qf_port.c` source file defines these three functions for μC/OS-II (see also Section 8.1.6).

(23-27)   Posting an event to a queue must always increment the reference counter of a dynamic event. This must happen exactly as shown.

(28)   The event pointer is posted to the μC/OS-II message queue with the μC/OS-II call `OSQPost()`, which uses the standard FIFO policy. Note that μC/OS-II message queues are designed to accept only pointer-size objects. This is exactly what QF needs. The function is called inside the assertion macro `Q_ALLEGE()`, to make sure that the operation always succeeds (this is part of the QF's event delivery guarantee). Note that `Q_ALLEGE()` evaluates its argument even if assertions are disabled (see Section 6.7.3 in Chapter 6).

(29)   The event pointer is posted to the μC/OS-II message queue with the μC/OS-II call `OSQPostFront()`, which uses the LIFO policy. Again, the assertion makes sure that the event is posted successfully.

> **NOTE**
>
> You should make sure that the message-posting operation you're using is callable from ISRs. The µC/OS-II functions `OSQPost()` and `OSQPostFront()` are callable from ISRs.

(30)   The event is retrieved from the message queue with the µC/OS-II call `OSQPend()`. The second argument to this function is the timeout, whereas timeout of 0 indicates indefinite waiting on an empty event queue.

(31)   The pending operation must not fail.

8.3.4   Building the µC/OS-II Port

The QF port to µC/OS-II comes with the build script `make.bat` located in the port directory `<qp>\qpc\ports\80x86\ucos2\tcpp101\l\`. The most important aspect of the script is that none of the files listed in Table 8.2 are included in the QF library build, because this functionality is provided in the source file `qf_port.c` (Listing 8.16).

8.3.5   The System Clock Tick (`QF_tick()`)

In µC/OS-II, the system clock tick interrupt is coded in assembly (see the µC/OS-II port file `os_cpu_a.asm`). Generally, you should not touch such assembly files. µC/OS-II provides, however, a customizable "hook" function `OSTimeTickHook()`, which is called from the system clock tick ISR and is the ideal place to invoke `QF_tick()`.

Rather than defining the `OSTimeTickHook()` in `qf_port.c`, I decided to let the application define the `OSTimeTickHook()` callback so that you can easily add some more processing to it. You should not forget to invoke `QF_tick()`. I'd like to remind you again that µC/OS-II requires you to actually start the clock tick interrupt only from the task level in your application.

```
void OSTimeTickHook(void) {
    QF_tick();
    /* optionally, do some application-specific work ... */
}
```

8.3.6 Idle Processing

Finally, I'd like to mention the idle processing and the proper use of MCUs low-power modes with a preemptive kernel, such as μC/OS-II, because it is fundamentally different than in a nonpreemptive case like the "vanilla" kernel. A preemptive kernel performs a context switch to a special *idle task* when all other tasks are blocked. Most kernels provide a way to customize the idle task (e.g., μC/OS-II provides a callback function `OSTaskIdleHook()`) so that you can conveniently implement the transition to a low-power MCU state from the idle task. The main difference between a preemptive kernel and a nonpreemptive foreground/background system or the "vanilla" kernel is that as long as tasks are ready to run, the preemptive kernel never switches the context back to the idle task. Consequently the transition to a low-power mode from the idle task is much simpler because it does not need to occur with interrupts locked. In fact, the `OSTaskIdleHook()` callback is always invoked with interrupts unlocked.

8.4 QF Port to Linux (Conventional POSIX-Compliant OS)

Programming for a general-purpose operating system (OS) such as Linux or Windows is very different from working with a typical RTOS or a "bare metal" embedded processor. A "big" OS strictly limits you to a given API, whether the POSIX API for Linux or the Win32 API for Windows.

In particular, the general-purpose APIs don't let you lock and unlock interrupts, so you need to employ a different mutual-exclusion mechanism to implement the QF critical section. Also surprisingly, the "big" APIs don't support a lightweight message queue, so you need to build you own out of the native QF event queue and a blocking mechanism supported in the given API.

In this section I describe a QF port to Linux, which should be also directly applicable on any POSIX-compliant OS, as it strictly adheres to the POSIX 1003.1cn1995 standard [Butenhof 97]. In this port, a QF application runs as a single process, with each QF active object executing in a separate lightweight POSIX thread (Pthread). The port uses a *Pthread mutex* to implement the QF critical section and the *Pthread condition variables* to provide the blocking mechanism for event queues of active objects.

8.4.1   The `qep_port.h` Header File

Listing 8.17 shows the `qep_port.h` header file for Linux. The GNU gcc compiler supports the C99 standard, so I simply include the `<stdint.h>` file. I have also increased the event signal size to 2 bytes, which gives you 64K different signals. With this configuration, the size of `QEvent` base structure is just 3 bytes, which most 32-bit compilers will pad to 4 bytes.

Listing 8.17 The `qep_port.h` header file for Linux

```
#ifndef qep_port_h
#define qep_port_h
                                         /* 2-byte (64K) signal space */
#define Q_SIGNAL_SIZE 2

#include <stdint.h>               /* C99-standard exact-width integers */
#include "qep.h"          /* QEP platform-independent public interface */

#endif                                                /* qep_port_h */
```

8.4.2   The `qf_port.h` Header File

Listing 8.18 shows the `qf_port.h` header file for Linux. You typically should not need to change this file as you move to a different POSIX-compliant OS.

Listing 8.18 The `qf_port.h` header file for Linux; boldface indicates elements of the Pthread API

```
    #ifndef qf_port_h
    #define qf_port_h
                                  /* Linux event queue and thread types */
(1) #define QF_EQUEUE_TYPE        QEQueue
(2) #define QF_OS_OBJECT_TYPE     pthread_cond_t
(3) #define QF_THREAD_TYPE        pthread_t

                 /* The maximum number of active objects in the application */
(4) #define QF_MAX_ACTIVE         63
                      /* various QF object sizes configuration for this port */
(5) #define QF_EVENT_SIZ_SIZE    4
(6) #define QF_EQUEUE_CTR_SIZE   4
(7) #define QF_MPOOL_SIZ_SIZE    4
(8) #define QF_MPOOL_CTR_SIZE    4
(9) #define QF_TIMEEVT_CTR_SIZE  4
```

```
                              /* QF critical section entry/exit for Linux, see NOTE01 */
(10)  /* QF_INT_KEY_TYPE not defined, "unconditional interrupt locking" policy */
(11)  #define QF_INT_LOCK(dummy)    pthread_mutex_lock(&QF_pThreadMutex_)
(12)  #define QF_INT_UNLOCK(dummy) pthread_mutex_unlock(&QF_pThreadMutex_)

(13)  #include <pthread.h>                              /* POSIX-thread API */
(14)  #include "qep_port.h"                                   /* QEP port */
(15)  #include "qequeue.h"                          /* Linux needs event-queue */
(16)  #include "qmpool.h"                          /* Linux needs memory-pool */
(17)  #include "qf.h"                 /* QF platform-independent public interface */


      /**********************************************************************
       * interface used only inside QF, but not in applications
       */
                                      /* OS-object implementation for Linux */
(18)  #define QACTIVE_EQUEUE_WAIT_(me_) \
          while ((me_)->eQueue.frontEvt == (QEvent *)0) \
          pthread_cond_wait(&(me_)->osObject, &QF_pThreadMutex_)

(19)  #define QACTIVE_EQUEUE_SIGNAL_(me_) \
          pthread_cond_signal(&(me_)->osObject)
(20)  #define QACTIVE_EQUEUE_ONEMPTY_(me_) ((void)0)
                                      /* native QF event pool operations */
(21)  #define QF_EPOOL_TYPE_        QMPool
(22)  #define QF_EPOOL_INIT_(p_, poolSto_, poolSize_, evtSize_) \
          QMPool_init(&(p_), poolSto_, poolSize_, evtSize_)
(23)  #define QF_EPOOL_EVENT_SIZE_(p_) ((p_).blockSize)
(24)  #define QF_EPOOL_GET_(p_, e_)     ((e_) = (QEvent *)QMPool_get(&(p_)))
(25)  #define QF_EPOOL_PUT_(p_, e_)    (QMPool_put(&(p_), e_))

(26)  extern pthread_mutex_t QF_pThreadMutex_;  /* mutex for QF critical section */
```

(1)   The Linux port employs the QF native QEQueue as the event queue for active objects.

(2)   The Pthread condition variable is used for blocking the QF native event queue. Note that each active object has its own private condition variable.

(3)   Each active object also holds a handle to its Pthread.

(4)   The Linux port is configured to use the maximum allowed number of active objects.

(5-9) Linux requires a 32-bit CPU, so I configure all sizes of internal QF objects to 4 bytes.

(10)   The `QF_INT_KEY_TYPE` macro is not defined. This means that the interrupt status is not preserved across the QF critical section.

(11)   The QF critical section is implemented with a single global Pthread mutex `QF_pThreadMutex_`. The mutex is locked upon entry to a critical section.

(12)   The global mutex `QF_pThreadMutex_` is unlocked upon exit from a critical section.

NOTE

The global mutex `QF_pThreadMutex_` is configured as a normal "fast" Pthread mutex that cannot handle nested locks. Consequently, the QF port to Linux does *not* support nesting of critical sections. This QF port is designed to never nest critical sections internally, but you should be careful not to call QF services from critical sections at the application level.

---

(13)   The system header file `<pthread.h>` contains the Pthread API.

(14)   This QF port uses the QEP event processor.

(15)   This QF port uses the native QF event queue `QEQueue`.

(16)   This QF port uses the native QF memory pool `QMPool`.

(17)   The platform-independent `qf.h` header file must be always included.

The following three macros `QACTIVE_EQUEUE_WAIT_()`, `QACTIVE_EQUEUE_SIGNAL_()`, and `QACTIVE_EQUEUE_ONEMPTY_()` customize the native QF event queue to use the Pthread condition variable for blocking and signaling the active object's thread. (See Section 7.8.3 in Chapter 7 for the context in which QF calls these macros.)

(18)   As long as the queue is empty, the private condition variable `osObject` blocks the calling thread. Note that the macro `ACTIVE_EQUEUE_WAIT_()` is called from critical section, that is, with the global mutex `QF_pThreadMutex_` locked.

The behavior of the `pthread_cond_wait()` function requires explanation. Here is the description from the POSIX-thread standard:

> "The function pthread_cond_wait() atomically releases the associated mutex and causes the calling thread to block on the condition variable. Atomically here means 'atomically with respect to access by another thread to the mutex and then the condition variable.' That is, if another thread is able to acquire the mutex after the about-to-block thread has released it, then a subsequent call to pthread_cond_signal() or pthread_cond_broadcast() in that thread behaves as if it were issued after the about-to-block thread has blocked."

The bottom line is that the global mutex QF_pThreadMutex_ remains unlocked only as long as pthread_cond_wait() blocks. The mutex gets locked again as soon as the function unblocks. This means that the macro ACTIVE_EQUEUE_WAIT_() returns within critical section, which is exactly what the intervening code in QActive_get_() expects.

The while -loop around the pthread_cond_wait() call is necessary because of the following comment in the POSIX-thread documentation:

> "Since the return from pthread_cond_wait() does not imply anything about the value of the predicate, the predicate should be re-evaluated upon such return."

(19)   The macro QACTIVE_EQUEUE_SIGNAL_() is called when an event is inserted into an empty event queue (so the queue becomes not-empty). Note that this macro is called from a critical section.

(20)   The macro QACTIVE_EQUEUE_ONEMPTY_() is called when the queue is becoming empty. This macro is defined to nothing in this port.

(21-25)   The Linux port uses QMPool as the QF event pool. The platform abstraction layer (PAL) macros are set to access the QMPool operations (see Section 7.9 in Chapter 7).

(26)   The global mutex QF_pThreadMutex_ is declared as an external variable.

8.4.3   The qf_port.c Source File

The qf_port.c source file shown in Listing 8.19 provides the "glue-code" between QF and the POSIX API. The general assumption I make here is that QF is going to be used in real-time applications (perhaps "soft real-time"). This means that I'm trying to use as much as possible the real-time features available in the standard POSIX API. Since some of these features require the "superuser" privileges, the actual real-time behavior of the application will depend on the privilege level at which it is launched. As always with a general-purpose OS used for real-time applications, your actual mileage may vary.

Listing 8.19 The `qf_port.c` header file for Linux; boldface indicates elements of the Pthread API

```
      #include "qf_pkg.h"
      #include "qassert.h"

      #include <sys/mman.h>                                    /* for mlockall() */
      #include <sys/select.h>                                  /* for select() */
      Q_DEFINE_THIS_MODULE(qf_port)

      /* Global objects --------------------------------------------------------*/
 (1)  pthread_mutex_t QF_pThreadMutex_ = PTHREAD_MUTEX_INITIALIZER;

      /* Local objects ---------------------------------------------------------*/
      static uint8_t l_running;

      /*..........................................................................*/
      void QF_init(void) {

                              /* lock memory so we're never swapped out to disk */
 (2)      /*mlockall(MCL_CURRENT | MCL_FUTURE);          uncomment when supported */
      }
      /*..........................................................................*/
 (3)  void QF_run(void) {
          struct sched_param sparam;
          struct timeval timeout = { 0 };                   /* timeout for select() */

 (4)      QF_onStartup();                                    /* invoke startup callback */

                  /* try to maximize the priority of the ticker thread, see NOTE01 */
 (5)      sparam.sched_priority = sched_get_priority_max(SCHED_FIFO);
 (6)      if (pthread_setschedparam(pthread_self(), SCHED_FIFO, &sparam) == 0) {
                      /* success, this application has sufficient privileges */
          }
          else {
              /* setting priority failed, probably due to insufficient privieges */
          }
          l_running = (uint8_t)1;
 (7)      while (l_running) {
 (8)          QF_tick();                                     /* process the time tick */

 (9)          timeout.tv_usec = 8000;
(10)          select(0, 0, 0, 0, &timeout);    /* sleep for the full tick, NOTE05 */
          }
```

```
(11)        QF_onCleanup();                                    /* invoke cleanup callback */
(12)        pthread_mutex_destroy(&QF_pThreadMutex_);
(13)   }
       /*..........................................................................*/
     void QF_stop(void) {
(14)        l_running = (uint8_t)0;                            /* stop the loop in QF_run() */
       }
       /*..........................................................................*/
(15) static void *thread_routine(void *arg) {       /* the expected POSIX signature */
(16)        ((QActive *)arg)->running = (uint8_t)1; /* allow the thread loop to run */
(17)        while (((QActive *)arg)->running) { /* QActive_stop() stopps the loop */
(18)            QEvent const *e = QActive_get_((QActive *)arg);/*wait for the event */
(19)            QF_ACTIVE_DISPATCH_(&((QActive *)arg)->super, e);/* dispatch to SM */
(20)            QF_gc(e);     /* check if the event is garbage, and collect it if so */
            }
(21)        QF_remove_((QActive *)arg);/* remove this object from any subscriptions */
            return (void *)0;                               /* return success */
(22)   }
       /*..........................................................................*/
     void QActive_start(QActive *me, uint8_t prio,
                        QEvent const *qSto[], uint32_t qLen,
                        void *stkSto, uint32_t stkSize,
                        QEvent const *ie)
       {
            pthread_attr_t attr;
            struct sched_param param;

(23)        Q_REQUIRE(stkSto == (void *)0); /* p-threads allocate stack internally */

(24)        QEQueue_init(&me->eQueue, qSto, (QEQueueCtr)qLen);
(25)        pthread_cond_init(&me->osObject, 0);

(26)        me->prio = prio;
(27)        QF_add_(me);                         /* make QF aware of this active object */
(28)        QF_ACTIVE_INIT_(&me->super, ie);     /* execute the initial transition */

            /* SCHED_FIFO corresponds to real-time preemptive priority-based scheduler
             * NOTE: This scheduling policy requires the superuser privileges
             */
(29)        pthread_attr_init(&attr);
(30)        pthread_attr_setschedpolicy(&attr, SCHED_FIFO);
                                                                 /* see NOTE04 */
(31)        param.sched_priority = prio
                                  + (sched_get_priority_max(SCHED_FIFO)
                                  - QF_MAX_ACTIVE - 3);
(32)        pthread_attr_setschedparam(&attr, &param);
(33)        pthread_attr_setdetachstate(&attr, PTHREAD_CREATE_DETACHED);

(34)        if (pthread_create(&me->thread, &attr, &thread_routine, me) != 0) {
```

Continued onto next page

```
                    /* Creating the p-thread with the SCHED_FIFO policy failed.
                     * Most probably this application has no superuser privileges,
                     * so we just fall back to the default SCHED_OTHER policy
                     * and priority 0.
                     */
(35)            pthread_attr_setschedpolicy(&attr, SCHED_OTHER);
(36)            param.sched_priority = 0;
(37)            pthread_attr_setschedparam(&attr, &param);
(38)            Q_ALLEGE(pthread_create(&me->thread, &attr, &thread_routine, me)== 0);
            }
(39)        pthread_attr_destroy(&attr);
        }
    /*.................................................................*/
    void QActive_stop(QActive *me) {
(40)        me->running = (uint8_t)0;          /* stop the event loop in QActive_run() */
(41)        pthread_cond_destroy(&me->osObject); /* cleanup the condition variable */
        }
```

(1)   The global `Pthread mutex QF_pThreadMutex_` variable for the QF critical section is defined.

(2)   On POSIX systems that support it, you might want to call the `mlockall()` function to lock in physical memory of all the pages mapped by the address space of a process. This prevents nondeterministic swapping of the process memory to disk and back. The standard desktop Linux does not support `mlockall()`, so it is commented out.

(3)   The `QF_run()` function is called from `main()` to let the framework execute the application. In this QF port, the `QF_run()` function is used as the "ticker thread" to periodically call the `QF_tick()` function.

(4)   The callback function `QF_onStartup()` is called to give the application a chance to perform startup.

(5,6)   These two lines of code attempt to set the current thread (the "ticker thread") to the `SCHED_FIFO` scheduling policy and to the maximum priority within that policy.

In Linux, the scheduler policy closest to real time is the `SCHED_FIFO` policy, available only with the "superuser" privileges. `QF_run()` attempts to set this policy to maximize its priority so that the system clock tick occurs in the most timely manner. However, setting the `SCHED_FIFO` policy might fail, most probably due to insufficient privileges.

(6)   The "ticker" thread runs in loop, as long as the `l_running` flag is set.

(7)   The "ticker" thread calls `QF_tick()` outside of any critical section.

(9,10)   The "ticker" thread is put to sleep for the rest of the time slice.

I use the `select()` system call as a fairly portable way to sleep because it seems to deliver the shortest sleep time of just one clock tick. The timeout value passed to `select()` is rounded up to the nearest tick (10 milliseconds on desktop Linux). The timeout cannot be too short, because the system might choose to busy-wait for very short timeouts. An obvious alternative—the POSIX `nanosleep` system call—seems to be unable to block for less than two clock ticks (20 milliseconds).

Also according to the man pages, the function `select()` on Linux *modifies* the timeout argument to reflect the amount of time not slept. Most other implementations do not do this. I handle this quirk in a portable way by always setting the microsecond part of the structure before each `select()` call (see line (9)).

(11)   When the loop exits, the callback function `QF_onCleanup()` is called to give the application a chance to perform cleanup.

(12)   The global Pthread mutex `QF_pThreadMutex_` is cleaned up before exit.

(13)   The `QF_run()` function exits, which causes the `main()` function to exit. The system terminates the process and shuts down all Pthreads spawned from `main()`.

(14)   The exit sequence just described in triggered when the application calls `QF_stop()`, which stops the loop in `QF_run()`.

The following static function `thread_routine()` specifies the thread function of all active objects.

(15)   In this POSIX port, all active object threads execute the same function `thread_routine()`, which has the structure shown in Figure 6.12(A) in Chapter 6. The thread routine has the exact signature expected by POSIX API `pthread_create()`. The parameter `arg` is set to the active object owning the thread.

(16)   The thread routine sets the `QActive.running` flag to continue the local event loop.

(17)   The event loop continues as long as the `QActive.running` flag is set.

(18-20)   These are the three steps of the active object thread (see Listing 7.8 in Chapter 7).

(21)   After the event loop terminates, the active object is removed from the framework.

(22)   The return from the thread routine cleans up the POSIX thread.

(23)   The `pthread_create()` function allocates the stack space for the thread internally. This assertion makes sure that the stack storage is not provided, because that would be wasteful.

(24)   The native QF event queue of the active object is initialized.

(25)   The Pthread condition variable is initialized.

(26)   The active object's priority is set.

(27)   The active object is registered with the QF framework.

(28)   The active object's state machine is initialized.

(29-33)   The attribute structure for the active object thread is initialized. In the first attempt, the thread is created with the `SCHED_FIFO` policy.

According to the man pages (for `pthread_attr_setschedpolicy()`) the only value supported in the Linux Pthread implementation is `PTHREAD_SCOPE_SYSTEM`, meaning that the threads contend for CPU time with all processes running on the machine. In particular, thread priorities are interpreted relative to the priorities of all other processes on the machine. This is good, because it seems that if we set the priorities high enough, no other process (or threads running within) can gain control over the CPU. However, QF limits the number of priority levels to `QF_MAX_ACTIVE`. Assuming that a QF application will be real time, this port reserves the three highest Linux priorities for the system threads (e.g., the ticker, I/O), and the rest highest-priorities for the active objects.

(34)   The active object Pthread is created. If the thread creation fails, it is most likely due to insufficient privileges to use the real-time policy `SCHED_FIFO`.

(35-37)   The thread attributes are modified to use the default scheduling policy `SCHED_OTHER` and priority zero.

(38) The Pthread creation is attempted again. This time it must succeed or the application cannot continue.

(39) The Pthread attribute structure is cleaned up.

(40) To stop an active object, the `QActive_stop()` function clears the `QActive.running` flag. This stops the active object event loop at line (17) and causes the thread routine to exit.

(41) The condition variable is cleaned up.

8.5 Summary

Portability of system-level software, such as the QF real-time framework, is critical to its usability, especially in the real-time embedded systems (RTES) domain. QF has been designed from the ground up to be highly adaptable to various CPU architectures, operating systems, and compilers. Without a doubt, the ease of portability has been the most difficult, tedious, and time-consuming aspect of the framework's design, implementation, and testing.

QF contains a clearly defined platform abstraction layer (PAL), which encapsulates all platform-specific code and cleanly separates it from the platform-independent code. The QF PAL must be more flexible than the hardware abstraction layers (HALs) found in various RTOSs. On one hand, the PAL must allow easy porting of the QF framework to "bare metal" CPUs and compilers. This kind of portability is used in the standalone QF configurations, such as the cooperative "vanilla" kernel, and the QK preemptive kernel that I describe in Chapter 10. On the other hand, the QF PAL must also allow integration between the QF and any RTOS/OS, which occurs at the level of the API provided in the external RTOS/OS.

The proper structure and completeness of the PAL becomes apparent only after the framework has been ported to a wide range of different actual targets, including peculiar CPUs and not always standard compilers. To date, the standalone QF configurations have been ported to over 10 different CPU architectures, ranging from 8-bit (e.g., 8051, PIC, AVR, 68HC(S)08, Cypress M8C/PSoC), through 16-bit (e.g., MSP430, M16C, x86-real mode), to 32-bit architectures (e.g., traditional ARM, ARM Cortex-M3, Altera Nios II, x86). QF has been also ported to six major operating systems and RTOSs, including Linux (POSIX), Windows (Win32), and VxWorks.

Check the Website www.quantum-leaps.com for a list of available QF ports because new ports are added frequently.

After downloading a QF port or developing your own, you need to test the port to verify that it works on your particular target system with your specific operating system and compiler. At this point, you will need a test application, simple enough so that you will essentially test the QF framework all by itself, yet not completely trivial so that it will put most of the QF mechanisms through the paces. In the next chapter I describe a simple test application that historically I use to test all QF ports.

CHAPTER 9

Developing QP Applications

Example is not the main thing in influencing others. It is the only thing.
—*Albert Schweitzer*

In the previous two chapters, I explained the internal workings of the QF real-time framework and issues related to porting QF to various CPUs, operating systems, and compilers. However, I want you to realize that the way the QF framework itself is implemented internally is very different from the way you develop *applications* running on top of the framework.

A real-time framework, as any piece of *system-level* software, must internally employ many low-level mechanisms, such as critical sections and various blocking APIs of the underlying RTOS, if you use an RTOS. These mechanisms are always tricky to use correctly and programmers often underestimate the true risks and costs of their use.

But the good news is that this traditional approach to concurrent programming is contained within the framework. Once the framework is built and thoroughly tested, it offers you a faster, safer, and more reliable way of developing concurrent, event-driven software. A QF application has no more need to fiddle directly with critical sections, semaphores, or other such mechanisms. You can program active objects effectively and safely without even knowing what a semaphore is. Yet your application as a whole can reap all the benefits of multitasking, such as optimal, deterministic responsiveness and good CPU utilization.

My goal in this chapter is to explain how to develop a QP application that uses both the QF real-time framework and the QEP event processor described in Part I of this book. I begin with some general rules and heuristics for developing robust and maintainable QP applications. Next I describe the test application that historically I have used to verify all QP ports. I walk you through all steps required to design and implement this application. As you go over these steps, you might also flip back to the "Fly 'n' Shoot" game in Chapter 1, which is a bit more advanced example than the one I use here. In the chapter, I explain how to adapt the test application to all three QF ports discussed in Chapter 8. The chapter concludes with guidelines for sizing event queues and event pools.

9.1   Guidelines for Developing QP Applications

The QP event-driven platform enables building efficient and maintainable event-driven applications in C and C++. However, it is also possible to use QP incorrectly, basically defeating its advantages. This section summarizes the main rules and heuristics for making the most out of active object computing implemented with QP.

9.1.1   Rules

When developing active object–based applications, you should try to heed the following two rules, without exception:

- Active objects should interact only through an asynchronous event exchange and should *not* share memory or other resources

- Active objects should *not* block or busy-wait for events in the middle of RTC processing.

I strongly recommend that you take these rules seriously and follow them religiously. In exchange, the QF real-time framework can guarantee that your application is free from the traditional perils of preemptive multitasking, such as race conditions, deadlocks, priority inversions, starvation, and nondeterminism. In particular, you will never need to use mutexes, semaphores, monitors, or other such troublesome mechanisms at the application level. Even so, your QP applications can be fully deterministic and can handle hard real-time deadlines efficiently.

The rules of using active objects impose a certain programming discipline. In developing your QP applications, you will certainly be tempted to circumvent the rules.

Occasionally, sharing a variable among different active objects or a mutually exclusive blocking active object threads might seem like the easiest solution. However, you should resist such quick fixes. First, you should convince yourself that the rules are there for a good reason (e.g., see Chapters 6 and 7). Second, you must trust that it is possible to arrive at a good solution without breaking the rules.

I repeatedly find that obeying the rules ultimately results in a better design and invariably pays dividends in the increased flexibility and robustness of the final software product. In fact, I propose that you treat every temptation to break the rules as an opportunity to discover something important about your application. Perhaps instead of sharing a variable, you will discover a new signal or a crucial event parameter that conveys some important information.

Many examples from other arts and crafts demonstrate that discipline can be good for art. Indeed, an artist's aphorism says, "Form is liberating." As Fred Brooks [Brooks 95] eloquently writes: "Bach's creative output hardly seems to have been squelched by the necessity of producing a limited-form cantata each week."

I am firmly convinced that the external provision of architecture such as the QF real-time framework enhances, not cramps, creativity.

9.1.2   Heuristics

Throughout Part II of this book, you can find several basic guidelines for constructing active object–based systems. Here is the quick summary.

- Event-driven programming requires a paradigm shift from traditional sequential programming. In the traditional approach, you concentrate on shared resources and various blocking mechanisms, such as semaphores, to signal events. Event-driven programming is all about writing *nonblocking* code and returning quickly to the event loop.

- Your main goal is to achieve *loose coupling* among active objects. You seek a partitioning of the problem that avoids resource sharing and requires minimal communication (in terms of number and size of exchanged events).

- The main strategy for avoiding resource sharing is to encapsulate the resources in dedicated active objects that manage the resources for the rest of the system.

- The responsiveness of an active object is determined by the longest RTC step of its state machine. To meet hard real-time deadlines, you need to either break up longer processing into shorter steps or move such processing to other, lower-priority active objects.

- A good starting point in developing an active object–based application is to draw sequence diagrams for the primary use cases. These diagrams help you discover signals and event parameters, which, in turn, determine the structure of active objects.

- As soon as you have the first sequence diagrams, you should build an executable model of it. The QP event-driven platform has been specifically designed to enable the construction and execution of vastly incomplete (virtually empty) prototypes. The high portability of QP enables you to build the models on a different platform than your ultimate target (e.g., your PC).

- Most of the time you can concentrate only on the internal state machines of active objects and ignore their other aspects (such as threads of execution and event queues). In fact, developing a QP application consists mostly of elaborating on the state machines of active objects. The generic QEP hierarchical event processor (Chapter 4) and the basic state patterns (Chapter 5) can help you with that part of the problem.

9.2   The Dining Philosopher Problem

The test application that I historically have been using to verify QF ports is based on the classic Dining Philosophers Problem (DPP) posed and solved by Edsger Dijkstra back in 1971 [Dijkstra 71]. The DPP application is simpler than the "Fly 'n' Shoot" game described in Chapter 1 and can be tested only with a couple of LEDs on your target board, as opposed to the graphic display required by the "Fly 'n' Shoot" game. Still, DPP contains six concurrent active objects that exchange events via publish-subscribe and direct event-posting mechanisms. The application uses five time events (timers) as well as dynamic and static events.

9.2.1   Step 1: Requirements

First, your always need to understand what your application is supposed to accomplish. In the case of a simple application, the requirements are conveyed through the problem specification, which for the DPP is as follows.

Five philosophers are gathered around a table with a big plate of spaghetti in the middle (see Figure 9.1). Between each two philosophers is a fork. The spaghetti is so slippery that a philosopher needs two forks to eat it. The life of a philosopher consists of alternate periods of thinking and eating. When a philosopher wants to eat, he tries to acquire forks. If successful in acquiring two forks, he eats for a while, then puts down the forks and continues to think. The key issue is that a finite set of tasks (philosophers) is sharing a finite set of resources (forks), and each resource can be used by only one task at a time. (An alternative Oriental version replaces spaghetti with rice and forks with chopsticks, which perhaps explains better why philosophers need two chopsticks to eat.)

Figure 9.1: The Dining Philosopher Problem.

9.2.2   Step 2: Sequence Diagrams

A good starting point in designing any event-driven system is to draw sequence *diagrams* for the main scenarios (main-use cases) identified from the problem specification. To draw such diagrams, you need to break up your problem into active objects with the main goal of minimizing the coupling among active objects. You seek a partitioning of the problem that avoids resource sharing and requires minimal communication in terms of number and size of exchanged events.

DPP has been specifically conceived to make the philosophers contend for the forks, which are the shared resources in this case. In active object systems, the generic design strategy for handling such shared resources is to encapsulate them inside a dedicated active object and to let that object manage the shared resources for the rest of the system (i.e., instead of directly sharing the resources, the rest of the application shares the dedicated active object). When you apply this strategy to DPP, you will naturally arrive at a dedicated active object to manage the forks. I named this active object Table.

The sequence diagram in Figure 9.2 shows the most representative event exchanges among any two adjacent Philosophers and the Table active objects.

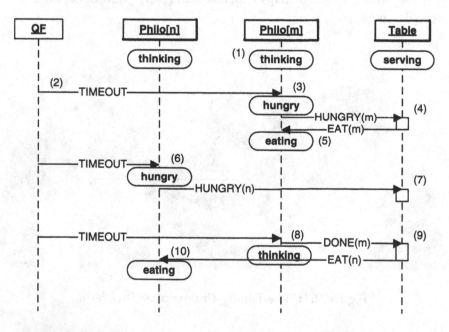

Figure 9.2: The sequence diagram of the DPP application.

(1)   Each Philosopher active object starts in the "thinking" state. Upon the entry to this state, the Philosopher arms a one-shot time event to terminate the thinking.

(2)   The QF framework posts the time event (timer) to Philosopher[m].

(3) Upon receiving the TIMEOUT event, Philosopher[m] transitions to "hungry" state and posts the HUNGRY(m) event to the Table active object. The parameter of the event tells the Table which Philosopher is getting hungry.

(4) The Table active object finds out that the forks for Philosopher[m] are available and grants it permission to eat by publishing the EAT(m) event.

(5) The permission to eat triggers the transition to "eating" in Philosopher[m]. Also, upon the entry to "eating," the Philosopher arms its one-shot time event to terminate the eating.

(6) The Philosopher[n] receives the TIMEOUT event and behaves exactly as Philosopher[m], that is, transitions to "hungry" and posts HUNGRY(n) event to the Table active object.

(7) This time the Table active object finds out that the forks for Philosopher[n] are *not* available, and so it does not grant the permission to eat. Philosopher[n] remains in the "hungry" state.

(8) The QF framework delivers the timeout for terminating the eating to Philosopher [m]. Upon the exit from "eating," Philosopher[m] publishes event DONE(m) to inform the application that it is no longer eating.

(9) The Table active object accounts for free forks and checks whether any direct neighbors of Philosopher[m] are hungry. Table posts event EAT(n) to Philosopher[n].

(10) The permission to eat triggers the transition to "eating" in Philosopher[n].

9.2.3 Step 3: Signals, Events, and Active Objects

Sequence diagrams like Figure 9.2 help you discover events exchanged among active objects. The choice of signals and event parameters is perhaps the most important design decision in any event-driven system. The signals affect the other main application components: events and state machines of the active objects.

In QP, signals are typically enumerated constants and events with parameters are structures derived from the QEvent base structure. Listing 9.1 shows signals and events used in the DPP application. The DPP sample code for the DOS version (in C) is located in the <qp>\qpc\examples\80x86\dos\tcpp101\1\dpp\ directory, where <qp> stands for the installation directory you chose to install the accompanying software.

Listing 9.1 Signals and events used in the DPP application (file dpp.h)

```
        #ifndef dpp_h
        #define dpp_h

(1)     enum DPPSignals {
(2)         EAT_SIG = Q_USER_SIG,      /* published by Table to let a philosopher eat */
            DONE_SIG,                  /* published by Philosopher when done eating */
            TERMINATE_SIG,             /* published by BSP to terminate the application */
(3)         MAX_PUB_SIG,                      /* the last published signal */

(4)         HUNGRY_SIG,       /* posted directly from hungry Philosopher to Table */
(5)         MAX_SIG                                   /* the last signal */
        };

        typedef struct TableEvtTag {
(6)         QEvent super;                              /* derives from QEvent */
            uint8_t philoNum;                       /* Philosopher number */
        } TableEvt;

        enum { N_PHILO = 5 };                       /* number of Philosophers */

(7)     void Philo_ctor(void);         /* ctor that instantiates all Philosophers */
(8)     void Table_ctor(void);

(9)     extern QActive * const AO_Philo[N_PHILO]; /* "opaque" pointers to Philo AOs */
(10)    extern QActive * const AO_Table;          /* "opaque" pointer to Table AO */

        #endif                                             /* dpp_h */
```

(1) For smaller applications such as the DPP, I define all signals in one enumeration (rather than in separate enumerations or, worse, as preprocessor #define macros). An enumeration automatically guarantees the uniqueness of signals.

(2) Note that the user signals must start with the offset Q_USER_SIG to avoid overlapping the reserved QEP signals.

(3)   I like to group all the globally published signals at the top of the enumeration, and I use the MAX_PUB_SIG enumeration to automatically keep track of the maximum published signals in the application.

(4)   I decided that the Philosophers will post the HUNGRY event directly to the Table object rather than publicly publish the event (perhaps a Philosopher is "embarrassed" to be hungry, so does not want other Philosophers to know about it). That way, I can demonstrate direct event posting and publish-subscribe mechanisms coexisting in a single application.

(5)   I use the MAX_SIG enumeration to automatically keep track of the total number of signals used in the application.

(6)   Every event with parameters, such as the TableEvt, derives from the QEvent base structure.

I like to keep the code and data structure of every active object strictly encapsulated within its own C-file. For example, all code and data for the active object Table are encapsulated in the file table.c, with the external interface consisting of the function Table_ctor() and the pointer AO_Table.

(7,8)   These functions perform an early initialization of the active objects in the system. They play the role of static "constructors," which in C you need to call explicitly, typically at the beginning of main().

(9,10)   These global pointers represent active objects in the application and are used for posting events directly to active objects. Because the pointers can be initialized at compile time, I like to declare them const so that they can be placed in ROM. The active object pointers are "opaque" because they cannot access the whole active object, but only the part inherited from the QActive structure.

9.2.4   Step 4: State Machines

At the application level, you can mostly ignore such aspects of active objects as the separate task contexts or private event queues and view them predominantly as state machines. In fact, your main job in developing your QP application consists of elaborating the state machines of your active objects.

Figure 9.3(A) shows the state machines associated with Philosopher active object, which clearly shows the life cycle consisting of states "thinking," "hungry," and "eating."

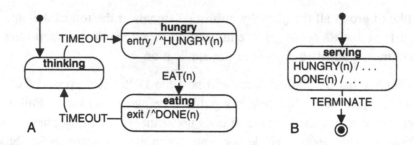

Figure 9.3: State machines associated with the Philosopher active object (A), and Table active object (B).

This state machine generates the HUNGRY event on entry to the "hungry" state and the DONE event on exit from the "eating" state, because this exactly reflects the semantics of these events. An alternative approach—to generate these events from the corresponding TIMEOUT transitions—would not guarantee the preservation of the semantics in potential future modifications of the state machine. This actually is the general guideline in state machine design.

GUIDELINE

Favor entry and exit actions over actions on transitions.

Figure 9.3(B) shows the state machine associated with the Table active object. This state machine is trivial because Table keeps track of the forks and hungry philosophers by means of extended state variables rather than by its state machine. The state diagram in Figure 9.3(B) obviously does not convey how the Table active object behaves, since the specification of actions is missing. I decided to omit the actions because including them required cutting and pasting most of the Table code into the diagram, which would make the diagram too cluttered. In this case, the diagram simply does not add much value over the code.

As I mentioned before, I like to strictly encapsulate each active object inside a dedicated source file (.C file). Listing 9.2 shows the declaration (active object structure) and complete definition (state-handler functions) of the Table active object in the file `table.c`. In the explanation section immediately following this listing, I focus on the techniques of encapsulating active objects and using QF services. I don't repeat here the recipes for coding state machine elements, which I already gave in Part I of this book (Chapters 1 and 4).

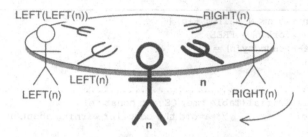

Figure 9.4: Numbering of philosophers and forks (see the macros LEFT() and RIGHT() in Listing 9.2).

Listing 9.2 Table active object (file `table.c`); boldface indicates the QF services

```
       #include "qp_port.h"
       #include "dpp.h"
       #include "bsp.h"

       Q_DEFINE_THIS_FILE

       /* Active object class ---------------------------------------------*/
(1)    typedef struct TableTag {
(2)        QActive super;                                /* derives from QActive */
(3)        uint8_t fork[N_PHILO];                        /* states of the forks */
(4)        uint8_t isHungry[N_PHILO];          /* remembers hungry philosophers */
       } Table;

       static QState Table_initial(Table *me, QEvent const *e);    /* pseudostate */
       static QState Table_serving(Table *me, QEvent const *e);   /* state handler */

(5)    #define RIGHT(n_)  ((uint8_t)(((n_) + (N_PHILO - 1)) % N_PHILO))
(6)    #define LEFT(n_)   ((uint8_t)(((n_) + 1) % N_PHILO))
       enum ForkState { FREE, USED };

       /* Local objects --------------------------------------------------*/
(7)    static Table l_table;          /* the single instance of the Table active object */

       /* Global-scope objects -------------------------------------------*/
(8)    QActive * const AO_Table = (QActive *)&l_table;              /* "opaque" AO pointer */

       /*................................................................*/
(9)    void Table_ctor(void) {
           uint8_t n;
           Table *me = &l_table;
(10)       QActive_ctor(&me->super, (QStateHandler)&Table_initial);
```

Continued onto next page

```
(11)            for (n = 0; n < N_PHILO; ++n) {
                    me->fork[n] = FREE;
                    me->isHungry[n] = 0;
                }
        }
        /*..................................................................*/
        QState Table_initial(Table *me, QEvent const *e) {
            (void)e;                /* avoid the compiler warning about unused parameter */

(12)        QActive_subscribe((QActive *)me, DONE_SIG);
(13)        QActive_subscribe((QActive *)me, TERMINATE_SIG);
(14)        /* signal HUNGRY_SIG is posted directly */

            return Q_TRAN(&Table_serving);
        }
        /*..................................................................*/
        QState Table_serving(Table *me, QEvent const *e) {
            uint8_t n, m;
            TableEvt *pe;

            switch (e->sig) {
                case HUNGRY_SIG: {
(15)                BSP_busyDelay();
                    n = ((TableEvt const *)e)->philoNum;
                                    /* phil ID must be in range and he must be not hungry */
(16)                Q_ASSERT((n < N_PHILO) && (!me->isHungry[n]));

(17)                BSP_displayPhilStat(n, "hungry ");
                    m = LEFT(n);
                    if ((me->fork[m] == FREE) && (me->fork[n] == FREE)) {
                        me->fork[m] = me->fork[n] = USED;
                        pe = Q_NEW(TableEvt, EAT_SIG);
                        pe->philoNum = n;
                        QF_publish((QEvent *)pe);
                        BSP_displayPhilStat(n, "eating ");
                    }
                    else {
                        me->isHungry[n] = 1;
                    }
                    return Q_HANDLED();
                }
                case DONE_SIG: {
                    BSP_busyDelay();
                    n = ((TableEvt const *)e)->philoNum;
                                    /* phil ID must be in range and he must be not hungry */
(18)                Q_ASSERT((n < N_PHILO) && (!me->isHungry[n]));

                    BSP_displayPhilStat(n, "thinking");
                    m = LEFT(n);
```

```
                                                /* both forks of Phil[n] must be used */
(19)                Q_ASSERT((me->fork[n] == USED) && (me->fork[m] == USED));

                    me->fork[m] = me->fork[n] = FREE;
                    m = RIGHT(n);                          /* check the right neighbor */
                    if (me->isHungry[m] && (me->fork[m] == FREE)) {
                        me->fork[n] = me->fork[m] = USED;
                        me->isHungry[m] = 0;
                        pe = Q_NEW(TableEvt, EAT_SIG);
                        pe->philoNum = m;
(20)                    QF_publish((QEvent *)pe);
                        BSP_displayPhilStat(m, "eating ");
                    }
                    m = LEFT(n);                           /* check the left neighbor */
                    n = LEFT(m);                    /* left fork of the left neighbor */
                    if (me->isHungry[m] && (me->fork[n] == FREE)) {
                        me->fork[m] = me->fork[n] = USED;
                        me->isHungry[m] = 0;
                        pe = Q_NEW(TableEvt, EAT_SIG);
                        pe->philoNum = m;
(21)                    QF_publish((QEvent *)pe);
                        BSP_displayPhilStat(m, "eating ");
                    }
                    return Q_HANDLED();
                }
            case TERMINATE_SIG: {
(22)                QF_stop();
                    return Q_HANDLED();
                }
            }
        return Q_SUPER(&QHsm_top);
    }
```

(1)   To achieve true encapsulation, I place the declaration of the active object structure in the source file (.C file).

(2)   Each active object in the application derives from the QActive base structure.

(3)   The Table active object keeps track of the forks in the array fork[]. The forks are numbered as shown in Figure 9.4.

(4)   Similarly, the Table active object needs to remember which philosophers are hungry, in case the forks aren't immediately available. Table keeps track of hungry philosophers in the array isHungry[]. Philosophers are numbered as shown in Figure 9.4.

(5,6) The helper macros LEFT() and RIGHT() access the left and right philosopher or fork, respectively, as shown in Figure 9.4.

(7)   I statically allocate the Table active object. By defining this object as static I make it inaccessible outside the .C file.

(8)   Externally, the Table active object is known only through the "opaque" pointer AO_Table. The pointer is declared 'const' (with the const *after* the '*'), which means that the pointer itself cannot change. This ensures that the active object pointer cannot change accidentally and also allows the compiler to allocate the active object pointer in ROM.

(9)   The function Table_ctor() performs the instantiation of the Table active object. It plays the role of the static "constructor," which in C you need to call explicitly, typically at the beginning of main().

> **NOTE**
>
> In C++, static constructors are invoked automatically before main(). This means that in the C++ version of DPP (found in <qp>\qpcpp\examples\80x86\dos\tcpp101\l\dpp\), you provide a regular constructor for the Table class and don't bother with calling it explicitly. However, you must make sure that the startup code for your particular embedded target includes the additional steps required by the C++ initialization.

(10)   The constructor must first instantiate the QActive superclass.

(11)   The constructor can then initialize the internal data members of the active object.

(12,13) The active object subscribes to all interesting signals in the topmost initial transition.

> **NOTE**
>
> I often see new QP users forget subscribing to events, and then the application appears "dead" when you first run it.

(14)   Note that Table does not subscribe to the HUNGRY event, because this event is posted directly.

(15)   I sprinkled the state machine with calls to the function `BSP_busyDelay()`
         to artificially prolong the RTC processing. The function
         `BSP_busyDelay()` busy-waits in a counted loop, whereas you can adjust
         the number of iterations of this loop from the command line or through
         a debugger. This technique lets me increase the probability of various
         preemptions and thus helps me use the DPP application for stress-testing
         various QP ports.

(16,18,19) The Table state machine extensively uses assertions to monitor correct
            execution of the DPP application. For example, in line (19) both forks of a
            philosopher who just finished eating must be used.

(17)   The output to the screen is a BSP (board support package) operation. The
         different BSPs implement this operation differently, but the code of the
         Table state machine does not need to change.

(20,21) It is possible that the Table active object publishes two events in a single
         RTC step.

(22)   Upon receiving the TERMINATE event, the Table active object calls
         `QF_stop()` to stop QF and return to the underlying operating system.

The Philosopher active objects bring no essentially new techniques, so I don't reproduce
the listing of the `philo.c` file here. The only interesting aspect of philosophers that
I'd like to mention is that all five philosopher active objects are instances of the same
active object class. The philosopher state machine also uses a few assertions to monitor
correct execution of the application according to the problem specification.

9.2.5   Step 5: Initializing and Starting the Application

Most of the system initialization and application startup can be written in a platform-
independent way. In other words, you can use essentially the same `main()` function for
the DPP application with many QP ports.

Typically, you start all your active objects from `main()`. The signature of the
`QActive_start()` function forces you to make several important decisions about
each active object upon startup. First, you need to decide the relative priorities of the
active objects. Second, you need to decide the size of the event queues you preallocate
for each active object. The correct size of the queue is actually related to the priority,
as I discuss in the upcoming Section 9.4. Third, in some QF ports, you need to give

each active object a separate stack, which also needs to be preallocated adequately. And finally, you need to decide the order in which you start your active objects.

The order of starting active objects becomes important when you use an OS or RTOS, in which a spawned thread starts to run immediately, possibly preempting the main() thread from which you launch your application. This could cause problems if, for example, the newly created active object attempts to post an event directly to another active object that has not been yet created. Such a situation does not occur in DPP, but if it is an issue for you, you can try to lock the scheduler until all active objects are started. You can then unlock the scheduler in the QF_onStartup() callback, which is invoked right before QF takes over control. Some RTOSs (e.g., µC/OS-II) allow you to defer the start of multitasking until after you start active objects. Another alternative is to start active objects from within other active objects, but this design increases coupling because the active object that serves as the launch pad must know the priorities, queue sizes, and stack sizes for all active objects to be started.

Listing 9.3 Initializing and starting the DPP application (file main.c)

```
      #include "qp_port.h"
      #include "dpp.h"
      #include "bsp.h"

      /* Local-scope objects --------------------------------------------------*/
(1)   static QEvent const *l_tableQueueSto[N_PHILO];
(2)   static QEvent const *l_philoQueueSto[N_PHILO][N_PHILO];
(3)   static QSubscrList   l_subscrSto[MAX_PUB_SIG];

(4)   static union SmallEvent {
(5)       void *min_size;
          TableEvt te;
(6)       /* other event types to go into this pool */
(7)   } l_smlPoolSto[2*N_PHILO];               /* storage for the small event pool */

      /*...................................................................*/
      int main(int argc, char *argv[]) {
          uint8_t n;

(8)       Philo_ctor();            /* instantiate all Philosopher active objects */
(9)       Table_ctor();                 /* instantiate the Table active object */

(10)      BSP_init(argc, argv);         /* initialize the Board Support Package */

(11)      QF_init();        /* initialize the framework and the underlying RT kernel */

(12)      QF_psInit(l_subscrSto, Q_DIM(l_subscrSto)); /* init publish-subscribe */
```

```
                                          /* initialize event pools... */
(13)    QF_poolInit(l_smlPoolSto, sizeof(l_smlPoolSto), sizeof(l_smlPoolSto[0]));

        for (n = 0; n < N_PHILO; ++n) {          /* start the active objects... */
(14)        QActive_start(AO_Philo[n], (uint8_t)(n + 1),
                          l_philoQueueSto[n], Q_DIM(l_philoQueueSto[n]),
                          (void *)0, 0,                    /* no private stack */
                          (QEvent *)0);
        }
(15)    QActive_start(AO_Table, (uint8_t)(N_PHILO + 1),
                      l_tableQueueSto, Q_DIM(l_tableQueueSto),
                      (void *)0, 0,                        /* no private stack */
                      (QEvent *)0);
(16)    QF_run();                                    /* run the QF application */

        return 0;
    }
```

(1,2)   The memory buffers for all event queues are statically allocated.

(3)   The memory space for subscriber lists is also statically allocated. The
        MAX_PUB_SIG enumeration comes in handy here.

(4)   The union SmallEvent contains all events that are served by the "small" event
        pool.

(5)   The union contains a pointer-size member to make sure that the union size will
        be at least that big.

(6)   You add all events that you want to be served from this event pool.

(7)   The memory buffer for the "small" event pool is statically allocated.

(8,9)   The main() function starts with calling all static "constructors" (see Listing 9.1
        (7-8)). This step is not necessary in C++.

(10)   The target board is initialized.

(11)   QF is initialized together with the underlying OS/RTOS.

(12)   The publish-subscribe mechanism is initialized. You don't need to call
        QF_psInit() if your application does not use publish-subscribe.

(13)   Up to three event pools can be initialized by calling QF_poolInit() up to
        three times. The subsequent calls must be made in the order of increasing

block sizes of the event pools. You don't need to call `QF_poolInit()` if your application does not use dynamic events.

(14,15)   All active objects are started using the "opaque" active object pointers (see Listing 9.1(9-10)). In this particular example, the active objects are started without private stacks. However, some RTOSs, such as µC/OS-II, require preallocating stacks for all active objects.

(16)   The control is transferred to QF to run the application. `QF_run()` might never return.

9.2.6   Step 6: Gracefully Terminating the Application

Terminating an application is not really a big concern in embedded systems because embedded programs almost never have a need to terminate gracefully. The job of a typical embedded system is never finished, and most embedded software runs forever or until the power is removed, whichever comes first.

> **NOTE**
>
> You still need to carefully design and test the fail-safe mechanism triggered by a CPU exception or assertion violation in your embedded system. However, such a situation represents a *catastrophic* shutdown, followed perhaps by a reset. The subject of this section is the *graceful* termination, which is part of the normal application life cycle.

However, in desktop programs, or when embedded applications run on top of a general-purpose operating system, such as Linux, Windows, or DOS, the shutdown of a QP application becomes important. The problem is that to terminate *gracefully*, the application must clean up all resources allocated by the application during its lifetime. Such a shutdown is always application-specific and cannot be preprogrammed generically at the framework level.

The DPP application uses the following mechanism to shut down: When the user decides to terminate the application, the global TERMINATE event is published. In DPP, only the Table active object subscribes to this event (Listing 9.2(13)), but in general all active objects that need to clean up anything before exiting should subscribe to the TERMINATE event. The last subscriber, which is typically the lowest-priority subscriber, calls the `QF_stop()` function (Listing 9.2(22)). As described in Chapter 8, `QF_stop()` is implemented in the QF port. Often, `QF_stop()` causes the `QF_run()` function to return. Right before transferring control to the underlying operating system, QF invokes the `QF_onCleanup()` callback. This callback gives the application the last chance to clean up globally (e.g., the DOS version restores the original DOS interrupt vectors).

Finally, you can also stop individual active objects and let the rest of the application continue execution. The cleanest way to end an active object's thread is to have it stop itself by calling `QActive_stop(me)`, which should cause a return from the active object's thread routine. Of course, to "commit a suicide" voluntarily, the active object must be running and cannot be waiting for an event. In addition, before disappearing, the active object should release all the resources acquired during its lifetime. Finally, the active object should unsubscribe from receiving all signals and somehow should make sure that no more events will be posted to it directly. Unfortunately, all these requirements cannot be preprogrammed generically and always require some work on the application programmer's part.

9.3   Running DPP on Various Platforms

I generally use the same DPP source code to test the QP ports on various CPUs, operating systems, and compilers. The only platform-dependent file is the board support package (BSP) definition and sometimes the `main()` function. In this section I describe what needs to be done to execute the DPP application with the "vanilla" kernel (I cover two versions: for 80×86 and Cortex-M3), as well as µC/OS-II on DOS and Linux.

9.3.1   "Vanilla" Kernel on DOS

The code for the DPP port to 80x86 with the "vanilla" kernel is located in the directory `<qp>\qpc\examples\80x86\dos\tcpp101\l\dpp\`. The directory contains the Turbo C++ 1.01 project files to build the application. You can execute the application by double-clicking the executables in the `dbg\`, `rel\`, or `spy\` subdirectories. Figure 9.5 shows the output generated by the DPP executable. Listing 9.4 shows the BSP for this version of DPP.

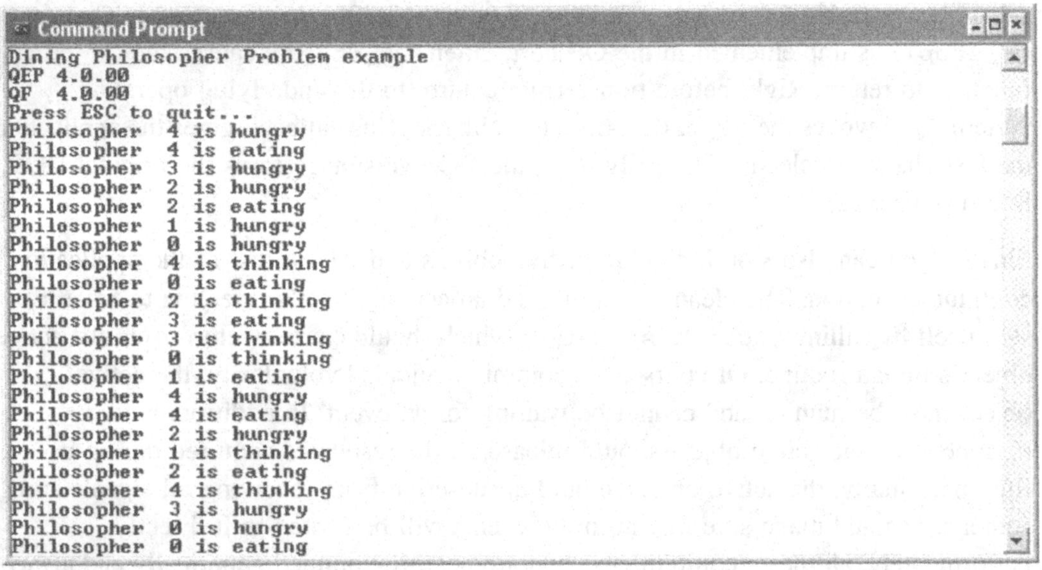

```
Command Prompt                                                    _ □ ×
Dining Philosopher Problem example
QEP 4.0.00
QF  4.0.00
Press ESC to quit...
Philosopher  4 is hungry
Philosopher  4 is eating
Philosopher  3 is hungry
Philosopher  2 is hungry
Philosopher  2 is eating
Philosopher  1 is hungry
Philosopher  0 is hungry
Philosopher  4 is thinking
Philosopher  0 is eating
Philosopher  2 is thinking
Philosopher  3 is eating
Philosopher  3 is thinking
Philosopher  0 is thinking
Philosopher  1 is eating
Philosopher  4 is hungry
Philosopher  4 is eating
Philosopher  2 is hungry
Philosopher  1 is thinking
Philosopher  2 is eating
Philosopher  4 is thinking
Philosopher  3 is hungry
Philosopher  0 is hungry
Philosopher  0 is eating
```

Figure 9.5: DPP test application running in a DOS console.

Listing 9.4 BSP for the DPP application with the "Vanilla" kernel on DOS (file `<qp>\qpc\examples\80x86\dos\tcpp101\l\dpp\bsp.c`)

```c
     #include "qp_port.h"
     #include "dpp.h"
     #include "bsp.h"


     /* Local-scope objects------------------------------------------------*/
     static void interrupt (*l_dosTmrISR)();
     static void interrupt (*l_dosKbdISR)();
     static uint32_t l_delay = 0UL;   /* limit for the loop counter in busyDelay() */


     #define TMR_VECTOR    0x08
     #define KBD_VECTOR    0x09


     /*..................................................................*/
(1)  void interrupt ISR_tmr(void) {
(2)      QF_INT_UNLOCK(dummy);                              /* unlock interrupts */
(3)      QF_tick();                    /* call QF_tick() outside of critical section */
(4)      QF_INT_LOCK(dummy);                          /* lock interrupts again */
(5)      outportb(0x20, 0x20);              /* write EOI to the master 8259A PIC */
     }
     /*..................................................................*/
```

```
         void interrupt ISR_kbd(void) {
             uint8_t key;
             uint8_t kcr;

             QF_INT_UNLOCK(dummy);                          /* unlock interrupts */
             key = inport(0x60);         /* key scan code from the 8042 kbd controller */
             kcr = inport(0x61);                      /* get keyboard control register */
             outportb(0x61, (uint8_t)(kcr | 0x80));  /* toggle acknowledge bit high */
             outportb(0x61, kcr);                     /* toggle acknowledge bit low */
             if (key == (uint8_t)129) {                       /* ESC key pressed? */
                 static QEvent term = {TERMINATE_SIG, 0};           /* static event */
                 QF_publish(&term);                 /* publish to all interested AOs */
             }
             QF_INT_LOCK(dummy);                       /* lock interrupts again */
             outportb(0x20, 0x20);         /* write EOI to the master 8259A PIC */
         }
         /*................................................................*/
         void QF_onStartup(void) {
                                            /* save the origingal DOS vectors ... */
(6)          l_dosTmrISR = getvect(TMR_VECTOR);
(7)          l_dosKbdISR = getvect(KBD_VECTOR);

             QF_INT_LOCK(dummy);
(8)          setvect(TMR_VECTOR, &ISR_tmr);
(9)          setvect(KBD_VECTOR, &ISR_kbd);
             QF_INT_UNLOCK(dummy);
         }
         /*................................................................*/
         void QF_onCleanup(void) {           /* restore the original DOS vectors ... */
             QF_INT_LOCK(dummy);
(10)         setvect(TMR_VECTOR, l_dosTmrISR);
(11)         setvect(KBD_VECTOR, l_dosKbdISR);
             QF_INT_UNLOCK(dummy);
             _exit(0);                                            /* exit to DOS */
         }
         /*................................................................*/
         void QF_onIdle(void) {              /* called with interrupts LOCKED */
(12)         QF_INT_UNLOCK(dummy);                    /* always unlock interrutps */
         }
         /*................................................................*/
         void BSP_init(int argc, char *argv[]) {
             if (argc > 1) {
(13)             l_delay = atol(argv[1]);      /* set the delay counter for busy delay */
             }
             printf("Dining Philosopher Problem example"
                     "\nQEP %s\nQF %s\n"
                     "Press ESC to quit...\n",
                     QEP_getVersion(),
                     QF_getVersion());
         }
```

Continued onto next page

```
    /*.....................................................................*/
    void BSP_busyDelay(void) {
        uint32_t volatile i = l_delay;
(14)    while (i-> 0UL) {                                      /* busy-wait loop */
        }
    }
    /*.....................................................................*/
    void BSP_displayPhilStat(uint8_t n, char const *stat) {
(15)    printf("Philosopher %2d is %s\n", (int)n, stat);
    }
    /*.....................................................................*/
    void Q_onAssert(char const Q_ROM * const Q_ROM_VAR file, int line) {
        QF_INT_LOCK(dummy);                                   /* cut-off all interrupts */
        fprintf(stderr, "Assertion failed in %s, line %d", file, line);
(16)    QF_stop();
    }
```

(1)   The compiler-supported Turbo C++ 1.01 compiler provides an extended keyword "interrupt" that enables you to program ISRs in C/C++. The compiler-supported ISRs are adequate for the "vanilla" kernel.

(2)   The 80x86 processor locks interrupts in hardware before vectoring to the ISR. The interrupts can be unlocked right away, though, because the 8259A programmable interrupt controller prioritizes interrupts before they reach the CPU.

(3)   The QF_tick() service is called outside of the critical section. You cannot call any QF services within a critical section, because this "vanilla" port uses the simple "unconditional interrupt locking and unlocking" policy, which precludes nesting critical sections.

(4)   Interrupts are locked before the interrupt is exited.

(5)   The end-of-interrupt (EOI) instruction is sent to the master 8259A PIC, so that it ends prioritization of this interrupt level.

(6,7)   The original DOS interrupts vectors are saved to be restored upon cleanup.

(8,9)   The customized interrupts are set for this port. This must happen in a critical section.

(10,11) Upon cleanup, the original DOS interrupts are restored.

(12)   In the "vanilla" kernel, the `QF_idle()` callback is invoked with interrupts locked and must always unlock interrupts (see Section 8.2.4 in Chapter 8).

(13,14)   The loop counter for the `BSP_busyDelay()` function is set from the first command-line parameter. You should not go overboard with this parameter, because you might overload the CPU by creating an unschedulable set of tasks. In this case QF will eventually assert on overflowing an event queue.

(15)   The output of the philosopher status is implemented as a `printf()` statement (see Figure 9.5). Note that the output occurs only from the context of the Table active object.

(16)   Upon an assertion failure, the application is stopped and cleanly exits to the DOS prompt.

9.3.2   "Vanilla" Kernel on Cortex-M3

The code for the DPP port to Cortex-M3 with the "vanilla" kernel is located in the directory `<qp>\qpc\examples\cortex-m3\vanilla\iar\dpp\`. The directory contains the IAR EWARM v5.11 project files to build the application and download it to the EV-LM3S811 board. Figure 9.6 shows the display of the board while it is executing the application. Listing 9.5 shows the BSP for this version of DPP.

Figure 9.6: DPP test application running on the EV-LM3S811 board (Cortex-M3). The status of each Philosopher is displayed as "t" (thinking), "e" (eating), or "h" (hungry).

Listing 9.5 BSP for the DPP application with the "vanilla" kernel on bare metal Cortex-M3 (file `<qp>\qpc\examples\cortex-m3\vanilla\iar\dpp-ev-lm3s811\bsp.c`**)**

```
       #include "qp_port.h"
       #include "dpp.h"
       #include "bsp.h"

(1)    #include "hw_ints.h"
       ...                         /* other Luminary Micro driver library include files */

       /* Local-scope objects -----------------------------------------------------*/
       static uint32_t l_delay = 0UL;    /* limit for the loop counter in busyDelay() */

       /*........................................................................*/
(2)    void ISR_SysTick(void) {
           QF_tick();                              /* process all armed time events */
           /* add any application-specific clock-tick processing, as needed */
       }
       ...
       /*........................................................................*/
       void BSP_init(int argc, char *argv[]) {
           (void)argc;                        /* unused: avoid the complier warning */
           (void)argv;                        /* unused: avoid the compiler warning */

           /* Set the clocking to run at 20MHz from the PLL. */
(3)        SysCtlClockSet(SYSCTL_SYSDIV_10 | SYSCTL_USE_PLL
                         | SYSCTL_OSC_MAIN | SYSCTL_XTAL_6MHZ);

           /* Enable the peripherals used by the application. */
(4)        SysCtlPeripheralEnable(SYSCTL_PERIPH_GPIOA);
           SysCtlPeripheralEnable(SYSCTL_PERIPH_GPIOC);

           /* Configure the LED, push button, and UART GPIOs as required. */
(5)        GPIODirModeSet(GPIO_PORTA_BASE, GPIO_PIN_0 | GPIO_PIN_1,
                         GPIO_DIR_MODE_HW);
           GPIODirModeSet(GPIO_PORTC_BASE, PUSH_BUTTON, GPIO_DIR_MODE_IN);
           GPIODirModeSet(GPIO_PORTC_BASE, USER_LED, GPIO_DIR_MODE_OUT);
           GPIOPinWrite(GPIO_PORTC_BASE, USER_LED, 0);

           /* Initialize the OSRAM OLED display. */
(6)        OSRAMInit(1);
(7)        OSRAMStringDraw("Dining Philos", 0, 0);
(8)        OSRAMStringDraw("0 ,1 ,2 ,3 ,4", 0, 1);
       }
       /*........................................................................*/
       void BSP_displayPhilStat(uint8_t n, char const *stat) {
           char str[2];
           str[0] = stat[0];
```

```
            str[1] = '\0';
 (9)        OSRAMStringDraw(str, (3*6*n + 6), 1);
       }
       /*.............................................................*/
       void BSP_busyDelay(void) {
            uint32_t volatile i = l_delay;
            while (i-- > 0UL) {                               /* busy-wait loop */
            }
       }
       /*.............................................................*/
       void QF_onStartup(void) {
            /* Set up and enable the SysTick timer.  It will be used as a reference
             * for delay loops in the interrupt handlers.  The SysTick timer period
             * will be set up for BSP_TICKS_PER_SEC.
             */
(10)        SysTickPeriodSet(SysCtlClockGet() / BSP_TICKS_PER_SEC);
(11)        SysTickEnable();
(12)        IntPrioritySet(FAULT_SYSTICK, 0xC0);      /* set the priority of SysTick */
(13)        SysTickIntEnable();                     /* Enable the SysTick interrupts */
(14)        QF_INT_UNLOCK(dummy);                 /* set the interrupt flag in PRIMASK */
       }
       /*.............................................................*/
       void QF_onCleanup(void) {
(15)   }
       /*.............................................................*/
       void QF_onIdle(void) {          /* entered with interrupts LOCKED, see NOTE01 */
            /* toggle the User LED on and then off, see NOTE02 */
            GPIOPinWrite(GPIO_PORTC_BASE, USER_LED, USER_LED);      /* User LED on */
            GPIOPinWrite(GPIO_PORTC_BASE, USER_LED, 0);           /* User LED off */

(16)   #ifdef NDEBUG
            /* Put the CPU and peripherals to the low-power mode.
             * you might need to customize the clock management for your application,
             * see the datasheet for your particular Cortex-M3 MCU.
             */
(17)        __asm("WFI");                                   /* Wait-For-Interrupt */
       #endif
(18)        QF_INT_UNLOCK(dummy);                       /* always unlock the interrupts */
       }
       /*.............................................................*/
       void Q_onAssert(char const Q_ROM * const Q_ROM_VAR file, int line) {
            (void)file;                                   /* avoid compiler warning */
            (void)line;                                   /* avoid compiler warning */
            QF_INT_LOCK(dummy);        /* make sure that all interrupts are disabled */
(19)        for (;;) {        /* NOTE: replace the loop with reset for the final version */
            }
       }
       /* error routine that is called if the Luminary library encounters an error */
(20)   void __error__(char *pcFilename, unsigned long ulLine) {
            Q_onAssert(pcFilename, ulLine);
       }
```

(1)   The BSP for Cortex-M3 relies on the driver library provided by Luminary Micro with the EV-LM3S811 board.

(2)   As described in Section 8.2.3 in Chapter 8, ISRs in Cortex-M3 are just regular C functions. The system clock tick is implemented with the Cortex-M3 SysTick interrupt, specifically designed for that purpose. Note that the Cortex-M3 enters ISRs with interrupts unlocked, so there is no need to unlock interrupts before calling QF services, such as `QF_tick()`.

(3-5)   The board initialization includes enabling all peripherals used in the DPP application.

(6-8)   The graphic OLED display driver is initialized and the screen is prepared for the DPP application.

(9)   The output of the philosopher status is implemented as drawing a single letter on the screen (see Figure 9.6). Note that the output occurs only from the context of the Table active object.

(10)   Upon startup, the hardware system clock tick rate is set.

(11)   The system clock tick hardware is enabled.

(12)   The Cortex-M3 performs prioritization of all interrupts in hardware, and it is highly recommended to explicitly set the priority of every interrupt used by the application. The Cortex-M3 represents an ISR priority in the three most significant bits of a byte, whereas 0xE0 is the lowest and 0x00 is the highest hardware priority. Priority 0xC0 corresponds to the second-lowest priority in the system.

(13)   The system clock tick interrupt is enabled in hardware.

(14)   The interrupts are enabled.

(15)   The DPP application running on the EV-LM3S811 board operates on "bare metal" and has no operating system to return to. The cleanup callback is not used in this case.

(16-18)   In Section 8.2.4, I have already discussed idle processing for the "vanilla" kernel running on Cortex-M3.

(19)   The assertion handler enters a forever loop in the DPP application. You need to replace this loop with the fail-safe shutdown, followed perhaps by a reset in the production version of your application.

(20)   The function __error__() is used inside the Luminary Micro driver library. This
function has the same purpose and signature as Q_onAssert().

9.3.3   μC/OS-II

The code for the DPP port to 80x86 with the μC/OS-II RTOS is located in the directory
<qp>\qpc\examples\80x86\ucos2\tcpp101\l\dpp\. The directory contains
the make.bat batch file to build the application. You can execute the DPP application
by double-clicking the executables in the dbg\, rel\, or spy\ subdirectories.
Figure 9.7 shows the output generated by the DPP executable.

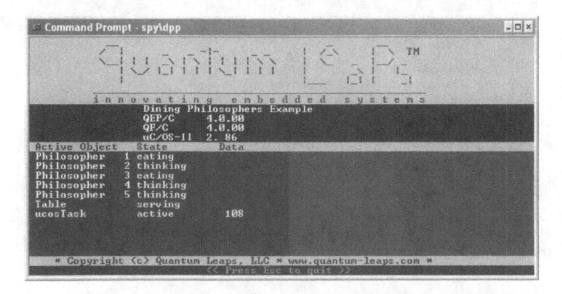

**Figure 9.7: DPP test application running in a DOS console on top
of μC/OS-II v2.86.**

As shown in Listing 9.6, in case of μC/OS-II, you need to modify the main.c source
file to supply the *private stacks* for the active object tasks. This is one of the big-ticket
items in terms of RAM usage required by a traditional preemptive kernel. You also
need to create a dedicated μC/OS-II task to start the interrupts, as described in the
Micro-C/OS-II book [Labrosse 02]. Listing 9.7 shows the customization of the
μC/OS-II hooks (callbacks) to call the QF clock tick processing.

Listing 9.6 `main()` **function for the DPP application with the μC/OS-II RTOS on DOS (file** `<qp>\qpc\examples\80x86\ucos2\tcpp101\l\dpp\main.c`**)**

```
      #include "qp_port.h"
      #include "dpp.h"
      #include "bsp.h"

      /* Local-scope objects -------------------------------------------------*/
      . . .
(1)   static OS_STK l_philoStk[N_PHILO][256];      /* stacks for the Philosophers */
(2)   static OS_STK l_tableStk[256];                       /* stack for the Table */
(3)   static OS_STK l_ucosTaskStk[256];              /* stack for the ucosTask */

      /*.....................................................................*/
      int main(int argc, char *argv[]) {
          . . .

          for (n = 0; n < N_PHILO; ++n) {
              QActive_start(AO_Philo[n], (uint8_t)(n + 1),
                        l_philoQueueSto[n], Q_DIM(l_philoQueueSto[n]),
(4)                     l_philoStk[n], sizeof(l_philoStk[n]), (QEvent *)0);
          }
          QActive_start(AO_Table, (uint8_t)(N_PHILO + 1),
                    l_tableQueueSto, Q_DIM(l_tableQueueSto),
(5)                 l_tableStk, sizeof(l_tableStk), (QEvent *)0);

              /* create a uC/OS-II task to start interrupts and poll the keyboard */
          OSTaskCreate(&ucosTask,
                    (void *)0,                                    /* pdata */
(6)                 &l_ucosTaskStk[Q_DIM(l_ucosTaskStk) - 1],
                    0);                          /* the highest uC/OS-II priority */
          QF_run();                              /* run the QF application */

          return 0;
      }
```

(1-3)   You need to statically allocate the private stacks for all μC/OS-II tasks that you use in the application. Here, I have oversized all stacks of to 256 of 16-bit stack entries (see definition of OS_STK in the μC/OS-II port file `os_cpu.h`). However, μC/OS-II allows each stack to have a different size.

(4,5)   The stack storage is passed to the active objects through the `stkSto` and `stkSize` parameters of the `QActive_start()` function.

(6)   I also create additional "raw" μC/OS-II task `ucosTask()` that starts all interrupts and polls the keyboard to find out when to terminate the application. The body of the `ucosTask()` function is shown in Listing 9.7.

**Listing 9.7 BSP for the DPP application for the µC/OS-II RTOS on DOS
(file `<qp>\qpc\examples\80x86\ucos2\tcpp101\1\dpp\bsp.c`)**

```
      #include "qp_port.h"
      #include "dpp.h"
      #include "bsp.h"
      #include "video.h"
      /*..........................................................*/
(1)   void ucosTask(void *pdata) {
          (void)pdata;         /* avoid the compiler warning about unused parameter */

(2)       QF_onStartup();      /* start interrupts including the clock tick, NOTE01 */

          for (;;) {
(3)           OSTimeDly(OS_TICKS_PER_SEC/10);              /* sleep for 1/10 s */
              if (kbhit()) {                               /* poll for a new keypress */
                  uint8_t key = (uint8_t)getch();
                  if (key == 0x1B) {                       /* is this the ESC key? */
(4)                   QF_publish(Q_NEW(QEvent, TERMINATE_SIG));
                  }
                  else {                                   /* other key pressed */
                      Video_printNumAt(30, 13 + N_PHILO, VIDEO_FGND_YELLOW, key);
                  }
              }
          }
      }
      /*..........................................................*/
      void OSTimeTickHook(void) {
(5)       QF_tick();
          /* add any application-specific clock-tick processing, as needed */
      }
      /*..........................................................*/
      void OSTaskIdleHook(void) {
(6)       /* put the MCU to sleep, if desired */
      }
      . . .
```

(1) The BSP contains a "raw" µC/OS-II task with the main responsibility of starting the interrupts, which in µC/OS-II must occur only after the OSStart() function is called from QF_run() (see [Labrosse 02]).

(2) The QF_onStartup() callback starts interrupts and is identical in this case as in Listing 9.4(6-9).

(3) As any conventional task, ucosTask() must call some blocking RTOS function. In this case, the task blocks on the timed delay.

(4) Every time the task wakes up, it polls the keyboard and checks whether the user hit the Esc key. If so, the μC/OS-II task publishes the static TERMINATE event. This call provides an example of how to generate QP events from external, third-party code.

(5) The QF_tick() processing is invoked from the μC/OS-II hook. Note that this particular μC/OS-II port uses the "saving and restoring interrupt status" policy (μC/OS-II critical section type 3). This means that it's safe to call a QF service, even though μC/OS-II calls OSTimeTickHook() with interrupts locked.

(6) Under a preemptive kernel such as μC/OS-II, a transition to a low-power sleep mode does not need to occur atomically (as it must in the nonpreemptive "vanilla" kernel). Refer to Section 8.3.6 in Chapter 8 for the discussion of idle processing under a preemptive kernel.

9.3.4 Linux

The code for the DPP port to Linux is located in the directory <qp>\qpc\examples \80x86\linux\gnu\dpp\. The directory contains the Makefile to build the application. You can execute the application from a console, as shown in Figure 9.8. The real-time behavior of the application depends on the privilege level. If you launch

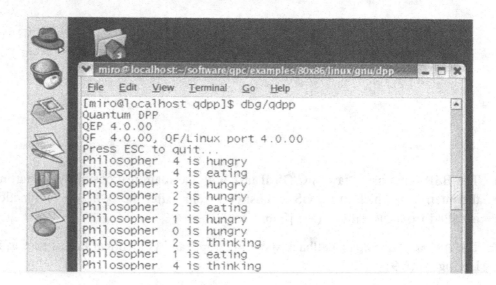

Figure 9.8: DPP test application running in Linux (Redhat 9).

the application with the "superuser" privileges, the QF port will use the SCHED_FIFO real-time scheduler and will prioritize active object threads high (see Section 8.4 in Chapter 8). Otherwise, the application will execute under the default SCHED_OTHER scheduler without a clear notion of priorities for active objects or the "ticker" task. Listing 9.8 shows the BSP for Linux.

Listing 9.8 BSP for the DPP application for Linux
(file `<qp>\qpc\examples\80x86\linux\gnu\dpp\bsp.c`**)**

```
      #include "qp_port.h"
      #include "dpp.h"
      #include "bsp.h"

      #include <sys/select.h>
      . . .

      Q_DEFINE_THIS_FILE
      /* Local objects -------------------------------------------------------*/
(1)   static struct termios l_tsav;     /* structure with saved terminal attributes */
      static uint32_t l_delay;          /* limit for the loop counter in busyDelay() */


      /*..........................................................................*/
(2)   static void *idleThread(void *me) {         /* the expected P-Thread signature */
          for (;;) {
              struct timeval timeout = {0};                 /* timeout for select() */
              fd_set con;                          /* FD set representing the console */
              FD_ZERO(&con);
              FD_SET(0, &con);
              timeout.tv_usec = 8000;

              /* sleep for the full tick or until a console input arrives */
(3)           if (0 != select(1, &con, 0, 0, &timeout)) {       /* any descriptor set? */
                  char ch;
                  read(0, &ch, 1);
                  if (ch == '\33') {                            /* ESC pressed? */
(4)                   QF_publish(Q_NEW(QEvent, TERMINATE_SIG));
                  }
              }
          }
          return (void *)0;                                  /* return success */
      }
      /*..........................................................................*/
      void BSP_init(int argc, char *argv[]) {
          printf("Dining Philosopher Problem example"
                 "\nQEP %s\nQF %s\n"
                 "Press ESC to quit...\n",
                 QEP_getVersion(),
```

Continued onto next page

```
                QF_getVersion());
        if (argc > 1) {
        l_delay = atol(argv[1]);                /* set the delay from the argument */
        }
    }
    /*.........................................................................*/
    void QF_onStartup(void) {                                /* startup callback */
        struct termios tio;                        /* modified terminal attributes */
        pthread_attr_t attr;
        struct sched_param param;
        pthread_t idle;

(5)     tcgetattr(0, &l_tsav);                /* save the current terminal attributes */
        tcgetattr(0, &tio);                  /* obtain the current terminal attributes */
(6)     tio.c_lflag &= ~(ICANON | ECHO);      /* disable the canonical mode & echo */
        tcsetattr(0, TCSANOW, &tio);                      /* set the new attributes */

        /* SCHED_FIFO corresponds to real-time preemptive priority-based scheduler
         * NOTE: This scheduling policy requires the superuser priviledges
         */
        pthread_attr_init(&attr);
        pthread_attr_setschedpolicy(&attr, SCHED_FIFO);
        param.sched_priority = sched_get_priority_min(SCHED_FIFO);

        pthread_attr_setschedparam(&attr, &param);
        pthread_attr_setdetachstate(&attr, PTHREAD_CREATE_DETACHED);

(7)     if (pthread_create(&idle, &attr, &idleThread, 0) != 0) {
                /* Creating the p-thread with the SCHED_FIFO policy failed.
                 * Most probably this application has no superuser privileges,
                 * so we just fall back to the default SCHED_OTHER policy
                 * and priority 0.
                 */
            pthread_attr_setschedpolicy(&attr, SCHED_OTHER);
            param.sched_priority = 0;
            pthread_attr_setschedparam(&attr, &param);
(8)         Q_ALLEGE(pthread_create(&idle, &attr, &idleThread, 0) == 0);
        }
        pthread_attr_destroy(&attr);
    }
    /*.........................................................................*/
    void QF_onCleanup(void) {                                /* cleanup callback */
        printf("\nBye! Bye!\n");
(9)     tcsetattr(0, TCSANOW, &l_tsav);  /* restore the saved terminal attributes */
        QS_EXIT();                                      /* perform the QS cleanup */
    }
    /*.........................................................................*/
    void BSP_displayPhilStat(uint8_t n, char const *stat) {
(10)    printf("Philosopher %2d is %s\n", (int)n, stat);
    }
```

```
/*.........................................................*/
void BSP_busyDelay(void) {
    uint32_t volatile i = l_delay;
    while (i-> 0UL) {
    }
}
/*.........................................................*/
void Q_onAssert(char const Q_ROM * const Q_ROM_VAR file, int line) {
    fprintf(stderr, "Assertion failed in %s, line %d", file, line);
(11)    QF_stop();
}
```

(1)   The standard configuration of a Linux console does not allow collecting user keystrokes asynchronously. The mode of the terminal can be changed but needs to be restored upon exit. The BSP uses the local variable l_tsav to save the terminal settings.

(2)   The BSP contains a "raw" POSIX-thread idleThread() with the main responsibility of polling the console for asynchronous input and terminating the application when the user presses the Esc key.

(3)   The idleThread() uses the select() POSIX call as the main blocking mechanism.

(4)   The idleThread() generates and publishes the TERMINATE event when it detects the Esc keypress.

(5)   Upon startup the terminal attributes are saved into the static variable t_sav.

(6)   The canonical mode of the terminal is switched off to allow collecting keystrokes asynchronously.

(7)   The idle thread is first created with the real-time SCHED_FIFO scheduling policy and the lowest possible priority. Using the SCHED_FIFO policy requires "superuser" privileges and might fail if the application is launched without these privileges.

(8)   If creating the thread under SCHED_FIFO fails, the thread is created under the default SCHED_OTHER policy. This time, the thread must be created successfully; otherwise, the application cannot continue, and hence the assertion.

(9)   The cleanup callback restores the saved terminal attributes.

(10) The output of the philosopher status is implemented as a `printf()` statement (see Figure 9.8). Note that the output occurs only from the context of the Table active object.

(11) Upon an assertion failure, the application is stopped and cleanly exits to Linux.

9.4 Sizing Event Queues and Event Pools

Event queues and event pools are the necessary burden you need to accept when you work within the event-driven paradigm. They are the price to pay for the convenience and speed of development.

The main problem with event queues and event pools is that they consume your precious memory. To minimize that memory, you need to size them appropriately. In this respect, event queues and pools are no different from execution stacks—these data structures all trade some memory for the convenience of programming.

The adequate sizing of event queues and event pools is especially important in QF applications because QF raises an assertion when an event queue overflows or an event pool runs out of events. QF treats both these situations as first-class bugs equally bad as overflowing the stack.

Note that the problem with sizing event queues and event pools is common to all active object–based frameworks, not specifically to QF. For instance, application frameworks that accompany design automation tools have this problem as well. However, the tools handle the problem behind the scenes by using massively oversized defaults. In fact, you should do exactly the same thing: Create oversized event queues and event pools in the early stages of development.

The minimization of memory consumed by event queues, event pools, and execution stacks is like shrink-wrapping your event-driven application. You should do it toward the end of application development because it stifles the flexibility you need in the earlier stages. Note that any change in processing time, interrupt load, or event production patterns can invalidate both your static analysis and the empirical measurements of queue and pool usage. However, that doesn't mean that you shouldn't care at all about event queues and event pools throughout the design and early implementation phase. To the contrary, understanding the general rules for sizing event queues and pools helps you conserve memory by avoiding unnecessary bursts in event production or by breaking up excessively long RTC steps. These techniques are analogous to the ways execution stack space is conserved by avoiding deep call nesting and big automatic variables.

9.4.1   In Sizing Event Queues

One basic fact that you need to understand about event queues is that they work only when the average event production rate $<P(t)>$ does not exceed the average event consumption rate $<C(t)>$. If this condition is not satisfied, the event queue is of no use and always eventually overflows, no matter how big you make it. This fact does not mean that the production rate $P(t)$ cannot occasionally exceed the consumption rate $C(t)$, but such a burst of event production can persist for only a short time. The bursts should also be sufficiently spread out over time to allow cleanup of the queue.

Some software designers try to work around these fundamental limitations by using message queues in a more "creative" way. For example, designers either allow blocking of the producer threads when the queue is full, effectively reducing the production rate $P(t)$, or allow messages to be lost, effectively boosting the consumption rate $C(t)$. The QF views both techniques as an abuse of event queues and simply asserts a contract violation. The basic premise behind this policy is that such a creative use of event queues destroys the event-delivery guarantee (see Chapter 6).

The empirical method is perhaps the simplest and most popular technique used to determine the required capacity of event queues, or any other buffers for that matter (e.g., execution stacks). This technique involves running the system for a while and then stopping it to examine how much of various buffers has been used. The QF implementation of the event queue (the QEQueue class) maintains the nMin data member specifically for this purpose (see Listing 7.24(12-14) in Chapter 7). You can inspect this low-watermark easily using a debugger or through a memory dump.

The alternative technique relies on a static analysis of event production and event consumption. The QF framework uses event queues in a rather specific way (e.g., there is only one consumer thread); consequently, the production rate $P(t)$ and the consumption rate $C(t)$ are strongly correlated.

For example, consider a QF application running under a preemptive, priority-based kernel.[1] Assume further that the highest-priority active object receives events only from other active objects (but not from ISRs). Whenever any of the lower-priority active objects posts or publishes an event for the highest-priority object, the kernel immediately assigns the CPU to the recipient. The kernel makes the context switch because, at this point, the recipient is the highest-priority thread ready to run. The highest-priority active object awakens and runs to completion, consuming any event posted to it. Therefore, the highest-priority active object really doesn't need to queue events (the maximum depth of its event queue is 1).

When the highest-priority active object receives events from ISRs, more events can queue up for it. In the most common arrangement, an ISR produces only one event per activation. In addition, the real-time deadlines are typically such that the highest-priority active object must consume the event before the next interrupt. In this case, the object's event queue can grow, at most, to two events: one from a task and the other from an ISR.

You can extend this analysis recursively to lower-priority active objects. The maximum number of queued events is the sum of all events that higher-priority threads and ISRs can produce for the active object within a given deadline. The deadline is the longest RTC step of the active object, including all possible preemptions by higher-priority threads and ISRs. For example, in the DPP application, all Philosopher active objects perform very little processing (they have short RTC steps). If the CPU can complete these RTC steps within one clock tick, the maximum length of the Philosopher queue would be three events: one from the clock-tick ISR and two from the Table active object (Table can sometimes publish two events in one RTC step).

The rules of thumb for the static analysis of event queue capacity are as follows.

- The size of the event queue depends on the priority of the active object. Generally, the higher the priority, the shorter the necessary event queue. In particular, the highest-priority active object in the system immediately consumes all events posted by the other active objects and needs to queue only those events posted by ISRs.

---

[1] The following discussion also pertains approximately to foreground/background systems with priority queues (see Section 7.11 in Chapter 7). However, the analysis is generally not applicable to desktop systems (e.g., Linux or Windows), where the concept of thread priority is much fuzzier.

- The queue size depends on the duration of the longest RTC step, including all potential (worst-case) preemptions by higher-priority active objects and ISRs. The faster the processing, the shorter the necessary event queue. To minimize the queue size, you should avoid very long RTC steps. Ideally, all RTC steps of a given active object should require about the same number of CPU cycles to complete.

- Any correlated event production can negatively affect queue size. For example, sometimes ISRs or active objects produce multiple event instances in one RTC step (e.g., the Table active object occasionally produces two permissions to eat). If minimal queue size is critical in your application, you should avoid such bursts by, for example, spreading event production over many RTC steps.

Remember also that the static analysis pertains to a steady-state operation after the initial transient. On startup, the relative priority structure and the event production patterns might be quite different. Generally, it is safest to start active objects in the order of their priority, beginning from the lowest-priority active objects because they tend to have the biggest event queues.

9.4.2 Sizing Event Pools

The size of event pools depends on how many events of different kinds you can sink in your system. The obvious sinks of events are event queues because as long as an event instance waits in a queue, the instance cannot be reused. Another potential sink of events is the event producer. A typical event-generation scenario is to create an event first (assigning a temporary variable to hold the event pointer), then fill in the event parameters and eventually post or publish the event. If the execution thread is preempted after event creation but before posting it, the event is temporarily lost for reuse.

In the simplest case of just one event pool (one size of events) in the system, you can determine the event pool size by adding the sizes of all the event queues plus the number of active objects in the system.

When you use more event pools (the QF allows up to three event pools), the analysis becomes more involved. Generally, you need to proceed as with event queues. For each event size, you determine how many events of this size can accumulate at any given time inside the event queues and can otherwise exist as temporaries in the system.

9.4.3 System Integration

An important aspect of QF-based applications is their integration with the rest of the embedded real-time software, most notably with the device drivers and the I/O system.

Generally, this integration must be based on the event-driven paradigm. QF allows you to post or publish events from any piece of software, not necessarily from active objects. Therefore, if you write your own device drivers or have access to the device driver source code, you can use the QF facilities for creating and publishing or posting events directly.

You should view any device as a shared resource and, therefore, restrict its access to only one active object. This method is safest because it evades potential problems with reentrancy. As long as access is strictly limited to one active object, the RTC execution within the active object allows you to use nonreentrant code. Even if the code is protected by some mutual exclusion mechanism, as is often the case for commercial device drivers, limiting the access to one thread avoids priority inversions and nondeterminism caused by the mutual blocking of active objects.

Accessing a device from just one active object does not necessarily mean that you need a separate active object for every device. Often, you can use one active object to encapsulate many devices.

9.5 Summary

The internal implementation of the QF real-time framework uses a lot of low-level mechanisms such as critical sections, mutexes, and message queues. However, after the infrastructure for executing active objects is in place, the development of QF-based applications can proceed much easier and faster. The higher productivity comes from encapsulated active objects that can be programmed without the troublesome low-level mechanisms traditionally associated with multitasking programs. Yet, the application as a whole can still take full advantage of multithreading.

Developing a QP application involves defining signals and event classes, elaborating state machines of active objects, and deploying the application on a concrete platform. The high portability of QP software components enables you to develop large portions of the code on a different platform than the ultimate target.

Programming with active objects requires some discipline on the part of the programmer because sharing memory and resources is prohibited. The experience of many software developers has shown that it is possible to write efficient applications without breaking this rule. Moreover, the discipline actually helps create software products that are safer, more robust, and easier to test and maintain.

You can view event queues and event pools as the costs inherently associated with event-driven programming paradigm. These data structures, like execution stacks, trade some memory for programming convenience. You should start application development with oversized queues, pools, and stacks and shrink them only toward the end of product development. You can combine basic empirical and analytical techniques for minimizing the size of event queues and event pools.

When integrating the QP application with device drivers and other software components, you should avoid sharing any nonreentrant or mutex-protected code among active objects. The best strategy is to localize access to such code in a dedicated active object.

Active object-based applications tend to be much more resilient to change than traditional blocking tasks because active objects never block and thus are more responsive than blocked tasks. Also, the higher adaptability of event-driven systems is rooted in the separation of concerns of signaling events and state of the object. In particular, active objects use state machines instead of blocking to represent modes of operation and use event passing instead of unblocking to signal interesting occurrences.

The active object-based computing model has been around long enough for programmers to accumulate a rich body of experience about how to best develop such systems. For example, the Real-Time Object-Oriented Modeling (ROOM) method of Selic and colleagues [Selic+ 94] provides a comprehensive set of related development strategies, processes, techniques, and tools. Douglass [Douglass 99, 02, 06] presents unique state patterns, safety-related issues, plenty of examples, and software process applicable to real-time development.

CHAPTER 10

Preemptive Run-to-Completion Kernel

Simplicity is the soul of efficiency.
—R. Austin Freeman (in The Eye of Osiris)

In Section 6.3.8 of Chapter 6 I mentioned a perfect match between the active object computing model and a super-simple, run-to-completion (RTC) *preemptive* kernel. In this chapter I describe such a kernel, called QK, which is part of the QP event-driven platform and is tightly integrated with the QF real-time framework.

I begin this chapter by enumerating good and bad reasons for choosing a preemptive kernel in the first place. I then follow with an introduction to RTC kernels. Next I describe the implementation of the QK preemptive kernel and how it integrates with the QF real-time framework. I then move on to the advanced QK features, such as the priority-ceiling mutex, extended context switch to support various coprocessors (e.g., the 80x87 floating point coprocessor), and thread-local storage (e.g., used in the Newlib standard library). Finally, I describe how to port the QK kernel to various CPUs and compilers. As usual, I illustrate all the features by means of executable examples that you can actually run on any x86-based PC.

10.1 Reasons for Choosing a Preemptive Kernel

Before I go into the details of QK, let me make absolutely clear that preemptive multitasking opens up an entirely new dimension of complexity in the design and debugging of the application, to say the least. It's simply much easier to analyze and debug a program in which tasks cannot preempt each other at every instruction and instead can only yield to one other after each RTC step. Allowing task preemptions can lead to a variety of tricky problems, all ultimately routed in *resource sharing* among tasks. You must be extremely careful because the resource sharing might be more camouflaged than you think. For example, without realizing it, you might be using some *nonreentrant code* from the standard libraries or other sources. Moreover, preemptive multitasking always costs more in terms of the stack usage (RAM) and CPU cycles for scheduling and context switching than does nonpreemptive scheduling, such as the "vanilla" cooperative kernel (see Section 7.11 in Chapter 7).

When you choose a preemptive kernel, such as QK or any other preemptive RTOS for that matter, I want you to do it for good reasons. Let me begin with the bad reasons for choosing a preemptive kernel. First, in active object computing model you don't need a preemptive kernel for partitioning the problem. Though this is by far the most common rationale for choosing an RTOS in the traditional sequential programming model, it's not a valid reason in a system of active objects. Active objects already divide the original problem, regardless of the underlying kernel or RTOS.

Second, you don't need a preemptive kernel to implement efficient blocking because event-driven systems generally don't block (see Chapter 6). Third, since event-driven systems don't poll or block, the RTC steps tend to be naturally quite short. Therefore, chances are that you can achieve adequate task-level response with the simple nonpreemptive kernel (I discuss the execution profile of a nonpreemptive kernel in Section 7.11 in Chapter 7). Often you can easily improve the task-level response of the vanilla kernel by breaking up long RTC steps into short enough pieces by using the "Reminder" state pattern described in Chapter 5. And finally, you don't need a preemptive kernel to take advantage of the low-power sleep modes of your MCU. As described in Section 7.11.1 in Chapter 7, the cooperative vanilla kernel allows you to use low-power sleep modes safely.

Having said all this, however, I must also say that a preemptive kernel can be a very powerful and an indispensable tool—for a specific class of problems. A preemptive kernel executes higher-priority tasks virtually independently of tasks of lower priority. When a high-priority task needs to run, it simply preempts right away any lower-priority task that might be currently running, so the low-priority processing becomes effectively transparent to all tasks of higher priority. Therefore, a preemptive, priority-based kernel *decouples* high-priority tasks from the low-priority tasks in the *time domain*. This unique ability is critical in control-type applications.

In Chapter 1, I mentioned a GPS receiver example in which the hard real-time control loops of GPS signal tracking must execute in parallel to slow, floating point-intensive number crunching, graphic LCD access, and other I/O. This type of application is not well served by a nonpreemptive kernel, in which the task-level response is the longest RTC step in the whole system. It is simply impractical to identify and break up all the low-priority RTC steps into short enough pieces to meet the tight timing constraints of the control loops. Using a nonpreemptive kernel would also result in a fragile design because any change in the low-priority task could impact the timing of high-priority control tasks. In contrast, with a preemptive kernel you can be sure that high-priority processing is virtually insensitive to changes in the low-priority tasks. If you have a control application like that, a preemptive kernel might be actually the simplest, most elegant, and most *robust* way to design and implement your system.

NOTE

The choice of a kernel type is really a tradeoff between the coupling in the time domain and sharing of resources. A nonpreemptive kernel permits you to share resources among tasks but couples the tasks in the time domain. A preemptive kernel decouples the tasks in the time domain but is unforgiving for sharing resources among tasks. Under a preemptive kernel, any mechanism that allows you to share resources *safely* (such as a mutex) introduces some coupling among tasks in the time domain.

10.2 Introduction to RTC Kernels

All event-driven systems handle events in discrete RTC steps. Ironically, most conventional RTOSs force programmers to model these simple, one-shot event reactions using tasks structured as continuous endless loops. This serious mismatch is caused by the sequential programming paradigm underlying all traditional blocking kernels (see Section 6.2.2 in Chapter 6).

Though the event-driven, active object computing model can be made to work with a traditional blocking kernel, as described in Section 6.3 in Chapter 6, it really does not use the capabilities of such a kernel efficiently. An active object task structured as an endless event loop (Figure 6.5(B)) blocks really in just one place in the loop and under one condition only—when the event queue is empty. Thus, it is obviously overkill to use sophisticated machinery capable of blocking at any number of places in the task's execution path to block at just one *a priori* known point.

The event-driven paradigm calls for a different, much simpler, type of truly event-driven, *run-to-completion* kernel. A kernel of this type breaks entirely with the loop structure of the tasks and instead uses tasks structured as one-shot, discrete, *run-to-completion functions*, very much like ISRs [Samek+ 06]. In fact, an RTC kernel views interrupts very much like tasks of a "super-high" priority, except that interrupts are prioritized in hardware by the interrupt controller, whereas tasks are prioritized in software by the RTC kernel (see Figure 6.9 in Chapter 6).

> **NOTE**
>
> The one-shot RTC tasks correspond directly to active objects, as described in Section 6.3.8 in Chapter 6. Therefore, in the following discussion I use the terms *active object* and *task* interchangeably.

10.2.1   Preemptive Multitasking with a Single Stack

To be able to efficiently block anywhere in the task code, a conventional real-time kernel maintains relatively complex execution contexts—including separate stack spaces—for each running task, as shown in Figure 6.2 in Chapter 6. Keeping track of the details of these contexts and switching among them requires lots of bookkeeping and sophisticated mechanisms to implement the context switch magic. In contrast, an RTC kernel can be ultra simple because it doesn't need to manage multiple stacks and all the associated bookkeeping.

By requiring that all tasks run to completion and enforcing fixed-priority scheduling, an RTC kernel can instead manage all context information using the machine's *natural* stack protocol. Whenever a task posts an event to a higher-priority task, an RTC kernel uses a regular C function call to build the higher-priority task context on top of the preempted-task context. Whenever an interrupt preempts a task and the interrupt posts an event to a higher-priority task, the RTC kernel uses the already established interrupt stack frame on top of which to build the higher-priority task context, again using a regular C function call.

This simple form of context management is adequate because every task, just like every ISR, runs to completion. Because the preempting task must also run to completion, the lower-priority context will never be needed until the preempting task (and any higher-priority tasks that might preempt it) has completed and returned—at which time the preempted task will, naturally, be at the top of the stack, ready to be resumed.

At this point, it is interesting to observe that most prioritized interrupt controllers (e.g., the 8259A inside the PC, the AIC in AT91-based ARM MCUs from Atmel, the NVIC in ARM Cortex-M3, and many others) implement in hardware the same asynchronous scheduling policy for interrupts as an RTC kernel implements in software for tasks. In particular, any prioritized interrupt controller allows only higher-priority interrupts to preempt currently serviced interrupt. All interrupts must run to completion and cannot "block." All interrupts nest on the same stack.

This close similarity should help you understand the operation of an RTC kernel because it is based on exactly the same principle widely used and documented in hardware design (just pick up a datasheet of any aforementioned microprocessors). Also, the similarity further reinforces the symmetry between RTC tasks and interrupts illustrated in Figure 6.9 in Chapter 6.

10.2.2 Nonblocking Kernel

One obvious consequence of the simplistic stack-management policy, and the most severe limitation of an RTC kernel, is that the RTC tasks *cannot block*. The kernel cannot leave a high-priority task context on the stack and at the same time resume a lower-priority task. The lower-priority task context simply won't be accessible on top of the stack unless the higher-priority task completes. Of course the inability to block disqualifies an RTC kernel for use with the traditional sequential programming paradigm, which is all about blocking and waiting for events at various points in the task's code.

But the inability to block in the middle of an RTC step is not really a problem for event-driven active objects because they don't need to block anyway. In other words, an active object computing model can benefit from the simplicity and excellent performance of an RTC kernel while being insensitive to the limitations of such a kernel.

10.2.3 Synchronous and Asynchronous Preemptions

As a fully preemptive multitasking kernel, an RTC kernel must ensure that at *all times* the CPU executes the highest-priority task that is ready to run. Fortunately, only two scenarios can lead to readying a higher-priority task:

1. When a lower-priority task posts an event to a higher-priority task, the kernel must immediately suspend the execution of the lower-priority task and start the higher-priority task. This type of preemption is called *synchronous preemption* because it happens synchronously with posting an event to the task's event queue.

2. When an interrupt posts an event to a higher-priority task than the interrupted task, upon completion of the ISR the kernel must start execution of the higher-priority task instead of resuming the lower-priority task. This type of preemption is called *asynchronous preemption* because it can happen asynchronously, any time interrupts are not explicitly locked.

Figure 10.1 illustrates the synchronous preemption scenario caused by posting an event from a low-priority task to a high-priority task.

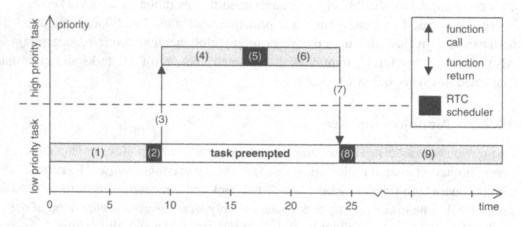

Figure 10.1: Synchronous preemption by a high priority task in an RTC kernel.

(1)   The low-priority task is executing.

(2)   At some point during normal execution, a low-priority task posts or publishes an event to a high-priority task, thus making it ready to run. Posting an event to a queue engages the scheduler of the RTC kernel.

(3)   The scheduler detects that a high-priority task becomes ready to run, so it calls the high-priority task function. Note that the scheduler does not return.

(4)   The high-priority task runs, but at some time it too posts an event to the lower-priority task than itself.

(5)   Event posting engages the RTC scheduler, but this time the scheduler does not find any higher-priority tasks than the current priority. The scheduler returns to the high-priority task.

(6)   The high-priority task runs to completion.

(7)   The high-priority task naturally returns to the RTC scheduler invoked at step 2.

(8)   The scheduler checks once more for a higher-priority task to start, but it finds none. The RTC scheduler returns to the low-priority task

(9)   The low-priority task continues.

Obviously, the synchronous preemption is not limited to only one level. If the high-priority task posts or publishes events to a still higher-priority task in point 5 of Figure 10.1, the high-priority task will be synchronously preempted and the scenario will recursively repeat itself at a higher level of nesting.

Figure 10.2 illustrates the asynchronous preemption scenario caused by an interrupt.

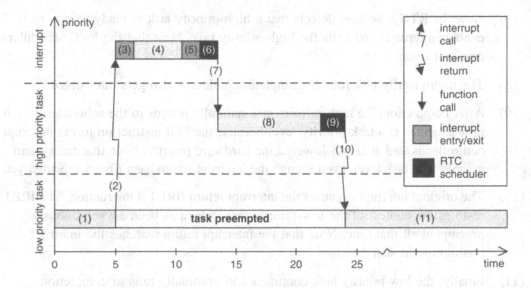

Figure 10.2: Asynchronous preemption by an interrupt and a high-priority task in an RTC kernel.

(1)   A low-priority task is executing and interrupts are unlocked.

(2)   An asynchronous event interrupts the processor. The interrupt immediately preempts any executing task.

(3)   The interrupt service routine (ISR) executes the RTC kernel-specific entry, which saves the priority of the interrupted task into a stack-based variable and raises the current priority of the RTC kernel to the ISR level (above any task).

The raising of the current priority informs the RTC kernel that it executes in the ISR context.

(4)   The ISR continues to perform some work and, among other things, posts or publishes an event to the high-priority task. Posting an event engages the RTC scheduler, which immediately returns because no task has a higher priority than the current priority.

(5)   The ISR continues and finally executes the RTC kernel-specific exit.

(6)   The RTC kernel-specific ISR exit sends the end-of-interrupt (EOI[1]) instruction to the interrupt controller, restores the saved priority of the interrupted task into the current priority, and invokes the RTC scheduler.

(7)   Now the RTC scheduler detects that a high-priority task is ready to run, so it enables interrupts and calls the high-priority task. Note that the RTC scheduler does not return.

(8)   The high-priority task runs to completion, unless it also gets interrupted.

(9)   After completion, the high-priority task naturally returns to the scheduler, which now executes at a task priority level because the EOI instruction to the interrupt controller issued at step 5 lowered the hardware priority. Note that the system priority is at task level, even though the interrupt return hasn't been executed yet.

(10)   The original interrupt executes the interrupt return (IRET[2]) instruction. The IRET restores the context of the low-priority task, which as been asynchronously preempted all that time. Note that the interrupt return matches the interrupt preemption in step 2.

(11)   Finally, the low-priority task continues and eventually runs to completion.

It is important to point out that conceptually the interrupt handling ends in the RTC kernel-specific interrupt exit (5), even though the interrupt stack frame still remains on the stack and the IRET instruction has not been executed yet. The interrupt ends because the EOI instruction is issued to the interrupt controller. Before the EOI instruction, the interrupt controller allows only interrupts of higher priority than the

---

[1] The EOI instruction is understood here generically and denotes a specific machine instruction to stop prioritizing the current interrupt nesting level.
[2] The IRET instruction is understood here generically and denotes a specific machine instruction for returning from an interrupt.

currently serviced interrupt. After the EOI instruction followed by the call to the RTC scheduler, the interrupts get unlocked and the interrupt controller allows all interrupt levels, which is exactly the behavior expected at the task level.

NOTE

Some processor architectures (e.g., ARM Cortex-M3) hardwire the EOI and the IRET instructions together, meaning that EOI cannot be issued independently from IRET. (Note that I treat the instructions EOI and IRET generically in this discussion.) In this case an extra dummy interrupt stack frame must be synthesized, so the EOI/IRET instruction will leave the original interrupt stack frame on the stack. However, such CPU architectures are actually rare, and most processors allow lowering the hardware interrupt priority level without issuing the IRET instruction.

Consequently, the asynchronous preemption is not limited to only one level. The high-priority task runs with interrupts unlocked (Figure 10.2(8)), so it too can be asynchronously preempted by an interrupt, including the same level interrupt as the low-priority task in step 2. If the interrupt posts or publishes events to a still higher-priority task, the high-priority task will be asynchronously preempted and the scenario will recursively repeat itself at a higher level of nesting.

10.2.4 Stack Utilization

Charting the stack utilization over time provides another, complementary view of the synchronous and asynchronous preemption scenarios depicted in Figures 10.1 and 10.2, respectively. To demonstrate the essential behavior, I ignore the irrelevant function calls and other unrelated stack activity.

Figure 10.3 illustrates the stack utilization across the synchronous preemption scenario. The timeline and labels used in Figure 10.3 are identical to those used in Figure 10.1 to allow you to easily correlate these two diagrams.

(1) Initially, the stack pointer points to the low-priority task stack frame.

(2) At some point during normal execution, a low-priority task posts or publishes an event to a high-priority task, which calls the RTC scheduler. A stack frame of the scheduler is pushed on the stack.

(3) The scheduler detects that a high-priority task becomes ready to run, so it calls the high-priority task. A stack frame of the high-priority task is pushed on the stack.

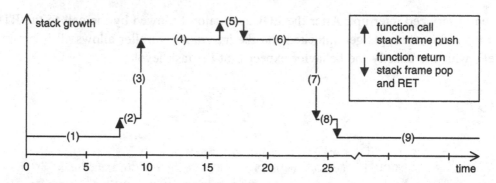

Figure 10.3: Stack utilization during the synchronous preemption scenario.

(4)   High-priority task executes and at some point posts an event to the low-priority task.

(5)   Event posting engages the RTC scheduler, so another scheduler stack frame is pushed on the stack. The scheduler does not find any higher-priority tasks ready to run, so it immediately returns.

(6)   The high-priority task runs to completion.

(7)   The high-priority task naturally returns to the RTC scheduler invoked at step 2, so the task's stack frame is popped off the stack.

(8)   The scheduler checks once more for a higher-priority task to start, but it finds none, so the RTC scheduler returns to the low-priority task popping off its stack frame.

(9)   The low-priority task continues.

Figure 10.4 illustrates the stack utilization during the asynchronous preemption scenario. The time-line and labels used in Figure 10.4 are identical to those used in Figure 10.2 to enable easy correlating of these two diagrams.

(1)   Initially, the stack pointer points to the low-priority task stack frame.

(2)   An asynchronous event interrupts the processor. The interrupt immediately preempts any executing task and the hardware arranges for pushing the interrupt stack frame onto the stack (zigzag arrow). The interrupt service routine (ISR) starts executing and possibly pushes some more context onto the stack (dashed up-arrow). The ISR stack frame is fully built.

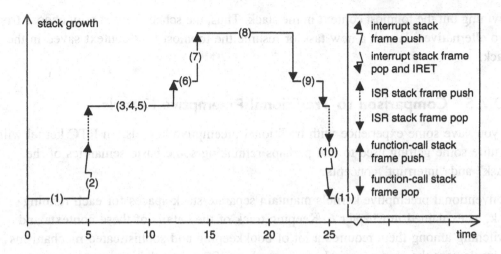

Figure 10.4: Stack utilization during the asynchronous preemption scenario.

(3-5)   The ISR runs to completion and executes the RTC kernel-specific exit, which sends the EOI command to the interrupt controller.

(6)   The RTC scheduler is called, which pushes its stack frame on the stack.

(7)   The scheduler detects that a high-priority task is ready to run, so it enables interrupts and calls the high-priority task. The call to high-priority task function pushes the task's stack frame on the stack.

(8)   The high-priority task runs to completion and returns to the scheduler. The return pops the task's function stack frame off the stack.

(9)   The scheduler resumes and checks for more high-priority tasks to execute but does not find any and returns popping its stack frame off the stack.

(10)   The ISR stack frame gets popped off the stack (the dashed down-arrow). Next the hardware executes the IRET instruction, which causes the interrupt stack frame to pop off the stack.

(11)   The interrupt return exposes the preempted low-priority stack, which is now resumed and continues to run.

As you can see, all context (both the interrupt and task contexts) are kept in a single stack. This forces the kernel to be nonblocking. The scheduler can never access anything but the topmost context in the stack. Thus, the scheduler can only choose from two alternatives: launch a new task or resume the topmost task context saved in the stack.

10.2.5 Comparison to Traditional Preemptive Kernels

If you have some experience with traditional preemptive kernels, an RTC kernel will require some getting used to and perhaps rethinking some basic semantics of the "task" and "interrupt" concepts.

Conventional preemptive kernels maintain separate stack spaces for each running task, as explained in Chapter 6. Keeping track of the details of these contexts and switching among them requires a lot of bookkeeping and sophisticated mechanisms to implement the context switch. In general, an ISR stores the interrupt context on one task's stack and restores the context from another task's stack. After restoring the task's context into the CPU registers, the traditional scheduler always issues the IRET[3] instruction. The key point is that the interrupt context remains saved on the preempted task's stack, so the saved interrupt context outlives the duration of the interrupt handler. Therefore, defining the duration of an interrupt from saving the interrupt context to restoring the context is problematic.

The situation is not really that much different under an RTC kernel, such as QK. An ISR stores the interrupt context on the stack, which happens to be common for all tasks and interrupts. After some processing, the ISR issues the EOI[4] instruction to the interrupt controller and calls the RTC scheduler. If no higher-priority tasks are ready to run, the scheduler exits immediately, in which case the ISR restores the context from the stack and executes the IRET instruction to return to the original task exactly at the point of preemption. Otherwise, the RTC scheduler unlocks interrupts and calls a higher-priority task. The interrupt context remains saved on the stack, just as in the traditional kernel.

The point here is that the ISR is defined from the time of storing interrupt context to the time of issuing the EOI instruction and enabling interrupts inside the RTC scheduler, not necessarily to the point of restoring the interrupt context via the IRET instruction.

---

[3] The IRET instruction is understood here generically and means the instruction that causes the return from interrupt.

[4] The EOI instruction is understood here generically and denotes a specific machine instruction to stop prioritizing the current interrupt nesting level.

This definition is more precise and universal because under any kernel the interrupt context remains stored on one stack or another and typically outlives the duration of an interrupt processing.

NOTE

The definition of ISR duration is not purely academic but has tangible practical implications. In particular, ROM monitor-based debugging at the ISR level is much more challenging than debugging at the task level. Even though all context nests on the same stack, debugging RTC tasks is as easy as debugging the main() task, because the interrupts are unlocked and the hardware interrupt priority at the interrupt controller level is set to the task level.

By managing all task and interrupt contexts in a single stack, an RTC kernel can run with far less RAM[5] than a typical blocking kernel. Because tasks don't have private stacks, there is no unused private stack space associated with suspended tasks. Furthermore, a traditional kernel does not distinguish between the synchronous and asynchronous preemptions and makes all preemptions look like the more stack-intensive asynchronous preemptions. Finally, an RTC kernel does not need to maintain the task control blocks (TCBs; see Figure 6.2 in Chapter 6) for each task.

Because of this simplicity, context switches in an RTC kernel (especially the "synchronous preemptions") can involve much less stack space and CPU overhead than in any traditional kernel. But even the "asynchronous preemptions" in an RTC kernel end up typically using significantly less stack space and fewer CPU cycles. A traditional kernel must typically save all CPU registers in strictly defined order and in "one swoop" onto the private task stack, to be able to restore the registers in an orderly fashion, also in "one swoop." In contrast, an RTC kernel doesn't really care about the order of registers stored and whether they are stored in "one swoop" or piecemeal. The only relevant aspect is that the CPU state be restored exactly to the previous status, but it's irrelevant how this happens. This means that the basic ISR entry and exit sequences that most embedded C compilers are capable of generating are

---

[5] In one case of a specialized GPS receiver application, an RTC kernel brought almost 80 percent reduction of the stack space compared to a traditional preemptive kernel running the same event-driven application [Montgomery 06].

typically adequate for an RTC kernel while being inadequate for most traditional kernels. The C compiler is in a much better position to optimize interrupt stack frames for specific ISRs by saving only the actually clobbered registers and not saving the preserved registers. In this respect, an RTC kernel can take advantage of the C compiler capabilities whereas a traditional kernel can't.

The last point is perhaps best illustrated by a concrete example. All C compilers for ARM processors (I mean the traditional ARM architecture[6]) adhere to the ARM Procedure Call Standard (APCS) that prescribes which registers must be preserved across a C function call and which can be clobbered. The C compiler-generated ISR entry initially saves only the registers that might be clobbered in a C function, which is less than half of all ARM registers. The rest of the registers get saved later, inside C functions invoked from the ISR, if and only if such registers are actually used. This is an example of a context save occurring "piecemeal," which is perfectly suitable for an RTC kernel. In contrast, a traditional kernel must save all ARM registers in "one swoop" upon ISR entry, and if an ISR calls C functions (which it typically does), many registers are saved again. Needless to say, such policy requires more RAM for the stacks and more CPU cycles for a context switch (perhaps by a factor of two) than an RTC kernel.

10.3   QK Implementation

QK is a lightweight, priority-based, RTC kernel specifically designed to provide preemptive multitasking capabilities to the QF real-time framework. QK is not a standalone kernel but rather is just an add-on to QF, similar to the "vanilla" kernel described in Chapter 7. QK is provided as one of the components of the QP event-driven platform.

In this section, I describe the platform-independent QK source code, whereas I focus on the basic kernel functions, such as keeping track of tasks and interrupts, scheduling, and context switching.

---

[6] The new ARMv7 architecture (e.g., Cortex-M3) saves registers in hardware upon interrupt entry, so a C compiler is not involved. However, even in this case the hardware-generated interrupt stack frame takes into account the APCS because the hardware pushes only the eight clobbered ARM registers on the stack.

10.3.1 QK Source Code Organization

Listing 10.1 shows the directories and files comprising the QK preemptive kernel in C. The structure of the C++ version is almost identical, except the implementation files have the .cpp extension. The general source code organization of all QP components is described in Section 8.1.3 in Chapter 8.

```
Listing 10.1  QK source code organization

<qp>\qpc\                    - QP/C root directory (<qp>\qpcpp for QP/C++)
  |
  +-include\                 - QP platform-independent header files
  |  +-qk.h                  - QK platform-independent interface
  |  +-..
  |
  +-qk\                      - QK preemptive kernel
  | +-source\                - QK platform-independent source code (*.C files)
  | | +-qk_pkg.h             - internal, interface for the QK implementation
  | | +-qk.c                 - definition of QK_getVersion() and QActive_start()
  | | +-qk_sched.c           - definition of QK_schedule_()
  | | +-qk_mutex.c           - definition of QK_mutexLock()/QK_mutexUnlock()
  | | +-qk_ext.c             - definition of QK_scheduleExt_()
  | |
  | +-lint\                  - QK options for lint
  |    +-opt_qk.lnt          - PC-lint options for linting QK
  |
  +-ports\                   - Platform-specific QP ports
  | +-80x86\                 - Ports to the 80x86 processor
  | | +-qk\                  - Ports to the QK preemptive kernel
  | | | +-tcpp101\           - Ports with the Turbo C++ 1.01 compiler
  | | |   +-l\               - Ports using the Large memory model
  | | |     +-dbg\           - Debug build
  | | |     | +-qf.lib        - QF  library
  | | |     | +-qep.lib       - QEP library
  | | |     +-rel\           - Release build
  | | |     +-spy\           - Spy build (with software instrumentation)
  | | |     +-make.bat        - batch script for building the QP libraries
  | | |     +-qep_port.h      - QEP platform-dependent include file
  | | |     +-qf_port.h       - QF  platform-dependent include file
  | | |     +-qk_port.h       - QK  platform-dependent include file
  | | |     +-qs_port.h       - QS  platform-dependent include file
  | | |     +-qp_port.h       - QP  platform-dependent include file
  | +-cortex-m3\             - Ports to the Cortex-M3 processor
```

Continued onto next page

```
| |  +-qk\                  - Ports to the QK preemptive kernel
| |  | +-iar\               - Ports with the IAR compiler
| |
+-examples\                 - Platform-specific QP examples
| +-80x88\                  - Examples for the 80x86 processor
| |  +-qk\                  - Examples for the QK preemptive kernel
| |  | +- . . .
| +-cortex-m3\              - Examples for the Cortex-M3 processor
| |  +-qk\                  - Examples for the QK preemptive kernel
| |  | +- . . .
| +- . . .
```

10.3.2 The qk.h Header File

The qk.h header file, shown in Listing 10.2, integrates the QK kernel with the QF framework. The structure of qk.h closely resembles the vanilla kernel header file qvanilla.h discussed in Section 7.11.2 in Chapter 7. The QK kernel uses many of the same basic building blocks provided in QF. Specifically, the QK kernel uses the native QF active object event queues (see Section 7.8.3 in Chapter 7), the QF native memory pool (see Section 7.9), and the QF priority set (see Section 7.10) to keep track of all active object event queues. Additionally, the central element of the QK design is the current systemwide priority, which is just a byte. Figure 10.5 shows the QK data elements.

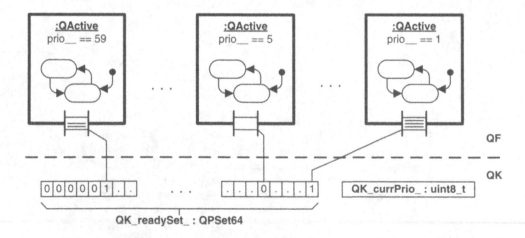

Figure 10.5: Data elements used by the QK preemptive kernel.

Listing 10.2 The QK preemptive kernel interface (<qp>\qpc\include\qk.h)

```
        #ifndef qk_h
        #define qk_h

(1)     #include "qequeue.h"        /* The QK kernel uses the native QF event queue */
(2)     #include "qmpool.h"         /* The QK kernel uses the native QF memory pool */
(3)     #include "qpset.h"          /* The QK kernel uses the native QF priority set */

        /* public-scope objects */
(4)     extern QPSet64 volatile QK_readySet_;              /**< QK ready-set */
(5)     extern uint8_t volatile QK_currPrio_;    /**< current task/interrupt priority*/
(6)     extern uint8_t volatile QK_intNest_;              /**< interrupt nesting level */

        /***************************************************************************/
        /* QF configuration for QK */


(7)     #define QF_EQUEUE_TYPE          QEQueue

        #if defined(QK_TLS) || defined(QK_EXT_SAVE)
(8)         #define QF_OS_OBJECT_TYPE   uint8_t
(9)         #define QF_THREAD_TYPE      void *
        #endif                                    /* QK_TLS || QK_EXT_SAVE */

        /* QK active object queue implementation.........................*/
(10)    #define QACTIVE_EQUEUE_WAIT_(me_) \
            Q_ASSERT((me_)->eQueue.frontEvt != (QEvent *)0)

(11)    #define QACTIVE_EQUEUE_SIGNAL_(me_) \
(12)        QPSet64_insert(&QK_readySet_, (me_)->prio); \
(13)        if (QK_intNest_ == (uint8_t)0) { \
(14)            QK_SCHEDULE_(); \
            } \
            else ((void)0)

(15)    #define QACTIVE_EQUEUE_ONEMPTY_(me_) \
            QPSet64_remove(&QK_readySet_, (me_)->prio)

        /* QK event pool operations...........................................*/
(16)    #define QF_EPOOL_TYPE_          QMPool
(17)    #define QF_EPOOL_INIT_(p_, poolSto_, poolSize_, evtSize_) \
            QMPool_init(&(p_), poolSto_, poolSize_, evtSize_)
(18)    #define QF_EPOOL_EVENT_SIZE_(p_)    ((p_).blockSize)
(19)    #define QF_EPOOL_GET_(p_, e_)       ((e_) = (QEvent *)QMPool_get(&(p_)))
(20)    #define QF_EPOOL_PUT_(p_, e_)       (QMPool_put(&(p_), (e_)))

(21)    void QK_init(void);                               /* QK initialization */
(22)    void QK_onIdle(void);                             /* QK idle callback */
(23)    char const Q_ROM * Q_ROM_VAR QK_getVersion(void);
```

Continued onto next page

```
(24)  typedef uint8_t QMutex;                        /* QK priority-ceiling mutex */
(25)  QMutex QK_mutexLock(uint8_t prioCeiling);
(26)  void QK_mutexUnlock(QMutex mutex);

      /* QK scheduler and extended scheduler */
(27)  #ifndef QF_INT_KEY_TYPE
(28)      void QK_schedule_(void);
(29)      void QK_scheduleExt_(void);                /* QK extended scheduler */
(30)      #define QK_SCHEDULE_()    QK_schedule_()
      #else
(31)      void QK_schedule_(QF_INT_KEY_TYPE intLockKey);
(32)      void QK_scheduleExt_(QF_INT_KEY_TYPE intLockKey); /* extended scheduler */
(33)      #define QK_SCHEDULE_()    QK_schedule_(intLockKey_)
      #endif

      #endif                                          /* qk_h */
```

(1)   The QK kernel uses the native QF event queue, so it needs to include the
      qequeue.h header file.

(2)   The QK kernel uses the native QF memory pool, so it needs to include the
      qmpool.h header file.

(3)   The QK kernel uses the native QF priority set, so it needs to include the
      qpset.h header file.

(4)   The global variable QK_readySet_ is a priority set that maintains the
      global status of all active object event queues, as shown in Figure 10.5.
      QK_readySet_ is declared as volatile because it can change asynchronously
      in ISRs.

(5)   The global variable QK_currPrio_ represents the global systemwide priority
      of the currently running task or interrupt. QK_currPrio_ is declared as
      volatile because it can change asynchronously in ISRs.

(6)   The global variable QK_intNest_ represents the global systemwide interrupt
      nesting level. QK_intNest_ is declared as volatile because it can change
      asynchronously in ISRs.

(7)   The QK kernel uses QEQueue as the event queue for active objects (see also
      Listing 7.7(8)).

(8)   In QK, the QActive data member osObject is used as a bitmask of flags
      representing various properties of the thread. For example, a bit of osObject

bitmask might contain the information whether the thread uses a particular coprocessor. (Refer to Section 10.4.3.)

(9) In QK, the `QActive` data member `thread` is used to point to the thread local storage for that thread. (Refer to Section 10.4.2.)

(10) The QK kernel never blocks. The QK scheduler calls `QActive_get_()` only when it knows for sure that the event queue contains at least one event (see Listing 10.4(22)). Since this is certainty in this type of kernel, the `QACTIVE_EQUEUE_WAIT_()` macro (see Listing 7.24(2) in Chapter 7) asserts that the event queue is indeed not empty.

(11) The macro `QACTIVE_EQUEUE_SIGNAL_()` is called from `QActive_postFIFO()` or `QActive_postLIFO()` when an event is posted to an empty queue (see Listing 7.25(5) in Chapter 7). Note that the macro is invoked inside a critical section. (Also, because I know exactly the context in which the macro is used, I don't bother surrounding the macro body with the `do {...} while(0)` loop.)

(12) The active object becomes ready to run, so its priority is inserted into the ready-set `QK_readySet_`.

(13) This `if` statement tests the QK interrupt nesting level because if the event posting occurs at the task level, the QK scheduler must be invoked to handle a potential synchronous preemption (see Section 10.2.3). The scheduler is not called from an interrupt, because a task certainly cannot preempt an interrupt.

(14) The QK scheduler is called via the macro `QK_SCHEDULE_()`, defined in lines (30) or (33), depending on the interrupt-locking policy used.

NOTE

The QK scheduler is always called from a critical section, that is, with interrupts locked. The scheduler might unlock interrupts internally, but always returns with interrupts locked.

---

(15) The macro `QACTIVE_EQUEUE_ONEMPTY_()` is called from `QActive_get_()` when the queue is becoming empty (see Listing 7.24(12) in Chapter 7). This is exactly when the priority of the active object needs to be removed from the ready-set `QK_readySet_` because the active object is no longer

ready to run. Note that QACTIVE_EQUEUE_ONEMPTY_() is called from a critical section.

(16-20)   The QK kernel uses QMPool as the QF event pool. The platform abstraction layer (PAL) macros are set to access the QMPool operations (see Section 7.9 in Chapter 7).

(21)   The QK kernel initialization is invoked from QF_init(). The QK_init() performs CPU-specific initialization and is defined at the QK port level.

(22)   The QK idle loop calls the QK_onIdle() callback to give the application a chance to customize the idle processing.

> **NOTE**
>
> The QK_onIdle() callback is distinctively different from the QF_onIdle() callback used by the cooperative vanilla kernel, because a preemptive kernel handles idle processing differently than a nonpreemptive one. Specifically, the QK_onIdle() callback is always called with interrupts *unlocked* and does not need to unlock interrupts.

(23)   The QK_getVersion() function allows you to obtain the current version of the QK kernel as a constant string "x.y.zz," where x is the one-digit major number (e.g., 3), y is the one-digit minor number (e.g., 5), and zz is the two-digit release number (e.g., 00).

(24)   This typedef defines the QMutex type for the priority-ceiling mutex. I describe the QK mutex implementation in Section 10.4.1.

(25,26)   The functions QK_mutexLock() and QK_mutexUnlock() perform mutex locking and unlocking, respectively. Again, I describe them in Section 10.4.1.

The QK kernel, just like any other real-time kernel, uses the simplest and most efficient way to protect critical sections of code from disruptions, which is to lock interrupts on entry to the critical section and unlock interrupts again on exit. QK uses the *same* critical section mechanism as the QF real-time framework, and in fact, QK defines the critical section mechanism for QF in the file qk_port.h. (See Section 7.3 in Chapter 7 for the description of the QF critical section policies and macros.)

(27)   As I mentioned at step 13, the QK scheduler is always invoked from a critical section but might need to unlock interrupts internally. Therefore, the signature

of the QK scheduler function depends on the interrupt-locking policy used, which is determined by the QF_INT_KEY_TYPE, as described in Section 7.3 in Chapter 7.

(28) When QF_INT_KEY_TYPE is not defined, the simple "unconditional interrupt locking and unlocking" policy is used, in which case the QK scheduler QK_schedule_() takes no parameters.

(29) Similarly, the extended QK scheduler QK_scheduleExt_() takes no parameters. I discuss the extended QK scheduler in the upcoming Section 10.4.3.

(30,33) The macro QK_SCHEDULE_() invokes the QK scheduler hiding the actual interrupt policy used.

(31) When QF_INT_KEY_TYPE is defined, the policy of "saving and restoring interrupt status" is used, in which case the QK scheduler QK_schedule_() takes the interrupt status key as a parameter.

(32) Similarly, he extended QK scheduler QK_scheduleExt_() takes the same interrupt status key as the parameter. I discuss the extended QK scheduler in the upcoming Section 10.4.3.

10.3.3 Interrupt Processing

Interrupt processing is always specific to your particular application, so obviously it cannot be programmed generically in a platform-independent manner. However, handling interrupts is critical to understanding how the QK kernel works, so here I explain it in general terms.

The most important thing you need to understand about interrupt processing under any *preemptive* kernel, not just QK, is that the kernel must be notified about entering and exiting an interrupt. Specifically, every interrupt must call the QK scheduler upon exit, to give the kernel a chance to handle the asynchronous preemption, as described in Section 10.2.3.

Unlike most conventional preemptive kernels, QK can typically work with interrupt service routines synthesized by the C compiler, which most embedded C cross-compilers support. Listing 10.3 shows the pseudocode for an ISR; Figure 10.6 shows the timeline for executing this code.

Listing 10.3 ISRs in QK; boldface indicates QK-specific interrupt entry and exit

```
(1)    void interrupt YourISR(void) {    /* typically entered with interrupts locked */

(2)       Clear the interrupt source, if necessary

(3)       ++QK_intNest_;              /* account for one more interrupt nesting level */
(4)       Unlock interrupts (depending on the interrupt policy used)

(5)       Execute ISR body, including calling QF services, such as:
          Q_NEW(), QActive_postFIFO(), QActive_postLIF(), QF_publish(), or QF_tick()

(6)       Lock interrupts, if they were unlocked in step (4)
(7)       Send the EOI instruction to the interrupt controller
(8)       -QK_intNest_;               /* account for one less interrupt nesting level */
(9)       if (QK_intNest_ == (uint8_t)0) {       /* coming back to the task level? */
(10)         QK_schedule_();           /* handle potential asynchronous preemption */
          }
       }
```

(1)   An ISR must usually be defined with a special extended keyword (such as " interrupt" in this case). Typically, an ISR is entered with interrupts locked, but some processor architectures (e.g., ARM Cortex-M3) don't lock interrupts. (Check your device's datasheet.)

(2)   If the interrupt source needs clearing, it's best to do it right away.

(3)   You need to tell QK that you are servicing an ISR, so that QK won't try to handle preemption at the ISR level. You inform the kernel by incrementing the global interrupt nesting level QK_intNest_. This must be done in a critical section, so if your processor does not lock interrupts automatically upon ISR entry (see line (1)), you need to explicitly lock interrupts before incrementing the nesting level.

(4)   Depending on the interrupt-locking policy used (see Section 7.3 in Chapter 7) and when an interrupt controller is present, you might need to unlock the interrupt at this point.

NOTE

Steps 3 and 4 constitute the QK-specific interrupt entry, and you can encapsulate them in a macro QK_ISR_ENTRY(), as shown in Section 10.5.

(5)   Execute the ISR body, including calling the indicated QF services. Note that all these services use critical sections internally. Therefore, if your interrupt-locking

policy does not support nesting of critical sections, you must make sure that interrupts are not locked.

(6) You need to lock interrupts if you unlocked them in line (4), because the following code must execute atomically.

(7) You need to send the EOI instruction to the interrupt controller to inform the hardware to stop prioritizing this interrupt level.

(8) The interrupt nesting level QK_intNest_ is decremented to account for leaving the interrupt.

(9) If the interrupt nesting level indicates that the interrupt returns to the task level, as opposed to another interrupt ...

(10) The QK scheduler is called to handle potential asynchronous preemption. Note that the scheduler is called with interrupts locked.

NOTE

Steps 6–10 constitute the QK-specific interrupt exit, and you can encapsulate them in a macro QK_ISR_EXIT(), as shown in Section 10.5.

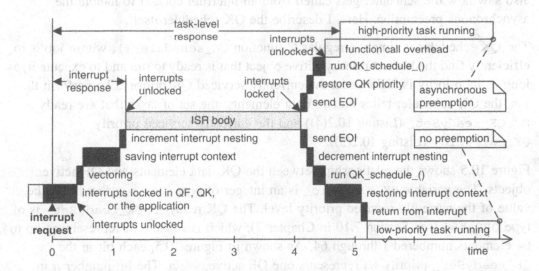

Figure 10.6: Timeline of servicing an interrupt and asynchronous preemption in QK. Black rectangles represent code executed with interrupts locked.

Figure 10.6 shows the timeline of interrupt servicing and asynchronous preemption under the QK preemptive kernel. I'd like to highlight two interesting points. First, the interrupt response under the QK kernel is as fast as under any other preemptive kernel and is mostly dominated by the longest critical section in the system and how long it takes the hardware to save the interrupt context to the stack. Second, the task-level response of the high-priority task is generally *faster* than any conventional preemptive kernel because the interrupt context does not need to be restored entirely from the stack and the interrupt return does not need to be executed to start the high-priority task. In the RTC kernel, all this is replaced by a function call, which typically is much faster than restoring the whole register set from the stack and executing the IRET instruction.

10.3.4   The `qk_sched.c` Source File (QK Scheduler)

The source file `qk_sched.c` implements the QK scheduler, which is the most important part of the QK kernel. As explained in Section 10.2.3, the QK scheduler is called at two junctures: (1) when an event is posted to an event queue of an active object (synchronous preemption), and (2) at the end of ISR processing (asynchronous preemption). In the `qk.h` header file (Listing 10.2(14)), you saw how the QK scheduler gets invoked to handle the synchronous preemptions. In the previous section, you also saw how the scheduler gets called from an interrupt context to handle the asynchronous preemption. Here, I describe the QK scheduler itself.

The QK scheduler is simply a regular C-function `QK_schedule_()`, whose job is to efficiently find the highest-priority active object that is ready to run and to execute it, as long as its priority is higher than the currently serviced QK priority. To perform this job, the QK scheduler relies on two data elements: the set of tasks that are ready to run `QK_readySet_` (Listing 10.2(4)) and the currently serviced priority `QK_currPrio_` (Listing 10.2(5)).

Figure 10.5 shows the relationship between the QK data elements and QF active objects. The variable `QK_currPrio_` is an integer of type `uint8_t` that holds the value of the currently serviced priority level. The QK ready-set `QK_readySet_` is of type `QPSet64` (see Section 7.10 in Chapter 7), which is capable of representing up to 64 elements numbered 1 through 64. As shown in Figure 10.5, each bit in the `QK_readySet_` priority set represents one QF active object. The bit number n in `QK_readySet_` is 1 if the event queue of the active object of priority n is not empty.

Conversely, bit number *m* in `QK_readySet_` is 0 if the event queue of the active object of priority *m* is empty or the priority level *m* is not used. Both variables `QK_currPrio_` and `QK_readySet_` are always accessed in a critical section to prevent data corruption.

Listing 10.4 shows the complete implementation of the `QK_schedule_()` function.

Listing 10.4 QK scheduler implementation
(`<qp>\qpc\qk\source\qk_sched.c`)

```
(1)   #include "qk_pkg.h"

      /* Public-scope objects -----------------------------------------------*/
(2)   QPSet64 volatile QK_readySet_;                            /* QK ready-set */
                                              /* start with the QK scheduler locked */
(3)   uint8_t volatile QK_currPrio_ = (uint8_t)(QF_MAX_ACTIVE + 1);
(4)   uint8_t volatile QK_intNest_;                   /* start with nesting level of 0 */

      /* ...................................................................*/
      /* NOTE: the QK scheduler is entered and exited with interrupts LOCKED */
(5)   #ifndef QF_INT_KEY_TYPE
(6)   void QK_schedule_(void) {
      #else
(7)   void QK_schedule_(QF_INT_KEY_TYPE intLockKey_) {
      #endif
          uint8_t p;
                              /* the QK scheduler must be called at task level only */
(8)       Q_REQUIRE(QK_intNest_ == (uint8_t)0);

(9)       if (QPSet64_notEmpty(&QK_readySet_)) {
              /* determine the priority of the highest-priority task ready to run */
(10)          QPSet64_findMax(&QK_readySet_, p);

(11)          if (p > QK_currPrio_) {                        /* do we have a preemption? */
(12)              uint8_t pin = QK_currPrio_;            /* save the initial priority */
                  QActive *a;
(13)  #ifdef QK_TLS                                     /* thread-local storage used? */
(14)              uint8_t pprev = pin;
      #endif
(15)              do {
                      QEvent const *e;
(16)                  a = QF_active_[p];                  /* obtain the pointer to the AO */

(17)                  QK_currPrio_ = p;       /* this becomes the current task priority */

      #ifdef QK_TLS                                     /* thread-local storage used? */
(18)                  if (p != pprev) {                    /* are we changing threads? */
```

Continued onto next page

```
(19)              QK_TLS(a);                    /* switch new thread-local storage */
(20)              pprev = p;
           }
   #endif
(21)           QK_INT_UNLOCK_();                        /* unlock the interrupts */

(22)           e = QActive_get_(a);              /* get the next event for this AO */
(23)           QF_ACTIVE_DISPATCH_(&a->super, e);        /* dispatch to the AO */
(24)           QF_gc(e);                  /* garbage collect the event, if necessary */

(25)           QK_INT_LOCK_();
                                 /* determine the highest-priority AO ready to run */
(26)           if (QPSet64_notEmpty(&QK_readySet_)) {
(27)               QPSet64_findMax(&QK_readySet_, p);
               }
               else {
(28)               p = (uint8_t)0;
               }
(29)       } while (p > pin);            /* is the new priority higher than initial? */
(30)       QK_currPrio_ = pin;                    /* restore the initial priority */

   #ifdef QK_TLS                                    /* thread-local storage used? */
(31)       if (pin != (uint8_t)0) {          /* no extended context for idle loop */
(32)           a = QF_active_[pin];
(33)           QK_TLS(a);                           /* restore the original TLS */
           }
   #endif
       }
(34) }
```

(1)   As every QK source file, the qk_sched.c file includes the wider
       "package-scope" QK interface qk_pkg.h, located in <qp>\qpc\qk\source\.
       The qk_pkg.h header file includes the platform-specific QK port header file
       qk_port.h, but it additionally defines some internal macros and objects
       shared only internally within QK.

(2)   The global variable QK_readySet_ is a priority set that maintains the global
       status of all active object event queues, as shown in Figure 10.5.

(3)   The global variable QK_currPrio_ represents the global systemwide priority of
       the currently running task or interrupt.

(4)   The global variable QK_intNest_ represents the global systemwide interrupt
       nesting level.

(5)   The QK scheduler is always invoked with *interrupts locked* but might need to
       unlock interrupts internally. Therefore, the signature of the QK scheduler

function depends on the interrupt-locking policy used, which is determined by the `QF_INT_KEY_TYPE`, as described in Section 7.3 in Chapter 7.

(6)   When `QF_INT_KEY_TYPE` is not defined, the simple "unconditional interrupt locking and unlocking" policy is used, in which case the QK scheduler `QK_schedule_()` takes no parameters.

(7)   When `QF_INT_KEY_TYPE` is defined, the policy of "saving and restoring interrupt status" is used, in which case the QK scheduler `QK_schedule_()` takes the interrupt status key as the parameter.

(8)   The QK scheduler should only be called at the task level.

(9)   If the ready-set `QK_readySet_` is not empty, the QK kernel has some events to process.

(10) The priority set quickly discovers the highest-priority, not-empty event queue, as I described in Section 7.10 in Chapter 7.

(11) The QK scheduler can preempt the currently running task only when the new priority is higher than the priority of the currently executing task.

NOTE

The QK scheduler is an indirectly recursive function. The scheduler calls task functions, which might post events to other tasks, which calls the scheduler. However, this recursion can continue only as long as the priority of the tasks keeps increasing. Posting an event to a lower- or equal-priority task (posting to self) stops the recursion because of the `if` statement in line (11).

---

(12) To handle the preemption, the QK scheduler will need to increase the current priority. However, before doing this, the current QK priority is saved into a stack variable `pin`.

(13) If the macro `QK_TLS` is defined, the QK kernel manages the thread-local storage (TLS). I discuss TLS management in QK in the upcoming Section 10.4.2.

(14) For TLS, the variable `pprev` holds the previous task priority to help QK determine when a task change occurs.

(15) The `do` loop continues as long as the QK scheduler finds ready-to-run tasks of higher priority than the initial priority `pin`.

(16)   The active object pointer ' a' is resolved through the `QF_active_[]` priority-to-active object lookup table maintained internally by QF.

(17)   The current QK priority is raised to the level of the highest-priority task that is about to be started.

(18)   If TLS management is enabled, the scheduler checks whether the task change is about to occur.

(19)   If so, the `QK_TLS()` macro changes the TLS to the new task.

(20)   Also, the `pprev` variable is updated so that QK can discover when the next task change occurs.

(21)   Interrupts are unlocked to run the RTC task.

(22-24)   These are the three steps of the active object thread (see Listing 7.8 in Chapter 7).

NOTE

Steps 22–24 represent the body of the one-shot, RTC task in QK. Note that the RTC task is executed with interrupts unlocked.

---

(25)   Interrupts are locked so that the scheduler can check again for highest-priority active objects ready to run. The status of the QK ready set could have change during the RTC step just executed.

(26)   If the ready-set `QK_readySet_` is not empty, the QK kernel has still some events to process.

(27)   The priority set quickly discovers the new highest-priority, not-empty event queue based on the potentially changed `QK_readySet_`.

(28)   If the `QK_readySet_` turns out to be empty, the QK kernel has nothing more to do. The variable `p` is set to zero to terminate the `do-while` loop in the next step.

(29)   The `while` condition loops back to step (13) as long as the QK scheduler still finds ready-to-run tasks of higher priority than the initial priority `pin`.

(30)   After the loop terminates, the current QK priority must go back to the initial level.

(31) The TLS needs to be restored only if a task has been preempted. The priority 'pin' of zero corresponds to the QK idle loop. I assume that the idle loop does not use the TLS.

(32) The pointer to the preempted active object is resolved through the `QF_active_[]` priority-to-active object lookup.

(33) The `QK_TLS()` macro restores the TLS of the original preempted task.

(34) The QK scheduler always returns with interrupts locked.

10.3.5 The `qk.c` Source File (QK Startup and Idle Loop)

The `qk.c` source file, shown in Listing 10.5, defines the QK initialization, cleanup, startup, and idle loop.

Listing 10.5 QK startup and idle loop (`<qp>\qpc\qk\source\qk.c`)

```
(1)   #include "qk_pkg.h"
      #include "qassert.h"

      Q_DEFINE_THIS_MODULE(qk)

      /*................................................................*/
      void QF_init(void) {
          /* nothing to do for the QK preemptive kernel */
(2)       QK_init();                                /* might be defined in assembly */
      }
      /*................................................................*/
      void QF_stop(void) {
(3)       QF_onCleanup();                                       /* cleanup callback */
          /* nothing else to do for the QK preemptive kernel */
      }
      /*................................................................*/
(4)   void QF_run(void) {
          QK_INT_LOCK_KEY_

          QK_INT_LOCK_();
(5)       QK_currPrio_ = (uint8_t)0;    /* set the priority for the QK idle loop */
(6)       QK_SCHEDULE_();               /* process all events produced so far */
          QK_INT_UNLOCK_();

(7)       QF_onStartup();                                       /* startup callback */

(8)       for (;;) {                                            /* the QK idle loop */
```

Continued onto next page

```
(9)            QK_onIdle();                    /* invoke the QK on-idle callback */
        }
    }
    /*................................................................*/
(10) void QActive_start(QActive *me, uint8_t prio,
                     QEvent const *qSto[], uint32_t qLen,
(11)                 void *tls,
(12)                 uint32_t flags,
                     QEvent const *ie)
    {
        Q_REQUIRE(((uint8_t)0 < prio) && (prio <= (uint8_t)QF_MAX_ACTIVE));

(13)    QEQueue_init(&me->eQueue, qSto, (QEQueueCtr)qLen);
(14)    me->prio = prio;
(15)    QF_add_(me);                    /* make QF aware of this active object */

    #if defined(QK_TLS) || defined(QK_EXT_SAVE)
(16)    me->osObject = (uint8_t)flags;      /* osObject contains the thread flags */
(17)    me->thread  = tls;   /* contains the pointer to the thread-local storage */
    #else
        Q_ASSERT((tls == (void *)0) && (flags == (uint32_t)0));
    #endif

(18)    QF_ACTIVE_INIT_(&me->super, ie);         /* execute initial transition */
    }
    /*................................................................*/
    void QActive_stop(QActive *me) {
        QF_remove_(me);                 /* remove this active object from the QF */
    }
```

(1) As every QK source file, the qk.c file includes to the wider "package-scope"
     QK interface qk_pkg.h, located in <qp>\qpc\qk\source\. The qk_pkg.h
     header file includes the platform-specific QK port header file qk_port.h, but
     it additionally defines some internal macros and objects shared only internally
     within QK.

(2) The function QF_init() initializes the QF framework and the underlying kernel.
     In case of the QK kernel, this function has nothing to do, except invoking the
     QK_init() function to give the QK kernel a chance to initialize. QK_init() is
     defined in the QK port.

(3) The function QF_stop() stops execution of the QF framework. In case of
     the QK kernel, this function has nothing to do except invoke the
     QF_onCleanup() callback function to give the application a chance to
     clean up and exit to the underlying operating system (e.g., consider QK

kernel running on top of DOS). All QF callback functions are summarized in Section 8.1.8 in Chapter 8.

(4) Applications call the function `QF_run()` from `main()` to transfer the control to the framework. This function implements the startup and idle loop of the QK kernel.

(5) The current QK priority is reduced from the initial value of `QF_MAX_ACTIVE+1` (see Listing 10.4(3)) to zero, which corresponds to the priority of the QK idle loop.

> **NOTE**
>
> The QK current priority value of `QF_MAX_ACTIVE+1` effectively locks the QK scheduler, so the scheduler is not even called upon event posting or exit from ISRs.

(6) After reducing the priority level, the scheduler is invoked to process all events that might have been posted during the initialization of active objects. Note that the scheduler is called with interrupts locked.

(7) The `QF_onStartup()` callback function configures and starts interrupts. This function is typically implemented at the application level (in the BSP). All QF callback functions are summarized in Section 8.1.8 in Chapter 8.

(8) This is the *idle loop* of the QK kernel.

> **NOTE**
>
> When no interrupts are running and all event queues are empty, the QK kernel has nothing to do. The kernel then executes the idle loop. The idle loop is the only "task" structured as an endless loop in QK. The QK priority associated with the idle loop is zero and is the absolute lowest priority level in the system, which is not accessible to the RTC tasks. The task priorities in QK start at 1.

(9) The idle loop continuously calls the `QK_onIdle()` callback function to give the application a chance to put the CPU to a low-power sleep mode or to perform other processing (e.g., software-tracing output, see Chapter 11). The `QK_onIdle()` function is typically implemented at the application level (in the BSP).

> **NOTE**
>
> As a preemptive kernel, QK handles idle processing differently than a nonpreemptive vanilla kernel. Specifically, the `QK_onIdle()` callback is always called with interrupts *unlocked* and does not need to unlock interrupts (as opposed to the `QF_onIdle()` callback). Furthermore, a transition to a low-power sleep mode inside `QK_onIdle()` does not need to occur with interrupts locked. Such a transition is safe and does not cause any race conditions, because a preemptive kernel never switches the context back to the idle loop as long as events are available for processing.

(10)   The `QActive_start()` function initializes the event queue and starts the active object QK task.

(11)   For conventional kernels, the fifth and sixth parameters of `QActive_start()` represent the private stack memory and the size of that memory. The QK kernel does not need a per-task stack. Instead, the fifth parameter of `QActive_start()` is used as a pointer to the thread-local storage (TLS) for the QK task.

(12)   The sixth parameter is used as a bitmask of flags representing properties of the task, such as whether the task uses a coprocessor. I discuss a generic coprocessor context switch in QK in the upcoming Section 10.4.3.

(13)   The QK kernel uses the native QF event queue `QEQueue`, which needs to be initialized with the function `QEQueue_init()`.

(14)   The QF priority of the active object is set inside the active object.

(15)   The active object is added to the QF framework.

(16)   The task flags are stored in the `osObject` data member. I show an example of using the task flags in the upcoming Section 10.4.3.

(17)   The pointer to the TLS for this task is stored in the `thread` data member.

(18)   The internal state machine of the active object is initialized.

10.4   Advanced QK Features

Simple as it is, the QK kernel supports quite advanced features, which you find only in the more sophisticated real-time kernels. In this section I cover mutual exclusion that is robust against priority inversions, thread-local storage useful for thread-safe libraries, and extended context switching to support various coprocessors. If you happen to know how other kernels implement these features, I hope you'll appreciate the simple elegance of the QK implementation.

NOTE

All advanced QK features covered in this section are only necessary when you must share resources among multiple QK tasks (active objects). If you use strict encapsulation (as advised in Chapter 9) and never share memory, nonreentrant libraries, or coprocessors among active objects, you don't need to use *any* of these advanced features.

10.4.1 Priority-Ceiling Mutex

QK is a preemptive kernel, and as with all such kernels, you must be very careful with any resource sharing among QK tasks. Ideally, the QF active objects (i.e., QK tasks) should communicate exclusively via events and otherwise should not share any resources, which I have been advocating all along (see Chapter 9). This ideal situation allows you to program all active objects without ever worrying about mutual exclusion mechanisms to protect shared resources.

However, at the cost of increased coupling among active objects, you might choose to share selected resources. If you go this path, you take the burden on yourself to interlock the access to such resources (shared memory or devices).

One powerful method of guaranteeing mutually exclusive access to resources at your disposal is the critical section mechanism implemented with the QF macros `QF_INT_LOCK()` and `QF_INT_UNLOCK()`, as described in Section 7.3 in Chapter 7. For very short accesses this might well be the most efficient synchronization mechanism.

However, you can also use a much less intrusive mechanism available in QK. QK supports a priority-ceiling mutex to prevent task-level preemptions while accessing a shared resource. Priority-ceiling mutex is *immune to priority inversions* [Kalinsky 05] but is a more selective mechanism than interrupt locking because all tasks (and interrupts) of priority higher than the priority ceiling run as usual. Listing 10.6 shows an example of using a QK mutex to protect a shared resource.

Listing 10.6 Protecting a shared resource with a QK priority-ceiling mutex

```
     void your_function(arguments) {
(1)      QMutex mutex;
         . . .
(2)      mutex = QK_mutexLock(PRIO_CEILING);
(3)      You can safely access the shared resource here
(4)      QK_mutexUnlock(mutex);
         . . .
     }
```

---

(1)   You need to provide a temporary `mutex` variable of type `QMutex` (which is just a byte).

(2)   You lock the mutex by calling `QK_mutexLock()`. This function requires the priority ceiling parameter, which you choose to be the priority of the highest-priority task that may use the shared resource you want to protect. Typically, this priority is known at compile time because all QK tasks have fixed priorities assigned statically usually at system startup.

(3)   You access the shared resource.

(4)   You unlock the mutex by calling `QK_mutexUnlock()`.

As you can see, the mutex variables are used only temporarily, and there is no limitation on how many mutexes you can use in your application. In principle, mutex locks can even nest, so your code in Listing 10.6(3) could use another priority-ceiling mutex. Note that I mention this only as a theoretical possibility, not necessarily as a good or recommended design.

Before explaining how the QK protects the resource and why it is a nonblocking mechanism, I simply show in Listing 10.7 how it is implemented.

Listing 10.7 QK mutex (`<qp>\qpc\qk\source\qk_mutex.c`)

```
     QMutex QK_mutexLock(uint8_t prioCeiling) {
         uint8_t mutex;
         QK_INT_LOCK_KEY_
         QK_INT_LOCK_();
(1)      mutex = QK_currPrio_;        /* the original QK priority to return */
(2)      if (QK_currPrio_ < prioCeiling) {
```

```
(3)           QK_currPrio_ = prioCeiling;            /* raise the QK priority */
          }
      QK_INT_UNLOCK_();
(4)       return mutex;
      }
      /*.................................................................*/
   void QK_mutexUnlock(QMutex mutex) {
      QK_INT_LOCK_KEY_
      QK_INT_LOCK_();
(5)       if (QK_currPrio_ > mutex) {
(6)           QK_currPrio_ = mutex;                  /* restore the saved priority */
(7)           QK_SCHEDULE_();
          }
      QK_INT_UNLOCK_();
   }
```

(1) Inside a critical section, the current QK priority is saved in the temporary variable mutex to be returned from the QK_mutexLock().

(2) If the priority ceiling provided as the function argument exceeds the current QK priority ...

(3) The current QK priority is raised to the priority ceiling.

(4) The original QK priority is returned to the caller.

(5) Inside a critical section, the current QK priority is compared to the mutex argument.

(6) If the current priority exceeds the mutex, the current QK priority is reduced to the level of the mutex.

(7) Reducing the current QK priority might "expose" some ready-to-run tasks that have a higher priority than the reduced QK_currPrio_ level. The QK scheduler is called to process these potential synchronous preemptions. Note that the scheduler is called with interrupts locked.

As you can see, locking the mutex boils down to raising the current QK priority to the priority ceiling level. Recall that the QK scheduler can only launch tasks with priorities higher than the initial priority with which the scheduler was entered (Listing 10.4(11)). This means that temporarily increasing the current QK priority prevents preemptions from all tasks with priorities lower than or equal to the priority ceiling. This is exactly what a priority ceiling mutex is supposed to do to protect your resource.

Note that the QK mutex is a *nonblocking* mechanism. If a task that needs to protect a shared resource is running at all, it means that all tasks of higher priority have no events to process. Consequently, simply preventing launch of higher-priority tasks that might access the resource is sufficient to guarantee the mutually exclusive access to the resource. Of course, you don't need to worry about any lower-priority tasks that might be preempted because they never resume until the current task runs to completion.

10.4.2   Thread-Local Storage

Thread-local storage (TLS) is a mechanism by which variables are allocated such that there is one instance of the variable per extant thread. The canonical example of when TLS could be useful is the popular Newlib[7] standard C runtime library intended for use in embedded devices. Newlib's facilities are *reentrant*, but only when properly integrated into a multithreaded environment [Gatliff 01]. Because QK is a *preemptive* kernel, care must be taken to preserve the reentrant character of Newlib.

For example, consider the `errno` facility specified in the ANSI C standard. The runtime library sets `errno` when an error occurs within the library. Once set, `errno`'s value persists until the application clears it, which simplifies error notification by the library but can create reentrancy problems when multiple threads are using the library at the same time. If `errno` would be just a single global variable shared among all threads, neither thread would know who generated the error.

Newlib addresses this problem by redefining `errno` as a macro that (indirectly) references a global pointer called `_impure_ptr` (see Figure 10.7). The Newlib's `_impure_ptr` points to a structure of type `struct _reent`. This structure contains the traditional `errno` value specified by ANSI, but it also contains a lot of other elements, including signal handler pointers and file handles for standard input, output, and error streams.

The central idea of the Newlib design is that *every* thread in the application has its own copy of the `_reent` structure (shown as TLS in Figure 10.7) and that the `_impure_ptr` pointer is switched during context switches to always point at the

---

[7] www.sourceware.org/newlib/

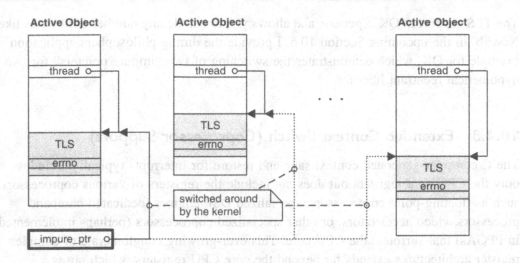

Figure 10.7: Pointer to thread-local storage (TLS) switched around by the kernel.

_reent structure of the *currently active thread*. Obviously, to perform the switching of the _impure_ptr, you need a helping hand from the kernel.

QK supports the TLS concept by providing a context-switch hook QK_TLS(), which is invoked every time a different task priority is processed (see Listing 10.4 (19,33)). The macro QK_TLS() receives from the kernel a pointer to the current active object. The following code fragment from qk_port.h defines the macro QK_TLS() for re-assigning the Newlib's _impure_ptr during context switches:

```
#define QK_TLS(act_)  (_impure_ptr = (struct _reent *)(act_)->thread)
```

Though the QK_TLS() macro will switch the _impure_ptr automatically, *you* are responsible for allocating the _reent structure in each active object. You also need to tell QK where the TLS is for every active object during startup by passing the pointer to the TLS as the fifth parameter of the QActive_start() function (see Listing 10.5(17)).

NOTE

The current implementation of the TLS in QK assumes that the thread-local storage is accessed neither in the ISRs nor in the idle loop (inside the QK_onIdle() callback function).

The TLS support in QK is generic and allows you to handle any number of libraries like Newlib. In the upcoming Section 10.6, I provide the dining philosophers application example for QK, which demonstrates the switching of two "impure pointers" for two hypothetical reentrant libraries.

10.4.3   Extended Context Switch (Coprocessor Support)

The C compiler-generated context save and restore for interrupts typically includes only the CPU core registers but does *not* include the registers of various coprocessors, such as floating-point coprocessors, specialized DSP engines, dedicated baseband processors, video accelerators, or other specialized coprocessors (perhaps implemented in FPGAs) that surround the CPU core. This ever-growing conglomerate of complex register architectures extends far beyond the core CPU registers, which poses a problem for a *preemptive* kernel if the various coprocessors are used by multiple tasks. The solution offered by advanced preemptive kernels is to include the various coprocessor registers in the context switch process, thus allowing sharing of the coprocessors among multiple tasks.

The QK kernel supports such *extended context switch* in a generic way, which you can easily customize for various coprocessors and hardware accelerators. The QK design of the extended context switch carefully minimizes the added overhead by saving and restoring the extended context only when necessary. The basic simplifying assumption is that neither ISRs, nor the QK idle loop use the coprocessor(s). Consequently, the extended context needs to be preserved only when a task preempts another task. Moreover, synchronous context switches generally don't need to be extended, because the context switch is in this case just a simple function call (see Section 10.2.3), which cannot happen in the middle of accessing a coprocessor.

This leaves only the asynchronous preemptions as really requiring the extended context switch. As described in Section 10.3.3, asynchronous preemptions are handled upon the exit from interrupts in the `QK_ISR_EXIT()` macro. Listing 10.8 shows pseudocode of the `QK_ISR_EXIT()` macro, which calls the *extended scheduler* `QK_scheduleExt_()` instead of `QK_scheduler_()`, as shown in Listing 10.3(10).

Listing 10.8 `QK_ISR_EXIT()` **macro with the extended context switch**

```
#define QK_ISR_EXIT() do { \
    Lock interrupts \
    Send the EOI instruction to the interrupt controller \
    --QK_intNest_; \
    if (QK_intNest_ == 0) { \
        QK_scheduleExt_(); \
    } \
} while (0)
```

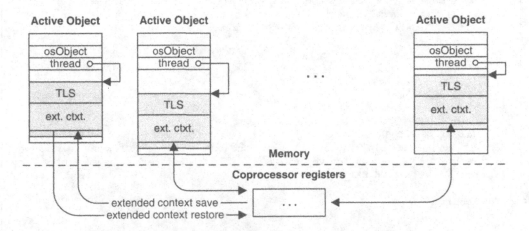

Figure 10.8: Extended context switch saves and restores coprocessor registers in the TLS area.

Figure 10.8 shows the additional extended context save and restore steps implemented in the extended scheduler `QK_scheduleExt_()`. The per-active object extended context is simply added to the TLS area, which is accessible via the `thread` data member of the `QActive` class.

Listing 10.9 shows the extended scheduler `QK_scheduleExt_()`. In the explanation section following this listing, I describe only the highlighted differences from the regular scheduler `QK_scheduler_()`, which I already explained in Listing 10.4.

Listing 10.9 QK extended scheduler implementation
(`<qp>\qpc\qk\source\qk_ext.c`)

```c
        #ifndef QF_INT_KEY_TYPE
(1)     void QK_scheduleExt_(void) {
        #else
(2)     void QK_scheduleExt_(QF_INT_KEY_TYPE intLockKey_) {
        #endif
            uint8_t p;
                                    /* the QK scheduler must be called at task level only */
            Q_REQUIRE(QK_intNest_ == (uint8_t)0);

            if (QPSet64_notEmpty(&QK_readySet_)) {
                /* determine the priority of the highest-priority task ready to run */
                QPSet64_findMax(&QK_readySet_, p);

                if (p > QK_currPrio_) {                       /* do we have a preemption? */
                    uint8_t pin = QK_currPrio_;            /* save the initial priority */
                    QActive *a;
        #ifdef QK_TLS                                     /* thread-local storage used? */
                    uint8_t pprev = pin;
        #endif
(3)     #ifdef QK_EXT_SAVE                               /* extended context-switch used? */
(4)                 if (pin != (uint8_t)0) {    /* no extended context for the idle loop */
(5)                     a = QF_active_[pin];          /* the pointer to the preempted AO */
(6)                     QK_EXT_SAVE(a);                   /* save the extended context */
                    }
        #endif
                    do {
                        QEvent const *e;
                        a = QF_active_[p];                  /* obtain the pointer to the AO */

                        QK_currPrio_ = p;      /* this becomes the current task priority */

        #ifdef QK_TLS                                       /* thread-local storage used? */
                        if (p != pprev) {                     /* are we changing threads? */
                            QK_TLS(a);                  /* switch new thread-local storage */
                            pprev = p;
                        }
        #endif
                        QK_INT_UNLOCK_();                        /* unlock the interrupts */

                        e = QActive_get_(a);             /* get the next event for this AO */
                        QF_ACTIVE_DISPATCH_(&a->super, e);        /* dispatch to the AO */
                        QF_gc(e);              /* garbage collect the event, if necessary */

                        QK_INT_LOCK_();
                                    /* determine the highest-priority AO ready to run */
                        if (QPSet64_notEmpty(&QK_readySet_)) {
```

```
                    QPSet64_findMax(&QK_readySet_, p);
                }
            else{
                p = (uint8_t)0;
                }
        } while (p > pin);             /* is the new priority higher than initial? */
        QK_currPrio_ = pin;                /* restore the initial priority */

(7)  #if defined(QK_TLS) || defined(QK_EXT_RESTORE)
(8)      if (pin != (uint8_t)0)             {/*no extended context for the idle loop */
(9)           a = QF_active_[pin];           /* the pointer to the preempted AO */
    #ifdef QK_TLS                            /* thread-local storage used? */
             QK_TLS(a);                      /* restore the original TLS */
    #endif
    #ifdef QK_EXT_RESTORE                     /* extended context-switch used? */
(10)          QK_EXT_RESTORE(a);              /* restore the extended context */
    #endif
             }
    #endif
        }
    }
}
```

(1,2)   The signature of the extended scheduler depends on the interrupt locking policy used, just like the regular scheduler.

(3)   If the macro QK_EXT_SAVE() is defined, the extended scheduler invokes the macro to save the extended context.

(4)   The extended context needs to be saved only if a task has been preempted. The priority 'pin' of zero corresponds to the QK idle loop. I assume that the idle loop does not to use the coprocessor(s).

NOTE

The idle loop does not correspond to an active object, so it does not have the TLS memory area to save the extended context.

---

(5)   The pointer to the preempted active object is resolved through the QF_active_[] priority-to-active object lookup.

(6)   The QK_EXT_SAVE() macro saves the extended context of the original preempted active object.

(7)   The following code is only needed when either TLS or extended context is used.

(8)   The TLS or the extended context needs to be restored only if a task has been preempted. The priority 'pin' of zero corresponds to the QK idle loop. I assume that the idle loop uses neither TLS nor the coprocessor(s).

(9)   The pointer to the preempted active object is resolved through the QF_active_[] priority-to-active object lookup.

(10) The QK_EXT_RESTORE() macro restores the extended context of the original preempted active object.

The QK_EXT_SAVE() and QK_EXT_RESTORE() macros allow you to save and restore as many coprocessor contexts as necessary for a given task. As shown in Figure 10.8, you need to provide per-task memory for all the extended contexts that you use. In the next section, I describe the QK port to 80x86 with the 80x87 floating point coprocessor (FPU) and I provide examples of the macros QK_EXT_SAVE() and QK_EXT_RESTORE() for the 80x87 FPU.

10.5   Porting QK

When you use QF with the QK preemptive kernel, you don't need to port the QF framework to the kernel because QF and QK are already integrated. However, you still need to port the QK kernel to the target CPU and compiler that you are using. Fortunately, this is quite easy due to the simplistic nature of the QK kernel. All you need to provide is the compiler-specific exact-width integer types in qep_porth.h, configure QF in qf_port.h, and finally provide the interrupt-locking policy and interrupt entry/exit in qk_port.h. You often don't need to write any platform-specific QK source files because most of the time QK can work with the ISRs generated by the C compiler.

Note that the preemptive QK kernel puts more demands on the target CPU and the compiler than the simple vanilla kernel described in Chapter 7. Generally, QK can be ported to a processor and compiler, if they satisfy the following requirements:

1.   The processor supports a hardware stack that can accommodate a fair amount of data (at least 256 bytes or more).

2.   The C or C++ compiler can generate reentrant code. In particular, the compiler must be able to allocate automatic variables on the stack.

3.   Interrupts can be locked and unlocked from C.

4.   The system provides a clock tick interrupt (typically 10 to 100Hz).

For example, some older CPU architectures, such as the 8-bit PIC microcontrollers, don't have a C-friendly stack architecture and consequently cannot easily run QK. Note, however, that in most cases you can use the nonpreemptive vanilla kernel.

In this section I show an example of QK kernel port to 80x86 CPU under DOS, with the legacy Turbo C++ 1.01 compiler configured to generate code for "large" memory model. The port will also demonstrate the advanced features, such as thread-local storage, and extended context switch for the 80x87 FPU. This port is located in `<qp>\qpc\ports\80x86\qk\tcpp101\l\`.

10.5.1 The `qep_port.h` Header File

Listing 10.10 shows the `qep_port.h` header file for 80x86/QK/Turbo C++ 1.01/Large memory model. The legacy Turbo C++ 1.01 is a prestandard compiler, so I `typedef` the six platform-specific exact-with integer types used in QP.

Listing 10.10 The `qep_port.h` header file for 80x86/QK/Turbo C++ 1.01/ Large memory model

```
#ifndef qep_port_h
#define qep_port_h

     /* Exact-width integer types for DOS/Turbo C++ 1.01/Large memory model */
typedef signed   char int8_t;
typedef signed   int  int16_t;
typedef signed   long int32_t;
typedef unsigned char uint8_t;
typedef unsigned int  uint16_t;
typedef unsigned long uint32_t;

#include "qep.h"         /* QEP platform-independent public interface */

#endif                                                  /* qep_port_h */
```

10.5.2 The `qf_port.h` Header File

Listing 10.11 shows the `qf_port.h` header file for 80x86/QK/Turbo C++ 1.01/Large memory model. You always need to configure the maximum number of active objects `QF_MAX_ACTIVE` and you need to include `qep_port.h`, `qk_porth.h`, and `qf.h` header files.

Listing 10.11 The `qf_port.h` header file for 80x86/QK/Turbo C++ 1.01/ Large memory model

```
#ifndef qf_port_h
#define qf_port_h

                /* The maximum number of active objects in the application */
#define QF_MAX_ACTIVE           63

#include "qep_port.h"                                          /* QEP port */
#include "qk_port.h"                                            /* QK port */
#include "qf.h"                         /* QF platform-independent interface */

#endif                                                      /* qf_port_h */
```

10.5.3   The `qk_port.h` Header File

The actual porting of QK to the CPU/Compiler of your choice happens in the `qk_port.h` header file. The first porting decision you need to make is the policy for locking and unlocking interrupts. To make this decision correctly, you need to learn a bit about your target CPU and the compiler to find out the most efficient way of enabling and disabling interrupts from C or C++. Generally, your first choice should be the safe policy of "saving and restoring the interrupt status" (Section 7.3.1 in Chapter 7). However, if you find out that it is safe to unlock interrupts inside ISRs because your target system can prioritize interrupts in hardware, you can use the simple and fast policy of "unconditional interrupt unlocking" (Section 7.3.2 in Chapter 7). With the fast policy you must always make sure that QF functions are invoked with interrupts unlocked, or more generally, that critical sections don't nest.

The next decision, related to the first, is the QK-specific interrupt entry and exit. Again, you need find out whether your CPU enters ISRs with interrupts locked or unlocked (most CPUs lock interrupts before vectoring to ISRs). If you decided to use the fast interrupt-locking policy, you *must* unlock interrupts in `QK_ISR_ENTRY()` and lock them again in `QK_ISR_EXIT()` to avoid nesting of critical sections when you call any QF services. If your system has an interrupt controller, you *might* decide to unlock interrupts inside ISRs even if you're using the safe policy of "saving and restoring interrupt context." I would generally recommend leaving interrupts locked throughout the whole ISR on systems that don't have interrupt controllers. Obviously, in the latter case you must be using the safe policy of "saving and restoring interrupt context," because most QF services that you call from ISRs use a critical section internally.

Finally, you need to customize the advanced features, such as the TLS and the extended context switch, if you plan to use them in your applications. Here, you need to find out which libraries require TLS support (e.g., Newlib). You also need to find what kind of coprocessors you want to support and how to save and restore their registers from C.

Listing 10.12 shows the `qk_port.h` header file for 80x86/QK/Turbo C++ 1.01/Large memory model. I decided to use the simple "unconditional interrupt unlocking" policy because the standard PC is equipped with the external 8259A Programmable Interrupt Controller (PIC) and the Turbo C++ 1.01 compiler provides the pair of functions `disable()` and `enable()`, to unconditionally lock and unlock interrupts, respectively. With this simple interrupt-locking policy, I must unlock interrupts in `QK_ISR_ENTRY()` and lock them again in `QK_ISR_EXIT()`. I also use the 80x87 floating point coprocessor (FPU) and two libraries that require TLS support.

Listing 10.12 The `qk_port.h` header file for 80x86/QK/Turbo C++ 1.01/ Large memory model

```
      #ifndef qk_port_h
      #define qk_port_h
                                              /* QF critical section entry/exit */
(1)   /* QF_INT_KEY_TYPE not defined */
(2)   #define QF_INT_LOCK(dummy)   disable()
(3)   #define QF_INT_UNLOCK(dummy) enable()

                                              /* QK-specific ISR entry and exit */
(4)   #define QK_ISR_ENTRY() do { \
          ++QK_intNest_; \
(5)       enable(); \
      } while (0)

(6)   #define QK_ISR_EXIT() do { \
(7)       disable(); \
(8)       outportb(0x20, 0x20); \
          --QK_intNest_; \
          if (QK_intNest_ == 0) { \
(9)           QK_scheduleExt_(); \
          } \
      } while (0)

      /* demonstration of advanced QK features: TLS and extended context switch   */
```

Continued onto next page

```
(10)   typedef struct Lib1_contextTag{              /* an example of a library context */
           double x;
       } Lib1_context;
(11)   extern Lib1_context * volatile impure_ptr1;

(12)   typedef struct Lib2_contextTag{              /* an example of a library context */
           double y;
       } Lib2_context;
(13)   extern Lib2_context * volatile impure_ptr2;

(14)   typedef union FPU_contextTag{
           uint32_t align;
(15)       uint8_t x87[108];                        /* the x87 FPU context takes 108-bytes */
       } FPU_context;

(16)   typedef struct ThreadContextTag{
           Lib1_context lib1;                                        /* library1 context */
           Lib2_context lib2;                                        /* library2 context */
           FPU_context fpu;                                             /* the FPU context */
       } ThreadContext;

(17)   enum QKTaskFlags{
           QK_LIB1_THREAD = 0x01,
           QK_LIB2_THREAD = 0x02,
           QK_FPU_THREAD  = 0x04
       };

                                                       /* QK thread-local storage */
(18)   #define QK_TLS(act_) \
(19)       impure_ptr1 = &((ThreadContext *)(act_)->thread)->lib1; \
(20)       impure_ptr2 = &((ThreadContext *)(act_)->thread)->lib2

                                          /* QK extended context (FPU) save/restore */
(21)   #define QK_EXT_SAVE(act_)   \
(22)       if (((act_)->osObject & QK_FPU_THREAD) != 0) \
(23)           FPU_save(&((ThreadContext *)(act_)->thread)->fpu)

(24)   #define QK_EXT_RESTORE(act_) \
(25)        if (((act_)->osObject & QK_FPU_THREAD) != 0) \
(26)            FPU_restore(&((ThreadContext *)(act_)->thread)->fpu)

(27)   void FPU_save   (FPU_context * fpu);               /* defined in assembly */
(28)   void FPU_restore(FPU_context * fpu);               /* defined in assembly */

       #include <dos.h>                                          /* see NOTE01 */
       #undef outportb    /*don't use the macro because it has a bug in Turbo C++ 1.01*/

(29)   #include "qk.h"              /* QK platform-independent public interface */
(30)   #include "qf.h"              /* QF platform-independent public interface */

       #endif                                              /* qk_port_h */
```

(1) The macro `QF_INT_KEY_TYPE` is *not* defined, meaning that the fast policy of "unconditional interrupt saving and restoring" is used. This is possible because the standard PC is equipped with the 8259A Programmable Interrupt Controller (PIC), which allows unlocking interrupts inside ISRs.

(2) The macro `QF_INT_LOCK()` is defined as the Turbo C++ function `disable()`.

(3) The macro `QF_INT_UNLOCK()` is defined as the Turbo C++ function `enable()`.

(4) As described in Section 10.3.3, the macro `QK_ISR_ENTRY()` is called upon the entry to every ISR.

(5) The 80x86 CPU enters ISRs with interrupts locked. However, interrupts must be unlocked before any QF or QK service can be used in the ISR body, because the fast interrupt-locking policy does not support nesting critical sections.

(6) As described in Section 10.3.3, the macro `QK_ISR_EXIT()` is called upon exit from every ISR.

(7) The interrupts unlocked upon entry must be locked again to prevent corruption of the QK variables.

(8) This output statement writes the EOI instruction to the master 8259A interrupt controller.

(9) As described in Section 10.3.3, the QK scheduler must be called at the exit from every interrupt to handle the asynchronous preemption. Here I use the extended QK scheduler because this port supports the extended context switch for the 80x87 FPU.

> **NOTE**
>
> If you don't define the macros `QK_EXT_SAVE()` and `QK_EXT_RESTORE()`, the extended QK scheduler is equivalent to the regular scheduler `QK_schedule_()`. You can always use the extended scheduler in the `QK_ISR_EXIT()` macro without any performance penalty, but if you want to save a little code space, you might want to use the regular scheduler.

(10)   This `typedef` specifies the per-thread context used by a hypothetical reentrant library `lib1`. I use this library to demonstrate the TLS switching capability of the QK kernel.

(11)   The `impure_ptr1` pointer points to the per-thread context of library `lib1`.

(12)   This `typedef` specifies the per-thread context used by a hypothetical reentrant library `lib2`. I use this library to demonstrate that the TLS implementation can handle multiple reentrant libraries.

(13)   The `impure_ptr2` pointer points to the per-thread context of library `lib2`.

(14)   This `typedef` specifies the per-thread FPU context.

(15)   The 80x87 FPU requires 108 bytes to store its context.

(16)   This `typedef` specifies the entire per-thread context, which includes the contexts of the library1, library2, and the FPU.

(17)   This enumeration defines the thread flags.

(18)   The macro `QK_TLS()` is the context-switch hook in which you customize TLS management.

(19)   The impure pointer for the reentrant library `lib1` is switched to the current active object.

(20)   The impure pointer for the reentrant library `lib2` is switched to the current active object.

(21)   The macro `QK_EXT_SAVE()` saves the extended context in asynchronous preemption.

(22)   The FPU context must only be saved for a task that actually uses the FPU.

(23)   The FPU context is saved to the active object's private location by calling the function `FPU_save()`.

(24)   The macro `QK_EXT_RESTORE()` restores the extended context in asynchronous preemption.

(25)   The FPU context must only be restored for a task that actually uses the FPU.

(26)   The FPU context is restored from the active object's private location by calling the function `FPU_restore()`.

(27,28) The prototypes for the functions `FPU_save()` and `FPU_restore()` are provided. These functions are defined in assembly.

(29) The `qk_port.h` header file must always include the QK platform-independent `qk.h` header file.

(30) The `qk_port.h` header file must always include the QF platform-independent `qf.h` header file.

10.5.4 Saving and Restoring FPU Context

The functions `FPU_save()` and `FPU_restore()`, declared in Listing 10.12(27,28), are part of the QK port. They are defined in the assembly file `<qp>\qpc\ports\ 80x86\qk\tcpp101\l\src\fpu.asm`. Both these functions are just shells for executing the 80x87 machine instructions `FSAVE` and `FRSTOR`, respectively.

10.6   Testing the QK Port

As usual, I use the dining philosopher problem (DPP) application to test the port. For QK, I have extended the basic DPP application discussed in Chapter 9 to demonstrate and test the advanced QK features, such as the priority-ceiling mutex, thread-local storage for multiple reentrant libraries, and the extended context switch for the 80x87 FPU. The DPP application for QK is located in the directory `<qp>\qpc\ examples\80x86\qk\tcpp101\l\dpp\`.

10.6.1   Asynchronous Preemption Demonstration

As it turns out, an interesting asynchronous preemption is not that easy to observe in the DPP application. By an interesting preemption, I mean a task asynchronously preempting another task, as opposed to simply a task asynchronously preempting the idle loop. Figure 10.9 illustrates why. The DPP application is mostly driven by the system clock tick interrupt (`ISR_tmr`), which posts the time events to the Philosopher active objects. Typically, the interrupts and state machines execute so quickly that all processing happens very close to the clock tick and the CPU goes quickly back to executing the QK idle loop. With the code executing so fast, the `ISR_tmr()` has no chance to actually preempt any QK task, just the idle loop. Consequently an asynchronous preemption cannot happen.

Therefore, to increase the odds of asynchronous preemptions in this application, I have added the second interrupt (the ISR_kbd triggered by the keyboard), which is asynchronous with respect to the clock tick and always posts an event to the Table active object. I also added some artificial CPU loading in the form of various busy-wait functions called from to the state machines and interrupts (I show examples of these functions later in this section). Finally, I've instrumented the ISRs to report preemptions caused by interrupts to the screen.

Since I cannot foresee the speed of your CPU, I have provided a command-line parameter to the DPP application that determines the delay incurred by the various busy-wait functions. On my 2GHz PC, I've been using the value of 100 iterations, which allowed me to easily catch several asynchronous preemptions. You should be careful not to go overboard with this parameter, though, because you can overload the CPU, or more scientifically stated, you can create an unschedulable set of tasks. In this case, QF will eventually overflow an event queue and assert.

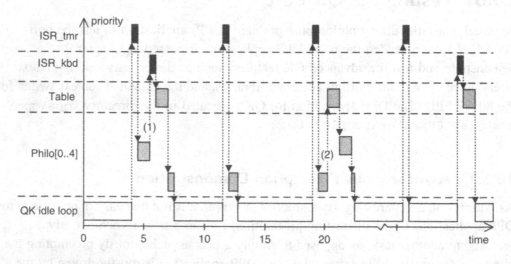

Figure 10.9: Execution profile of the DPP application with QK.

With all this scaffolding, you actually have a chance to observe an interesting asynchronous preemption, such as the instance shown in Figure 10.9(1). You need to run the DPP application (with the command-line parameter of 100 or so). As explained before, you will never get asynchronous preemptions unless you start typing on the keyboard. When the keyboard interrupt happens to come close enough after the clock tick, it might just manage to preempt one of the philosopher tasks. The keyboard ISR always posts an event to the Table object, and because Table has the highest priority in the system, upon the exit from ISR_kbd(), QK performs an asynchronous context switch to the Table active object. When this happens, you'll see that the preemption counter for one of the Philosopher tasks will increment on the screen (see Figure 10.10).

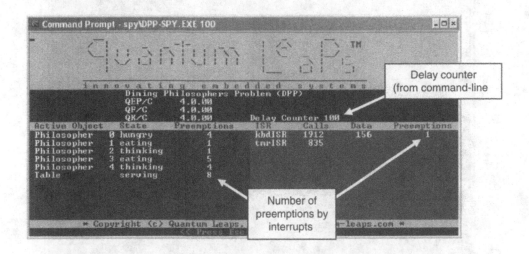

Figure 10.10: DPP application with QK running in a DOS console.

If you want to examine an asynchronous preemption closer, you can use the debugger built into the Turbo C++ IDE. You load the project DPP-DBG.PRJ (located in <qp>\qpc\examples\80x86\qk\tcpp101\l\dpp\) into the IDE and open the file qk_ext.c that I have specifically added to this project, even though it is already included in the qk.lib library. You set a breakpoint inside QK_scheduleExt_() indicated as label (1) in Figure 10.11 (see also Listing 10.12(9)). Next, select Run | Arguments... to define a command-line argument around 100. Now you can run the program. When you start typing on the keyboard, eventually you should hit the breakpoint (asynchronous preemption). You can step through the code from there. Figure 10.11 shows an example of my debug session.

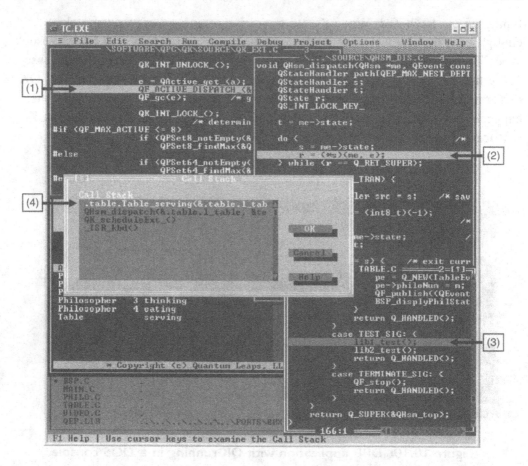

Figure 10.11: Asynchronous preemption examined in the Turbo C++ debugger.

(1)   The original breakpoint is set at the instruction that is only executed when a task preempts another task, but not the idle loop (inside `QK_scheduleExt_()`).

(2)   When you step into the `QF_ACTIVE_DISPATCH_()` macro, you get inside the function `QHsm_dispatch()`. Keep stepping (F7) until you reach the highlighted line. This line calls the current state-handler function of the high-priority active object.

(3)   When you step into again, you end up inside the Table active object.

(4)   The Call Stack window shows the call tree information, which is a nice byproduct of QK using just a single stack. You see that the extended QK scheduler is engaged and detects preemption, so it calls `QHsm_dispatch()` for the Table active object. Finally, the `Table_serving()` state handler processes the `TEST` event.

> **NOTE**
>
> This example should convince you that debugging QK tasks is straightforward, even though they nest on the interrupt stack frame, as shown in the Call Stack window (see Section 10.2.5). In contrast, debugging ISRs is hard because the Turbo C++ debugger freezes when interrupts are locked at the CPU or the 8259A PIC level.

10.6.2 Priority-Ceiling Mutex Demonstration

To demonstrate the QK priority-ceiling mutex, I've extended the Philosopher active object to allow random think and eat timeouts for philosophers rather than fixed timeouts used in the basic implementation. To implement the feature, I use the pseudorandom number (PRN) generator provided in Turbo C++ 1.01 (random()). This generator, like most PRN generators, is not reentrant because it must preserve its state from one call to the next. To prevent corruption of this internal state, I protect the generator with the QK mutex, as shown in Listing 10.13.

Listing 10.13 Protecting PRN generator with the priority-ceiling mutex (file `<qp>\qpc\examples\80x86\qk\tcpp101\1\dpp\philo.c`**)**

```
QState Philo_thinking(Philo *me, QEvent const *e) {
    switch (e->sig) {
        case Q_ENTRY_SIG: {
            QTimeEvtCtr think_time;
            QMutex mutex;

            mutex = QK_mutexLock(N_PHILO);
            think_time = (QTimeEvtCtr)(random(THINK_TIME) + 1);
            QK_mutexUnlock(mutex);

            QTimeEvt_postIn(&me->timeEvt, (QActive *)me, think_time);
            return (QState)0;
        }
        . . .
    }
    return (QState)&QHsm_top;
}
```

Note that I use the number of philosophers N_PHILO as the priority ceiling to lock the mutex. This ceiling corresponds to the highest-priority Philosopher active object that can access the PRN generator (Philosopher active objects have priorities

1..N_PHILO). Since the priority of the Table active object (N_PHILO + 1) is above the ceiling, the mutex does not affect Table, which is exactly what I wanted to achieve. Table does not use the resource (PRN generator in this case), so it should not be affected by the mutex.

10.6.3 TLS Demonstration

In the QK port header file qk_port.h (Listing 10.12), you saw the two contexts of two hypothetical libraries lib1 and lib2, which need TLS support in the same way as Newlib does. The QK port implemented the switching of the two "impure pointers" in the macro QK_TLS (Listing 10.12(18)). At the application level, I need to add the contexts to the active objects and I need to inform QK where these TLS contexts are located. I also need to call the library functions so that they are shared among all Philosopher active objects. Listing 10.14 shows these steps.

Listing 10.14 Incorporating the TLS context inside active objects (file `<qp>\qpc\examples\80x86\qk\tcpp101\1\dpp\philo.c`**)**

```
      typedef struct PhiloTag {
          QActive super;                    /* derives from the QActive base class */
(1)       ThreadContext context;                        /* thread context */
          QTimeEvt timeEvt;                 /* for timing out thining or eating */
      } Philo;
      . . .
      /*.........................................................*/
      void Philo_start(uint8_t n,
                   uint8_t p, QEvent const *qSto[], uint32_t qLen)
      {
          Philo *me = &l_philo[n];
          Philo_ctor(me);                                 /* instantiate */
(2)       impure_ptr1 = &me->context.lib1;    /* initialize reentrant library1 */
(3)       lib1_reent_init(p);
(4)       impure_ptr2 = &me->context.lib2;    /* initialize reentrant library2 */
(5)       lib2_reent_init(p);

          QActive_start((QActive *)me, p, qSto, qLen,
(6)                    &me->context,
(7)                    (uint8_t)(QK_LIB1_THREAD | QK_LIB2_THREAD | QK_FPU_THREAD),
                       (QEvent *)0);
      }
      QState Philo_thinking(Philo *me, QEvent const *e) {
          switch (e->sig) {
          . . .
              case TIMEOUT_SIG: {
```

```
(8)                 lib1_test();
(9)                 lib2_test();
                return (QState)0;
            }
            . . .
    }
    return Q_SUPER(&QHsm_top);
}
/*................................................................*/
QState Philo_hungry(Philo *me, QEvent const *e) {
    switch (e->sig) {
        . . .
        case EAT_SIG: {
            if (((TableEvt const *)e)->philoNum == PHILO_ID(me)) {
(10)                lib1_test();
(11)                lib2_test();
                return (QState)0;
            }
            break;
        }
        . . .
    }
    return Q_SUPER(&QHsm_top);
}
/*................................................................*/
QState Philo_eating(Philo *me, QEvent const *e) {
    switch (e->sig) {
        . . .
        case TIMEOUT_SIG: {
(12)            lib1_test();
(13)            lib2_test();
            return Q_TRAN(&Philo_thinking);
        }
        . . .
    }
    return Q_SUPER(&QHsm_top);
}
```

(1)   I place the data member context of ThreadContext type (defined in Listing 10.12(16)) directly inside the Philo class. That way I can be sure that every Philo object has the private ThreadContext area.

(2)   Upon the Philo active object initialization, I aim the "impure pointer" of library lib1 at the TLS context for this library.

(3)   I then let the library initialize the context.

(4,5) I repeat the same two steps for the library lib2.

(6)   I pass the pointer to the TLS as the fifth parameter to the `QActive_start()` function, to inform QK about the location of TLS for each active object (see also Listing 10.5(17)).

(7)   I set the thread attributes to inform QK that this active object uses library1, library2, and the FPU.

(8-13)   I pepper the `Philo` state machine with the calls to the libraries. These calls take a long time to run and provide the CPU loading that was necessary to test asynchronous preemptions.

NOTE

I made similar changes to the Table active object, so that it too shares the libraries with all Philosophers.

Finally, I need to define the library functions `lib1_test()` and `lib2_test()` that actually use the "impure pointers." Listing 10.15 shows the test code.

Listing 10.15 Using the TLS context inside the libraries (file `<qp>\qpc\examples\80x86\qk\tcpp101\1\dpp\bsp.c`)

```
      #include <math.h>
      . . .
      /* -----------------------------------------------------------------*/
      void lib1_reent_init(uint8_t prio) {
(1)       impure_ptr1->x = (double)prio * (M_PI / 6.0);
      }
      /* ................................................................*/
      void lib1_test(void) {
(2)       uint32_t volatile i = l_delay;
(3)       while (i-->OUL) {
(4)           volatile double r = sin(impure_ptr1->x) * sin(impure_ptr1->x)
                               + cos(impure_ptr1->x) * cos(impure_ptr1->x);
(5)           Q_ASSERT(fabs(r - 1.0) < 1e-99);          /* assert the identity */
          }
      }
      /* -----------------------------------------------------------------*/
      void lib2_reent_init(uint8_t prio) {
(6)       impure_ptr2->y = (double)prio * (M_PI / 6.0) + M_PI;
      }
      /* ................................................................*/
```

```
void lib2_test(void) {
    uint32_t volatile i = l_delay;
    while (i-->OUL) {
        volatile double r = sin(impure_ptr2->y) * sin(impure_ptr2->y)
                          + cos(impure_ptr2->y) * cos(impure_ptr2->y);
        Q_ASSERT(fabs(r - 1.0) < 1e-99);              /* assert the identity */
    }
}
```

(1)   I initialize the per-thread context of library lib1 (the x variable)
      so it depends on the priority of the task, which is different for each task.

(2,3) The parameter l_delayCtr is set from the command line (see Section 10.6.1)
      and determines the number of iterations performed in this function (the
      CPU loading that the function causes).

(4)   I use a mathematical identity $\sin^2(x)+\cos^2(x) == 1.0$ to compute the value of
      r based on the impure pointer impure_ptr1. This expression makes an
      extensive use of the 80x87 FPU.

(5)   I assert the identity. This assertion would fail if the impure pointer where
      switched incorrectly or the FPU would compute the expression incorrectly.

(6)   I initialize the per-thread context of library lib2 (the y variable) similarly
      as lib1, except I add a phase shift, so that the per-thread values are
      different that for lib1.

10.6.4   Extended Context Switch Demonstration

As discussed in Section 10.5.4, the 80x87 FPU context saving and restoring is
handled automatically in the QK extended scheduler. At the application level, you
need to include the per-thread FPU context in every active object, which is done in
Listing 10.14(1). You also need to set the QK_FPU_FLAG for every task that uses the
FPU (see Listing 10.14(7)). And finally, you must use the FPU to test it. Though
Listing 10.15 perform a lot of floating-point operations, it is also important to use
the correct compiler options to select the FPU. I've compiled both the QP libraries
and the DPP applications with the -f287 option, which instructs the Turbo C++
compiler to generate FPU hardware instructions.

10.7 Summary

A certain class of real-time embedded (RTE) stems, such as control applications, can vastly benefit from preemptive multitasking. The QP event-driven platform contains a lightweight, priority-based, preemptive kernel called QK.

QK is a special kind of a preemptive kernel, called a run-to-completion (RTC) or single-stack kernel, in which tasks are one-shot, RTC functions as opposed to endless loops as in most conventional RTOSs. The biggest limitation of RTC kernels is inability to block in the middle of a task, but this limitation is irrelevant for executing event-driven active objects, because active objects don't block in the middle of RTC steps anyway.

When applied in active object applications, the QK kernel provides the same execution profile as any other conventional, priority-based, preemptive kernel or RTOS. In fact, QK most likely *outperforms* all conventional preemptive RTOSs in all respects, such as speed, stack usage, code size (ROM footprint), complexity, ease of use, or any other metrics you want to apply.

QK supports advanced features, such as priority-ceiling mutex, thread-local storage, and extended context switch, which you can find only in sophisticated RTOSs. All these advanced features are helpful when you need to share memory, libraries, or devices among active object threads.

QK is also easier to port to new CPUs and compilers than most RTOSs, mainly because QK can work with compiler-generated interrupts that virtually all embedded C/C++ compilers support. Most of the time, you can complete a QK port without writing assembly code. In this chapter, I discussed the QK port to 80x86 CPU with the legacy C compiler running in real mode. Although this port can be valuable in itself, here I used it mostly to demonstrate QK capabilities. This book's accompanying Website at `www.quantum-leaps.com/psicc2` contains links to many other QK ports to popular embedded processors and compilers.

CHAPTER 11

Software Tracing for Event-Driven Systems

There has never been an unexpectedly short debugging period in the history of computers.
—Steven Levy

In any real-life project, getting the code written, compiled, and successfully linked is only the first step. The system still needs to be tested, validated, and tuned for best performance and resource consumption. A single-step debugger is frequently not helpful because it stops the system and exactly hinders seeing *live* interactions within the application. Clogging up high-performance code with `printf()` statements is usually too intrusive and simply unworkable in most embedded systems, which typically don't have adequate screens to print to. So the questions are: How can you monitor the behavior of a running real-time system without degrading the system itself? How can you discover and document elusive, intermittent bugs that are caused by subtle interactions among concurrent components? How do you design and execute repeatable unit and integration tests of your system? How do you ensure that a system runs reliably for long periods of time and gets top processor performance?

Techniques based on *software tracing* can answer many of these questions. Software tracing is a method for obtaining diagnostic information in a *live* environment without the need to stop the application to get the system feedback. Software tracing always involves some form of a target system *instrumentation* to log interesting discrete events for subsequent retrieval from the system and analysis.

Due to the inversion of a control, software tracing is particularly effective and powerful in combination with the event-driven programming model. An instrumented event-driven framework can provide much more comprehensive and detailed information than any traditional RTOS.

In this chapter, I describe the software-tracing system called Quantum Spy, which is part of the QP event-driven platform. I begin with a quick introduction to software tracing concepts. Next I walk you through an example of a software tracing session. I then describe the target-resident software-tracing component, called QS, explaining in detail the generation of trace data, the various filters, buffering, transmission protocol, and porting QS. Subsequently, I present the QSPY host application for receiving, displaying, storing, and analyzing the trace data. Finally I explain the steps required to add the QS software tracing component to a QP application.

11.1 Software Tracing Concepts

In a nutshell, software tracing is similar to peppering the code with `printf()` statements for logging and debugging, except that software tracing is much less intrusive and more selective than the primitive `printf()`. This quick overview introduces the basic concepts and describes some features you can expect from a commercial-grade software-tracing system.

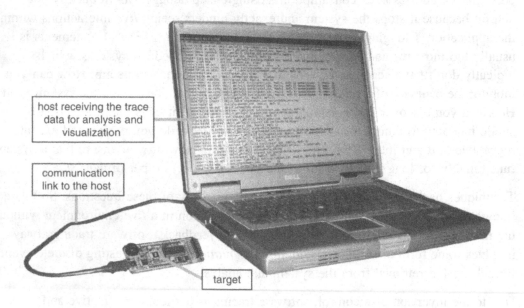

Figure 11.1: Typical setup for collecting software trace data.

Figure 11.1 shows a typical setup for software tracing. The embedded target system is executing instrumented code, which logs the trace data into a RAM buffer inside the target. From that buffer the trace data is sent over a data link to a host computer that stores, displays, and analyzes the information. This configuration means that a software tracing always requires two components: a *target-resident component* for collecting and sending the trace data and a *host-resident component* to receive, decompress, visualize, and analyze the data.

> **NOTE**
>
> Software-tracing instrumentation logs interesting discrete events that occur in the target system. I will call these discrete events *trace records*, to avoid confusing them with the application-level events.

A good tracing solution is minimally intrusive, which means that it can provide visibility into the running code with minimal impact on the target system behavior. Properly implemented and used, it will let you diagnose a live system without interrupting or significantly altering the behavior of the system under investigation.

Of course, it's always possible that the overhead of software tracing, no matter how small, will have some effect on the target system behavior, which is known as the *probe effect* (a.k.a. the *Heisenberg effect*). To help you determine whether that is occurring, you must be able to configure the instrumentation in and out both at compile time as well as at runtime.

To minimize the probe effect, a good trace system performs efficient, selective logging of trace records using as little processing and memory resources of the target as possible. Selective logging means that the tracing system provides user-definable, fine-granularity *filters* so that the target-resident component only collects events of interest—you can filter as many or as few instrumented events as you need. That way you can make the tracing as noninvasive as necessary.

To minimize the RAM usage, the target-resident trace component typically uses a circular trace buffer that is continuously updated, and new data overwrites the old when the buffer "wraps around" due to limited size or transmission rate to the host. This reflects the typically applied *last-is-best* policy in collecting the trace data. To focus on certain periods of time, software tracing provides configurable software triggers that can start and stop trace collection before the new data overwrites the old data of interest in the circular buffer.

To further maximize the amount of data collected in the trace buffer, the target-resident component typically applies some form of data compression to squeeze more trace information into the buffer and to minimize the bandwidth required to uplink the data to the host.

However, perhaps the most important characteristic of a flexible software-tracing system is the separation of trace logging (*what* information is being traced) from the data transmission mechanism (*how* and *when* exactly the data is sent to the host). This separation of concerns allows the transmissions to occur in the least time-critical paths of the code, such as the idle loop. Furthermore, clients should be able to employ any data transmission mechanism available on the target, meaning both the physical transport layer (e.g., serial port, SPI, USB, Ethernet, etc.) as well as implementation strategy (polling, interrupt, DMA, etc.). The tracing facility should tolerate and be able to detect any RAM buffer overruns due to bursts of tracing data production rate or insufficient transmission rate to the host.

Finally, the tracing facility must allow consolidating data from all parts of the system, including concurrently executing threads and interrupts. This means that the instrumentation facilities must be reentrant (i.e., both thread-safe and interrupt-safe). Also, to be able to correlate all this data, most tracing systems provide precise *timestamping* of the trace records.

11.2   Quantum Spy Software-Tracing System

As I mentioned in the introduction to this chapter, software tracing is especially effective and powerful in combination with the event-driven active object computing model (see Chapter 6). A running application built of active objects is a highly structured affair where all important system interactions funnel through the real-time framework and the state-machine engine. This offers a unique opportunity to instrument these relatively small parts of the overall code to gain unprecedented insight into the entire system.

Quantum Spy is a software-tracing system that enables live monitoring of event-driven QP applications with minimal target system resources and without stopping or significantly slowing down the code. The Quantum Spy system consists of the target-resident component, called QS, and the application running on a host workstation, called QSPY.

Many operating systems provide software-tracing capabilities. However, Quantum Spy takes software tracing to the entirely *new level*. Due to inversion of control, an instrumented state machine framework, as opposed to merely an RTOS, is capable of providing incomparably more comprehensive information about the running system, even without adding any instrumentation to the application code. For example, the QS trace data is thorough enough to produce complete sequence diagrams and detailed state machine activity for all state machines in the system. You can selectively monitor all event exchanges, event queues, event pools, and time events because all these elements are controlled by the framework. Additionally, if you use one of the kernels built into QP (the vanilla kernel or the preemptive QK kernel), you can obtain all the data available to a traditional RTOS as well, such as context switches and mutex activity.

11.2.1 Example of a Software-Tracing Session

To show you how software-tracing works in practice, I present an example of a software-tracing session. I use the dining philosophers problem (DPP) test application, which I introduced in Chapter 9. All versions of the DPP application included in the code accompanying this book contain the QS instrumentation. The tracing instrumentation becomes active when you build the "Spy" configuration.

Figure 11.2 shows how to collect the software trace data from the QK/DOS version of the DPP application located in the directory <qp>\qpc\examples\80x86\qk\ tcpp101\l\dpp\spy\dpp-spy.exe. You can rebuild the "Spy" configuration by loading the DPP-SPY.PRJ project into the Turbo C++ IDE. You need to run the DPP-SPY.EXE executable on a target PC with a serial port. You connect the serial port of the target machine to the serial port of a Windows or a Linux host workstation via a NULL-modem cable. On the host workstation, you need to start the QSPY host application that decompresses and visualizes the QS trace data.

The Windows executable of the QSPY host application is located in the directory <qp>\qpc\tools\qspy\win32\vc2005\Release\. Assuming that this directory is your current directory or is in your path, you invoke this console application by typing the following command at the command prompt:

```
qspy -c COM1 -b 115200
```

The first command-line parameter -c COM1 tells the QSPY host application to receive the trace data from COM1. If your target is connected to a different COM port, you

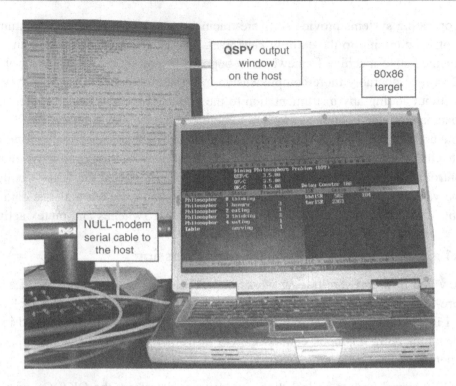

Figure 11.2: Collecting software trace data from a 80x86 target.

need to adjust the COM number. The second parameter configures the baud rate of the serial port to 115200.

NOTE

In the particular case of a Windows PC, you can use the same machine as the target and the host at the same time. You need to use a machine with two serial ports, which you connect with a NULL modem cable. You can use one serial port for the DPP target application running in a DOS-window and the other for the QSPY host application.

You might also use a Linux host machine. In case of Linux, you must first build the executable by running the Makefile located in the directory `<qp>/qpc/tools/qspy/linux/gnu/`. You invoke the Linux executable by typing the following command at the command prompt:

```
qspy -c /dev/ttyS0 -b 115200
```

The first parameter `-c /dev/ttyS0` tells the QSPY application to receive the trace data from the ttyS0 serial device. If you connected a different serial port to the target, you need to adjust the `ttyS` number.

As I mentioned before, all DPP applications included in the code accompanying this book are instrumented for software tracing, and I encourage you to try them all. For example, you can collect trace data from the EV-LM3S811 board (see Figure 11.1). The EV-LM3S811 board sends the QS trace data through the UART0 connected to the Virtual COM Port (VCP) provided by the USB debugger, so the QSPY host application can conveniently receive the trace data on the host PC. No additional serial cable is needed.

11.2.2   The Human-Readable Trace Output

The QSPY host application is just a simple console-type program without any fancy user interface. QSPY application displays the trace data in a human-readable textual format. Listing 11.1 shows fragments of such a data log generated from the DOS/QK version of the DPP application.

NOTE

The QSPY host application supports also exporting data to the powerful MATLAB environment, as described in Section 11.5. MATLAB is a registered trademark of The Mathworks, Inc.

Listing 11.1 Fragments of the software trace log from the DOS/QK version of the DPP application

```
qspy host application 3.5.00
Copyright (c) Quantum Leaps, LLC.
Mon Feb 25 12:20:13 2008

-T 4
-O 4
-F 4
-S 1
-E 2
-Q 1
-P 2
-B 2
-C 2
```

Continued onto next page

```
          Obj Dic: 16CA18D8->l_smlPoolSto
          Obj Dic: 16CA1900->l_tableQueueSto
          Obj Dic: 16CA1914->l_philoQueueSto[0]
. . . . . .
          EQ.INIT: Obj=l_tableQueueSto Len= 5
0000000000 AO.ADD : Active=16CA1CB8 Prio= 6
          Obj Dic: 16CA1CB8->&l_table
          Fun Dic: 141E0006->&QHsm_top
          Fun Dic: 12DA00C9->&Table_initial
          Fun Dic: 12DA020B->&Table_serving
          Sig Dic: 00000008,Obj=16CA1CB8 ->HUNGRY_SIG
0000000000 AO.SUB : Active=l_table Sig=DONE_SIG
0000000000 AO.SUB : Active=l_table Sig=TERMINATE_SIG
          Q_INIT : Obj=l_table Source=QHsm_top Target=Table_serving
0000000000 ==>Init: Obj=l_table New=Table_serving
0000070346 QF_isrE: IsrNest= 1, CurrPrio=255
          TICK   : Ctr=        1
0000070367 QF_isrX: IsrNest= 1, CurrPrio=255
0000135566 QF_isrE: IsrNest= 1, CurrPrio=255
          TICK   : Ctr=        2
0000135581 QF_isrX: IsrNest= 1, CurrPrio=255
. . . . .
0000461783 QF_isrE: IsrNest= 1, CurrPrio=255
          TICK   : Ctr=        7
          TE.ADRM: Obj=l_philo[1].timeEvt Act=l_philo[1]
0000461797 TE.POST: Obj=l_philo[1].timeEvt Sig=TIMEOUT_SIG Act=l_philo[1]
0000461808 AO.FIFO: Obj=l_philo[1] Evt(Sig=TIMEOUT_SIG, Pool=0, Ref= 0) Queue(nUsed= 5, nMax= 5)
0000461824 QF_isrX: IsrNest= 1, CurrPrio=255
0000461836 AO.GETL: Active= l_philo[1] Evt(Sig=TIMEOUT_SIG, Pool=0, Ref= 0)
0000461850 NEW    : Evt(Sig=HUNGRY_SIG, size=   3)
0000461862 MP.GET : Obj=l_smlPoolSto nFree=   9 nMin=  9
0000461874 AO.FIFO: Obj=l_table Evt(Sig=HUNGRY_SIG, Pool=1, Ref= 0) Queue(nUsed= 5, nMax= 5)
0000461886 AO.GETL: Active= l_table Evt(Sig=HUNGRY_SIG, Pool=1, Ref= 1)
0000461906 NEW    : Evt(Sig=EAT_SIG, size=   3)
0000461917 MP.GET : Obj=l_smlPoolSto nFree=  8 nMin=  8
0000461929 PUBLISH: Evt(Sig=EAT_SIG, Pool=1, Ref= 0)
0000461941 AO.FIFO: Obj=l_philo[4] Evt(Sig=EAT_SIG, Pool=1, Ref= 1) Queue(nUsed= 5, nMax= 5)
0000461953 AO.FIFO: Obj=l_philo[3] Evt(Sig=EAT_SIG, Pool=1, Ref= 2) Queue(nUsed= 5, nMax= 5)
0000461965 AO.FIFO: Obj=l_philo[2] Evt(Sig=EAT_SIG, Pool=1, Ref= 3) Queue(nUsed= 5, nMax= 5)
0000461977 AO.FIFO: Obj=l_philo[1] Evt(Sig=EAT_SIG, Pool=1, Ref= 4) Queue(nUsed= 5, nMax= 5)
0000461987 AO.FIFO: Obj=l_philo[0] Evt(Sig=EAT_SIG, Pool=1, Ref= 5) Queue(nUsed= 5, nMax= 5)
0000462001 GC-ATT : Evt(Sig=EAT_SIG, Pool=1, Ref= 5)
0000462018 Intern : Obj=l_table Sig=HUNGRY_SIG Source=Table_serving
0000462030 GC     : Evt(Sig=HUNGRY_SIG, Pool=1, Ref= 1)
0000462042 MP.PUT : Obj=l_smlPoolSto nFree=    9
0000462054 AO.GETL: Active= l_philo[4] Evt(Sig=EAT_SIG, Pool=1, Ref= 5)
0000462065 Intern : Obj=l_philo[4] Sig=EAT_SIG Source=Philo_thinking
0000462077 GC-ATT : Evt(Sig=EAT_SIG, Pool=1, Ref= 4)
0000462089 AO.GETL: Active= l_philo[3] Evt(Sig=EAT_SIG, Pool=1, Ref= 4)
0000462101 Intern : Obj=l_philo[3] Sig=EAT_SIG Source=Philo_thinking
```

```
0000462111 GC-ATT : Evt(Sig=EAT_SIG, Pool=1, Ref= 3)
0000462123 AO.GETL: Active= l_philo[2] Evt(Sig=EAT_SIG, Pool=1, Ref= 3)
0000462135 Intern : Obj=l_philo[2] Sig=EAT_SIG Source=Philo_thinking
0000462146 GC-ATT : Evt(Sig=EAT_SIG, Pool=1, Ref= 2)
           Q_ENTRY: Obj=l_philo[1] State=Philo_hungry
0000462159 ==>Tran: Obj=l_philo[1] Sig=TIMEOUT_SIG Source=Philo_thinking New=Philo_hungry
0000462171 AO.GETL: Active= l_philo[1] Evt(Sig=EAT_SIG, Pool=1, Ref= 2)
0000462183 QK_muxL: OrgPrio= 2, CurrPrio= 5
0000462195 QK_muxU: OrgPrio= 2, CurrPrio= 5
0000462207 TE.ARM : Obj=l_philo[1].timeEvt Act=l_philo[1] nTicks=  8 Interval=  0
           Q_ENTRY: Obj=l_philo[1] State=Philo_eating
0000462219 ==>Tran: Obj=l_philo[1] Sig=EAT_SIG Source=Philo_hungry New=Philo_eating
0000462231 GC-ATT : Evt(Sig=EAT_SIG, Pool=1, Ref= 1)
0000462243 AO.GETL: Active= l_philo[0] Evt(Sig=EAT_SIG, Pool=1, Ref= 1)
0000462255 Intern : Obj=l_philo[0] Sig=EAT_SIG Source=Philo_thinking
0000462265 GC     : Evt(Sig=EAT_SIG, Pool=1, Ref= 1)
0000462277 MP.PUT : Obj=l_smlPoolSto nFree= 10
0000527134 QF_isrE: IsrNest= 1, CurrPrio=255
           TICK   : Ctr=      8
0000527153 QF_isrX: IsrNest= 1, CurrPrio=255
0000592283 QF_isrE: IsrNest= 1, CurrPrio=255
. . .    . . .
```

The QS trace log shown in Listing 11.1 contains quite detailed information because most QS records are enabled (not blocked in the QS filters). The following bullet items highlight the most interesting parts of the trace and illustrate how you can interpret the trace data:

- The QS log always contains the QSPY application version number, the date and time of the run, and all the configuration options used by the QSPY host application.

- A log typically starts with the *dictionary records* that provide a mapping between addresses of various objects in memory and their symbolic names. The dictionary entries don't have timestamps.

- After the dictionaries, you see the active object initialization. For example, the EQ.INIT record indicates event queue initialization with the l_tableQueueSto buffer. After this the AO.ADD trace record you see adding the Table object with priority 6. At this point, the time-tick interrupt is not configured, so all timestamps are 0000000000 (timestamps are always placed in the first eight columns).

- Active object initialization can contain dictionary entries for items that are encapsulated within the active object. For example, initialization of Table inserts an object dictionary entry for l_table object and three function dictionary entries for state handlers QHsm_top, Table_initial and Table_serving. Finally, the topmost initial transition is taken from QHsm_top to Table_serving.

- After the active object initialization, interrupts are enabled, and the first Tick interrupt arrives at the timestamp 0000070346. You can find out the type of the interrupt by the unique priority number. For example, the priority of the Tick interrupt is 0xFF == 255.

- The Tick interrupt occurs seven times. You can determine the ticking rate by comparing the timestamps between interrupt entry of Tick 1 and 2, which is ((0000135566 − 0000070346)/7 = 65220 ~= 0x10000). In the case of the DPP application, the timestamp is provided from counter-0 of the 8254 timer/counter, which is driven from the oscillator running at 1.193182MHz. The same counter-0 of the 8254 also generates time tick interrupts every 0x10000 number of counts (the 18.2Hz DOS tick).

- In the Tick 7 interrupt entered at timestamp 0000461783, you see that a time event posts TIMEOUT_SIG events to the l_philo[1] active objects. This triggers a lot of activity in the application. In fact, over 42 trace records occur before the next Tick 8.

> **NOTE**
>
> The QSPY human-readable format contains many cryptic names for various trace records. The "QSPY Reference Manual" available in the code accompanying this book (see Section 11.6.6) contains documentation of all predefined QS trace records and their parameters.

11.3   QS Target Component

The target-resident component of the Quantum Spy tracing system is called QS. The QS target component consists of the ring buffer, the QS filters, and the instrumentation added to QEP, QF, QK, and the application, as shown in Figure 11.3.

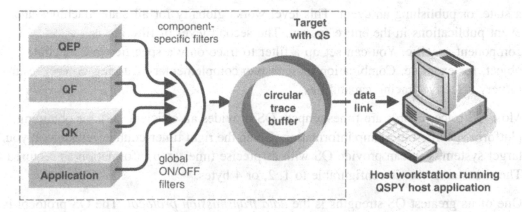

Figure 11.3: Structure of the QS target component.

Software tracing with QS is incomparably less intrusive than the primitive printf() statements because all the data formatting is *removed* from the target system and is done after the fact in the host. Additionally, the data-logging overhead incurred in the time-critical path of the target code is reduced to just storing the data into the trace buffer but typically does not include the overhead of sending the data out of the target device. In QS, data logging and sending to the host are separated so that the target system can typically perform the transmission outside of the time-critical path—for example, in the idle loop of the target CPU.

A nice byproduct of removing the data formatting from the target is a natural *data compression* compared to a formatted output. For example, ASCII representation of a single byte takes two hexadecimal digits (and three decimal digits), so avoiding the formatting gives at least a factor of two in data density. On top of this natural compression, QS uses such techniques as data dictionaries, and compressed format information, which in practice result in a compression factor of 4 to 5 compared to the expanded human-readable format.

Obviously, QS cannot completely eliminate the overhead of software tracing. But with the fine-granularity *filters* available in QS, you can make this impact as small as necessary. For greatest flexibility, QS uses two complementary levels of filters (see Figure 11.3). The first level is filtering based on trace record type, such as entry to a state, or publishing an event. This level works globally for all state machines and event publications in the entire system. The second level of filtering is component-specific. You can set up a filter to trace only a specific state machine object, for example. Combination of such two complementary filtering criteria results in very selective tracing capabilities.

Most QS trace records are timestamped. QS provides an efficient API for obtaining platform-specific timestamp information. Given the right timer-counter resource in your target system, you can provide QS with as precise timestamp information as required. The timestamp size is configurable to 1, 2, or 4 bytes.

One of its greatest QS strengths is the *data transmission protocol*. The QS protocol is very lightweight but has many the elements of the High Level Data Link Control (HDLC) protocol [HDLC] defined by the International Standards Organization (ISO). The protocol has provisions for detecting transmission errors and allows for instantaneous resynchronization after any error, such as data dropouts due to RAM buffer overruns.

Finally, QS contains a lightweight API for implementing data transmission to the host. The API supports any implementation strategy (polling, interrupt, DMA, etc.) and any physical transport layer (e.g., serial port, SPI, USB, Ethernet, data file, etc.)

11.3.1   QS Source Code Organization

Listing 11.2 shows the platform-independent directories and files comprising the QS software-tracing component in C. The structure of the C++ version is almost identical, except the implementation files have the .cpp extension.

Listing 11.2 Platform-independent QS source code organization

```
<qp>\qpc\              - QP/C root directory (<qp>\qpcpp for QP/C++)
  |
  +-include/           - QP platform-independent header files
  | +-qs.h             - QS platform-independent active interface
  | +-qs_dummy.h       - QS platform-independent inactive interface
  |
  +-qs/                - QS target component
  | +-source/          - QS platform-independent source code (*.C files)
```

```
| | +-qs_pkg.h      - internal, packet-scope interface for QS implementation
| | +-qs.c          - internal ring buffer and formatted output functions
| | +-qs_.c         - definition of basic unformatted output functions
| | +-qs_blk.c      - definition of block-oriented interface QS_getBlock()
| | +-qs_byte.c     - definition of byte-oriented interface QS_getByte()
| | +-qs_f32.c      - definition of 32-bit floating point output QS_f32()
| | +-qs_f64.c      - definition of 64-bit floating point output QS_f64()
| | +-qs_mem.c      - definition of memory-block output
| | +-qs_str.c      - definition of zero-terminated string output
|
+-ports\            - Platform-specific QP ports
| +- . . .
+-examples\         - Platform-specific QP examples
| +- . . .
```

The QS source files contain typically just one function or a data structure definition per file. This design aims at deploying QS as a fine-granularity library that you statically link with your applications. Fine granularity means that the QS library consists of several small loosely coupled modules (object files) rather than a single module that contains all functionality.

11.3.2 The QS Platform-Independent Header Files qs.h **and** qs_dummy.h

As most software tracing systems for C or C++, QS relies heavily on the C preprocessor for the tracing instrumentation to be enabled or disabled at compile time without changing the instrumented source code.

NOTE

Most QS facilities are provided in form of preprocessor macros. Depending on the global macro Q_SPY, the QS facilities are either defined to provide actual QS services or are "dummied-out" to prevent any code generation when the global macro Q_SPY is not defined. That way, the QS instrumentation can be *left in the code* at all times but becomes active only when the code is compiled with the macro Q_SPY defined. You typically define the macro Q_SPY externally through the compiler option (usually -D).

Listing 11.3 shows the platform-independent header file <qp>\qpc\include\qs.h, which specifies the *active* interface to all QS facilities. The platform-independent header file <qp>\qpc\include\qs_dummy.h, shown in Listing 11.4, specifies the *inactive* QS interface. Typically, you never need to explicitly include either of these header files in your application, because they are already included by all instrumented QP components. If the macro Q_SPY is defined, the QP components include the qs.h header file; otherwise they include the qs_dummy.h header file.

Listing 11.3 Active QS interface (fragments of the header file `<qp>\qpc\include\qs.h`)

```
        #ifndef qs_h
        #define qs_h

        #ifndef Q_SPY
 (1)        #error "Q_SPY must be defined to include qs.h"
        #endif

 (2)    enum QSpyRecords {
            /* QEP records */
 (3)        QS_QEP_STATE_ENTRY,                        /**< a state was entered */
            QS_QEP_STATE_EXIT,                         /**< a state was exited */
            . . .
            /* QF records */
 (4)        QS_QF_ACTIVE_ADD,            /**< an AO has been added to QF (started) */
            QS_QF_ACTIVE_REMOVE,      /**< an AO has been removed from QF (stopped) */
            QS_QF_ACTIVE_SUBSCRIBE,         /**< an AO subscribed to an event */
            QS_QF_ACTIVE_UNSUBSCRIBE,      /**< an AO unsubscribed to an event */
            QS_QF_ACTIVE_POST_FIFO,   /**< an event was posted (FIFO) directly to AO */
            . . .
            /* QK records */
 (5)        QS_QK_MUTEX_LOCK,                        /**< the QK mutex was locked */
            QS_QK_MUTEX_UNLOCK,                    /**< the QK mutex was unlocked */
            QS_QK_SCHEDULE,                /**< the QK scheduled a new task to execute */
            . . .
            /* Miscellaneous QS records */
 (6)        QS_SIG_DICTIONARY,                        /**< signal dictionary entry */
            QS_OBJ_DICTIONARY,                        /**< object dictionary entry */
            QS_FUN_DICTIONARY,                      /**< function dictionary entry */
            QS_ASSERT,                                    /** assertion failed */
            . . .
            /* User records */
 (7)        QS_USER              /**< the first record available for user QS records */
        };
        . . .
        /* Macros for adding QS instrumentation to the client code . . . . . . . . . . . . . . . . */
 (8)    #define QS_INIT(arg_)             QS_onStartup(arg_)
 (9)    #define QS_EXIT()                 QS_onCleanup()
(10)    #define QS_FILTER_ON(rec_)        QS_filterOn(rec_)
(11)    #define QS_FILTER_OFF(rec_)       QS_filterOff(rec_)
(12)    #define QS_FILTER_SM_OBJ(obj_)    (QS_smObj_ = (obj_))
```

```
      #define QS_FILTER_AO_OBJ(obj_)  (QS_aoObj_ = (obj_))
      #define QS_FILTER_MP_OBJ(obj_)  (QS_mpObj_ = (obj_))
      #define QS_FILTER_EQ_OBJ(obj_)  (QS_eqObj_ = (obj_))
      #define QS_FILTER_TE_OBJ(obj_)  (QS_teObj_ = (obj_))
      #define QS_FILTER_AP_OBJ(obj_)  (QS_apObj_ = (obj_))


      /* Macros to generate user QS records (formatted data output) .............*/
(13)  #define QS_BEGIN(rec_, obj_)     ...
(14)  #define QS_END()                 ...
(15)  #define QS_BEGIN_NOLOCK(rec_, obj_) ...
(16)  #define QS_END_NOLOCK()          ...
      ...
(17)  #define QS_I8 (w_, d_)      QS_u8((uint8_t) (((w_) << 4)) | QS_I8_T,  (d_))
(18)  #define QS_U8 (w_, d_)      QS_u8((uint8_t) (((w_) << 4)) | QS_U8_T,  (d_))
      #define QS_I16(w_, d_)      QS_u16((uint8_t)(((w_) << 4)) | QS_I16_T, (d_))
      #define QS_U16(w_, d_)      QS_u16((uint8_t)(((w_) << 4)) | QS_U16_T, (d_))
      #define QS_I32(w_, d_)      QS_u32((uint8_t)(((w_) << 4)) | QS_I32_T, (d_))
      #define QS_U32(w_, d_)      QS_u32((uint8_t)(((w_) << 4)) | QS_U32_T, (d_))
(19)  #define QS_F32(w_, d_)      QS_f32((uint8_t)(((w_) << 4)) | QS_F32_T, (d_))
(20)  #define QS_F64(w_, d_)      QS_f64((uint8_t)(((w_) << 4)) | QS_F64_T, (d_))
(21)  #define QS_STR(str_)        QS_str(str_)
(22)  #define QS_STR_ROM(str_)    QS_str_ROM(str_)
(23)  #define QS_MEM(mem_, size_)  QS_mem((mem_), (size_))


      #if (QS_OBJ_PTR_SIZE == 1)
(24)      #define QS_OBJ(obj_)    QS_u8(QS_OBJ_T, (uint8_t)(obj_))
      #elif (QS_OBJ_PTR_SIZE == 2)
(25)      #define QS_OBJ(obj_)    QS_u16(QS_OBJ_T, (uint16_t)(obj_))
      #elif (QS_OBJ_PTR_SIZE == 4)
(26)      #define QS_OBJ(obj_)    QS_u32(QS_OBJ_T, (uint32_t)(obj_))
      #else
(27)      #define QS_OBJ(obj_)    QS_u32(QS_OBJ_T, (uint32_t)(obj_))
      #endif
      #if (QS_FUN_PTR_SIZE == 1)
(28)      #define QS_FUN(fun_)    QS_u8(QS_FUN_T, (uint8_t)(fun_))
      #elif (QS_FUN_PTR_SIZE == 2)
          ...
      #endif

      #if (Q_SIGNAL_SIZE == 1)
(29)      #define QS_SIG(sig_, obj_) QS_u8 (QS_SIG_T, (sig_)); QS_OBJ_(obj_)
      #elif (Q_SIGNAL_SIZE == 2)
          ...

      #endif

      /* Dictionary records ......................................................*/
(30)  #define QS_OBJ_DICTIONARY(obj_) ...
(31)  #define QS_FUN_DICTIONARY(fun_) ...
(32)  #define QS_SIG_DICTIONARY(sig_, obj_) ...
```

Continued onto next page

```
        . . .
        /* Macros used only internally in the QP code .............................*/
(33)    #define QS_BEGIN_(rec_, obj_)   . . .
(34)    #define QS_END_()              . . .
(35)    #define QS_BEGIN_NOLOCK_(rec_, obj_) . . .
(36)    #define QS_END_NOLOCK_()       . . .


        /* QS functions for managing the QS trace buffer .........................*/
(37)    void QS_initBuf(uint8_t sto[], uint32_t stoSize);
(38)    uint16_t QS_getByte(void);                      /* byte-oriented interface */
(39)    uint8_t const *QS_getBlock(uint16_t *pNbytes); /* block-oriented interface */


        /* QS callback functions, typically implemented in the BSP ................*/
(40)    uint8_t QS_onStartup(void const *arg);
(41)    void QS_onCleanup(void);
(42)    void QS_onFlush(void);
(43)    QSTimeCtr QS_onGetTime(void);

        #endif                                                           /* qs_h */
```

(1)   A compile-time error is reported if the qs.h header file is included without
        defining the Q_SPY macro (see also Listing 11.4(1)).

(2)   The enumeration QSpyRecords defines all the standard QS record types.

(3-5)   Each QP component generates specific QS record types. For example,
        standard QS records are designed for entering a state (3), adding an active
        object to the framework (4), or locking a QK mutex (5).

(6)   Standard QS records include also miscellaneous records, like the dictionary
        records (see Section 11.3.8).

(7)   The list QS records can be extended by adding user-defined records. The user
        records must start at the numerical value determined by QS_USER. Currently,
        QS supports up to 256 records, from which the first 70 are reserved for the
        standard, predefined records. This leaves 186 records for application-specific
        records. I discuss application-specific records in Section 11.3.9.

(8,9)   As I mentioned before, all QS services are defined as preprocessor macros.
        That way, you can leave them in the code, even if software tracing is disabled.
        Here the services for initializing and terminating QS are specified.

(10,11) These two macros implement the global QS filter, which turns tracing of a
        given QS trace record on or off. I discuss QS filters in Section 11.3.5.

(12)   This macro implements the local QS filter. This filter type allows you selectively trace only specified state machine object. I discuss QS filters in Section 11.3.5.

(13,14)   These macros open and close an application-specific QS trace record. I discuss the application-specific records in Section 11.3.9.

(15,16)   These macros also open and close an application-specific QS trace record, except they don't lock and unlock interrupts. These macros are supposed to be used inside an already established critical section.

> **NOTE**
>
> The QS trace buffer is obviously a shared resource, which must be protected against corruption. QS uses interrupt locking (critical section) as the mutual exclusion mechanism. The macro QS_BEGIN() locks interrupts and the macro QS_END() unlocks interrupts, so the whole QS record is saved to the QS buffer in one critical section. You should avoid producing big trace records because this could extend interrupt latency.

(17,18)   These macros are used to output an unsigned 8-bit integer and a signed 8-bit integer in an application-specific trace record.

(19,20)   These two macros output a 32-bit and 64-bit IEEE floating-point numbers, respectively, to an application-specific trace record.

(21)   This macro outputs a zero-terminated string to an application-specific trace record.

(22)   This macro outputs a zero-terminated string allocated in ROM to an application-specific trace record.

> **NOTE**
>
> Some Harvard CPU architectures use different instructions to access data in program memory (ROM) than in RAM.

(23)   This macro outputs a memory block of specified length to an application-specific trace record. The block size cannot exceed 255 bytes.

(24-27)   The macro QS_OBJ() outputs an object pointer to an application-specific trace record. Note how the actual macro definition depends on the object pointer size defined by the macro `QS_OBJ_PTR_SIZE`. This idiom is used quite often in QS.

(28)   The macro `QS_FUN()` outputs a function pointer to an application-specific trace record. Note how the actual macro definition depends on the function pointer size defined by the macro `QS_FUN_PTR_SIZE`.

(29)   The macro `QS_SIG()` outputs a signal value to an application-specific trace record. Note how the actual macro definition depends on the signal size defined by the macro `Q_SIGNAL_SIZE`.

NOTE

The macro `QS_SIG()` outputs both the signal value and state machine object pointer. This is done to avoid ambiguities, when numerical signal values are reused in different state machines.

(30-32)   These macros output various dictionary trace records to the QS trace buffer. I discuss dictionary trace records in Section 11.3.8.

(33,34)   These internal QS macros open and close an internal QS trace record.

(35,36)   These internal macros also open and close an internal QS trace record, except they don't lock and unlock interrupts. These macros are supposed to be used inside an already established critical section.

(37)   The function `QS_initBuf()` initializes the QS buffer. The caller must provide the storage for the buffer and its size. The function must be called before QS trace buffer can be used, typically from `QS_onStartup()`.

(38)   The function `QS_getByte()` obtains 1 byte from the QS trace buffer (see Section 11.3.7).

(39)   The function `QS_getBlock()` obtains a contiguous block of data in QS trace buffer (see Section 11.3.7).

(40)   The `QS_onStartup()` callback function initializes the QS tracing output and the trace buffer (see `QS_initBuf()`). The function returns the status of initialization. It is called from the macro `QS_INIT()`.

(41)   The `QS_onCleanup()` callback function cleans up the QS tracing output. The function is called from the macro `QS_EXIT()`.

(42) The `QS_onFlush()` callback function flushes the QS trace buffer to the host. The function typically busy-waits until the whole buffer is transmitted to the host. `QS_onFlush()` is called at the end of each dictionary record (see Section 11.3.8), but you can also call it explicitly via macro `QS_FLUSH()`. I provide an example of the `QS_onFlush()` callback implementation in Section 11.6.2.

(43) The `QS_onGetTime()` callback function returns the timestamp for a QS record. I provide an example of the `QS_onGetTime()` callback implementation in Section 11.6.3.

Listing 11.4 Inactive QS interface (fragments of the header file `<qp>\qpc\include\qs_dummy.h`)

```
       #ifndef qs_dummy_h
       #define qs_dummy_h

       #ifdef Q_SPY
(1)        #error "Q_SPY must NOT be defined to include qs_dummy.h"
       #endif

(2)    #define QS_INIT(arg_)                ((uint8_t)1)
(3)    #define QS_EXIT()                    ((void)0)
       #define QS_DUMP()                    ((void)0)
       #define QS_FILTER_ON(rec_)           ((void)0)
       #define QS_FILTER_OFF(rec_)          ((void)0)
       #define QS_FILTER_SM_OBJ(obj_)       ((void)0)
       . . .
(4)    #define QS_GET_BYTE(pByte_)          ((uint16_t)0xFFFF)
(5)    #define QS_GET_BLOCK(pSize_)         ((uint8_t *)0)

(6)    #define QS_BEGIN(rec_, obj_)         if (0) {
(7)    #define QS_END()                     }
       #define QS_BEGIN_NOLOCK(rec_, obj_)  QS_BEGIN(rec_, obj_)
       #define QS_END_NOLOCK()              QS_END()
       #define QS_I8(width_, data_)         ((void)0)
       #define QS_U8(width_, data_)         ((void)0)
       . . .
       #define QS_SIG(sig_, obj_)           ((void)0)
       #define QS_OBJ(obj_)                 ((void)0)
       #define QS_FUN(fun_)                 ((void)0)

       #define QS_SIG_DICTIONARY(sig_, obj_) ((void)0)
       #define QS_OBJ_DICTIONARY(obj_)      ((void)0)
```

Continued onto next page

```
    #define QS_FUN_DICTIONARY(fun_)          ((void)0)
    #define QS_FLUSH()                       ((void)0)

    . . .
    #endif                                                    /* qs_dummy_h */
```

(1) A compile-time error is reported if the qs_dummy.h header file is included when the Q_SPY macro is defined (see also Listing 11.3(1)).

(2) The dummy QS initialization always returns 1, meaning successful initialization.

(3) Most other QS dummy macros are defined as ((void)0), which is a valid empty C expression that can be terminated with a semicolon.

(4) The dummy QS macro for obtaining a byte from the trace buffer always returns 0xFFFF, which means that end of data is reached. I discuss QS API for accessing the trace buffer in Section 11.3.7.

(5) The dummy QS macro for obtaining a block of data to output always returns a NULL pointer, which means that there is no data in the buffer. I discuss QS API for accessing the trace buffer in Section 11.3.7.

(6,7) The dummy QS macros for opening and closing a trace record compile as if (0) {...}. Any active code between the braces is eliminated because of the FALSE condition of the if statement.

NOTE

Some trace records might contain temporary variables and expressions that are only used for the trace output. The "if (0) {" statement establishes a new scope to define such temporary variables.

11.3.3 QS Critical Section

The QS target component must protect the internal integrity of the trace buffer, which is shared among concurrently running tasks and interrupts (see Figure 11.3). To guarantee mutually exclusive access to the trace buffer, QS uses the same mechanism as the rest of the QP platform, that is, QS locks interrupts on entry to the critical section of code and unlocks interrupts again on exit.

When QS detects that the QF critical section macros `QF_INT_LOCK()`/
`QF_INT_UNLOCK()` are defined, QS uses the provided definitions for its own critical
section. However, when you use QS without the QF real-time framework, you need
to define the platform-specific interrupt locking/unlocking policy of QS in the
`qs_port.h` header file, as shown in Listing 11.5.

NOTE

QS can be used with just the QEP component or even completely standalone, without any
other QP components. In these cases, QS must provide its own, independent critical section
mechanism.

Listing 11.5 QS macros for interrupt locking and unlocking (file `qs_port.h`)

```
#define QS_INT_KEY_TYPE       . . .
#define QS_INT_LOCK(key_)     . . .
#define QS_INT_UNLOCK(key_)   . . .
```

The QS macros are exactly analogous to the QF macros `QF_INT_KEY_TYPE`,
`QF_INT_LOCK()`, and `QF_INT_UNLOCK()`, respectively. Refer to Section 7.3 in
Chapter 7 for more details.

11.3.4 General Structure of QS Records

Like all software-tracing systems, QS logs the tracing data in discrete chunks called QS
trace records. These trace records have the general structure shown in Listing 11.6.

Listing 11.6 General structure of a QS record

```
QS_BEGIN_xxx(record_type)                      /* trace record begin */
    QS_yyy(data);                                /* QS data element */
    QS_zzz(data);                                /* QS data element */
    . . .                                        /* QS data element */
QS_END_xxx()                                   /* trace record end */
```

Each trace record always begins with one variant of the macro `QS_BEGIN_xxx()` and
ends with the matching macro `QS_END_xxx()`.

> **NOTE**
>
> The macros QS_BEGIN_xxx() and QS_END_xxx() are *not* terminated with the semicolon.

Sandwiched between these two macros are the data-generating macros that actually insert individual data elements into the QS trace buffer. QS provides four variants of the begin/end macro pairs for different purposes (Listing 11.3(13-16 and 33-36)).

The first two variants (Listing 11.3(13-16)) are for creating application-specific QS records (see Section 11.3.9). The QS_BEGIN()/QS_END() pair locks interrupts at the beginning of the record and unlocks at the end. The pair QS_BEGIN_NOLOCK()/ QS_END_NOLOCK() is for application-specific records without entering the QS critical and should be used only within an already established critical section.

The third and fourth variants of the begin-record/end-record QS macros (Listing 11.3 (33-36)) are for internal use within QP components to generate the predefined QS records. Such predefined records are generated with QS_BEGIN_()/QS_END_() or QS_BEGIN_NOLOCK_()/QS_END_NOLOCK_() macro pairs, depending whether a critical section must be entered or not.

11.3.5   QS Filters

One of the main roles of the begin-record macros QS_BEGIN_xxx() is to implement the filtering of QS records before they reach the trace buffer. QS provides two complementary levels of filtering: the global on/off filter and local filters (see Figure 11.3).

Global On/Off Filter

The global on/off filter is based on the record IDs. The qs.h header file provides the enumeration of all the predefined internal trace records IDs that are already instrumented into the QP components (Listing 11.3(2)). The enumeration of the predefined records ends with the QS_USER value, which is the first numerical value available for application-specific trace records. I discuss the application-specific trace records in Section 11.3.9.

The global on/off filter is efficiently implemented by means of an array of bitmasks QS_glbFilter_[], where each bit represents one trace record. Currently the QS_glbFilter_[] array contains 32 bytes for a total of 32*8 bits for 256 different trace records. A little more than a quarter of these records are already taken by the predefined QP trace records. The remaining three quarters are available for application-specific use.

The macro QS_BEGIN() for opening a trace record shows how the global on/off filter is implemented:

```
#define QS_BEGIN(rec_, obj_) \
    if ((((QS_glbFilter_[(uint8_t)(rec_) >> 3U] \
        & (1U << ((uint8_t)(rec_) & 7U))) != 0) . . .\
```

The global on/off filter works by checking the state of the bit corresponding to the given trace record argument "rec_." This check is accomplished by the familiar expression (bimask & bit) != 0, where the bitmask is QS_glbFilter_[(uint8_t) (rec_) >> 3U] and the bit is (1U << ((uint8_t)(rec_) & 7U)). Note that for any constant value of the argument rec_, both the bitmask and the bit are compile-time constants. For example, a global filter check for a record ID of 46, say, costs as much as the expression (QS_glbFilter_[5] & 0x40) != 0).

NOTE
The global filter is specifically implemented to use byte-size computations only to be efficient even on 8-bit machines.

QS provides a simple interface for setting and clearing individual bits in the QS_glbFilter_[] bitmask array. The macro QS_FILTER_ON(rec_) turns on the bit corresponding to record "rec_." Conversely, the macro QS_FILTER_OFF(rec_) turns off the bit corresponding to record "rec_." In both cases, the special constant QS_ALL_RECORDS affects all records. Specifically, QS_FILTER_ON (QS_ALL_RECORDS) turns on all records, and QS_FILTER_OFF(QS_ALL_RECORDS) turns off all records. Examples of these macros are provided in Listing 11.16.

Just after QS initialization, the global on/off filter is set to OFF for all records types. You need to explicitly turn the filter ON for some records to enable the tracing.

> **NOTE**
>
> Globally disabling all records through QS_FILTER_OFF(QS_ALL_RECORDS) is a useful way
> of implementing a software-tracing *trigger*. You can use this trigger to rapidly stop the
> tracing after an interesting event, to prevent new trace data from overwriting interesting data
> in case the data uplink to the host cannot keep up with the production rate of new trace data.

Local Filters

The local filters allow generation of trace records only for specified objects. For
example, you might set up a local filter to log only activities of a given state machine
object. Independently, you might set up another local filter to log only activities of a
given memory pool.

The Table 11.1 summarizes all specified local filters and the predefined QS records
controlled by these filters.

Table 11.1: Local filter summary

Local Filter	Object Type	Example	Applies to QS Records
QS_FILTER_SM_OBJ()	State machine	QS_FILTER_SM_OBJ (&l_qhsmTst);	QS_QEP_STATE_EMPTY, QS_QEP_STATE_ENTRY, QS_QEP_STATE_EXIT, QS_QEP_STATE_INIT, QS_QEP_INIT_TRAN, QS_QEP_INTERN_TRAN, QS_QEP_TRAN, QS_QEP_IGNORED
QS_FILTER_AO_OBJ()	Active object	QS_FILTER_AO_OBJ (&l_philo[3]);	QS_QF_ACTIVE_ADD, QS_QF_ACTIVE_REMOVE, QS_QF_ACTIVE_SUBSCRIBE, QS_QF_ACTIVE_UNSUBSCRIBE, QS_QF_ACTIVE_POST_FIFO, QS_QF_ACTIVE_POST_LIFO, QS_QF_ACTIVE_GET, QS_QF_ACTIVE_GET_LAST
QS_FILTER_MP_OBJ()[1]	Memory pool	QS_FILTER_MP_OBJ (l_regPoolSto);	QS_QF_MPOOL_INIT, QS_QF_MPOOL_GET, QS_QF_MPOOL_PUT,

Table 11.1: Local filter summary—Cont'd

Local Filter	Object Type	Example	Applies to QS Records
QS_FILTER_EQ_OBJ()[2]	Event queue	QS_FILTER_EQ_OBJ (l_philQueueSto[3]);	QS_QF_EQUEUE_INIT, QS_QF_EQUEUE_POST_FIFO, QS_QF_EQUEUE_POST_LIFO, QS_QF_EQUEUE_GET, QS_QF_EQUEUE_GET_LAST
QS_FILTER_TE_OBJ()	Time event	QS_FILTER_TE_OBJ (&l_philo[3]. timeEvt);	QS_QF_TICK, QS_QF_TIMEEVT_ARM, QS_QF_TIMEEVT_AUTO_DISARM, QS_QF_TIMEEVT_DISARM_ATTEMPT, QS_QF_TIMEEVT_DISARM, QS_QF_TIMEEVT_REARM, QS_QF_TIMEEVT_POST, QS_QF_TIMEEVT_PUBLISH
QS_FILTER_AP_OBJ()	Generic application object	QS_FILTER_AP_OBJ (&myAppObject);	Application-specific records starting with QS_USER

[1] Memory pool is referenced by the memory buffer managed by the pool.
[2] Event queue is referenced by the ring buffer managed by the queue.

The first column of Table 11.1 enlists the QS macros you need to use to set/clear the local filters. For example, you specify the state machine local filter by invoking:

```
QS_FILTER_SM_OBJ(aStateMachinePointer);
```

where `aStateMachinePointer` is the pointer to the state machine object you want to trace.

You deactivate any local filter by passing the NULL pointer to the appropriate QS macro. For example, to open up the local filter for all state machine objects, you write the following code:

```
QS_FILTER_SM_OBJ(0);
```

Just after QS initialization, all local filters is set to NULL, meaning that the local filters are open for all objects.

The highlighted code in the QS_BEGIN() macro definition shows the actual implementation of the local filter for the application-specific objects:

```
#define QS_BEGIN(rec_, obj_) \
    if (((QS_glbFilter_[(uint8_t)(rec_) >> 3U] \
        & (1U << ((uint8_t)(rec_) & 7U))) != 0) \
      && ((QS_apObj_ == (void *)0) || (QS_apObj_ == (obj_)))) \
    { \
        . . .
```

The QS local filters are closely related to the object dictionary records (see Section 11.3.8) and both facilities consistently use the same conventions.

11.3.6   QS Data Protocol

The data transmission protocol used in QS to transmit trace data from the target to the host is one of its greatest strengths. The protocol is very lightweight but has many elements of the HDLC protocol defined by the ISO.

The QS protocol has been specifically designed to simplify the data management overhead in the target yet allow detection of any data dropouts due to the trace buffer overruns. The protocol has not only provisions for detecting gaps in the data and other errors but allows for instantaneous resynchronization after any error, to minimize data loss.

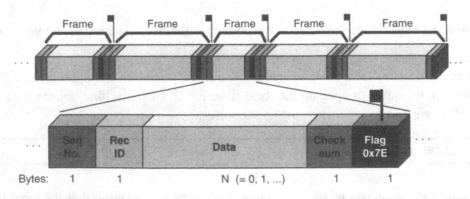

Figure 11.4: QS transmission protocol.

The QS protocol transmits each trace record in an HDLC-like frame. The upper part of Figure 11.4 shows the serial data stream transmitted from the target containing frames of different lengths. The bottom part of Figure 11.4 shows the details of a single frame:

1. Each frame starts with the Frame Sequence Number byte. The target QS component increments the Frame Sequence Number for every frame inserted into the circular buffer. The Sequence Number naturally rolls over from 255 to 0. The Frame Sequence Number allows the QSPY host component to detect any data discontinuities.

2. Following the Fame Sequence Number is the Record ID byte, which is one of the predefined QS records (see Listing 11.3(2)) or an application-specific record (see Section 11.3.9).

3. Following the Record ID is zero or more data bytes.

4. Following the data is the Checksum. The Checksum is computed over the frame Sequence Number, the Record ID, and all the data bytes. The next section gives the detailed checksum computation formula.

5. Following the Checksum is the HDLC Flag, which delimits the frame. The HDLC flag is the 01111110 binary string (0x7E hexadecimal). Note that the QS protocol uses only one HDLC Flag at the end of each frame and no HDLC Flag at the beginning of a frame. In other words, only one Flag is inserted between frames.

The QS target component performs the HDLC-like framing described above at the time the bytes are inserted into the circular trace buffer. This means that the data in the buffer is already cleanly divided into frames and can be transmitted in any chunks, typically *not* aligned with the frame boundaries.

Transparency

One of the most important characteristics of HDLC-type protocols is establishing very easily identifiable frames in the serial data stream. Any receiver of such a protocol can instantaneously synchronize to the frame boundary by simply finding the Flag byte. This is because the special Flag byte can never occur within the content of a frame. To avoid confusing unintentional Flag bytes that can naturally occur in the data stream with an intentionally sent Flag, HDLC uses a technique known as *transparency* (a.k.a. *byte stuffing* or *escaping*) to make the Flag bytes transparent during the transmission. Whenever the transmitter encounters a Flag byte in the data, it inserts a 2-byte escape sequence to the output stream. The first byte is the Escape byte, defined as binary 01111101 (hexadecimal 0x7D). The second byte is the original byte XOR-ed with 0x20.

Of course, now the Escape byte itself must also be transparent to avoid interpreting an unintentional Escape byte as the 2-byte escape sequence. The procedure of escaping the Escape byte is identical to that of escaping the Flag byte.

The transparency of the Flag and Escape bytes complicates slightly the computation of the Checksum. The transmitter computes the Checksum over the Fame Sequence Number, the Record ID, and all data bytes before performing any byte stuffing. The receiver must apply the exact reversed procedure of performing the byte unstuffing before computing the Checksum.

An example might make this clearer. Suppose that the following trace record needs to be inserted to the trace buffer (the transparent bytes are shown in bold):

```
Record ID = 0x7D, Record Data = 0x7D 0x08 0x01
```

Assuming that the current Frame Sequence Number is, say, 0x7E, the Checksum will be computed over the following bytes:

```
Checksum == (uint8_t)(~(0x7E + 0x7D + 0x7D + 0x08 + 0x01)) == 0x7E
```

and the actual frame inserted into the QS trace buffer will be as follows:

```
0x7D 0x5E 0x7D 0x5D 0x7D 0x5D 0x08 0x01 0x7D 0x5E 0x7E
```

Obviously, this is a degenerated example, where the Frame Sequence Number, the Record ID, a data byte, and the Checksum itself turned out to be the transparent bytes. Typical overhead of transparency with real trace data is one escape sequence per several trace records.

Endianness

In addition to the HDLC-like framing, the QS transmission protocol specifies the endianness of the data to be *little endian*. All multibyte data elements, such as 16- and 32-bit integers, pointers, and floating-point numbers, are inserted into the QS trace buffer in the little-endian byte order (least significant byte first). The QS data inserting macros (see Listing 11.3(17-23)) place the data in the trace buffer in a platform-neutral manner, meaning that the data is inserted into the buffer in the little-endian order regardless of the endianness of the CPU. Also, the data-inserting macros copy the data to the buffer 1 byte at a time, thus avoiding any potential data misalignment problems. Many embedded CPUs, such as ARM, require certain alignment of 16- and 32-bit quantities.

11.3.7 QS Trace Buffer

As described in the previous section, the QS target component performs the HDLC-like framing at the time the bytes are inserted into the QS trace buffer. This means that only complete frames are placed in the buffer, which is the pivotal point in the design of the QS target component and has two important consequences.

First, the use of HDLC-formatted data in the trace buffer allows decoupling the data insertion into the trace buffer from the data removal out of the trace buffer. You can simply remove the data in whichever chunks you like, without any consideration for frame boundaries. You can employ just about any repetition physical data link available on the target for transferring the trace data from the target to the host.

Second, the use of the formatted data in the buffer enables the "last is best" tracing policy. The QS transmission protocol maintains both the Frame Sequence Number and the Checksum over each trace record, which means that any data corruption caused by overrunning the old data with the new data can be always *reliably* detected. Therefore, the new trace data is simply inserted into the circular trace buffer, regardless of whether it perhaps overwrites the old data that hasn't been sent out yet or is in the process of being sent. The burden of detecting any data corruption is placed on the QSPY host component. When you start missing the frames (which the host component easily detects by discontinuities in the Frame Sequence Number), you have several options. Your can apply some additional filtering, increase the size of the buffer, or improve the data transfer throughput.

Initializing the QS Trace Buffer `QS_initBuf()`

Before you can start producing trace data, you must initialize the QS trace buffer by calling the `QS_initBuf()` function. Typically, you invoke this function from the `QS_onStartup()` callback, which you typically define in your application. Listing 11.7 shows an example of initializing the QS trace buffer.

Listing 11.7 Initializing QS trace buffer with `QS_initBuf(`

```
(1)  #ifdef Q_SPY                                   /* define QS callbacks */

(2)  uint8_t QS_onStartup(void const *arg) {
(3)      static uint8_t qsBuf[2*1024];              /* buffer for Quantum Spy */
(4)      QS_initBuf(qsBuf, sizeof(qsBuf));

         Initialize the QS data link

         return success;            /* return 1 for success and 0 for failure */
     }
     #endif                                         /* Q_SPY */
```

(1) The QS callback functions (such as `QS_onStartup()`) are defined only when QS tracing is enabled.

(2) At a minimum, the `QS_onStartup()` callback function must initialize the QS trace buffer.

(3) You need to statically allocate the storage for the QS trace buffer. The size of the buffer depends on the nature of your application and the data link to the host. Obviously, a bigger buffer is needed if you want to trace events occurring at a high rate and producing a higher volume of trace data. On the other hand, using a higher-bandwidth data link to the host allows you to reduce the size of the trace buffer.

(4) The `QS_initBuf()` function initializes internal QS variables to use the provided trace buffer.

NOTE

QS can work with a trace buffer of any size, but smaller buffers will lose data if the buffer "wraps around." You will always know when any data loss occurs, however, because the QS data protocol maintains a sequence number in every trace record (see Section 11.3.6). When the QSPY host application detects a discontinuity in the sequence numbers, it produces the following message:

```
*** Incorrect record past seq=xxx
*** Dropped yy records
```

You have several options to avid losing trace records due to data overruns. You can increase the size of the trace buffer (Listing 11.7(3)) or apply more filtering to reduce the amount of trace data produced (see Section 11.3.5). You can also employ a faster data link to the host. Finally, sometimes you can improve the data throughput by changing the policy of sending trace data to the host. For example, using only the idle processing might not utilize the full available bandwidth of the data link if the idle processing executes too infrequently.

Byte-Oriented Interface: `QS_getByte()`

The lack of any constraints on removing the data from the trace buffer means that you can remove 1 byte at a time at arbitrary time instances. QS provides the function `QS_getByte()` for such byte-oriented interfaces. The signature of `QS_getByte()` is shown in Listing 11.3(38). Listing 11.8 shows an example of how to use this function.

The `QS_getByte()` function returns the byte in the least significant 8 bits of the 16-bit return value if the byte is available. If the trace buffer has no more data, the function returns `QS_EOD` (end-of-data), which is defined in `qs.h` as `((uint16_t) 0xFFFF)`.

NOTE

The function `QS_getByte()` does *not* lock interrupts internally and is not reentrant. You should always design your software such that `QS_getByte()` is called with interrupts locked. In addition, an application should consistently use either the byte-oriented interface `QS_getByte()` or the block-oriented interface `QS_getBlock()` (see the next subsection), but never both at the same time.

Listing 11.8 Using `QS_getByte()` **to output data to a 16550-compatible UART**

```
(1)   void QF_onIdle(void) {                           /* called with interrupts LOCKED */
(2)       QF_INT_UNLOCK(dummy);                         /* always unlock interrupts */
(3)   #ifdef Q_SPY
(4)       if ((inportb(l_uart_base + 5) & (1 << 5)) != 0) {          /* THR Empty? */
(5)           uint8_t fifo = UART_16550_TXFIFO_DEPTH; /*depth of the 16550 Tx FIFO */
(6)           uint16_t b;
(7)           QF_INT_LOCK(dummy);
```

Continued onto next page

```
(8)             while ((fifo != 0)
(9)                 && ((b = QS_getByte()) != QS_EOD))        /* get the next byte */
               {
(10)               QF_INT_UNLOCK(dummy);
(11)               outportb(l_base + 0, (uint8_t)b);   /* insert byte into TX FIFO */
(12)               --fifo;
(13)               QF_INT_LOCK(dummy);
               }
(14)            QF_INT_UNLOCK(dummy);
           }
       #endif
       }
```

(1)   Idle processing is ideal for implementing trace data output. In this example, I use the QF_onIdle() idle callback of the cooperative vanilla kernel (see Section 8.2.4 in Chapter 8).

(2)   As explained at the end of Section 7.11.1 in Chapter 7, the QF_onIdle() idle callback is invoked with interrupts locked and it always must unlock interrupts.

(3)   The QS trace buffer output is performed only when QS is active, that is, when the macro Q_SPY is defined.

(4)   The Transmitter Holding Register Empty bit of the 16550 UART is checked.

(5)   The 16550 UART can accept up to the TX FIFO depth bytes (typically 16).

(6)   The temporary variable 'b' will hold the return value from QS_getByte(). Note that it is 2 bytes wide.

(7)   Interrupts are locked before calling QS_getByte().

(8)   The loop continues until there is room in the TX FIFO.

(9)   The QS_getByte() function is called to obtain the next trace byte to transmit. The return value of QS_EOD indicates end of data.

(10)   Interrupts can be unlocked.

(11)   The trace byte is written to the Transmitter Holding Register.

(12)   One less byte is available in the TX FIFO.

(13)   Interrupts are locked to make another call to QS_getByte().

(14)   Interrupts are unlocked before returning to the caller.

Block-Oriented Interface: `QS_getBlock()`

QS also provides an alternative block-oriented interface for obtaining a contiguous block of data at a time. QS provides the function `QS_getBlock()` for such block-oriented interface. The signature of `QS_getBlock()` is shown in Listing 11.3(39). Such a block-oriented interface is very useful for DMA-type transfers. Listing 11.9 shows an example of how to use this function.

If any bytes are available at the time of the call, the function returns the pointer to the beginning of the data block within the QS trace buffer and writes the number of *contiguous* bytes in the block to the location pointed to by `pNbytes`. The value of `*pNbytes` is also used *as input* to limit the maximum size of the data block that the caller can accept. Note that the bytes are not copied from the trace buffer.

If no bytes are available in the QS buffer when the function is called, the function returns a NULL pointer and sets the value pointed to by `pNbytes` to zero.

You should not assume that the QS trace buffer becomes empty after `QS_getBlock()` returns a data block with fewer bytes than the initial value of `*pNbytes`. Sometimes the data block falls close to the end of the trace buffer and you need to call `QS_getBlock()` again to obtain the rest of the data that "wrapped around" to the beginning of the QS data buffer. After the `QS_getBlock()` returns a memory block to the caller, the caller must transfer all the bytes in the returned block before calling `QS_getBlock()` again.

NOTE

The function `QS_getBlock()` does not lock interrupts internally and is not reentrant. You should always design your software such that `QS_getBlock()` is called with interrupts locked.

Listing 11.9 Using `QS_getBlock()` **to output data to a 16550-compatible UART**

```
(1)   void QF_onIdle(void) {                          /* called with interrupts LOCKED */
(2)       QF_INT_UNLOCK(dummy);                        /* always unlock interrupts */
(3)   #ifdef Q_SPY
(4)       if ((inportb(l_uart_base + 5) & (1 << 5)) != 0) {          /* THR Empty? */
(5)           uint16_t fifo = UART_16550_TXFIFO_DEPTH;      /* 16550 Tx FIFO depth */
(6)           uint8_t const *block;
```

Continued onto next page

```
(7)            QF_INT_LOCK(dummy);
(8)            block = QS_getBlock(&fifo);      /* try to get next block to transmit */
(9)            QF_INT_UNLOCK(dummy);
(10)           while (fifo-- != 0) {                        /* any bytes in the block? */
(11)               outportb(l_uart_base + 0, *block++);
               }
        }
    #endif
    }
```

(1) Idle processing is ideal for implementing trace data output. In this example, I use the QF_onIdle() idle callback of the cooperative vanilla kernel (see Section 8.2.4 in Chapter 8).

(2) As explained at the end of Section 7.11.1 in Chapter 7, the QF_onIdle() idle callback is invoked with interrupts locked and it always must unlock interrupts.

(3) The QS trace buffer output is performed only when QS is active, that is, when the macro Q_SPY is defined.

(4) The Transmitter Holding Register Empty bit of the 16550 UART is checked.

(5) The 16550 UART can accept up to the TX FIFO depth bytes (typically 16).

(6) The temporary pointer 'block' will hold the return value from QS_getBlock().

(7) Interrupts are locked before calling QS_getBlock().

(8) The QS_getBlock() function is called to obtain the contiguous block of trace data to transmit.

(9) Interrupts can be unlocked.

(10) The loop continues while there is room in the TX FIFO.

(11) The trace byte is written to the Transmitter Holding Register.

11.3.8   Dictionary Trace Records

By the time you compile and load your application image to the target, the symbolic information about the object names, function names, and signal names is stripped from the code. Therefore, if you want to have the symbolic information available to the QSPY host-resident component, you need to supply it somehow to the software-tracing system.

QS provides special trace records designed expressly for including the symbolic information about the target code in the trace itself. The dictionary records included in the trace for the QSPY host application are very much like the symbolic information embedded in the object files for the traditional single-step debugger.

The dictionary trace records are not absolutely required to generate the trace in the same way as the symbolic information in the object files is not absolutely required to debug code. However, in both cases, the availability of the symbolic information greatly improves productivity in working with the software trace or the debugger.

QS supports three types of dictionary trace records: object dictionary, function dictionary, and signal dictionary. The following subsections cover these types in detail.

NOTE
As all QS trace records, the dictionary trace records are generated in a critical section of code, that is, interrupts are locked for the time the data is inserted into the QS trace buffer. Additionally, after unlocking interrupts, the callback function OS_onFlush() is invoked at the end of each dictionary record. This callback function typically busy-waits until all data are sent out to the host, which might take considerable time. For that reason, dictionary entries should be generated only during the system initialization, when the real-time constraints do not yet apply.

Object Dictionaries

Object dictionaries are generated with the macro QS_OBJ_DICTIONARY() that associates the address of the object in memory with its symbolic name. Listings 11.19 and 11.20 provide some examples of how you use this macro. The QS_OBJ_DICTIONARY() macro takes only one argument, the address of the object, and uses internally the "stringizing" preprocessor operator to convert the provided argument to a C string. Therefore, you should invoke the QS_OBJ_DICTIONARY() macro with meaningfully named persistent objects, such as &l_table, or &l_philo[3], and not generic pointers, such as "me" (or "this" in C++), because the latter will not help you much in recognizing the object name in the trace.

Table 11.2 enlists object dictionaries you can provide to furnish the symbolic information used by the QSPY data output. Note that QS identifies memory pools by the memory buffer managed by the memory pool, because the actual memory pool objects are buried inside the QF framework and are not accessible to the application developer. In addition, event queues are identified by the ring buffer managed by the queue, not by the queue object itself.

Table 11.2: Object dictionaries required for the predefined QS records

Object Type	Example(s)	QS Records
State machine	QS_OBJ_DICTIONARY (&l_table); See Listing 11.19(3)	QS_QEP_STATE_EMPTY, QS_QEP_STATE_ENTRY, QS_QEP_STATE_EXIT, QS_QEP_STATE_INIT, QS_QEP_INIT_TRAN, QS_QEP_INTERN_TRAN, QS_QEP_TRAN, QS_QEP_IGNORED
Active object	QS_OBJ_DICTIONARY (&l_philo[0]); See Listing 11.20(4)	QS_QF_ACTIVE_ADD, QS_QF_ACTIVE_REMOVE, QS_QF_ACTIVE_SUBSCRIBE, QS_QF_ACTIVE_UNSUBSCRIBE, QS_QF_ACTIVE_POST_FIFO, QS_QF_ACTIVE_POST_LIFO, QS_QF_ACTIVE_GET, QS_QF_ACTIVE_GET_LAST
Memory pool[1]	QS_OBJ_DICTIONARY (l_smlPoolSto); See Listing 11.16(7)	QS_QF_MPOOL_INIT, QS_QF_MPOOL_GET, QS_QF_MPOOL_PUT,
Event queue[2]	QS_OBJ_DICTIONARY (l_philQueueSto[0]); See Listing 11.16(9)	QS_QF_EQUEUE_INIT, QS_QF_EQUEUE_POST_FIFO, QS_QF_EQUEUE_POST_LIFO, QS_QF_EQUEUE_GET, QS_QF_EQUEUE_GET_LAST
Time event	QS_OBJ_DICTIONARY (&l_philo[0].timeEvt); See Listing 11.20(5)	QS_QF_TICK, QS_QF_TIMEEVT_ARM, QS_QF_TIMEEVT_AUTO_DISARM, QS_QF_TIMEEVT_DISARM_ATTEMPT, QS_QF_TIMEEVT_DISARM, QS_QF_TIMEEVT_REARM, QS_QF_TIMEEVT_POST, QS_QF_TIMEEVT_PUBLISH

[1]Memory pool is referenced by the memory buffer managed by the pool.
[2]Event queue is referenced by the ring buffer managed by the queue.

The object dictionary records are closely related to the QS local filters (see Section 11.3.5). Both facilities consistently use the same conventions. For example, a local filter for a specific memory pool is selected by means of the QS_FILTER_MP_OBJ() macro, which accepts a pointer to the memory buffer managed by the memory pool. Similarly, a local filter for a specific event queue is selected by means of the QS_FILTER_EQ_OBJ() macro, which accepts a pointer to the ring buffer managed by the event queue.

Function Dictionaries

Function dictionaries are generated with the macro QS_FUN_DICTIONARY(), which associates the address of the function in memory with its symbolic name. Listing 11.19 (4-6) provides examples of how you use this macro. The main purpose of the function dictionaries is to provide symbolic names for *state-handler functions*.

The QS_FUN_DICTIONARY() macro takes only one argument: the address of the function, and uses internally the "stringization" preprocessor operator to convert the provided argument to a C string.

Signal Dictionaries

Signal dictionaries are generated with the macro QS_SIG_DICTIONARY() that associates the numerical value of the event signal and the state machine object to the symbolic name of the signal.

The reason for using both the signal value and the state machine object rather than just the signal value is that a signal value alone is not sufficient to uniquely identify the symbolic signal. Only the globally published signals are required to be systemwide unique. Other signals, used only locally, can have completely different meanings for different state machines in the system.

The QS_SIG_DICTIONARY() macro takes two arguments: the numerical value of the signal and the address of the state machine object. The macro uses internally the "stringization" preprocessor operator to convert the provided signal argument to a string. The state machine object is not converted to a string, so the actual variable name you use is irrelevant.

Listing 11.19(7-10) provides examples of how you use the QS_SIG_DICTIONARY() macro. Listing 11.19(7-9) shows how to specify globally published signals that are associated with multiple state machines. In this case, you specify NULL as the state machine object. In contrast, Listing 11.19(10) shows a dictionary entry for the local signal TIMEOUT_SIG. This signal is associated only with the Philosopher state machines.

11.3.9   Application-Specific QS Trace Records

The application-specific QS records allow you to generate tracing information from the application-level code. You can think of the application-specific records as an equivalent to `printf()` but with much less overhead. Listing 11.10 shows an example of an application-specific QS record.

Listing 11.10 Example of an application-specific trace record

```
QS_BEGIN(MY_QS_RECORD, myObjectPointer)           /* trace record begin */
    QS_STR("Hello");                              /* string data element */
    QS_U8(3, n);                      /* uint8_t data, 3-decimal digits format */
    . . .                                                   /* QS data */
    QS_MEM(buf, sizeof(buf));             /* memory block of a given size */
QS_END()                                            /* trace record end */
```

In most cases, the application-specific records are enclosed with the `QS_BEGIN()`/ `QS_END()` pair of macros. This pair of macros locks interrupts at the beginning and unlocks at the end of each record (see Section 11.3.3). Occasionally you would want to generate trace data from within already established critical sections or ISRs. In such rare occasions, you would use the macros `QS_BEGIN_NOLOCK()`/`QS_END_NOLOCK()` to avoid nesting of critical sections.

The record-begin macro `QS_BEGIN()` takes two arguments. The first argument (e.g., `MY_QS_RECORD`) is the enumerated record type, which is used in the global on/off filter (Section 11.3.5) and is part of the each record header. The application-specific record types must start with the value `QS_USER` to avoid overlap with the predefined QS records already instrumented into the QP components.

The second argument (e.g., `myObjectPointer`) is used for the local filter, which allows you to selectively log only specific application-level objects. Listing 11.21 shows an example of an application-specific trace record, including the use of the second parameter of the `QS_BEGIN()` macro.

NOTE

If you don't want to use the local filter for a given application-specific trace record, you can use `NULL` as the second argument to the macros `QS_BEGIN()` or `QS_BEGIN_NOLOCK()`. That way, the trace record will always be produced, regardless of the setting of the application-specific local filter.

Sandwiched between the QS_BEGIN()/QS_END() macros are data elements that you want to store in the trace record. The macros for generating the data elements are shown in Listing 11.3(17-23). The supported data elements include signed and unsigned integers of 8-bit, 16-bit, and 32-bit size; floating-point numbers of 32-bit and 64-bit size; zero-terminated strings; and variable-size memory blocks. Special macros are also provided for inserting platform-dependent elements, such as event signals, object pointers, and function pointers. For these configurable or platform-specific data elements, QS logs only the minimal number of bytes required on the given platform.

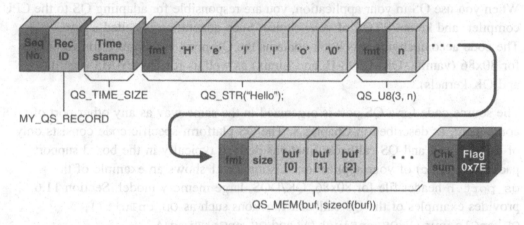

Figure 11.5: Encoding of the application-specific trace record from Listing 11.10 (escaping bytes are omitted for clarity).

The biggest challenge in supporting arbitrary trace records is that the host-resident component (the QSPY application) doesn't "know" the structure of such records, so the data type information must be stored with the data itself. Figure 11.5 shows the encoding of the application-specific trace record from Listing 11.10. The application-specific trace record, like all QS records, starts with the Sequence Number and the Record ID (see Section 11.3.6). Every application-specific trace record also contains the timestamp immediately following the Record ID. The number of bytes used by the timestamp is configurable by the macro QS_TIME_SIZE. After the timestamp, you see the data elements, such as the "Hello" string, an unsigned byte 'n,' some other data, and finally a memory block. Each of these data elements starts with a format byte, which actually contains both the data-type information (in the lower nibble) and the format width for displaying that element (in the upper nibble). For example, the data element QS_U8(3, n) will cause the value 'n' to be encoded as uint8_t with the format width of 3 decimal digits. The maximum allowed format width is 15 decimal digits.

As shown in Listing 11.10, you can place many data elements of any kind in any order inside an application-specific record. The only limitation is that a complete record must fit in the QS trace buffer. Of course, you should avoid big trace records anyway, to keep the critical sections short (QS records are always placed in the buffer in a critical section of code). Furthermore, you might want to conserve the buffer space.

11.3.10   Porting and Configuring QS

When you use QS in your application, you are responsible for adapting QS to the CPU, compiler, and kernel/RTOS of your choice. Such adaptation is called a *port*. The code accompanying this book contains the QS ports and application examples for 80x86 (vanilla, QK, C/OS-II, and Linux) as well as for Cortex-M3 (vanilla and QK kernels).

The source code for a QS port is organized in the same way as any other port of a QP component, as described in Chapter 8. The QS platform-specific code consists only of qs_port.h and QS callback functions defined typically in the board support package (bsp.c) of your application. Listing 11.11 shows an example of the qs_port.h header file for 80x86, QK/DOS, large memory model. Section 11.6 provides examples of the QS callback functions such as QS_onInit(), QS_onCleanup(), QS_onFlush(), and QS_onGetTime().

Listing 11.11 QS port header file for 80x86, QK/DOS, large memory model
(`<qp>\qpc\ports\80x86\qk\tcpp101\l\qp_port.h`)

```
    #ifndef qs_port_h
    #define qs_port_h

(1) #define QS_OBJ_PTR_SIZE     4
(2) #define QS_FUN_PTR_SIZE     4
(3) #define QS_TIME_SIZE        4

(4) #include "qf_port.h"                              /* use QS with QF */
(5) #include "qs.h"         /* QS platform-independent public interface */

    #endif                                           /* qs_port_h */
```

---

(1)   The macro QS_OBJ_PTR_SIZE specifies the size (in bytes) of an object pointer on the particular platform.

(2) The macro `QS_FUN_PTR_SIZE` specifies the size (in bytes) of a function pointer on the particular platform.

(3) The macro `QS_TIME_SIZE` configures the size (in bytes) of the QS time stamp `QSTimeCtr` (see also Listing 11.3(43)).

(4) The QF port header file `qf_port.h` is included if QS is used together with the QF real-time framework.

> **NOTE**
>
> When QS is combined with QF, the QS critical section is the same as it is defined in the `qf_port.h` header file. However, QS can also be used with just the QEP component, or even completely standalone. In these cases, QS must provide its own, independent critical section mechanism by defining the macros `QS_INT_KEY_TYPE`, `QS_INT_LOCK()`, and `QS_INT_UNLOCK()` (see Section 11.3.3).

(5) The platform-independent `qs.h` header file must always be included in the `qf_port.h` header file.

11.4 The QSPY Host Application

As described in Section 11.2, the host-resident component for the Quantum Spy software-tracing system is the QSPY host application. QSPY is a simple console application without any fancy GUI because its purpose is to provide only the QS data parsing, storing, and exporting to such powerful tools as MATLAB. QSPY has been designed from the ground up to be platform-neutral. The application is written in portable C++ and ports to Linux and Windows with various compilers are provided.

QSPY is easily adaptable to various target-host communication links. Out of the box, the QSPY host application supports serial (RS232), TCP/IP, and file communication links. Adding other communication links is easy because the data link is accessed only through a generic hardware abstraction layer (HAL).

The QSPY application accepts several command-line parameters to configure the data link and all target dependencies, such as pointer sizes, signal sizes, and the like. This means that the single QSPY host application can process data from any embedded target. The application has been tested with a wide range of 8-, 16-, or 32-bit CPUs.

QSPY provides a simple consolidated, human-readable textual output to the screen. If the QS trace data contains dictionary trace records (see Section 11.3.8), QSPY applies this symbolic information to output the provided identifiers for objects, signals, and states. Otherwise, QSPY outputs the hexadecimal values of various pointers and signals.

Finally, QSPY can export the trace data in the matrix format readable by MATLAB. A special MATLAB script to import QSPY trace data to MATLAB is provided. Once the data is available in MATLAB matrices, it can be conveniently manipulated and visualized with this powerful tool.

QSPY comes with a Reference Manual in electronic format (see Section 11.6.6), which contains detailed explanations of all command-line options, the human-readable format, and the MATLAB interface.

11.4.1   Installing QSPY

The QSPY host application is included in the code accompanying this book in the directory `<qp>\qpc\tools\qspy\` (for QP/C) and also in `<qp>\qpcpp\tools\qspy\` (for QP/C++). The two versions are actually identical except that QSPY for QP/C includes the C-version of the `qs.h` header file, whereas the QP/C++ version includes the C++ version of `qs.h`. Listing 11.12 shows the contents of the QP Root Directory after the installation of QS component.

> **NOTE**
>
> The QSPY host application includes the header file `<qp>\qpc\include\qs.h`. The `qs.h` header file provides the link between the QS target-resident component and the QSPY host-resident component.

Listing 11.12 Source code organization for the QSPY host application

```
<qp>\qpc\              - QP/C root directory (<qp>\qpcpp for QP/C++)
  |
  +-doxygen\           - QP/C documentation generated with Doxygen
  | +-html\            - "QP/C Reference Manual" in HTML format
  | | +-index.html     - The starting HTML page for the "QP/C Reference Manual"
  | | |                   (contains the "QSPY Reference Manual")
  | | +- . . .
  | +-qpc.chm          - "QP/C Reference Manual" in CHM Help format
  |                      (contains the "QSPY Reference Manual")
```

```
|
+-include/           - QP platform-independent header files
| +-qs.h             - QS platform-independent header file (used by QSPY)
|
+-tools\             - Tools directory
| +-qspy\            - QSPY host application
| | +-include\       - platform-independent include
| | | +-dict.h       - dictionary class header file
| | | +-getopt.h     - command-line option parser
| | | +-hal.h        - Hardware Abstraction Layer header file
| | | +-qspy.h       - QSPY parser header file
| | +-source\        - platform-independent sources (C++)
| | | +-dict.cpp     - dictionary class implementation
| | | +-getopt.c     - command-line option parser
| | | +-main.cpp     - main() entry point
| | | +-qspy.cpp     - QSpy parser
| | |
| | +-linux\         - Linux version of QSPY
| | | +-gnu\         - GNU compiler
| | | | +-dbg\       - debug build directory
| | | | +-rel\       - release build directory
| | | | +-com.cpp    - serial port HAL for Linux
| | | | +-tcp.cpp    - TCP/IP port HAL for Linux
| | | | +-Makefile   - make file to build QSPY for Linux
| | |
| | +-win32\         - Win32 (Windows) version of QSPY
| | | +-mingw\       - MinGW compiler (GNU)
| | | | +-dbg\       - debug build directory
| | | | +-rel\       - release build directory
| | | | | +-qspy.exe - QSPY executable
| | | | +-com.cpp    - serial port HAL for Win32
| | | | +-tcp.cpp    - TCP/IP port HAL for Win32
| | | | +-make.bat   - Simple batch script to build QSPY
| | |
| | | +-vc2005\      - Visual C++ 2005 toolset
| | | | +-Debug\     - debug build directory
| | | | +-Release\   - release build directory
| | | | | +-qspy.exe - QSPY executable
| | | | +-com.cpp    - serial port HAL for Win32
| | | | +-tcp.cpp    - TCP/IP port HAL for Win32
| | | | +-qspy.sln   - Visual C++ Solution to build QSPY for Win32
| | |
| | +-matlab\        - MATLAB scripts
| | | +-qspy.m       - MATLAB script to import the QS data into MATLAB
| | | +-dpp.spy      - Example of a QS binary file from DPP application
| | | +-philo_timing.m - example MATLAB script to generate timing diagrams
| | |                       for the DPP example
```

11.4.2   Building QSPY Application from Sources

The QSPY source code is written in portable C++, with ports to Windows and Linux already provided (see Listing 11.12). Note that the QSPY host application is coupled with the QS target component through the header file `<qp>\qpc\include\qs.h`, which enumerates the predefined QS records.

Building QSPY for Windows with Visual C++ 2005

The Win32 executable of the QSPY application is provided in the file `<qp>\qpc\tools\qspy\win32\vc2005\Release\qspy.exe`. This executable should run on any version of 32-bit Windows.

If you want to rebuild the application, the directory `<qp>\qpc\tools\qspy\win32\vc2005\` contains the Microsoft Visual C++ 2005 solution file `qspy.sln` to build the QSPY application. You simply load the solution file to the Visual C++ 2005 IDE and start the build by pressing F7.

Building QSPY for Windows with MinGW

Alternatively, you can use the open source MinGW (Minimalist GNU for Windows) toolset available from `www.mingw.org` to build the QSPY executable. The directory `<qp>\qpc\tools\qspy\win32\mingw\` contains a simple batch file `make.bat` to build the QSPY application. You probably need to modify the definition of the `MINGW` symbol at the top of the batch file to point it to the location where you installed the MinGW toolset. By default, `make.bat` produces the debug version of the application in the directory `<qp>\qpc\tools\qspy\win32\mingw\dbg\`. To produce the release version, add the 'rel' parameter to the `make.bat` script (`make rel`). The release version is produced in the release directory: `<qp>\qpc\tools\qspy\win32\mingw\rel\`.

Building QSPY for Linux

The directory `<qp>\qpc\tools\qspy\linux\gnu\` contains the `Makefile` for building QSPY for Linux. By default, the `Makefile` produces the debug version of the application in the directory `<qp>\qpc\tools\qspy\linux\gnu\dbg\`. To produce the release version, add the 'rel' target to the make (`make rel`). The release version is produced in the release directory: `<qp>\qpc\tools\qspy\linux\gnu\rel\`.

11.4.3   Invoking QSPY

The QSPY host application is designed to work with all possible target CPUs and data links, which requires a wide range of configurability. For example, for any given target CPU, the QSPY application must "know" the size of object pointers, function pointers, event signals, timestamp size, and so on. You provide this information to QSPY by means of command-line parameters, which are summarized in Table 11.3 and also in the "QSPY Reference Manual" (see Section 11.6.6). Note that the options are *case sensitive*.

Table 11.3: Summary of QSpy command-line options

Option	Example	Default	Must Match QP Macro (QP Port Header File)	Comments
-h	-h			Help; prints the summary of options
-q	-q			Quiet mode (no stdout output)
-o	-o qs.txt			Produces output to the specified file
-s	-s qs.spy			Saves the binary input to the specified file; not compatible with -f
-m	-m qs.mat			Generates MATLAB output to the specified file
-c	-c COM2	COM1		COM port selection; not compatible with −t, -p, -f
-b	-b 115200	38400		Baud rate selection; not compatible with −t, -p, -f
-t	-t			TCP/IP input selection; not compatible with −c, -b, -f
-p	-p 6602	6601		TCP/IP server port number; not compatible with −c, -b, -f

Continued onto next page

Table 11.3: Summary of QSpy command-line options—Cont'd

Option	Example	Default	Must Match QP Macro (QP Port Header File)	Comments
-f	-f qs.spy			File input selection; not compatible with –c, -b, -t, -p
-T	-T 2	4	QS_TIME_SIZE (qs_port.h)	Time stamp size in bytes; valid values: 1, 2, 4
-O	-O 2	4	QS_OBJ_PTR_SIZE (qs_port.h)	Object pointer size in bytes; valid values: 1, 2, 4
-F	-F 2	4	QS_FUN_PTR_SIZE (qs_port.h)	Function pointer size in bytes; valid values: 1, 2, 4
-S	-S 2	1	Q_SIGNAL_SIZE (qep_port.h)	Signal size in bytes; valid values: 1, 2, 4
-E	-E 1	2	QF_EVENT_SIZ_SIZE (qf_port.h)	Event-size size in bytes (i.e., the size of variables that hold event size); valid values: 1, 2, 4
-Q	-Q 1	2	QF_EQUEUE_CTR_SIZE (qf_port.h)	Queue counter size in bytes; valid values 1, 2, 4
-P	-P 4	2	QF_MPOOL_CTR_SIZE (qf_port.h)	Pool counter size in bytes; valid values: 1, 2, 4
-B	-B 1	2	QF_MPOOL_SIZ_SIZE (qf_port.h)	Block size size in bytes (i.e., the size of variables that hold memory block size); valid values 1, 2, 4
-C	-C 4	2	QF_TIMEEVT_CTR_SIZE (qf_port.h)	Time event counter size; valid values: 1, 2, 4

Your main concern when invoking QSPY is to *match* exactly the target system you are using. The fourth column of Table 11.3 lists the configuration macros used by the target system as well as the platform-specific QP header files where those macros are defined. You need to use the corresponding QSPY command-line option only when the QP macro differs from the default. The default values assumed by QSPY are consistent with the defaults used in QP.

NOTE

When you do not match the QSPY host application with the QS target component, the QSPY application will be unable to correctly parse the mismatched trace records and will start generating the following errors:

```
********** 028: Error xx bytes unparsed
********** 014: Error -yy bytes unparsed
```

The number in front of the error indicates the Record ID of the trace record that could not be parsed.

11.5   Exporting Trace Data to MATLAB

The QSPY host application can also export trace data to MATLAB, which is a popular numerical computing environment and a high-level technical programming language. Created by The MathWorks, Inc., MATLAB allows easy manipulation and plotting of data represented as *matrices*.

Figure 11.6 summarizes the interface between the QSPY host application and MATLAB. The interface consists of the QSPY MATLAB output file, the qspy.m MATLAB script, and MATLAB matrices generated by the script in the current MATLAB workspace. The following sections explain these elements.

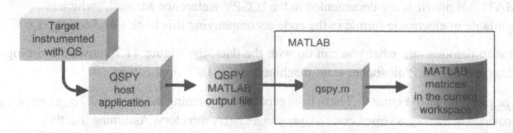

Figure 11.6: Exporting trace data to MATLAB.

11.5.1   Analyzing Trace Data with MATLAB

When you invoke QSPY with the −m <file name> option, the QSPY application generates a MATLAB-readable file of the specified name in addition to the human-readable format discussed in the previous section.

The MATLAB output file is an ASCII file that contains all the trace records formatted for MATLAB. However, the various trace records in the MATLAB file are still in the same order as they were produced in the target and don't yet form proper MATLAB matrices, which are the most natural way of representing data within MATLAB. You can find an example of a QSPY MATLAB output in the file `<qp>\qpc\ examples\80x86\qk\tcpp101\l\dpp\dpp.mat`.

The directory `<qp>\qpc\tools\qspy\matlab\` contains the MATLAB script `qspy.m`, which reads in the QSPY MATLAB file and converts the data into several MATLAB matrices in the current workspace. Assuming that the directory `<qp>\qpc\ tools\qspy\matlab\` is included in the MATLAB path, you invoke the script from the MATLAB command window as follows:

```
Q_FILE='<qp>\qpc\examples\80x86\qk\tcpp101\l\dpp\dpp.mat'; qspy
```

The variable `Q_FILE` is set to the file name of the QSPY MATLAB file. Note that the `qspy.m` script is intentionally not a MATLAB function because its main purpose is to fill the current workspace with matrices that remain after the script is done, which is not possible with a function that runs in a separate temporary workspace.

At this point you have all the data conveniently represented in MATLAB matrices. After filling in the matrices, the `qspy.m` script executes the 'whos' command to show the created objects. The matrices with the prefix `Q_` contain the time-ordered trace data. All MATLAB matrices are documented in the "QSPY Reference Manual," which is available in electronic format in the code accompanying this book (see Section 11.6.6).

Just to demonstrate what you can do with the data, the Figure 11.7 shows the timing diagrams for all Philosopher state machines in the DPP application.

The plots shown in Figure 11.7 have been generated by running the script `philo_timing.m` provided in the `<qp>\qpc\tools\qspy\matlab\` directory. Assuming that this directory is in the MATLAB path, you simply type the script's name at the MATLAB prompt:

```
» philo_timing
```

The `philo_timing.m` script displays the data from the MATLAB matrix `Q_STATE`, which you generated by running the `qspy.m` script. The `Q_STATE` matrix contains all the state machine information. Section 11.5.4 explains how this plot has been generated.

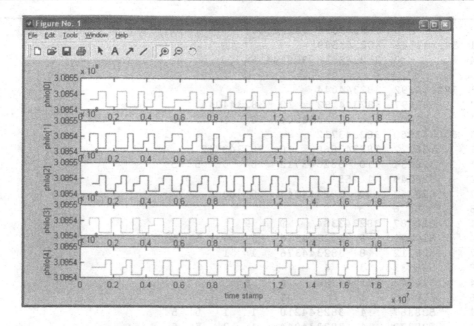

Figure 11.7: MATLAB plot showing timing diagrams of the five Philosophers generated from the QS trace data. The vertical axis represents states "thinking" (lowest), "hungry" (middle), and "eating" (top).

11.5.2 MATLAB Output File

The QSPY MATLAB file is in ASCII format and Listing 11.13 shows a snippet of the QSPY MATLAB file generated from the DPP application (see also Section 11.2.1).

Listing 11.13 Fragment of the QSPY MATLAB file for the DPP application

```
. . . . . .
62 Philo_initial= 308543675;
62 Philo_thinking= 308544589;
62 Philo_hungry= 308544835;
62 Philo_eating= 308545073;
60 HUNGRY_SIG=[   8  382343546];
60 TIMEOUT_SIG=[ 10  382343546];
12       0   4  382343546
 3  382343546  337510406  308544589
50       0  64   64
51       0  64   64
```

Continued onto next page

```
 32          0   382343694   382343546                9            0
  1   382343546   308544589
  4          0   382343546   308544589
. . . . . .
 33   382344192   382344044
 37      528197   382344192     10   382344044
 14      528208     10   382344044     0     0     5     5
 33   382343860   382343712
 37      528222   382343860     10   382343712
 14      528234     10   382343712     0     0     5     5
 42      528249   1 255
 17      528262     10   382344044     0     0
 28      528276      3      8
 24      528287   382343384      9     9
 14      528299      8   382344376     1     0     5     5
 17      528312      8   382344376     1     1
 28      528333      3      4
 24      528343   382343384      8     8
 26      528355      4             1
 14      528367      4   382344210     1     1     5     5
 14      528379      4   382344044     1     2     5     5
 14      528391      4   382343878     1     3     5     5
 14      528403      4   382343712     1     4     5     5
 14      528415      4   382343546     1     5     5     5
. . . . . .
```

The QSPY MATLAB file is stored in portable ASCII format for cross-platform portability, but is really not intended to be human-readable. The purpose of Listing 11.13 is simply to demonstrate that the data is mostly numerical, with the only exception of the "dictionary" entries, which actually are stored as MATLAB commands.

The MATLAB output file shown in Listing 11.13 contains all the trace records formatted for MATLAB. However, the various records at this stage are still in the same order as they were produced in the target and don't yet form proper MATLAB matrices, which are the most natural way of representing data within MATLAB.

11.5.3   MATLAB Script `qspy.m`

The MATLAB script `qspy.m`, located in the directory `<qp>\qpc\tools\qspy\ matlab\`, is designed to read the QSPY MATLAB file and sort the different records into various MATLAB matrices for subsequent analysis. In Section 11.5.1, I described how to invoke the script from MATLAB. Here I discuss the `qspy.m` script itself, which is shown in Listing 11.14.

Listing 11.14 Script `qspy.m` **for Importing QSPY trace data to MATLAB**

```
(1)   % the string Q_FILE must be defined
(2)   fid = fopen(Q_FILE, 'r');
      if fid == -1
(3)       error('file not found')
      end

(4)   Q_STATE     = []; % sate entry/exit, init, tran, internal tran, ignored
      Q_EQUEUE    = []; % QEQueue
      Q_MPOOL     = []; % QMPool
      Q_NEW       = []; % new/gc
      Q_ACTIVE    = []; % active add/remove, subscribe/unsubscribe
      Q_PUB       = []; % publish/publish attempt
      Q_TIME      = []; % time event arm/disarm/rearm, clock tick
      Q_INT_LOCK  = []; % interrupt locking/unlocking
      Q_ISR_LOCK  = []; % ISR entry/exit
      Q_MUTEX     = []; % QK mutex locking/unlocking
      Q_SCHED     = []; % QK scheduler events

      Q_TOT       = 0;  % total number of records processed

(5)   while feof(fid) == 0
          line = fgetl(fid);
          Q_TOT = Q_TOT+1;

(6)       rec = sscanf(line, '%d', 1);    % extract the record type
(7)       switch rec     % discriminate based on the record type

              % QEP trace records
(8)           case 1    %  QS_QEP_STATE_ENTRY
                  Q_STATE(size(Q_STATE,1)+1,:) = ...
                      [NaN 1 sscanf(line, '%*u %u %u')' NaN 1];

              case 2    %  QS_QEP_STATE_EXIT
                  Q_STATE(size(Q_STATE,1)+1,:) = ...
                      [NaN 2 sscanf(line, '%*u %u %u')' NaN 1];

              case 3    %  QS_QEP_STATE_INIT
                  Q_STATE(size(Q_STATE,1)+1,:) = ...
                      [NaN 3 sscanf(line, '%*u %u %u %u')' 1];

              case 4    %  QS_QEP_INIT_TRAN
                  tmp = sscanf(line, '%*u %u %u %u')';
                  Q_STATE(size(Q_STATE,1)+1,:) = ...
```

Continued onto next page

```
                    [tmp(1) 3 tmp(2) NaN tmp(3) 1];
            case 5    %   QS_QEP_INTERN_TRAN
                Q_STATE(size(Q_STATE,1)+1,:) = ...
                    [sscanf(line, '%*u %u %u %u %u')' NaN 1];

            case 6    %   QS_QEP_TRAN
                Q_STATE(size(Q_STATE,1)+1,:) = ...
                    [sscanf(line, '%*u %u %u %u %u')' 1];

            case 7    %   QS_QEP_IGNORED
                Q_STATE(size(Q_STATE,1)+1,:) = ...
                    [sscanf(line, '%*u %u %u %u %u')' NaN 0];

            % QF trace records
            case 10    %   QS_QF_ACTIVE_ADD
                tmp = sscanf(line,'%*u %u %u %u %u')';
                Q_ACTIVE(size(Q_ACTIVE,1)+1,:) = [tmp(1) NaN tmp(2) tmp(3) 1];
            . . . .

            % Miscallaneous QS records
            case 60    %   QS_SIG_DICTIONARY
                eval(line(5:end));

            case 61    %   QS_OBJ_DICTIONARY
                eval(line(5:end));

            case 62    %   QS_FUN_DICTIONARY
                eval(line(5:end));


            . . .
            % User records
(9)         % . . .
        end
    end

    % cleanup ...
(10) fclose(fid);
    clear fid;
    clear line;
    clear rec;
    clear tmp;

    % display status information...
(11) Q_TOT
(12) whos
```

(1)   This is a simple MATLAB script, not a MATLAB function, because the main purpose of qspy.m is to create new data that *remain* in the workspace after the script finishes so that you can use them for further computations. (A MATLAB function runs in a separate workspace, which disappears after the function returns.)

(2)   The variable Q_FILE contains the name of the QSPY MATLAB file and must be defined before starting the qspy.m script.

(3)   The specified MATLAB file must exist.

(4)   The QSPY MATLAB matrices are cleared to be filled with trace data.

(5)   The entire MATLAB file is processed one line at a time.

(6)   The first number in each line is the QS record type (see Listing 11.3(2)).

(7)   The switch statement discriminates based on the record type.

(8)   Different case statements read specific record types and place the data into MATLAB matrices described in the "QSPY Reference Manual."

(9)   Here you can extend the script to include user-specific (your) trace records.

(10)   In this section the temporary data is cleaned up (this is a simple script, not a function).

(11,12)   The total number of records and the contents of the current workspace are displayed.

11.5.4   MATLAB Matrices Generated by qspy.m

The qspy.m script generates 11 MATLAB matrices (see Listing 11.14). Each of these matrices contains different group of related QS trace records. The "QSPY Reference Manual," available in electronic format in the code accompanying this book (see Section 11.6.6), contains documentation for all MATLAB matrices generated by the qspy.m script. Here I give you only one example, the matrix Q_STATE, which stores all QS records pertaining to state machine activities in the system. Table 11.4 summarizes how the QS records are stored in the Q_STATE matrix.

By MATLAB convention, the different variables such as timestamp, event signal, and the like are put into columns, allowing observations to vary down through the rows.

Table 11.4: Q_STATE **matrix (N-by-6) produced by the** qspy.m **script**

MATLAB Index	1	2	3	4	5	6
QS Record	Timestamp	Signal	State Machine Object	Source State	New State	Event Handled
QS_QEP_STATE_ENTRY	NaN	1	$\sqrt{}^2$	$\sqrt{}$	NaN	1
QS_QEP_STATE_EXIT	NaN	2	$\sqrt{}^2$	$\sqrt{}$	NaN	1
QS_QEP_STATE_INIT	NaN	3	$\sqrt{}^2$	$\sqrt{}$	$\sqrt{}$	1
QS_QEP_STATE_INIT_TRAN	$\sqrt{}$	3	$\sqrt{}^2$	NaN	$\sqrt{}$	1
QS_QEP_STATE_INTERN_TRAN	$\sqrt{}$	$\sqrt{}^1$	$\sqrt{}^2$	$\sqrt{}$	NaN	1
QS_QEP_STATE_TRAN	$\sqrt{}$	$\sqrt{}^1$	$\sqrt{}^2$	$\sqrt{}$	$\sqrt{}$	1
QS_QEP_STATE_IGNORED	$\sqrt{}$	$\sqrt{}^1$	$\sqrt{}^2$	$\sqrt{}$	NaN	0

[1] The valid user signal is > 3.
[2] Per inheritance, an active object is a state machine object as well.

Therefore, a data set consisting of N time samples of six variables is stored in a matrix of size N-by-6. The checkmark '$\sqrt{}$' in a given cell of the matrix represents data available from the QSPY file. Other values represent data added by the qspy.m script to fill all the matrix cells and to allow unambiguous identification of the trace records. For example, the following six index matrices unambiguously select the QS trace record from the matrix:

QS Record	MATLAB Index Matrix
QS_QEP_STATE_ENTRY	Q_STATE(:,2) == 1
QS_QEP_STATE_EXIT	Q_STATE(:,2) == 2
QS_QEP_STATE_INIT	Q_STATE(:,2) == 3
QS_QEP_STATE_INIT_TRAN	isnan(Q_STATE(:,4))
QS_QEP_STATE_INTERN_TRAN	Q_STATE(:,2) > 3 & isnan(Q_STATE(:,5))
QS_QEP_STATE_TRAN	Q_STATE(:,2) > 3 & ~isnan(Q_STATE(:,5))
QS_QEP_STATE_IGNORED	~Q_STATE(:,6)

As an example of using the information contained in the matrix Q_STATE, consider the timing diagrams for the Philosopher active objects shown in Figure 11.7. These timing diagrams have been generated with the script `philo_timing.m` shown in Listing 11.15.

Listing 11.15 MATLAB script `philo_timing.m` that generates timing diagrams shown in Figure 11.7

```
(1)  t=Q_STATE(:,2)>3 & ~isnan(Q_STATE(:,5)); % QS_QEP_STATE_TRAN
(2)  o=Q_STATE(:,3) == l_philo_0_;
(3)  subplot(5,1,1); stairs(Q_STATE(o & t,1),Q_STATE(o & t,5),'r')
(4)  o=Q_STATE(:,3) == l_philo_1_;
     subplot(5,1,2); stairs(Q_STATE(o & t,1),Q_STATE(o & t,5),'r')
     o=Q_STATE(:,3) == l_philo_2_;
     subplot(5,1,3); stairs(Q_STATE(o & t,1),Q_STATE(o & t,5),'r')
     o=Q_STATE(:,3) == l_philo_3_;
     subplot(5,1,4); stairs(Q_STATE(o & t,1),Q_STATE(o & t,5),'r')
     o=Q_STATE(:,3) == l_philo_4_;
     subplot(5,1,5); stairs(Q_STATE(o & t,1),Q_STATE(o & t,5),'r')
     xlabel('time stamp'); zoom on
```

---

(1) The index matrix 't' selects only the rows of the Q_STATE matrix that correspond to state transitions (the QS_QEP_STATE_TRAN trace record). The conditions used for a transition are: signal >3 (user-defined signal) and new state is available (not a NaN).

(2) The index matrix 'o' (object) selects only the Philosopher 0 state machine object. This line of code makes use of the dictionary entry l_pholo_0_.

(3) The timing diagram for Philosopher 0 is drawn using the index matrices 't' and 'o.'

(4) The index matrix 'o' is created for Philosopher 1, 2, 3, and 4 state machine objects and timing diagrams for these objects are drawn.

Obviously, this short demonstration barely scratches the surface of the possibilities. Refer to the "QSPY Reference Manual" for the description of other MATLAB matrices. The rest is MATLAB.

11.6   Adding QS Software Tracing to a QP Application

In this section I show you how to add QS software tracing to the DPP application, which I already used as an example of a software-tracing session at the beginning of this chapter (Section 11.2.1).

The DPP example for 80x86 with the QK preemptive kernel demonstrates the use of QS with all QP components: QEP, QF, and QK. The example is located in the directory <qp>\qpc\examples\80x86\qk\tcpp101\l\dpp\. You can rebuild the "Spy" configuration by loading the DPP-SPY.PRJ project into the Turbo C++ IDE. The application links to the "Spy" versions of the QP libraries located in the directory <qp>\qpc\ports\80x86\qk\tcpp101\l\spy\.

The DPP example demonstrates all aspects of QS setup. In particular it demonstrates how to send the QS trace data over a serial port (UART) and how to timestamp QS records with submicrosecond precision using the 8254 timer/counter found in every x86-based PC.

11.6.1   Initializing QS and Setting Up the Filters

Listing 11.16 Initialization of QS, setting up the filters, and generating dictionary entries (file <qp>\qpc\examples\80x86\qk\tcpp101\l\dpp\main.c**)**

```
(1)  #include "qp_port.h"
     #include "dpp.h"
     #include "bsp.h"

     /* Local-scope objects --------------------------------------------*/
     static QEvent const *l_tableQueueSto[N_PHILO];
     static QEvent const *l_philoQueueSto[N_PHILO][N_PHILO];
     static QSubscrList   l_subscrSto[MAX_PUB_SIG];

     static union SmallEvent {
         void *min_size;
         TableEvt te;
         /* other event types to go into this pool */
     } l_smlPoolSto[2*N_PHILO];              /* storage for the small event pool */

     /*................................................................*/
     int main(int argc, char *argv[]) {
         uint8_t n;
```

```
        Philo_ctor();                      /* instantiate all Philosopher active objects */
        Table_ctor();                           /* instantiate the Table active object */

(2)     BSP_init(argc, argv);                   /* initialize the BSP (including QS) */

        QF_init();        /* initialize the framework and the underlying RT kernel */

                                                    /* setup the QS filters ... */
(3)     QS_FILTER_ON (QS_ALL_RECORDS);
(4)     QS_FILTER_OFF(QS_QF_INT_LOCK);
(5)     QS_FILTER_OFF(QS_QF_INT_UNLOCK);
(6)     QS_FILTER_OFF(QS_QK_SCHEDULE);
                                            /* provide object dictionaries... */
(7)     QS_OBJ_DICTIONARY(l_smlPoolSto);
(8)     QS_OBJ_DICTIONARY(l_tableQueueSto);
(9)     QS_OBJ_DICTIONARY(l_philoQueueSto[0]);
        QS_OBJ_DICTIONARY(l_philoQueueSto[1]);
        QS_OBJ_DICTIONARY(l_philoQueueSto[2]);
        QS_OBJ_DICTIONARY(l_philoQueueSto[3]);
        QS_OBJ_DICTIONARY(l_philoQueueSto[4]);

        QF_psInit(l_subscrSto, Q_DIM(l_subscrSto));  /* init publish-subscribe */

                                            /* initialize event pools... */
        QF_poolInit(l_smlPoolSto, sizeof(l_smlPoolSto), sizeof(l_smlPoolSto[0]));

        for (n = 0; n < N_PHILO; ++n) {          /* start the active objects... */
            QActive_start(AO_Philo[n], (uint8_t)(n + 1),
                    l_philoQueueSto[n], Q_DIM(l_philoQueueSto[n]),
                    (void *)0, 0, (QEvent *)0);
        }
        QActive_start(AO_Table, (uint8_t)(N_PHILO + 1),
                    l_tableQueueSto, Q_DIM(l_tableQueueSto),
                    (void *)0, 0, (QEvent *)0);
        QF_run();                                      /* run the QF application */

        return 0;
    }
```

(1)   When the QS tracing is enabled (i.e., the macro Q_SPY is defined), the header file qp_port.h includes the QS active interface qs.h (see Listing 11.3).

(2)   The BSP initialization also initializes QS (see Listing 11.17).

(3)   Right after initialization, all QS global filters are disabled. Here I enable all global filters.

(4-6) I disable the high-volume trace records to avoid overrunning the QS trace buffer (see Section 11.3.7).

(7-9)   I provide object dictionary entries (see Section 11.3.8) for all local-scope objects defined in this module. Note that I need to do it here because these objects are only known in this translation unit.

11.6.2   Defining Platform-Specific QS Callbacks

Listing 11.17 shows the platform-specific QS callback functions for QK/DOS. The QS tracing uses one of the standard 16550 UARTs (COM1-COM4) for data output at 115200 baud rate. I defer the discussion of the QS timestamp (the QS_onGetTime() callback) to the next section.

Listing 11.17 QS callbacks in the board support package (file `<qp>\qpc\examples\80x86\qk\tcpp101\1\dpp\bsp.c`**)**

```
        #include "qp_port.h"
        #include "dpp.h"
        #include "bsp.h"
        . . .

        /* Local-scope objects ---------------------------------------------*/
        #ifdef Q_SPY
            static uint16_t l_uart_base;        /* QS data uplink UART base address */
            . . .
            #define UART_16550_TXFIFO_DEPTH 16
        #endif

        . . .
        /*.................................................................*/
        void BSP_init(int argc, char *argv[]) {
            char const *com = "COM1";
            uint8_t n;

            if (argc > 1) {
                l_delay = atol(argv[1]);     /* set the delay counter for busy delay */
            }
            if (argc > 2) {
                com = argv[2];
                (void)com;                /* avoid compiler warning if Q_SPY not defined */
            }
(1)         if (!QS_INIT(com)) {                          /* initialize QS */
(2)             Q_ERROR();
            }
            . . .
        }
        /*.................................................................*/
```

```
(3)   void QK_onIdle(void) {
      #ifdef Q_SPY
          if ((inportb(l_uart_base + 5) & (1 << 5)) != 0) {          /* Tx FIFO empty? */
              uint16_t fifo = UART_16550_TXFIFO_DEPTH;        /* 16550 Tx FIFO depth */
              uint8_t const *block;
              QF_INT_LOCK(dummy);
              block = QS_getBlock(&fifo);        /* try to get next block to transmit */
              QF_INT_UNLOCK(dummy);
              while (fifo-- != 0) {                        /* any bytes in the block? */
                  outportb(l_uart_base + 0, *block++);
              }
          }
      #endif
      }
      . . .
      /*-------------------------------------------------------------------------*/
      #ifdef Q_SPY                                          /* define QS callbacks */
      /*.........................................................................*/
(4)   static uint8_t UART_config(char const *comName, uint32_t baud) {
          switch (comName[3]) {              /* Set the base address of the COMx port */
              case '1': l_uart_base = (uint16_t)0x03F8; break;            /* COM1 */
              case '2': l_uart_base = (uint16_t)0x02F8; break;            /* COM2 */
              case '3': l_uart_base = (uint16_t)0x03E8; break;            /* COM3 */
              case '4': l_uart_base = (uint16_t)0x02E8; break;            /* COM4 */
              default: return (uint8_t)0;        /* COM port out of range failure */
          }
          baud = (uint16_t)(115200UL / baud);              /* divisor for baud rate */
          outportb(l_uart_base + 3, (1 << 7));   /* Set divisor access bit (DLAB) */
          outportb(l_uart_base + 0, (uint8_t)baud);              /* Load divisor */
          outportb(l_uart_base + 1, (uint8_t)(baud >> 8));
          outportb(l_uart_base + 3, (1 << 1) | (1 << 0)); /* LCR:8-bits,no p,1stop */
          outportb(l_uart_base + 4, (1 << 3) | (1 << 1) | (1 << 0)); /*DTR,RTS,Out2*/
          outportb(l_uart_base + 1, 0);     /* Put UART into the polling FIFO mode */
          outportb(l_uart_base + 2, (1 << 2) | (1 << 0));  /* FCR: enable, TX clear */

          return (uint8_t)1;                                         /* success */
      }
      /*.........................................................................*/
(5)   uint8_t QS_onStartup(void const *arg) {
          static uint8_t qsBuf[2*1024];                    /* buffer for Quantum Spy */
(6)       QS_initBuf(qsBuf, sizeof(qsBuf));
          return UART_config((char const *)arg, 115200UL);
      }
      /*.........................................................................*/
(7)   void QS_onCleanup(void) {
      }
      /*.........................................................................*/
(8)   void QS_onFlush(void) {
          uint16_t fifo = UART_16550_TXFIFO_DEPTH;            /* 16550 Tx FIFO depth */
```

Continued onto next page

```
          uint8_t const *block;
          QF_INT_LOCK(dummy);
          while ((block = QS_getBlock(&fifo)) != (uint8_t *)0) {
              QF_INT_UNLOCK(dummy);
                                                      /* busy-wait until TX FIFO empty */
 (9)          while ((inportb(l_uart_base + 5) & (1 << 5)) == 0) {
              }

              while (fifo- != 0) {                          /* any bytes in the block? */
                  outportb(l_uart_base + 0, *block++);
              }
              fifo = UART_16550_TXFIFO_DEPTH;       /* re-load 16550 Tx FIFO depth */
              QF_INT_LOCK(dummy);
          }
          QF_INT_UNLOCK(dummy);
      }
      /*.............................................................*/
(10)  QSTimeCtr QS_onGetTime(void) {                          /* see Listing 11.18 */
          . . .
      }
      #endif                                                  /* Q_SPY */
      /*-------------------------------------------------------------*/
```

(1) The macro QS_INIT() initializes the QS component. The macro returns FALSE
     when the initialization fails (e.g., the specified COM port cannot be opened).

(2) Failing to initialize QS causes an error. Note that the assertion only fires when
     QS is active because the dummy version of the QS_INIT() macro always
     returns TRUE (see Listing 11.4(2)).

(3) The QK preemptive kernel calls the QK_onIdle() callback function from the
     idle loop (see Section 10.3.5 in Chapter 10). This callback is an ideal place to
     perform output of the trace data in the least intrusive way. The QS output in
     the QK_onIdle() function uses the QS_getBlock() API, which I already
     discussed in Listing 11.9 in Section 11.3.7.

(4) The function UART_config() configures one of the standard UARTs of the
     80x86-based PC (COM1-COM4).

(5) The callback function QS_onStartup() initializes the QS component.

(6) The function QS_onStartup() must always initialize the QS trace buffer by
     calling QS_initBuf() (see Section 11.3.7).

(7) The callback QS_onCleanup() function performs cleanup of QS. This function
     has nothing to clean up in this case.

(8)   The callback function QS_onFlush() flushes the entire trace buffer to the host. This function is called after each dictionary trace record to avoid overrunning the trace buffer during the system initialization.

(9)   The function QS_onFlush() busy-waits until the data is sent out. Note that this policy might be only appropriate during the initial transient.

(10)   The callback function QS_onGetTime() provides the timestamp for the QS trace records. I discuss this function in the next section.

11.6.3   Generating QS Timestamps with the QS_onGetTime() Callback

Most QS trace records are timestamped, which ties all the trace records to the common timeline. To be truly useful, the QS timestamps should have microsecond-level resolution or better, which is only possible with a dedicated clock or a timer device. The callback function QS_onGetTime() encapsulates the particular method of obtaining the timestamp. QS always calls this callback function inside a critical section of code.

In case of a standard PC, the time can be obtained from the 8284 timer/counter. The counter-0 of the 8254 chip is a 16-bit down-counter that is set up to generate the standard 18.2Hz clock-tick interrupt when it underflows from 0 to 0xFFFF. The counting rate is 1.193182MHz, which works out to approximately 0.838 microseconds per count.

The basic idea of using this 16-bit down-counter for 32-bit timestamping of the QS records is shown in Figure 11.8. The system clock-tick interrupt maintains the 32-bit-wide timestamp at tick l_tickTime. The clock-tick interrupt increments l_tickTime by 0x10000 counts to account for the 16-bit rollover. The complete 32-bit timestamp is constructed by adding l_tickTime and the on-chip count, or actually (0x10000 – the on-chip count), because the 8284 timer/counter counts down.

The method just described provides fine-granularity timestamp most of the time but occasionally can be off by the full period of the 16-bit counter (0x10000 counts). This can happen because the once-per-period "sampling" rate of the system tick interrupt is not sufficient to completely resolve the timer-cycle ambiguity. This "undersampling" allows a small time window in which the 16-bit clock rolls over but before the system clock interrupt increments l_tickTime by 0x10000 counts. This can happen due to interrupt locking. As a remedy, the QS_onGetTime() contains a protection against the

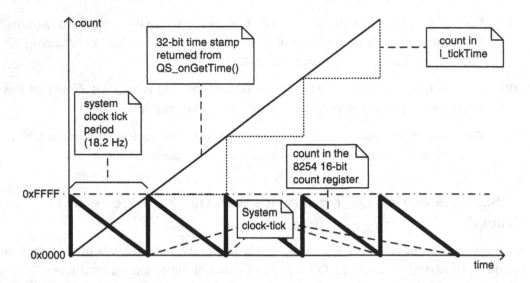

Figure 11.8: Using channel 0 of the 8254 PIT to generate 32-bit QS timestamp.

time going "backward." This solution assumes that the function QS_getTime() is called at least once per system clock tick. Listing 11.18 shows the QS timestamp implementation.

Listing 11.18 Generating QS timestamp (file <qp>\qpc\examples\80x86\ qk\tcpp101\l\dpp\bsp.c**)**

```
    /* Local-scope objects -------------------------------------------*/
    #ifdef Q_SPY
        static QSTimeCtr l_tickTime;              /* keeps timestamp at tick */
        static uint32_t l_lastTime;                      /* last timestamp */
    #endif
    . . .
    void interrupt ISR_tmr(void) {
        uint8_t pin;

    #ifdef Q_SPY
(1)     l_tickTime += 0x10000;                          /* add 16-bit rollover */
    #endif
        QK_ISR_ENTRY(pin, TMR_ISR_PRIO);   /* inform QK about entering the ISR */
        QF_tick();                   /* call QF_tick() outside of critical section */
        QK_ISR_EXIT(pin);                      /* inform QK about exiting the ISR */
    }
```

```
      /*................................................................*/
      #ifdef Q_SPY                                    /* define QS callbacks */
      ...
(2)   QSTimeCtr QS_onGetTime(void) {             /* invoked with interrupts locked */
          uint32_t now;
          uint16_t count16;                       /* 16-bit count from the 8254 */
(3)       if (l_tickTime != 0) {                        /* time tick has started? */
(4)           outportb(0x43, 0);               /* latch the 8254's counter-0 count */
(5)           count16 = (uint16_t)inportb(0x40);/* read the low byte of counter-0 */
(6)           count16 += ((uint16_t)inportb(0x40) << 8);    /* add on the hi byte */

(7)           now = l_tickTime + (0x10000 - count16);

(8)           if (l_lastTime > now) {           /* are we going "back" in time? */
(9)               now += 0x10000;               /* assume that there was one rollover */
              }
(10)          l_lastTime = now;
          }
          else {
(11)          now = 0;
          }
(12)      return (QSTimeCtr)now;
      }
      #endif                                                      /* Q_SPY */
```

(1)   The system clock tick interrupt increments the timestamp at tick l_tickTime by 0x10000 to account for the 16-bit counter rollover.

(2)   The QS_onGetTime() is always called inside a critical section of code. The QSTimeCtr type is defined according to the macro QS_TIME_SIZE (see Listing 11.11(3)).

(3,11) During the initial transient, before the clock-tick interrupt is enabled, the QS_onGetTime() returns 0.

(4)   The counter-0 in the 8254 is latched.

(5,6) The two halves of the counter can be now safely read, starting with the least significant byte.

(7)   The complete 32-bit timestamp 'now' is constructed.

(8)   Due to undersampling of the counter, the check is performed to find out if a counter rollover has been missed. (*Note:* This check assumes that QS_onGetTime() is called at least once per rollover period.)

(9)   If so, the counter is corrected by 0x10000.

(10) The last timestamp value is updated for the next time around.

(12) The timestamp value is returned to the caller.

11.6.4   Generating QS Dictionary Records from Active Objects

The few dictionary records generated from `main()` (Listing 11.16) provide only
the symbolic information available at the global level. However, the encapsulated
application components, such as the Philosopher and Table active objects, also
need to provide the symbolic information in the trace to the QSPY host
application. Listings 11.19 and 11.20 show how to generate dictionary trace
records from individual components without compromising their
encapsulation.

**Listing 11.19 Generating dictionary trace records from the Table active
object (file** `<qp>\qpc\examples\80x86\qk\tcpp101\1\dpp\table.c`**)**

```
(1)    static Table l_table;       /* the single instance of the Table active object */
       . . .
(2)    QState Table_initial(Table *me, QEvent const *e) {
           (void)e;          /* suppress the compiler warning about unused parameter */

(3)        QS_OBJ_DICTIONARY(&l_table);

(4)        QS_FUN_DICTIONARY(&QHsm_top);
(5)        QS_FUN_DICTIONARY(&Table_initial);
(6)        QS_FUN_DICTIONARY(&Table_serving);

(7)        QS_SIG_DICTIONARY(DONE_SIG,      0);                     /* global signals */
(8)        QS_SIG_DICTIONARY(EAT_SIG,       0);
(9)        QS_SIG_DICTIONARY(TERMINATE_SIG, 0);

(10)       QS_SIG_DICTIONARY(HUNGRY_SIG,    me);          /* signal just for Table */

           /* signal HUNGRY_SIG is posted directly */
           QActive_subscribe((QActive *)me, DONE_SIG);
           QActive_subscribe((QActive *)me, TERMINATE_SIG);

           return Q_TRAN(&Table_serving);
       }
```

(1) The Table active object instance is declared at file scope and is strictly encapsulated inside `table.c`.

(2) Generally, the topmost initial pseudostate (see Section 4.5.3 in Chapter 4) is the best place for generating the dictionary records. The QF real-time framework executes the topmost initial transition directly from `QActive_start()`; therefore, the dictionary records are generated during the initial transient, before the real-time constraints start to apply.

(3) Note that the object dictionary `QS_OBJ_DICTIONARY()` is generated using the address of the static variable `l_table` rather than the generic pointer "`me.`" Obviously, the name "`l_table`" is more descriptive than "`me.`"

(4-6) The function dictionary records provide symbolic names for the state-handler functions. Note that a function dictionary record for `QHsm_top()` should be also provided.

(7-9) The signal dictionary records for globally published signals must be associated with all state machines in the system. You achieve this by using zero as the second parameter to the macro `QS_SIG_DICTIONARY()`.

(10) The `HUNGRY_SIG` signal is posted directly to the Table active object, so it does need to have the same meaning globally. You can associate such signals with a particular state machine object by providing the address of the state machine as the second parameter to the macro `QS_SIG_DICTIONARY()`.

Listing 11.20 Generating dictionary trace records from the Philosopher active objects (file `<qp>\qpc\examples\80x86\qk\tcpp101\l\dpp\ philo.c`)

```
(1)   static Philo l_philo[N_PHILO];                    /* storage for all Philos */
      . . .
      /*........................................................................*/
(2)   QState Philo_initial(Philo *me, QEvent const *e) {
          static uint8_t registered;        /* starts off with 0, per C-standard */
(3)       if (!registered) {
              registered = (uint8_t)1;

(4)           QS_OBJ_DICTIONARY(&l_philo[0]);
(5)           QS_OBJ_DICTIONARY(&l_philo[0].timeEvt);
              QS_OBJ_DICTIONARY(&l_philo[1]);
```

Continued onto next page

```
            QS_OBJ_DICTIONARY(&l_philo[1].timeEvt);
            QS_OBJ_DICTIONARY(&l_philo[2]);
            QS_OBJ_DICTIONARY(&l_philo[2].timeEvt);
            QS_OBJ_DICTIONARY(&l_philo[3]);
            QS_OBJ_DICTIONARY(&l_philo[3].timeEvt);
            QS_OBJ_DICTIONARY(&l_philo[4]);
            QS_OBJ_DICTIONARY(&l_philo[4].timeEvt);

(6)         QS_FUN_DICTIONARY(&Philo_initial);
            QS_FUN_DICTIONARY(&Philo_thinking);
            QS_FUN_DICTIONARY(&Philo_hungry);
            QS_FUN_DICTIONARY(&Philo_eating);

            randomize();    /* initialize the random number generator just once */
        }
(7)     QS_SIG_DICTIONARY(HUNGRY_SIG, me);              /* signal for each Philos */
(8)     QS_SIG_DICTIONARY(TIMEOUT_SIG, me);             /* signal for each Philos */

        QActive_subscribe((QActive *)me, EAT_SIG);

        return Q_TRAN(&Philo_thinking);             /* top-most initial transition */
    }
```

(1)   All Philosopher active object instances are declared at file scope and are strictly encapsulated inside philo.c.

(2)   The Philo_initial() initial pseudostate is invoked once for every Philosopher active object.

(3)   The test of the static variable registered ensures that the code inside the if statement runs only once.

(4)   An object dictionary is generated for the Philosopher active object. Note that the object dictionary QS_OBJ_DICTIONARY() is generated using the address of the static variable l_philo[0] rather than the generic pointer "me". Obviously, the name "l_philo[0]" is more descriptive than "me."

(5)   An object dictionary entry is also generated for the private time event member of the Philosopher active object.

(6)   The function dictionary entries for all Philosopher state-handler functions are generated only once.

(7,8) In contrast, the signal dictionary records for local signals are generated for every Philosopher instance.

11.6.5 Adding Application-Specific Trace Records

Although QS generates a very detailed trace even without adding any instrumentation to the application, it also allows inserting application-specific records into the trace (see Section 11.3.9). Listing 11.21 provides an example of a simple application-specific trace record that reports the Philosopher status.

Listing 11.21 Generating application-specific trace records (file `<qp>\qpc\examples\80x86\qk\tcpp101\1\dpp\bsp.c`**)**

```
      #ifdef Q_SPY
        . . .
        enum AppRecords {                    /* application-specific trace records */
(1)         PHILO_STAT = QS_USER
        };
      #endif
        . . .
      /*.................................................................*/
      void BSP_displayPhilStat(uint8_t n, char const *stat) {
          Video_printStrAt(17, 12 + n, VIDEO_FGND_YELLOW, stat);

(2)       QS_BEGIN(PHILO_STAT, AO_Philo[n]) /* application-specific record begin */
(3)           QS_U8(1, n);                            /* Philosopher number */
(4)           QS_STR(stat);                           /* Philosopher status */
(5)       QS_END()
      }
```

---

(1) The application-specific trace record types need to be enumerated. Note that the user-level record types do not start from zero but rather are offset by the constant `QS_USER`.

(2) An application-specific trace record starts with `QS_BEGIN()`. The first parameter is the record type. The second parameter is the object pointer corresponding to this trace record, which is set to `AO_philo[n]`. This means that you have an option to use the application-specific local filter `QS_FILTER_AP_OBJ(AO_philo[<n>])` to selectively trace only the `AO_philo[<n>]` object, where `<n>=0..4`. As described in Section 11.3.9, you can also set the second parameter to `NULL`, which will disable the local filter.

(3) The `QS_U8()` macro outputs the byte 'n' to the trace record to be formatted as 1 using one digit.

(4) The `QS_STR()` macro outputs the string 'stat' to the trace record.

(5) An application-specific trace record ends with `QS_END()`.

The following QSPY trace output shows the application-specific trace records generated by the DPP application:

```
            . . .
0000525113 User000: 4 eating
            . . .
0000591471 User000: 3 hungry
            . . .
0000591596 User000: 2 hungry
            . . .
0000591730 User000: 0 hungry
            . . .
0000852276 User000: 4 thinking
            . . .
0000852387 User000: 3 eating
            . . .
0000983937 User000: 1 thinking
            . . .
0000984047 User000: 0 eating
            . . .
0001246064 User000: 3 thinking
```

11.6.6 "QSPY Reference Manual"

The source code available from the companion Website to this book at www.quantum-leaps.com/psicc2/ contains the "QSPY Reference Manual" in HTML and CHM-Help formats (see Figure 11.9). The "QSPY Reference Manual" contains descriptions of the command-line options, human-readable format, and all MATLAB matrices generated by the qspy.m script.

The "QSPY Reference Manual" is part of the bigger manual "QP Reference Manual" (see Section 7.12 in Chapter 7), which is located in <qp>\qpc\doxygen\ directory. The HTML documentation is found in <qp>\qpc\doxygen\html\, whereas the CHM Help format is located in <qp>\qpc\qpc.chm.

11.7 Summary

Testing and debugging your system can often take more calendar time and effort than analysis, design, and coding combined. The biggest problem, especially in embedded systems domain, is the very limited visibility into the target system.

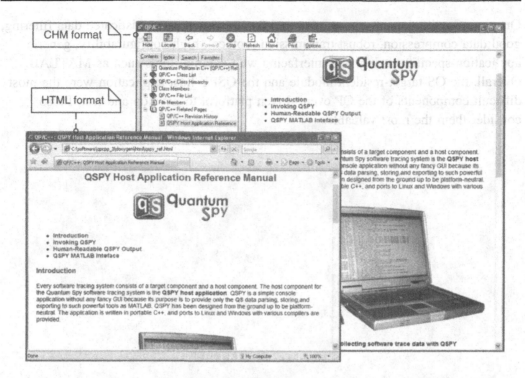

Figure 11.9: Screen shots of the "QSPY Reference Manual," which is available in HTML and CHM-help formats.

Software tracing is a method for obtaining diagnostic information in a *live* environment without the need to stop the application to get the system feedback. Software-tracing techniques are especially effective and powerful in combination with the event-driven paradigm because all important system interactions funnel through the event-driven infrastructure. The inversion of control so common in event-driven architectures offers a unique opportunity to instrument just the event-driven infrastructure, to gain unprecedented insight into the entire system.

The Quantum Spy software-tracing system allows you to monitor the behavior of QP applications in a live environment without degrading the application itself. It allows you to discover and document elusive, intermittent bugs that are caused by subtle interactions among concurrent components. It enables executing repeatable unit and integration tests of your system. It can help you ensure that your system runs reliably for long periods of time and gets top processor performance.

Quantum Spy implements many advanced features, such as sophisticated data filtering, good data compression, robust transmission protocol, high configurability, generic application-specific tracing, and interfacing with external tools such as MATLAB. Overall, the QS target-resident module and the QSPY host application were the most difficult components of the QP event-driven platform to develop and test. I also consider them the most valuable.

CHAPTER 12

QP-nano: How Small Can You Go?

All life on Earth is insects...
—Scientific American, July 2001

In this chapter I describe a reduced version of the event-driven infrastructure called QP-nano, which has been specifically designed to enable active object computing with UML-style hierarchical state machines on *low-end* 8- and 16-bit single-chip microcontrollers (MCUs). By low-end MCUs I mean devices such as 8051, PIC, AVR, MSP430, 68HC08/11/12, R8C/Tiny, and others alike, with a few hundred bytes of RAM and a few kilobytes of ROM. Embedded in myriads of products, these "invisible computers" far outnumber all other processor types in a similar way as countless species of insects far outnumber all other life forms on Earth [Turely 02].

Even though the QP event-driven platform is by no means big, a minimal QP application still requires around 1KB of RAM and some 10KB of ROM (see Figure 7.2 in Chapter 7), which is comparable to the footprint of a very small, bare-bones conventional RTOS. In comparison, a minimal QP-nano application can fit in a system with just 100 bytes of RAM and 2KB of ROM. This tiny footprint, especially in RAM, makes QP-nano ideal for high-volume, cost-sensitive, event-driven applications such as motor control, lighting control, capacitive touch sensing, remote access control, RFID, thermostats, small appliances, toys, power supplies, battery chargers, or just about any custom system on a chip (SOC or ASIC) that contains a small processor inside. Also, because the event-driven paradigm naturally uses the CPU only when handling events and otherwise can very easily switch the CPU into a low-power sleep mode (see Section 6.3.7 in Chapter 6), QP-nano is particularly suitable for ultra-low power applications, such as wireless sensor networks or implantable medical devices.

I begin this chapter by describing the key features of QP-nano. I then walk you through the QP-nano version of the "Fly 'n' Shoot" game, which I introduced in Chapter 1, so that you can easily compare how QP-nano differs from the full-version QP. Next I describe the QP-nano source code. I conclude with some more QP-nano examples for a very small, ultra-low-power MSP430F2013 MCU [TI 07].

12.1 Key Features of QP-nano

QP-nano is a generic, portable, *ultra-lightweight*, event-driven infrastructure designed specifically for low-end 8- and 16-bit MCUs. As shown in Figure 12.1, QP-nano consists of a hierarchical event processor called QEP-nano, a minimal real-time framework called QF-nano, and a choice between a preemptive run-to-completion kernel called QK-nano or a cooperative "vanilla" kernel. The key QP-nano features are:

- Full support for hierarchical state nesting, including guaranteed entry/exit action execution on arbitrary state transition topology with up to four levels of state nesting

- Support for up to eight concurrently executing active objects[1] with deterministic, thread-safe event queues

- Support for events with a byte-wide signal (255 signals) and one scalar parameter, configurable to 0 (no parameter), 1, 2, or 4 bytes

- Direct event delivery mechanism with first-in, first-out (FIFO) queuing policy

- One single-shot time event (timer) per active object with configurable dynamic range of 0 (no time events), 1, 2, or 4 bytes

- Built-in cooperative "vanilla" kernel (see Section 6.3.7 in Chapter 6)

- Built-in preemptive RTC kernel called QK-nano (see Section 6.3.8 in Chapter 6)

- Low-power architecture with idle callback function for easy implementation of power-saving modes

---

[1] This does not mean that your application is limited to eight state machines. Each active object can manage any number of stateful components, as described in the "Orthogonal Component" state pattern in Chapter 5.

- Provisions in the code to handle *nonstandard extensions* in the C compilers for popular low-end CPU architectures (e.g., allocation of constant objects in the code space, reentrant functions, etc.)

- Assertion-based error handling policy

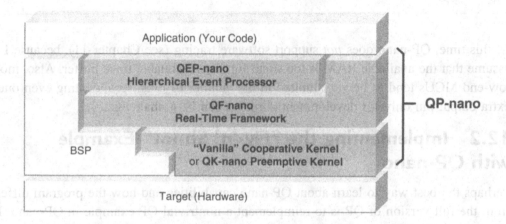

Figure 12.1: QP-nano components (in gray) and their relationship with the target hardware, board support package (BSP), and application.

By far the biggest challenge in QP-nano design is the extremely tight RAM budget, which I assumed to be only around 100 bytes, including the C stack. Obviously, with RAM in such short supply I was forced to carefully count every byte of RAM. This is in contrast to the full-version QP, where I was not trying to save every last byte of RAM if this would reduce programming convenience, flexibility, or performance.

Perhaps the most important implication of the severely limited RAM is that QP-nano does *not* support events with arbitrary-sized parameters. Instead, QP-nano allows only fixed-size events with one scalar parameter, configurable to 1, 2, or 4 bytes (or 0 bytes, which means no event parameter). This has far-reaching simplifying consequences. First, event queues in QP-nano hold entire events, not just pointers to events, as in the full-version QP. Small, fixed-size events are simply copied by value into and out of event queues in an inherently thread-safe manner. Second, the copy-by-value policy eliminates the need for event pools, which would not fit into the available RAM anyway. Finally, reference counting of events is unnecessary in this design.

> **NOTE**
>
> A single scalar event parameter means that QP-nano always associates the configured number of bytes with every event, but it does not mean that you can have only one event parameter. In fact, each event can have as many event parameters as you can squeeze into the available bits.

At this time, QP-nano does *not* support software tracing (see Chapter 11), because I assume that the available RAM is too small for any reasonable trace buffer. Also, most low-end MCUs tend to be very limited in the number of pins, so allocating even one extra output pin only for development purposes can be a challenge.

12.2   Implementing the "Fly 'n' Shoot" Example with QP-nano

Perhaps the best way to learn about QP-nano capabilities and how the program differs from the full version of QP is to reimplement a nontrivial QP example in QP-nano. In this section, I'll walk you through the QP-nano version of the "Fly 'n' Shoot" game introduced in Chapter 1. I recommend that you flip back to Chapter 1 and refresh your understanding of that application and its implementation based on the full-version QP.

The code accompanying this book contains four QP-nano implementations of the "Fly 'n' Shoot" game for two embedded targets and two different kernels. Here I'll use the version for DOS with the nonpreemptive kernel compiled with the legacy Turbo C++ 1.01 compiler, which you can run directly on any Windows PC. You can find this version in the directory <qp>\qpn\examples\80x86\tcpp101\game\. The same application code (except for the BSP) is also available for the Cortex-M3 EV-LM3S811 board (see Figure 1.2 in Chapter 1). The Cortex-M3 code is located in the directory <qp>\qpn\examples\cortex-m3\iar\game-ev-lm3s811\.

> **NOTE**
>
> The LM3S811 MCU (32-bit ARM Cortex-M3) with 8KB of RAM and 64KB of ROM is certainly a very big machine for QP-nano. I use it in this section only to provide a direct comparison to the same application implemented with full-version QP. In the upcoming Section 12.7, I describe QP-nano examples for the ultra-low power Texas Instruments board called eZ430-F2013, which is based on the MSP430F2013 MCU with only 128 bytes of RAM and 2KB of ROM [TI 07].

12.2.1 The `main()` function

Listing 12.1 shows the `main.c` source file for the "Fly 'n' Shoot" application, which contains the `main()` function along with some important data structures required by QP-nano.

Listing 12.1 The file `main.c` **of the "Fly 'n' Shoot" game application**

```
(1)   #include "qpn_port.h"                              /* QP-nano port */
(2)   #include "bsp.h"                          /* Board Support Package (BSP) */
(3)   #include "game.h"                             /* application interface */

      /*.................................................................*/
(4)   static QEvent 1_tunnelQueue[GAME_MINES_MAX + 4];
(5)   static QEvent 1_shipQueue[2];
(6)   static QEvent 1_missileQueue[2];

      /* QF_active[] array defines all active object control blocks ----*/
(7)   QActiveCB const Q_ROM Q_ROM_VAR QF_active[] = {
(8)       { (QActive *)0,              (QEvent *)0,       0             },
(9)       { (QActive *)&AO_tunnel, 1_tunnelQueue, Q_DIM(1_tunnelQueue) },
(10)      { (QActive *)&AO_ship,    1_shipQueue,    Q_DIM(1_shipQueue)   },
(11)      { (QActive *)&AO_missile, 1_missileQueue, Q_DIM(1_missileQueue) }
      };

      /* make sure that the QF_active[] array matches QF_MAX_ACTIVE in qpn_port.h */
(12)  Q_ASSERT_COMPILE(QF_MAX_ACTIVE == Q_DIM(QF_active) - 1);

      /*.................................................................*/
      void main (void) {
(13)      Tunnel_ctor ();
(14)      Ship_ctor   ();
(15)      Missile_ctor(GAME_MISSILE_SPEED_X);

(16)      BSP_init();                               /* initialize the board */

(17)      QF_run();                           /* transfer control to QF-nano */
      }
```

---

(1) Every application C file that uses QP-nano must include the `qpn_port.h` header file. This header file contains the specific adaptation of QP-nano to the given processor and compiler, which is called a *port*. The QP-nano port is typically located in the application directory.

(2) The `bsp.h` header file contains the interface to the board support package and is located in the application directory.

(3)   The game.h header file contains the declarations of events and other facilities shared among the components of the "Fly 'n' Shoot" game. I will discuss this header file in the upcoming Section 12.2.3. This header file is located in the application directory.

(4-6)   The application must provide storage for the event queues of all active objects used in the application. In QP-nano the storage is provided at compile time through the statically allocated arrays of events. Events are represented as instances of the QEvent structure declared in the <qp>\qpn\include\qepn.h header file, included from qpn_port.h. Each event queue of an active object can have a different length, and you need to decide this length based on your knowledge of the application. Refer to Chapters 6 and 7 for the discussion of sizing event queues.

(7)   Every QP-nano application *must* provide the constant array QF_active[], which defines all active object *control blocks* in the application. The control block QActiveCB structure groups together (1) the pointer to the corresponding active object instance, (2) the pointer to the event queue buffer of the active object, and (3) the length of the queue buffer.

In QP-nano, I use every opportunity to place data in ROM rather than in precious RAM. The QActiveCB structure contains data elements known at compile time so that these elements can be placed in ROM as opposed to placing them in the active object structure (RAM). That way I save anywhere from 10 to 80 bytes of RAM, depending on the number of active objects and the pointer size of the target CPU.

The Q_ROM macro is necessary on some CPU architectures to enforce placement of constant objects, such as the QF_active[] array, in ROM. On Harvard architecture CPUs (such as 8051 or AVR), the code and data spaces are separate and are accessed through different CPU instructions. The const keyword is not sufficient to place data in ROM, and various compilers often provide specific extended keywords to designate the code space for placing constant data, such as the "__code" extended keyword in the IAR 8051 compiler. The macro Q_ROM hides such nonstandard extensions. If you don't define Q_ROM in qepn_port.h, it will be defined to nothing in the qepn.h platform-independent header file.

The Q_ROM_VAR macro defines the compiler-specific directive for accessing a constant object in ROM. Many compilers for 8-bit MCUs provide variously sized pointers for accessing objects in various memories. Constant objects allocated in ROM often mandate the use of specific-size pointers (e.g., far pointers) to get access to ROM objects. The macro Q_ROM_VAR specifies the kind of the pointer to be used to access the ROM objects. An example of valid Q_ROM_VAR macro definition is __far (Freescale HC(S)08 compiler).

(8) The first entry (QF_active[0]) corresponds to an active object priority of zero, which is reserved for the idle task and cannot be used for any active object.

(9-11) The QF_active[] entries starting from one define the active object control blocks in the order of their relative *priorities*. The maximum number of active objects in QP-nano cannot exceed eight.

NOTE

The order or the active object control blocks in the QF_active[] array defines the *priorities* of active objects. This is the only place in the code where you assign active object priorities.

---

(12) This compile-time assertion (see Section 6.7.3 in Chapter 6) ensures that the dimension of the QF_active[] array matches the number of active objects QF_MAX_ACTIVE defined in the qpn_port.h header file.

In QP-nano, QF_MAX_ACTIVE denotes the *exact* number of active objects used in the application, as opposed to the full-version QP, where QF_MAX_ACTIVE denotes just the configurable maximum number of active objects.

NOTE

All active objects in QP-nano must be defined at compile time. This means that all active objects exist from the beginning and cannot be started (or stopped) later, as is possible in the full-version QP.

The macro QF_MAX_ACTIVE must be defined in the qpn_port.h header file because QP-nano uses the macro to optimize the internal algorithms based on the number of active objects. The compile-time assertion in line 12 makes sure that the configured number of active objects does indeed match exactly the number of active object control blocks defined in the QF_active[] array.

(13-15)   The `main()` function must first explicitly calls all active object constructors.

(16)   The board support package (BSP) is initialized.

(17)   At this point, you have initialized all components and have provided to the QF-nano framework all the information it needs to manage your application. The last thing you must do is to call the function `QF_run()` to pass the control to the QF-nano framework.

Overall, the application startup is much simpler in QP-nano than in full-version QP. Neither event pools nor publish-subscribe lists are supported, so you don't need to initialize them. You also don't start active objects explicitly. The QF-nano framework starts all active objects defined in the `QF_active[]` array automatically just after it gets control in `QF_run()`.

12.2.2   The `qpn_port.h` Header File

The `qpn_port.h` header file defines the QP-nano port and all configuration parameters for the particular application. Unlike in the full-version QP, QP-nano ports are typically defined at the application level. Also typically, the whole QP-nano port consists of just the `qpn_port.h` header file. Listing 12.2 shows the complete `qpn_port.h` file for the DOS version of the "Fly 'n' Shoot" game.

Listing 12.2 The `qpn_port.h` header file for the "Fly 'n' Shoot" game

```
      #ifndef qpn_port_h
      #define qpn_port_h

(1)   #define Q_PARAM_SIZE          4
(2)   #define QF_TIMEEVT_CTR_SIZE   2
(3)   #define Q_NFSM

      /* maximum # active objects--must match EXACTLY the QF_active[] definition   */
(4)   #define QF_MAX_ACTIVE         3

                                           /* interrupt locking policy for task level */
(5)   #define QF_INT_LOCK()         disable()
(6)   #define QF_INT_UNLOCK()       enable()

      /* Exact-width types (WG14/N843 C99 Standard) for Turbo C++/large model      */
(7)   typedef signed   char int8_t;
      typedef signed   int  int16_t;
```

```
        typedef signed   long  int32_t;
        typedef unsigned char  uint8_t;
        typedef unsigned int   uint16_t;
        typedef unsigned long  uint32_t;

        #include <dos.h>                                           /* DOS API */
        #undef outportb    /*don't use the macro because it has a bug in Turbo C++ 1.01*/

(8)     #include "qepn.h"          /* QEP-nano platform-independent header file */
(9)     #include "qfn.h"           /* QF-nano platform-independent header file */

        #endif                                                  /* qpn_port_h */
```

(1)   The macro Q_PARAM_SIZE defines the size (in bytes) of the scalar event parameter. The allowed values are 0 (no parameter), 1, 2, or 4 bytes. If you don't define this macro in qpn_port.h, the default of 0 (no parameter) will be assumed.

(2)   The macro QF_TIMEEVT_CTR_SIZE defines the size (in bytes) of the time event down-counter. The allowed values are 0 (no time events), 1, 2, or 4 bytes. If you don't define this macro in qpn_port.h, the default of 0 (no time events) will be assumed.

(3)   Defining the macro Q_NFSM eliminates the code for the simple nonhierarchical FSMs.

(4)   You must define the QF_MAX_ACTIVE macro as the *exact* number of active objects used in the application. The provided value must be between 1 and 8 and must be consistent with the definition of the QF_active[] array (see Listing 12.1(12)).

(5,6) The macros QF_INT_LOCK()/QF_INT_UNLOCK() define the task-level interrupt-locking policy for QP-nano. I discuss the QP-nano critical section in Section 12.3.2.

(7)   Just like the full-version QP, QP-nano uses a subset of the C99-standard exact-width integer types. The legacy Turbo C++ 1.01 compiler, which I'm using here, is a prestandard compiler and does not provide the <stdint.h> header file. In this case I just typedef the six exact-width integer types used in QP-nano.

(8)   The qpn_port.h must include the QEP-nano event processor interface qepn.h.

(9)   The qpn_port.h must include the QF-nano real-time framework interface qfn.h.

NOTE

The `qpn_port.h` header file in Listing 12.2 implicitly configures QP-nano to use the built-in cooperative "vanilla" kernel. The other alternative, which is the preemptive QK-nano kernel, is configured automatically when you include the `qkn.h` QK-nano interface in the `qpn_port.h` header file.

12.2.3   Signals, Events, and Active Objects in the "Fly 'n' Shoot" Game

In QP-nano, event signals are enumerated just as in the full-version QP. The only limitation is that signal values in QP-nano cannot exceed 255, because signals are always represented in a single byte.

In QP-nano, you cannot specify arbitrary event parameters, so you don't derive events as you do in full-version QP. Instead, all events in QP-nano are simply instances of the `QEvent` structure, which contains the fixed-size scalar parameter configured according to your definition of `Q_PARAM_SIZE` (see Listing 12.2(2)).

On the other hand, active objects in QP-nano are derived from the `QActive` base structure, just like they are in the full-version QP. One of the main concerns with respect to active object structures is to keep them encapsulated. In all QP-nano examples, including the "Fly 'n' Shoot" game, I demonstrate a technique to keep the active object structures and state machines completely opaque. I describe this technique in the explanation section following Listing 12.3, which shows the header file `game.h` included by all components of the "Fly 'n' Shoot" application.

Listing 12.3 Signals and active objects for the "Fly 'n' Shoot" game (the `game.h` header file)

```
    #ifndef game_h
    #define game_h

(1)  enum GameSignals {                              /* signals used in the game */
(2)      TIME_TICK_SIG = Q_USER_SIG,                 /* published from tick ISR */
         PLAYER_TRIGGER_SIG,  /* published by Player (ISR) to trigger the Missile */
         PLAYER_QUIT_SIG,          /* published by Player (ISR) to quit the game */
         GAME_OVER_SIG,         /* published by Ship when it finishes exploding */
         PLAYER_SHIP_MOVE_SIG,    /* posted by Player (ISR) to the Ship to move it */
```

```
              BLINK_TIMEOUT_SIG,              /* signal for Tunnel's blink timeout event */
              SCREEN_TIMEOUT_SIG,             /* signal for Tunnel's screen timeout event */
              TAKE_OFF_SIG,         /* from Tunnel to Ship to grant permission to take off */
              HIT_WALL_SIG,                   /* from Tunnel to Ship when Ship hits the wall */
              HIT_MINE_SIG,         /* from Mine to Ship or Missile when it hits the mine */
              SHIP_IMG_SIG,         /* from Ship to the Tunnel to draw and check for hits */
              MISSILE_IMG_SIG,    /* from Missile the Tunnel to draw and check for hits */
              MINE_IMG_SIG,                   /* sent by Mine to the Tunnel to draw the mine */
              MISSILE_FIRE_SIG,               /* sent by Ship to the Missile to fire */
              DESTROYED_MINE_SIG,  /* from Missile to Ship when Missile destroyed Mine */
              EXPLOSION_SIG,        /* from any exploding object to render the explosion */
              MINE_PLANT_SIG,                 /* from Tunnel to the Mine to plant it */
              MINE_DISABLED_SIG,        /* from Mine to Tunnel when it becomes disabled */
              MINE_RECYCLE_SIG,             /* sent by Tunnel to Mine to recycle the mine */
              SCORE_SIG      /* from Ship to Tunnel to adjust game level based on score */
          };

          /* active objects ......................................................*/
(3)       extern struct TunnelTag  AO_Tunnel;
(4)       extern struct ShipTag    AO_Ship;
(5)       extern struct MissileTag AO_Missile;

(6)       void Tunnel_ctor (void);
(7)       void Ship_ctor   (void);
(8)       void Missile_ctor(uint8_t speed);

          /* common constants and shared helper functions .........................*/
          . . .
          #endif                                                         /* game_h */
```

(1) All signals are defined in one enumeration, which automatically guarantees the uniqueness of signals.

(2) Note that the user signals must start with the offset Q_USER_SIG to avoid overlapping the reserved QEP-nano signals.

(3-5) I declare all active object instances in the system as extern variables. These declarations are necessary for the initialization of the QF_active[] array (see Listing 12.1(12)).

NOTE

The active object structures (e.g., struct TunnelTag) do *not* need to be defined globally in the application header file. The QF_active[] array needs only *pointers* to the active objects (see Listing 12.1(9-11)), which the compiler can resolve without knowing the full definition of the active object structure.

I never declare active object structures globally. Instead, I declare the active object structures in the file scope of the specific active object module (e.g., struct TunnelTag is declared in the tunnel.c file scope). That way I can be sure that each active object remains fully encapsulated.

(6-8)   Every active object in the system must provide a "constructor" function, which initializes the active object instance. These constructors don't take the "me" pointers, because they have access to the global active object instances (see (3-5)). However, the constructors can take some other initialization parameters. For instance, the Missile_ctor() takes the Missile speed parameter. Listing 12.1(13-15) shows that the constructors are called right at the beginning of main().

12.2.4   Implementing the Ship Active Object in QP-nano

Implementing active objects with QP-nano is very similar to the full-version QP. As before, you derive the concrete active object structures from the QActive base structure provided in QP-nano. Your main job is to elaborate the state machines of the active objects, which is also very similar to the full-version QP. The only important difference is that state-handler functions in QP-nano do *not* take the event pointer as the second argument. In fact, QP-nano state handlers take only one argument—the "me" pointer. The current event is embedded inside the state machine itself and is accessible via the "me" pointer. QP-nano provides macros Q_SIG() and Q_PAR() to conveniently access the signal and the scalar parameter of the current event, respectively.

Listing 12.4 shows the implementation of the Ship active object from the "Fly 'n' Shoot" game, which illustrates all aspects of implementing active objects with QP-nano. Correlate this implementation with the Ship state diagram in Figure 1.6 as well as with the QP implementation described in Section 1.7 in Chapter 1.

Listing 12.4 The Ship active object definition (file ship.c); boldface indicates QP-nano facilities

```
#include "qpn_port.h"
#include "bsp.h"
#include "game.h"
```

```
       /* local objects --------------------------------------------------*/
(1)    typedef struct ShipTag {
(2)        QActive super;                              /* extend the QActive class */
           uint8_t x;
           uint8_t y;
           uint8_t exp_ctr;
           uint16_t score;
       } Ship;                                         /* the Ship active object */

(3)    static QState Ship_initial   (Ship *me);
(4)    static QState Ship_active    (Ship *me);
       static QState Ship_parked    (Ship *me);
       static QState Ship_flying    (Ship *me);
       static QState Ship_exploding (Ship *me);

       /* global objects --------------------------------------------------*/
(5)    Ship AO_Ship;

       /*..................................................................*/
       void Ship_ctor(void) {
           Ship *me = &AO_Ship;
(6)        QActive_ctor(&me->super, (QStateHandler)&Ship_initial);
           me->x = GAME_SHIP_X;
           me->y = GAME_SHIP_Y;
       }

       /* HSM definition --------------------------------------------------*/
       QState Ship_initial(Ship *me) {
(7)        return Q_TRAN(&Ship_active);                /* top-most initial transition */
       }
       /*..................................................................*/
       QState Ship_active(Ship *me) {
(8)        switch (Q_SIG(me)) {
               case Q_INIT_SIG: {                      /* nested initial transition */
(9)                return Q_TRAN(&Ship_parked);
               }
               case PLAYER_SHIP_MOVE_SIG: {
(10)               me->x = (uint8_t)Q_PAR(me);
(11)               me->y = (uint8_t)(Q_PAR(me) >> 8);
(12)               return Q_HANDLED();
               }
           }
(13)       return Q_SUPER(&QHsm_top);
       }
       /*..................................................................*/
       QState Ship_flying(Ship *me) {
           switch (Q_SIG(me)) {
               case Q_ENTRY_SIG: {
                   me->score = 0;                      /* reset the score */
(14)               QActive_post((QActive *)&AO_Tunnel, SCORE_SIG, me->score);
                   return Q_HANDLED();
               }
               case TIME_TICK_SIG: {
```

Continued onto next page

```
                      /* tell the Tunnel to draw the Ship and test for hits */
(15)                  QActive_post((QActive *)&AO_Tunnel, SHIP_IMG_SIG,
                                   ((QParam)SHIP_BMP << 16)
                                   | (QParam)me->x
                                   | ((QParam)me->y << 8));

                      ++me->score; /* increment the score for surviving another tick */
                      if ((me->score % 10) == 0) {          /* is the score "round"? */
                          QActive_post((QActive *)&AO_Tunnel, SCORE_SIG, me->score);
                      }
                      return Q_HANDLED();
                  }
                  case PLAYER_TRIGGER_SIG: {                    /* trigger the Missile */
                      QActive_post((QActive *)&AO_Missile, MISSILE_FIRE_SIG,
                                   (QParam)me->x
                                   | (((QParam)me->y + SHIP_HEIGHT - 1) << 8));
                      return Q_HANDLED();
                  }
                  case DESTROYED_MINE_SIG: {
                      me->score += Q_PAR(me);
                      /* the score will be sent to the Tunnel by the next TIME_TICK */
                      return Q_HANDLED();
                  }
                  case HIT_WALL_SIG:
                  case HIT_MINE_SIG: {
(16)                  return Q_TRAN(&Ship_exploding);
                  }
              }
(17)      return Q_SUPER(&Ship_active);
      }
```

(1) This structure defines the Ship active object.

(2) The Ship active object structure derives from the framework structure QActive, as described in the sidebar "Single Inheritance in C" in Chapter 1.

(3) The Ship_initial() function defines the topmost initial transition in the Ship state machine. The only difference from the full-version QP is that the initial pseudostate function does not take the initial event parameter.

(4) The state-handler functions in QP-nano also don't take the event parameter. (In QP-nano, the current event is embedded in the state machine.) As in the full-version QP, a state-handler function in QP-nano returns the status of the event handling.

(5) In this line I allocate the global AO_Ship active object. Note that actual structure definition for the Ship active object is accessible only locally at the file scope of the ship.c file.

> **NOTE**
>
> QP-nano assumes that all global or static variables without explicit initialization value are initialized to zero upon system startup, which is a requirement of the ANSI-C standard. You should make sure that your startup code clears the static variables data section (a.k.a. the Block Started by Symbol section, or BSS) before calling `main()`.

(6)   As always, the derived structure is responsible for initializing the part inherited from the base structure. The "constructor" of the base class `QActive_ctor()` puts the state machine in the initial pseudostate `&Ship_initial`. The constructor also initializes the priority of the active object based on the `QF_active[]` array.

(7)   The topmost initial transition to state `Ship_active` is specified with the `Q_TRAN()` macro.

(8)   Every state handler is structured as a `switch` statement that discriminates based on the signal of the event, which in QP-nano is obtained by the macro `Q_SIG(me)`.

(9)   You designate the target of a nested initial transition with the `Q_TRAN()` macro.

(10,11)   You access the data members of the `Ship` state machine via the "me" parameter of the state-handler function. You access the event parameters via the `Q_PAR(me)` macro. Note that in this case two logical event parameters are actually from the scalar QP-nano parameter. The *x* coordinate of the Ship is sent in the least significant byte, and the *y* coordinate in the next byte.

> **NOTE**
>
> Each event can have as many event parameters as you can squeeze into the available bits.

(12)   You terminate the `case` statement with "`return Q_HANDLED()`" which informs QEP-nano that the internal transition has been handled.

(13)   The final `return` from a state-handler function designates the *superstate* of that state by means of the macro `Q_SUPER()`, which is exactly the same as in the full-version QP. QEP-nano provides the "top" state as a state-handler

function `QHsm_top()`, and therefore the `Ship_active()` state handler uses the pointer `&QHsm_top` as the argument to the `Q_SUPER()` macro (see the Ship state diagram in Figure 1.6 in Chapter 1).

(14)   The function `QActive_post()` posts the specified event signal and parameter directly to the recipient active object. Direct event posting is the only event delivery mechanism supported in QP-nano.

(15)   The event posting demonstrates how to combine several logical event parameters into the single scalar parameter managed by QP-nano. Of course, you must be careful not to overflow the dynamic range of the QP-nano parameter configured with the `Q_PARAM_SIZE` macro (see Listing 12.2(1)).

(16)   You designate the target of a transition with the `Q_TRAN()` macro.

(17)   The state "flying" (see Figure 1.6 in Chapter 1) nests in the state "active," so the state handler `Ship_flying()` designates the superstate by returning `Q_SUPER(&Ship_active)`.

12.2.5   Time Events in QP-nano

QP-nano maintains a single private time event (timer) for each active object in the system. These timers can be programmed (armed) to generate the reserved `Q_TIMEOUT` events after the specified number of clock ticks. Internally, QP-nano represents a time event only as a down-counter (typically 2 bytes of RAM). Only single-shot time events are supported because a periodic time event would require twice as much RAM to store the period. The `Q_TIMEOUT` signal is predefined in QP-nano and is one of the reserved signals (similar to `Q_ENTRY`, `Q_EXIT`, and `Q_INIT`).

Listing 12.5 shows a fragment of the Tunnel active object state machine that uses the QP-nano time event for blinking the "Game Over" text in the "game_over" state.

Listing 12.5 Using a QP-nano Time Event (file `tunnel.c`)

```
      QState Tunnel_game_over(Tunnel *me) {
         switch (Q_SIG(me)) {
            case Q_ENTRY_SIG: {
(1)            QActive_arm((QActive *)me, BSP_TICKS_PER_SEC/2);    /* 1/2 sec */
               me->blink_ctr = 5*2;                                /* 5s timeout */
               BSP_drawNString((GAME_SCREEN_WIDTH - 6*9)/2, 0, "Game Over");
               return (QState)0;
            }
```

```
                case Q_EXIT_SIG: {
(2)                 QActive_disarm((QActive *)me);
                    BSP_updateScore(0);              /* clear the score on the display */
                    return (QState)0;
                }
(3)             case Q_TIMEOUT_SIG: {
(4)                 QActive_arm((QActive *)me, BSP_TICKS_PER_SEC/2);     /* 1/2 sec */
                    BSP_drawNString((GAME_SCREEN_WIDTH - 6*9)/2, 0,
                                    (((me->blink_ctr & 1) != 0)
                                     ? "Game Over"
                                     : "        "));
                    if ((--me->blink_ctr) == 0) {         /* blinked enough times? */
                        Q_TRAN(&Tunnel_demo);
                    }
                    return (QState)0;
                }
            }
        return (QState)&Tunnel_active;
    }
```

(1) The time event associated with the active object is armed to expire after the specified number of clock ticks. Usually, arming of the time event occurs in the entry action to a state.

> **NOTE**
>
> While arming a time event, you must be careful not to exceed the preconfigured dynamic range of the internal down-counter. The dynamic range is configurable by means of the macro QF_TIMEEVT_CTR_SIZE in the qpn_port.h header file (see Listing 12.2(2)).

(2) The time event can be disarmed. The disarming usually occurs in the exit action from a state. At any moment you can also *rearm* a running timer by calling QActive_arm(), which simply replaces the value of the timer down-counter.

> **NOTE**
>
> A QP-nano time event can be easily used as a watchdog timer. You constantly rearm such time event by calling QActive_arm() so that it never expires.

(3) After the preprogrammed timeout, the timer generates the Q_TIMEOUT event, which you can handle just like any other event dispatched to the state machine.

(4)   You achieve a periodic timeout by arming the one-shot time event every time it expires. Note that often the period is a compile-time constant, which takes no precious RAM.

12.2.6   Board Support Package for "Fly 'n' Shoot" Application in QP-nano

QP-nano calls several platform-specific callback functions that you must define, typically in the board support package (BSP). Apart from the callbacks, you must also define all the interrupt service routines (ISRs), explained in more detail in Sections 12.5.1 and 12.6.4. Listing 12.6 shows the most important elements of the BSP for the "Fly 'n' Shoot" game in the DOS environment.

Listing 12.6 BSP for the "Fly 'n' Shoot" game under DOS (file `bsp.c`)

```
      #include "qpn_port.h"
      #include "game.h"
      #include "bsp.h"

      /* Local-scope objects -----------------------------------------------*/
      static void interrupt (*l_dosTmrISR)(void);
      static void interrupt (*l_dosKbdISR)(void);

      #define TMR_VECTOR      0x08
      #define KBD_VECTOR      0x09

      /*.............................................................*/
(1)   static void interrupt tmrISR(void) {     /* 80x86 enters ISRs with int. locked */

(2)       QF_tick();                              /* process all armed time events */

(3)       QActive_postISR((QActive *)&AO_Tunnel,    TIME_TICK_SIG, 0);
(4)       QActive_postISR((QActive *)&AO_Ship,      TIME_TICK_SIG, 0);
(5)       QActive_postISR((QActive *)&AO_Missile,   TIME_TICK_SIG, 0);
          outportb(0x20, 0x20);                    /* write EOI to the master PIC */
      }
      /*.............................................................*/
      void BSP_init(void) {
          . . .
      }
      /*.............................................................*/
      void BSP_drawBitmap(uint8_t const *bitmap, uint8_t width, uint8_t height) {
          Video_drawBitmapAt(0, 8, bitmap, width, height);
      }
      /*.............................................................*/
```

```
        void BSP_drawNString(uint8_t x, uint8_t y, char const *str) {
            Video_drawStringAt(x, 8 + y*8, str);
        }
        /*.................................................................*/
        void BSP_updateScore(uint16_t score) {
            if (score == 0) {
                Video_clearRect(68, 24, 72,   25, VIDEO_BGND_RED);
            }
            Video_printNumAt(68, 24, VIDEO_FGND_YELLOW, score);
        }
        /*.................................................................*/
(6)     void QF_onStartup(void) {
                                                /* save the original DOS vectors ... */
            l_dosTmrISR    = getvect(TMR_VECTOR);
            l_dosKbdISR    = getvect(KBD_VECTOR);

            QF_INT_LOCK();
            setvect(TMR_VECTOR, &tmrISR);
            setvect(KBD_VECTOR, &kbdISR);
            QF_INT_UNLOCK();
        }
        /*.................................................................*/
(7)     void QF_stop(void) {
                                            /* restore the original DOS vectors ... */
            if (l_dosTmrISR != (void interrupt (*)(void))0) {  /* DOS vectors saved? */
                QF_INT_LOCK();
                setvect(TMR_VECTOR, l_dosTmrISR);
                setvect(KBD_VECTOR, l_dosKbdISR);
                QF_INT_UNLOCK();
            }
            _exit(0);                                          /* exit to DOS */
        }
        /*.................................................................*/
(8)     void QF_onIdle(void) {                                 /* see NOTE01 */
            QF_INT_UNLOCK();
        }
        /*-----------------------------------------------------------------*/
(9)     void Q_onAssert(char const Q_ROM * const Q_ROM_VAR file, int line) {
            ...
            QF_stop();                                         /* stop QF and cleanup */
        }
```

(1)   Usually you can use the compiler-generated ISRs with QP-nano. Here I use the capability of the Turbo C++ 1.01 compiler to generate ISRs, which are designated with the extended keyword "interrupt."

(2)   Just as in the full-version QP, you need to call QF_tick() from the system clock-tick ISR.

(3-5)   QP-nano does not support publishing events. Instead, you directly post the TIME_TICK event to all active objects that need to receive it.

> **NOTE**
>
> QP-nano provides different services for ISRs and for the task level. You can only call two QP-nano functions from interrupts: `QF_tick()` and `QActive_postISR()`. Conversely, you should *never* call these two functions from the task level. This separation of APIs is closely related to the separate interrupt-locking policies for tasks and interrupts in QP-nano. I discuss the QP-nano interrupt-locking policy in Section 12.3.2.

(6)   The callback function `QF_onStartup()` is invoked by QP-nano after all active object state machines are started but before the event loop starts executing. `QF_onStartup()` is intended for configuring and starting interrupts.

(7)   The function `QF_stop()` stops QF-nano and returns control back to the operating system. This function is rarely used in deeply embedded systems, because a bare-metal system simply has no operating system to return to. In the DOS version, however, the `QF_stop()` function restores the original interrupts and exits to DOS.

(8)   This version of the "Fly 'n' Shoot" application has been implicitly configured to use the "vanilla" cooperative kernel (see the final note in Section 12.2.2). The vanilla kernel works in QP-nano exactly as described in Section 6.3.7 in Chapter 6. In particular, the `QF_onIdle()` callback is invoked with interrupts locked and must always unlock interrupts.

(9)   Finally, in every QP-nano application you need to provide the assertion-failure callback `Q_onAssert()`. QP-nano uses the same embedded-systems-friendly assertions as the full-version QP (see Section 6.7.3 in Chapter 6).

12.2.7 Building the "Fly 'n' Shoot" QP-nano Application

As shown in Figure 12.2, building a QP-nano application is simpler than the full-version QP. You merely need to add two QP-nano source files `qepn.c` and `qfn.c` to the project, and you need to instruct the compiler to search for the header files in the `<qp>\qpn\include\` directory. (If you use QK-nano, you additionally need to add the `qkn.c` source file.) The project file for building the "Fly 'n' Shoot" game for DOS with the Turbo C++ 1.01 compiler is found in `<qp>\qpn\examples\80x86\tcpp101\game\GAME.PRJ`. Similarly, the project file to build the game for Cortex-M3 with the IAR compiler is found in `C:\software\qpn\examples\cortex-m3\iar\game-ev-lm3s811\game.ewp`.

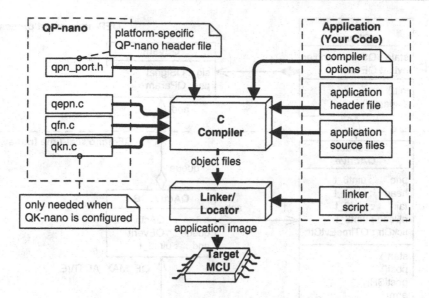

Figure 12.2: Building a QP-nano application.

12.3 QP-nano Structure

Figure 12.3 shows the main QP-nano elements and their relation to the application-level code, such as the "Fly 'n' Shoot" game example. As all real-time frameworks, QP-nano provides the central base class QActive for derivation of concrete[2] active objects. The QActive class is abstract, which means that it is not intended for direct instantiation but rather only for derivation of active object structures, such as Ship, Missile, and Tunnel shown in Figure 12.3 (see also the sidebar "Single Inheritance in C" in Chapter 1).

By default, the QActive class derives from the QHsm hierarchical state machine class defined in the QEP-nano event processor. This means that by virtue of inheritance, active objects are HSMs and inherit the init() and dispatch() state machine interface. QActive also contains the active object priority, which identifies the active object thread of execution, as well as an event queue and a time event counter.

As an option, you can configure QP-nano to derive QActive from the simpler, nonhierarchical state machine class QFsm. By doing this, you eliminate the hierarchical state machine code, which can save you some 300-500 bytes of ROM, depending on the code density of your target CPU.

---

[2] *Concrete class* is the OOP term and denotes a class that has no abstract operations or protected constructors. Concrete class can be instantiated, as opposed to *abstract class,* which cannot be instantiated.

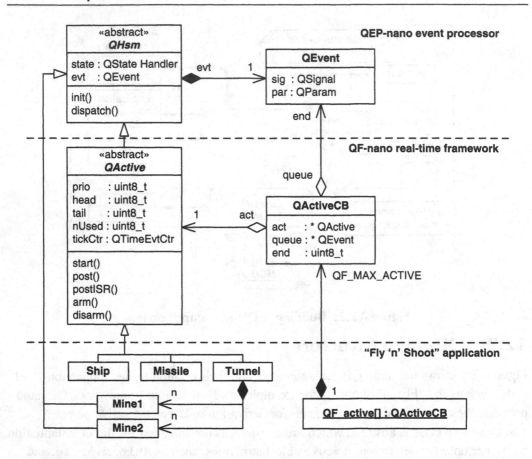

Figure 12.3: QEP-nano event processor, QF-nano real-time framework, and the "Fly 'n' Shoot" application.

The QEvent class represents events in QP-nano. The event signal QSignal is typedef'ed to a byte (uint8_t). Also embedded directly in the event is the single scalar event parameter. You cannot add any other parameters to the event by derivation. The size of the event parameter is configurable with the macro Q_PARAM_SIZE. In QP-nano, the state machine class QHsm (as well as the simpler QFsm) contains the current event. The event inside the state machine counts towards the event queue length.

QP-nano also introduces the active object control block structure QActiveCB, which contains read-only data elements known at compile time (see Listing 12.1(7)). Every QP-application must define the constant array QF_active[] that contains initialized active object control blocks for all active objects in the application.

The length of the `QF_active[]` array must match exactly the `QF_MAX_ACTIVE` macro in the `qpn_port.h` header file. The ordering of the active object control blocks in the `QF_active[]` array determines the priorities of the active objects.

12.3.1 QP-nano Source Code, Examples, and Documentation

Listing 12.7 shows the directories and files comprising QP-nano. Note that the `<qp>\qpn\examples\` directory contains several QP-nano executable examples for various embedded targets. In particular, I provide all the state design patterns described in Chapter 5 in the `<qp>\qpn\examples\80x86\tcpp101\` directory. The QP-nano directory `<qp>\qpn\doxygen\` also contains the "QP-nano Reference Manual" generated from the source code by the Doxygen utility. The reference manual is available in HTML and CHM Help formats.

Listing 12.7 QP-nano code organization

```
<qp>\qpn\              - QP-nano root directory
  |
  +-include\           - Platform independent QP header files
  | +-qassert.h        - QP-nano assertions (Section 6.7.3 in Chapter 6)
  | +-qepn.h           - QEP-nano platform-independent interface
  | +-qfn.h            - QF-nano  platform-independent interface
  | +-qkn.h            - QK-nano  platform-independent interface
  |
  +-source\            - QP-nano  platform-independent source code (*.C files)
  | +-qepn.c           - QEP-nano platform-independent source code
  | +-qfn.c            - QF-nano  platform-independent source code
  | +-qkn.c            - QK-nano  platform-independent source code
  |
  +-examples\          - Platform-specific QP examples
  | +-80x86\           - Examples for the 80x86 processor
  | | +-tcpp101\       - Examples with the Turbo C++ 1.01 compiler
  | |   +-comp\        - "Orthogonal Component" pattern (Chapter 5)
  | |   +-defer\       - "Deferred Event" pattern (Chapter 5)
  | |   +-dpp\         - DPP application (Chapter 9)
  | |   +-dpp-qk\      - DPP application with QK-nano
  | |   +-game\        - "Fly 'n' Shoot" game example
  | |   | +-dbg\       - Debug build
  | |   | | +-GAME.EXE - Debug executable
  | |   | +-GAME.PRJ   - Turbo C++ project to build the Debug version
  | |   | +-qpn_port.h - QP-nano port
  | |   | +-game.h     - The application header file
  | |   | +-bsp.c      - BSP for the application
  | |   | +-main.c     -
  | |   +-. . .
```

Continued onto next page

```
|  |    +-game-qk\      - "Fly 'n' Shoot" game example with QK-nano
|  |    +-history\      - "Transition to History" pattern (Chapter 5)
|  |    +-hook\         - "Ultimate Hook" pattern (Chapter 5)
|  |    +-pelican\      - PELICAN crossing example (see Section 12.7)
|  |    +-pelican-qk\   - PELICAN crossing example with QK-nano
|  |    +-qhsmtst\      - QHsmTst example (Section 2.3.15 in Chapter 2)
|  |    +-reminder\     - "Reminder" pattern (Chapter 5)
|  |
|  +-cortex-m3\       - Examples for the Cortex-M3 processor
|  | +-iar\           - Examples with the IAR compiler
|  |    +-game-ev-lm3s811       - "Fly 'n' Shoot" game example
|  |    +-game-qk-ev-lm3s811    - "Fly 'n' Shoot" game example with QK-nano
|  |    +-pelican-ev-lm3s811    - PELICAN crossing example (see Section 12.7)
|  |    +-pelican-qk-ev-lm3s811 - PELICAN crossing example with QK-nano
|  |
|  +-msp430\        - Examples for the MSP430 processor
|  | +-iar\         - Examples with the IAR compiler
|  |    +-bomb-eZ430         - Time bomb example (Section 3.6 in Chapter 3)
|  |    +-bomb-qk-eZ430      - Time bomb with QK-nano
|  |    +-dpp-eZ430          - Simplified DPP application
|  |    +-pelican-eZ430      - PELICAN crossing example (see Section 12.7)
|  |    +-pelican-qk-eZ430   - PELICAN crossing example with QK-nano
|  |    +-qhsmtst-eZ430      - QHsmTst example (Section 2.3.15 in Chapter 2)
|
+-doxygen\         - QP-nano documentation generated with Doxygen
| +-html\          - "QP-nano Reference Manual" in HTML format
| | +-index.html   - The starting HTML page for "QP-nano Reference Manual"
| | +- . . .
| +-Doxyfile       - Doxygen configuration file to generate the Manual
| +-qpn.chm        - QP-nano Reference Manual" in CHM Help format
| +-qpn_rev.h      - QP-nano revision history
```

12.3.2 Critical Sections in QP-nano

QP-nano, just like the full-version QP, achieves atomic execution of critical sections by briefly locking and unlocking interrupts. However, unlike the full-version QP, QP-nano uses a separate interrupt-locking policy for the task-level code and a different policy for ISRs. The ISR policy is used in the QP-nano functions QActive_postISR() and QF_tick(). The task-level policy is used in all other QP-nano services.

NOTE

Because the interrupt-locking policies for active objects and ISRs are *different*, you should never call the QP-nano functions intended for ISRs (QActive_postISR() and QF_tick()) inside tasks, and conversely, you should never call task-level QP-nano functions inside ISRs.

Task-Level Interrupt Locking

For the task level (code called from active objects), QP-nano employs the simple unconditional interrupt locking and unlocking policy, as described in Section 7.3.2 of Chapter 7. In this policy, interrupts are always unconditionally unlocked upon exit from a critical section, regardless of whether they were locked or unlocked before entry to the critical section. The pair of QP-nano macros QF_INT_LOCK()/QF_INT_UNLOCK(), shown in Listing 12.8, encapsulates the actual mechanism the compiler provides to lock and unlock interrupts from C. You need to consult the documentation of your CPU and the compiler to find how to achieve interrupt locking and unlocking in your particular system.

Listing 12.8 Example of QP-nano interrupt locking and unlocking macros for the task level

```
#define QF_INT_LOCK()      disable()
#define QF_INT_UNLOCK()    enable()
```

The simple task-level policy of "unconditional locking and unlocking interrupts" matches very well the architecture of low-end MCUs. The critical section is fast and straightforward, but does *not* allow nesting of critical sections. QP-nano never nests critical sections internally, but you should be careful not to nest critical sections in your own code.

NOTE

Since most of QP-nano functions lock interrupts internally, you should never call a QP-nano function from within an already established critical section.

ISR-Level Interrupt Locking

For the ISR level, QP-nano offers three critical section implementation options. The first default option is to do *nothing*, meaning that interrupts are neither locked nor unlocked inside ISRs. This policy is appropriate when interrupts cannot nest, because in this case the whole ISR represents a critical section of code, unless of course you explicitly unlock interrupts inside the ISR body. But unlocking interrupts inside ISRs is often not advisable, because most low-end micros aren't really designed to handle interrupt nesting. Low-end MCUs typically lack a fully prioritized interrupt controller and cannot afford the bigger stack requirements caused by nesting interrupts.

> **NOTE**
>
> Unlocking interrupts inside ISRs without an interrupt controller can cause all sorts of priority inversions, including the pathological case of an interrupt preempting itself recursively.

In case you can allow interrupt nesting, QP-nano offers two more interrupt-locking policies for ISRs. The first of these options is to use the task-level interrupt policy inside ISRs. As shown in Listing 12.9, you select this policy by defining the macro `QF_ISR_NEST`. In this case the function `QActive_postISR()` (as well as `QF_tick()`, which calls `QActive_postISR()`), unconditionally locks interrupts with the macro `QF_INT_LOCK()` and unlocks with `QF_INT_UNLOCK()`. Of course, to avoid nesting critical sections, you are responsible for making sure that interrupts are unlocked before calling `QActive_postISR()` or `QF_tick()` from your ISRs.

Listing 12.9 The ISR-level policy of "unconditional interrupt locking and unlocking"

```
#define QF_ISR_NEST
/* QF_ISR_KEY_TYPE not defined */
```

> **NOTE**
>
> You should always define the macro `QF_ISR_NEST` if interrupts can nest for some reason.

Finally, QP-nano also supports the more advanced policy of "saving and restoring interrupt status" (see Section 7.3.1 in Chapter 7). Listing 12.10 shows an example of this policy. This policy is the most expensive but also the most robust because interrupts are only unlocked upon exit from the critical section if they were unlocked upon entry.

Listing 12.10 Example of ISR-level policy of "saving and restoring interrupt status"

```
(1)  #define QF_ISR_NEST
(2)  #define QF_ISR_KEY_TYPE       int
(3)  #define QF_ISR_LOCK(key_)     ((key_) = int_lock())
(4)  #define QF_ISR_UNLOCK(key_)   int_unlock(key_)
```

(1)   The macro QF_ISR_NEST tells QP-nano that interrupts can nest.

(2)   The macro QF_ISR_KEY_TYPE indicates a data type of the "interrupt key" variable, which holds the interrupt status. Defining this macro in the qpn_port.h header file indicates to the QF-nano framework that the policy of "saving and restoring interrupt status" is used for ISRs.

(3)   The macro QF_ISR_LOCK() encapsulates the mechanism of interrupt locking. The macro takes the parameter key_, into which it saves the interrupt lock status.

(4)   The macro QF_ISR_UNLOCK() encapsulates the mechanism of restoring the interrupt status. The macro restores the interrupt status from the argument key_.

As an example, where the "saving and restoring interrupt status" interrupt-locking policy for ISRs is useful, consider the MCUs based on the popular ARM7 or ARM9 cores such as the AT91 family from Atmel, the LPC family from NXP, the TMS470 family from TI, the STR7 and STR9 families from ST, and others. The ARM7/ARM9 architecture supports two types of interrupts called FIQ and IRQ. In most ARM-based MCUs, the IRQ interrupt is typically prioritized in a vectored interrupt controller, but the FIQ typically is not. This means that you should *never* unlock interrupts inside the FIQ interrupt, but you should unlock interrupts inside IRQs, to allow the priority controller do its job. The advanced policy of "saving and restoring interrupt status" can be used safely in both FIQ and IRQ interrupts (see also [Samek 07a]).

12.3.3   State Machines in QP-nano

Just like the full-version QP, QP-nano contains an hierarchical event processor called QEP-nano. QEP-nano supports both hierarchical and nonhierarchical state machines. The only difference between the QEP-nano and full-version QEP is that the current event in QEP-nano is directly embedded in the state machine. That way, the current event is accessible to the state machine via the "me" pointer and does not need to be passed as a parameter to the state-handler functions. Other than that, however, QEP-nano supports the same set of features as the full-version QEP, including fully hierarchical state machines with entry and exit actions and nested initial transitions. Listing 12.11 shows the declaration of the QHsm base structure (class) that serves for derivation of HSMs in QP-nano.

Listing 12.11 QHsm **structure and related functions**
(file <qp>\qpn\include\qepn.h**)**

```
(1)  typedef uint8_t QState;    /* status returned from a state-handler function */
(2)  typedef QState (*QStateHandler)(struct QHsmTag *me);

     typedef struct QHsmTag {
(3)      QStateHandler state;        /* current active state of the HSM (private) */
(4)      QEvent evt;           /* currently processed event in the HSM (protected) */
     } QHsm;

(5)  #define QHsm_ctor   (me_, initial_) ((me_)->state  = (initial_))
(6)  void    QHsm_init   (QHsm *me);
     #ifndef QK_PREEMPTIVE
(7)      void QHsm_dispatch(QHsm *me);
     #else
(8)      void QHsm_dispatch(QHsm *me) Q_REENTRANT;
     #endif
(9)  QState   QHsm_top         (QHsm *me);

(10) #define Q_SIG(me_)        (((QFsm *)(me_))->evt.sig)

     #if (Q_PARAM_SIZE != 0)
(11)     #define Q_PAR(me_)    (((FHsm *)(me_))->evt.par)
     #endif

(12) #define Q_RET_HANDLED     ((QState)0)
(13) #define Q_RET_IGNORED     ((QState)1)
(14) #define Q_RET_TRAN        ((QState)2)
(15) #define Q_RET_SUPER       ((QState)3)

(16) #define Q_HANDLED()       (Q_RET_HANDLED)
(17) #define Q_IGNORED()       (Q_RET_IGNORED)
(18) #define Q_TRAN(target_)   \
         (((QFsm *)me)->state  = (QStateHandler)(target_), Q_RET_TRAN)
(19) #define Q_SUPER(super_)   \
         (((QFsm *)me)->state  = (QStateHandler)(super_),  Q_RET_SUPER)
```

(1)   This typedef defines QState as a byte that conveys the status of the event handling to the event processor (see also lines (12-15)).

(2)   This typedef defines the QStateHandler as a pointer to state-handler function. As you can see, the state-handler functions in QP-nano take only the "me" pointer but no event parameter.

(3)   The QHsm structure stores the state-variable state, which is a pointer to state-handler function. Typically, a pointer to function requires just 2 or 4 bytes of RAM depending on given CPU and C compiler options.

(4) In QP-nano, the QHsm structure also stores the current event evt, which can take between 1 and 5 bytes of RAM, depending on the configuration macro Q_PARAM_SIZE (see Listing 12.2(1)).

(5) The QHsm "constructor" function-like macro initializes the state variable to the initial-pseudostate function that defines the initial transition. Note that the initial transition is not actually executed at this point.

(6) The QHsm_init() function triggers the initial transition in the state machine.

(7,8) The QHsm_dispatch() function dispatches one event to the state machine.

NOTE

Some compilers for 8-bit MCUs, most notably the Keil C51 compiler for 8051, don't generate ANSI-C compliant reentrant functions by default due to the limited stack architecture in 8051. These compilers allow dedicating specific functions to be reentrant with a special extended keyword (such as "reentrant" for Keil C51). The macro Q_REENTRANT is defined to nothing by default, to work with ANSI-C compliant compilers, but can be defined to "reentrant" to work with Keil C51 and perhaps other embedded compilers.

---

(9) The QHsm_top() function is the hierarchical state handler for the *top state*. The application-level state-handler functions that don't explicitly nest in any other state return the &QHsm_top pointer to the event processor.

(10) The Q_SIG() macro provides access to the signal of the current event embedded in the state machine.

(11) The Q_PAR() macro provides access to the scalar parameter of the current event embedded in the state machine. The macro is only defined if the event parameter is configured.

(12-15) These constants define the status returned from state-handler functions to the QEP-nano event processor. The status values are identical in QEP-nano as in the full-version QEP.

(16) A state-handler function returns the macro Q_HANDLED() whenever it handles the current event.

(17) A state-handler function returns the macro Q_IGNORED() whenever it ignores (does not handle) the current event.

(18) The Q_TRAN() macro encapsulates the transition, exactly as it is done in the full-version QEP. The Q_TRAN() macro is defined using the *comma expression*. A comma expression is evaluated from left to right, whereas the type and value of the whole expression is the rightmost operand. The rightmost operand is in this case the status of the operation (transition), which is returned from the state-handler function. The pivotal aspect of this design is that the Q_TRAN() macro can be used with respect to structures derived (inheriting) from QFsm or QHsm, which in C requires explicit casting (upcasting) to the QFsm base structure (see the sidebar "Single Inheritance in C" in Chapter 1).

(19) The Q_SUPER() macro serves for specifying the superstate, exactly as it is done in the full-version QEP. The Q_SUPER() macro is very similar to the Q_TRAN() macro except Q_SUPER() returns the different status to the event processor.

12.3.4   Active Objects in QP-nano

As shown in Figure 12.3, the QF-nano real-time framework provides the base structure QActive for deriving application-specific active objects. QActive combines the following three essential elements:

- It is a state machine (derives from QHsm or QFsm).

- It has an event queue.

- It has an execution thread with a unique priority.

Listing 12.12 shows the declaration of the QActive base structure and related functions.

Listing 12.12 The QActive **base class for derivation of active objects (file** <qp>\qpn\include\qfn.h**)**

```
      typedef struct QActiveTag {
      #ifndef QF_FSM_ACTIVE
(1)       QHsm super;                        /* derives from the QHsm base structure */
      #else
(2)       QFsm super;                        /* derives from the QFsm base structure */
      #endif
(3)       uint8_t prio;              /* active object priority 1..QF_MAX_ACTIVE */
(4)       uint8_t head;                        /* index to the event queue head */
(5)       uint8_t tail;                        /* index to the event queue tail */
```

```
(6)        uint8_t nUsed;        /* number of events currently present in the queue */
      #if (QF_TIMEEVT_CTR_SIZE != 0)
(7)        QTimeEvtCtr tickCtr;                         /* time event down-counter */
      #endif
      } QActive;

      #ifndef QF_FSM_ACTIVE
(8)        #define QActive_ctor(me_, initial_)   QHsm_ctor(me_, initial_)
      #else
(9)        #define QActive_ctor(me_, initial_)   QFsm_ctor(me_, initial_)
      #endif

      #if (Q_PARAM_SIZE != 0)
(10)       void QActive_post    (QActive *me, QSignal sig, QParam par);
(11)       void QActive_postISR(QActive *me, QSignal sig, QParam par);
      #else
(12)       void QActive_post    (QActive *me, QSignal sig);
(13)       void QActive_postISR(QActive *me, QSignal sig);
      #endif

      #if (QF_TIMEEVT_CTR_SIZE != 0)
(14)       void QF_tick(void);
      #if (QF_TIMEEVT_CTR_SIZE == 1)                    /* single-byte tick counter? */
(15)       #define QActive_arm(me_, tout_) ((me_)->tickCtr = (QTimeEvtCtr)(tout_))
(16)       #define QActive_disarm(me_)     ((me_)->tickCtr = (QTimeEvtCtr)0)
      #else                                             /* multi-byte tick counter */
(17)       void QActive_arm(QActive *me, QTimeEvtCtr tout);
(18)       void QActive_disarm(QActive *me);
      #endif                                    /* (QF_TIMEEVT_CTR_SIZE == 1) */
      #endif                                    /* (QF_TIMEEVT_CTR_SIZE != 0) */
```

(1) By default (when the macro `QF_FSM_ACTIVE` is *not* defined), the `QActive` structure derives from the `QHsm` base structure, meaning that active objects are hierarchical state machines in QP-nano.

(2) However, when you define the macro `QF_FSM_ACTIVE` in `qpn_port.h`, the `QActive` structure derives from the `QFsm` base structure. In this case, active objects are traditional "flat" state machines.

(3) Active object remembers its unique priority, which in QP-nano is the index into the `QF_active[]` array. Priority numbering in QP-nano is identical as in full-version QP. The lowest possible task priority is 1 and higher-priority values correspond to higher-urgency active objects. The maximum allowed active object priority is determined by the macro `QF_MAX_ACTIVE`, which in QP-nano cannot exceed 8. Priority level zero is reserved for the idle loop.

(4,5) These are the head and tail indices for the event queue buffer.

(6)   The `nUsed` data member represents the number of events currently present in the queue. This number includes the extra event embedded in the state machine itself, not just the number of events in the ring buffer, so that `nUsed` of zero indicates an empty queue.

(7)   The time event down-counter is only present when you define `QF_TIMEEVT_CTR_SIZE` in `qpn_port.h`. The member 'tickCtr' is the internal down-counter decremented in every `QF_tick()` invocation (see the next section). The time event is posted when the down-counter reaches zero.

(8,9)   The active object constructor boils down to calling the base class constructor, which is either `QHsm_ctor()` or `QFsm_ctor()`, depending on the definition of the macro `QF_FSM_ACTIVE`.

(10-13)   The signatures of functions `QActive_post()` and `QActive_postISR()` depend on the presence or absence of the event parameter.

(14)   The QP-nano function `QF_tick()` (see the next section) handles the timeout events. This function is only provided when time events are configured.

(15,16)   On any CPU with the word size of at least 8 bits, setting a single-byte variable is atomic. In this case the `QActive_arm()` and `QActive_disarm()` operations don't need to use a critical section. For speed, they are implemented as macros.

> **NOTE**
>
> The tick counters inside active objects are simultaneously accessed from the task level and from `QF_tick()`, which is called from the ISR level. To prevent corruption of the tick counters, they must be always accessed atomically.

(17,18)   For multibyte tick counters I assume that they are updated by multiple machine instructions. In this case the `QActive_arm()` and `QActive_disarm()` operations are declared as functions that use critical sections inside.

12.3.5   The System Clock Tick in QP-nano

To manage time events QP-nano requires that you invoke the `QF_tick()` function from a periodic time source called the system clock tick (see Chapter 6, "System Clock Tick"). The system clock tick typically runs at a rate between 10Hz and 100Hz.

Listing 12.13 shows the implementation of QF_tick(). In QP-nano, this function can be called only from the clock-tick ISR. QF_tick() must always run to completion and never preempt itself. In particular the clock-tick ISR that calls QF_tick() must not be allowed to preempt itself. In addition, QF_tick() should never be called from two different ISRs, which potentially could preempt each other.

Listing 12.13 QF_tick() **function (file** <qp>\qpn\source\qfn.c**)**

```
      void QF_tick(void) {
(1)       static uint8_t p;                    /* declared static to save stack space */
(2)       p = (uint8_t)QF_MAX_ACTIVE;
(3)       do {
(4)           static QActive *a;               /* declared static to save stack space */
(5)           a = (QActive *)Q_ROM_PTR(QF_active[p].act);
(6)           if (a->tickCtr != (QTimeEvtCtr)0) {
(7)               if ((--a->tickCtr) == (QTimeEvtCtr)0) {
#if (Q_PARAM_SIZE != 0)
(8)                   QActive_postISR(a, (QSignal)Q_TIMEOUT_SIG, (QParam)0);
#else
(9)                   QActive_postISR(a, (QSignal)Q_TIMEOUT_SIG);
#endif
              }
          }
(10)      } while ((--p) != (uint8_t)0);
      }
```

---

(1) The temporary variable 'p' (priority of an active object) is declared static to save the stack space.

NOTE

Many older, but still immensely popular low-end micros (e.g., 8051 and PIC) have very limited stack. For these CPUs, trading regular RAM to save stack space is very desirable.

---

(2) All active objects are scanned starting from the highest-priority QF_MAX_ACTIVE, which is also the exact number of active objects in the application (see Listing 12.1(12)).

(3) I use the do loop to avoid checking the loop condition the first time through. The number of active objects QF_MAX_ACTIVE is guaranteed to be at least one, so I don't need to check it.

(4)   The temporary variable 'a' (a pointer to active object) is declared `static` to save the stack space.

(5)   The active object pointer is loaded from the ROM array `QF_active[]`, which maps active object priorities to active object pointers.

(6)   The time event of a given active object is running if the tick counter is nonzero.

(7)   The tick counter is decremented and tested against zero. By reaching zero, the time event automatically disarms itself.

(8,9) `QF_tick()` posts the `Q_TIMEOUT_SIG` event to the active object that owns the counter by means of the `QActive_postISR()` function. For that reason `QF_tick()` can be called only from the ISR context.

(10)   The loop continues for all active object priorities above zero.

12.4   Event Queues in QP-nano

Each active object in QP-nano has its own event queue. The queue consists of one event located inside the `QActive` structure from which all active objects derive (see Section 12.3.4), plus a ring buffer of events that is allocated outside of the active object.

Figure 12.4 shows the data structures that QP-nano uses to manage event queues of active objects. The constant array `QF_active[]` stores the active object "control blocks," which are instances of the `QActiveCB` structure. Each `QF_active[]` element contains the pointer to the active object `act`, the pointer to the ring buffer `queue`, and the index of the last element of the ring buffer `end`. All these elements are initialized at compile time for each active object.

The `QActive` structure stores the event queue elements that are changing at runtime. The queue storage consists of the external, user-allocated ring buffer `queue` plus the current event `evt` stored inside the state machine (see Listing 12.11(4)). All events dispatched to the state machine must go through the "current event" `evt` data member, which is indicated as dashed lines in Figure 12.4. This extra location outside the ring buffer optimizes queue operation by frequently bypassing buffering, because in most cases queues alternate between empty and nonempty states with just one event present in the queue at a time.

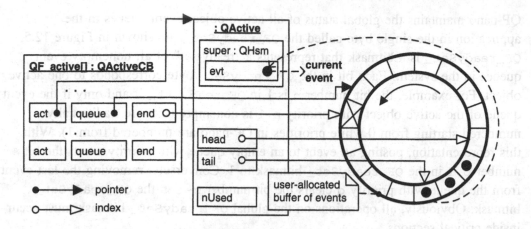

Figure 12.4: The relationship between the `QF_active[]` **array, the** `QActive` **struct and the ring buffer of event queue in QP-nano.**

NOTE

In extreme situations when the adequate queue depth is only one event, the whole ring buffer can be eliminated entirely. In such a case, you don't allocate the ring buffer and you use NULL as the queue pointer and zero as a valid queue length to initialize the `QF_active[]` array (see Listing 12.1(9-11)).

The ring-buffer indices `head` and `tail` as well as `end` from `QF_active[]` are relative to the `queue` pointer. These indices manage a ring buffer queue that the clients must preallocate as a contiguous array of events of type `QEvent`. Events are always extracted from the buffer at the `tail` index. New events are inserted at the `head` index, which corresponds to FIFO queuing. The `tail` index is always decremented when the event is extracted, as is the `head` index when an event is inserted. The `end` index limits the range of the `head` and `tail` indices that must "wrap around" to end once they reach zero. The effect is a counterclockwise movement of the `head` and `tail` indices around the ring buffer, as indicated by the arrow in Figure 12.4.

12.4.1 The Ready-Set in QP-nano (`QF_readySet_`)

QP-nano contains a cooperative "vanilla" kernel and a preemptive RTC kernel called QK-nano. To perform efficient scheduling in either one of these kernels,

QP-nano maintains the global status of all active object event queues in the application in the single byte called the QF_readySet_. As shown in Figure 12.5, QF_readySet_ is a bitmask that represents a "ready-set" of all nonempty event queues in the system. Each bit in the QF_readySet_ byte corresponds to one active object. For example, the bit number n is 1 in QF_readySet_ if and only if the event queue of the active object with priority n+1 is nonempty (bits are traditionally numbered starting from 0 while priorities in QP-nano are numbered from 1). With this representation, posting an event to an empty queue with priority p sets the bit number p-1 in the QF_readySet_ bitmask to 1. Conversely, removing the last event from the queue with priority q clears the bit number q-1 in the QF_readySet_ bitmask. Obviously, all operations on the global QF_readySet_ bitmask must occur inside critical sections.

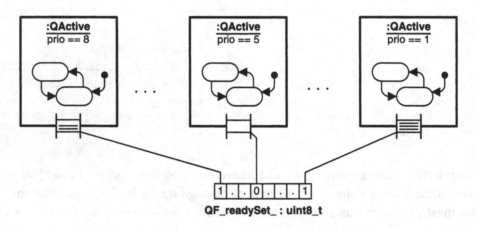

Figure 12.5: Representing state of all event queues in the QF_readySet_ **priority set.**

12.4.2   Posting Events from the Task Level (QActive_post())

Listing 12.14 shows the implementation of the QActive_post() function, which is used for posting events from one active object to another. You should *never* use this function to post events from the ISRs, because it uses task-level interrupt locking policy.

Listing 12.14 `QActive_post()` **function**
(file `<qp>\qpn\source\qfn.c`)

```
        #if (Q_PARAM_SIZE != 0)
(1)     void QActive_post(QActive *me, QSignal sig, QParam par) {
        #else
(2)     void QActive_post(QActive *me, QSignal sig) {
        #endif
(3)         QF_INT_LOCK();
            if (me->nUsed == (uint8_t)0) {                      /* is the queue empty? */
                ++me->nUsed;                                    /* update number of events */

(4)             Q_SIG(me) = sig;                                /* deliver the event directly */
        #if (Q_PARAM_SIZE != 0)
(5)             Q_PAR(me) = par;
        #endif
(6)             QF_readySet_ |= Q_ROM_BYTE(l_pow2Lkup[me->prio]);   /* set the bit */

        #ifdef QK_PREEMPTIVE
(7)             QK_schedule_();                        /* check for synchronous preemption */
        #endif
            }
            else {
(8)             QF_pCB_ = &QF_active[me->prio];
                        /* the queue must be able to accept the event (cannot overflow) */
(9)             Q_ASSERT(me->nUsed <= Q_ROM_BYTE(QF_pCB_->end));
                ++me->nUsed;                                    /* update number of events */
                                            /* insert event into the ring buffer (FIFO) */
(10)            ((QEvent *)Q_ROM_PTR(QF_pCB_->queue))[me->head].sig = sig;
        #if (Q_PARAM_SIZE != 0)
(11)            ((QEvent *)Q_ROM_PTR(QF_pCB_->queue))[me->head].par = par;
        #endif
(12)            if (me->head == (uint8_t)0) {
(13)                me->head = Q_ROM_BYTE(QF_pCB_->end);            /* wrap the head */
                }
(14)            --me->head;
            }
(15)        QF_INT_UNLOCK();
        }
```

(1,2)   The signature of the `QActive_post()` function depends whether you've
configured events with or without parameters.

(3)   The task-level posting to the event queue always happens in the task-level
critical section.

(4,5)   When the event queue is empty, the new event is copied directly to the current
event inside the state machine.

(6)   The bit corresponding to the priority of the active object is set in the QF_readySet_ bitmask. The constant lookup table l_pow2Lkup[] is initialized as follows: l_pow2Lkup[p] == (1 << (p-1)), for all priorities p = 1..8.

(7)   When the preemptive QK-nano kernel is configured (see Section 12.6), the preemptive scheduler is called to handle a potential synchronous preemption. (A synchronous preemption occurs when an active object posts an event to a higher-priority task.)

(8)   The global QF_pCB_ variable holds a pointer to the active object control block &QF_active[me->prio] located in ROM.

QF_pCB_ is defined as follows at the top of the qpn.c source file:

```
QActiveCB const Q_ROM * Q_ROM_VAR QF_pCB_;
```

The QF_pCB_ variable is used only locally within QP-nano functions, but I employ it to avoid loading the stack with a temporary variable. Such as global variable can be safely shared because all usage happens inside critical sections of code, anyway.

(9)   The assertion makes sure that the queue can accept this event.

NOTE

QP-nano treats the inability to post an event to a queue as an *error*. This assertion is part of the *event delivery guarantee policy* (see Section 6.7.6 in Chapter 6). It is the application designer's (your) responsibility to size the event queues adequately for the job at hand.

(10,11)   The new event is copied to the ring buffer at the head index.

(12)   The head index is checked for a wrap around.

(13)   If wrap around is required the head index is moved to the end of the buffer. This makes the buffer circular.

(14)   The head index is always decremented, including just after the wraparound. I've chosen to decrement the head (and also the tail) index because it leads to a more efficient implementation than incrementing the indices. The wraparound occurs in this case at zero rather than at the end. Comparing a variable to a constant zero is more efficient than any other comparison.

(15)   Interrupts are unlocked to leave the critical section.

12.4.3 Posting Events from the ISR Level (`QActive_postISR()`)

Listing 12.15 shows the implementation of the `QActive_postISR()` function, which is used for posting events from ISRs to active objects. You should *never* use this function to post events from active objects, because it uses the ISR-specific critical section mechanism (see Section 12.3.2).

> **Listing 12.15** `QActive_postISR()` **function**
> **(file** `<qp>\qpn\source\qfn.c`**)**

```
      #if (Q_PARAM_SIZE != 0)
      void QActive_postISR(QActive *me, QSignal sig, QParam par)
      #else
      void QActive_postISR(QActive *me, QSignal sig)
      #endif
      {
(1)   #ifdef QF_ISR_NEST
(2)   #ifdef QF_ISR_KEY_TYPE
(3)       QF_ISR_KEY_TYPE key;
(4)       QF_ISR_LOCK(key);
      #else
(5)       QF_INT_LOCK();
      #endif
      #endif
          if (me->nUsed == (uint8_t)0) {
              ++me->nUsed;                                  /* update number of events */

              Q_SIG(me) = sig;                              /* deliver the event directly */
      #if (Q_PARAM_SIZE != 0)
              Q_PAR(me) = par;
      #endif
              QF_readySet_ |= Q_ROM_BYTE(l_pow2Lkup[me->prio]);     /* set the bit */
          }
          else {
              QF_pCB_ = &QF_active[me->prio];
                      /* the queue must be able to accept the event (cannot overflow) */
              Q_ASSERT(me->nUsed <= Q_ROM_BYTE(QF_pCB_->end));
              ++me->nUsed;                                  /* update number of events */
                                          /* insert event into the ring buffer (FIFO) */
              ((QEvent *)Q_ROM_PTR(QF_pCB_->queue))[me->head].sig = sig;
      #if (Q_PARAM_SIZE != 0)
              ((QEvent *)Q_ROM_PTR(QF_pCB_->queue))[me->head].par = par;
      #endif
              if (me->head == (uint8_t)0) {
                  me->head = Q_ROM_BYTE(QF_pCB_->end);          /* wrap the head */
              }
              --me->head;
          }
      #ifdef QF_ISR_NEST
      #ifdef QF_ISR_KEY_TYPE
          QF_ISR_UNLOCK(key);
```

Continued onto next page

```
        #else
            QF_INT_UNLOCK();
        #endif
        #endif
        }
```

(1)   Interrupts are only locked when interrupt nesting is allowed.

(2)   If you define `QF_ISR_KEY_TYPE`, QP-nano uses the advanced policy of "saving and restoring interrupt status."

(3)   The advance policy requires a temporary variable `key`.

(4)   The interrupt status is saved into the `key` variable and interrupts are locked.

(5)   If you don't define `QF_ISR_KEY_TYPE`, QP-nano uses the simple policy of "unconditional interrupt unlocking," exactly the same as at the task level.

> **NOTE**
>
> The ISR-level event posting operation `QActive_postISR()` does *not* call the QK-nano scheduler, because a task can never synchronously preempt an ISR (compare Listing 12.14(7)).

12.5   The Cooperative "Vanilla" Kernel in QP-nano

By default, QP-nano uses the simple, cooperative "vanilla" scheduler, which works exactly as I described in Section 6.3.7 in Chapter 6. Listing 12.16 shows the `QF_run()` function in the `qfn.c` source file, which implements the whole "vanilla" kernel.

Listing 12.16 The cooperative "vanilla" kernel
(file `<qp>\qpn\source\qfn.c`)

```
(1)   #ifndef QK_PREEMPTIVE

      void QF_run(void) {
(2)       static uint8_t const Q_ROM Q_ROM_VAR log2Lkup[] = {
              0, 1, 2, 2, 3, 3, 3, 3, 4, 4, 4, 4, 4, 4, 4, 4
          };
(3)       static uint8_t const Q_ROM Q_ROM_VAR invPow2Lkup[] = {
              0xFF, 0xFE, 0xFD, 0xFB, 0xF7, 0xEF, 0xDF, 0xBF, 0x7F
          };
(4)       static QActive *a;                /* declared static to save stack space */
```

```
(5)          static uint8_t p;                          /* declared static to save stack space */

             /* trigger initial transitions in all registered active objects... */
(6)          for (p = (uint8_t)1; p <= (uint8_t)QF_MAX_ACTIVE; ++p) {
(7)              a = (QActive *)Q_ROM_PTR(QF_active[p].act);
(8)              Q_ASSERT(a != (QActive *)0);   /* QF_active[p] must be initialized */
(9)              a->prio = p;                 /* set the priority of the active object */

     #ifndef QF_FSM_ACTIVE
(10)             QHsm_init((QHsm *)a);         /* take the initial transition in HSM */
     #else
(11)             QFsm_init((QFsm *)a);         /* take the initial transition in FSM */
     #endif
             }

(12)     QF_onStartup();                               /* invoke startup callback */

(13)     for (;;) {                            /* the event loop of the vanilla kernel */
(14)         QF_INT_LOCK();
(15)         if (QF_readySet_ != (uint8_t)0) {
     #if (QF_MAX_ACTIVE > 4)
(16)             if ((QF_readySet_ & 0xF0) != 0U) {     /* upper nibble used? */
(17)                 p = (uint8_t)(Q_ROM_BYTE(log2Lkup[QF_readySet_ >> 4]) + 4);
                 }
                 else                        /* upper nibble of QF_readySet_ is zero */
     #endif
                 {
(18)                 p = Q_ROM_BYTE(log2Lkup[QF_readySet_]);
                 }
(19)             QF_INT_UNLOCK();

(20)             a = (QActive *)Q_ROM_PTR(QF_active[p].act);

     #ifndef QF_FSM_ACTIVE
(21)             QHsm_dispatch((QHsm *)a);                         /* dispatch to HSM */
     #else
(22)             QFsm_dispatch((QFsm *)a);                         /* dispatch to FSM */
     #endif

(23)             QF_INT_LOCK();
(24)             if ((--a->nUsed) == (uint8_t)0) {      /* queue becoming empty? */
(25)                 QF_readySet_ &= Q_ROM_BYTE(invPow2Lkup[p]);/* clear the bit */
                 }
                 else {
(26)                 QF_pCB_ = &QF_active[a->prio];

(27)                 Q_SIG(a) = ((QEvent *)Q_ROM_PTR(QF_pCB_->queue))[a->tail].sig;
     #if (Q_PARAM_SIZE != 0)
(28)                 Q_PAR(a) = ((QEvent *)Q_ROM_PTR(QF_pCB_->queue))[a->tail].par;
     #endif
(29)                 if (a->tail == (uint8_t)0) {                 /* wrap around? */
```

Continued onto next page

```
(30)                        a->tail = Q_ROM_BYTE(QF_pCB_->end);    /* wrap the tail */
                       }
(31)                   --a->tail;
                   }
(32)               QF_INT_UNLOCK();
               }
(33)           else {
(34)               QF_onIdle();                                    /* see NOTE01 */
               }
           }
       }

       #endif                                          /* #ifndef QK_PREEMPTIVE */
```

(1)   The cooperative vanilla kernel is only compiled when the preemptive QK-nano kernel is *not* configured.

(2)   The constant array `log2Lkup[]` is the binary-logarithm (log-base-2) lookup table, defined as `log2Lkup[bitmask] == log2(bmask)`, for `0 <= bitmask <= 15` (see Figure 7.6 in Chapter 7). The log-base-2 lookup quickly determines the most significant 1-bit in the bitmask.

(3)   The constant array `invPow2Lkup[]` is a bitwise negated power-2 lookup table, defined as `invPow2Lkup[p] == ~(1 << (p-1))`, for all priorities `p = 1..8`. This lookup table is used for masking off bits in the `QF_readSet_` bitmask.

(4,5) The temporary variables 'a' and 'p' are defined as `static` to save stack space.

NOTE

The static variable 'p' inside `QF_run()` is different than the analogous variable 'p' inside `QF_tick()`. You cannot reuse the same static variable for both, because these functions execute concurrently.

---

(6)   This `for` loop triggers initial transitions in all active objects, starting from the lowest-priority active object.

NOTE

Generally, active objects should be initialized in the order of priority because the lowest-priority active objects tend to have the longest event queues. This might be important if active objects post events to each other from the initial transitions.

---

(7)   The active object pointer is obtained from the active object control block in ROM.

(8)   This assertion makes sure that the QF_active[] array has been initialized correctly.

(9)   This internal priority of the active object is initialized consistently with the definition in the QF_active[] array.

NOTE

An active object instance (QActive) needs to know its priority to quickly access the corresponding control block (QActiveCB) by indexing into the QF_active[] array.

(10,11)   The initial transition in the active object state machine is triggered.

(12)   The QF_onStartup() callback function configures and starts interrupts. This function is implemented at the application level (in the BSP).

(13)   This is the *event loop* of the "vanilla" kernel.

(14)   Interrupts are locked to access the QF_readySet_ ready-set.

(15)   If the ready-set QF_readySet_ is not empty, the "vanilla" kernel has some events to process.

At this point, the vanilla kernel must quickly find the highest-priority active object with a nonempty event queue, which is achieved via a binary logarithm lookup table (see Section 7.11.1 in Chapter 7). However, to conserve the ROM, the log2Lkup[] lookup table in QP-nano can only handle values 0..15, which covers just four least significant bits of the QF_readySet_ bitmask.

(16)   When the number of active objects is greater than the range of the log2Lkup[] lookup table, I first test the upper nibble of the QF_readySet_ bitmask.

(17)   If the upper nibble is not zero, I shift the upper nibble to the lower 4 bits and apply the log2Lkup[] lookup table. I then need to add 4 to the priority of the active object.

(18)   Otherwise, I simply apply the log2Lkup[] lookup table to the lower nibble of the QF_readySet_ bitmask.

(19)   Interrupts can be unlocked.

(20)   The active object pointer is loaded from the ROM array QF_active[], which maps active object priorities to active object pointers.

(21,22)   The current event is dispatched to the active object state machine.

> **NOTE**
>
> Dispatching of the event into the state machine represents the *run-to-completion step* of the active object thread.

(23)   Interrupts are locked again to update the status of the active object event queue after processing the current event.

(24)   The number of events in the queue is decremented and tested against zero.

(25)   If the queue is becoming empty, the corresponding bit in the QF_readySet_ bitmask is cleared by masking it off with the invPow2Lkup[] lookup table.

(26)   Otherwise, the queue is not empty, so the next event must be copied from the ring buffer to the current event inside the active object state machine. The global pointer QF_pCB_ is set to point to the control block of the active object &QF_active[me->prio] located in ROM.

(27,28)   The next event is copied from the ring buffer at the tail index to the current event inside the state machine.

(29,30)   The tail index is checked for wraparound.

(31)   The tail index is always decremented.

(32)   Interrupts can be unlocked.

(33)   The else branch is taken when all active object event queues are empty, which is by definition the *idle condition* of the "vanilla" kernel.

(34)   The "vanilla" kernel calls the QF_onIdle() callback function to give the application a chance to put the MCU to a low-power sleep mode. The QF_onIdle() function is typically implemented at the application level (in the BSP).

> **NOTE**
>
> Most MCUs provide software-controlled low-power sleep modes, which are designed to reduce power dissipation by gating the clock to the CPU and various peripherals. To ensure a safe transition to a sleep mode, the "vanilla" kernel calls QF_onIdle() with interrupts *locked*. The QF_onIdle() function *must* always unlock interrupts internally, ideally atomically with the transition to a sleep mode.

12.5.1 Interrupt Processing Under the "Vanilla" Kernel

Interrupt processing under the "vanilla" kernel is as simple as in the foreground/background architecture. Typically, you can use the ISRs generated by the C compiler. The only special consideration for QP-nano is that you need to be consistent with respect to the interrupt-nesting policy. In particular, if you configured nesting interrupts by defining the macro QF_ISR_NEST in the qpn_port.h header file, you need to be consistent and unlock interrupts before calling the QP-nano functions QActive_postISR() or QF_tick().

12.5.2 Idle Processing under the "Vanilla" Kernel

The idle callback QF_onIdle() works in QP-nano exactly the same way as in the full-version QP. Refer to Section 8.2.4 in Chapter 8 for examples of defining this callback function for various CPUs.

12.6 The Preemptive Run-to-Completion QK-nano Kernel

QP-nano contains a preemptive run-to-completion (RTC) kernel called QK-nano, which works very similarly to the QK preemptive kernel available in the full-version QP and described in Chapter 10. I strongly recommend that you read Section 10.1 in Chapter 10 before you decide to use a preemptive kernel such as QK-nano.

Note that the preemptive QK-nano kernel puts more demands on the target CPU and the compiler than the nonpreemptive vanilla kernel. Generally, QK-nano can be used with a given processor and compiler if they satisfy the following requirements:

* The processor supports a hardware stack that can accommodate stack variables (not just return addresses).

- The MCU has enough stack memory to allow at least two levels of task nesting so that a high-priority active object can preempt a low-priority active object at least once. If you don't have even that much stack, you'll not be able to take advantage of the preemptive kernel.

- The C or C++ compiler can generate reentrant code. In particular, the compiler must be able to allocate automatic variables on the stack.

For example, some older CPU architectures, such as the 8-bit PIC MCUs, don't have a C-friendly stack architecture, even though they might have enough RAM, and consequently cannot easily run QK-nano.

12.6.1   QK-nano Interface `qkn.h`

You configure your QP-nano application to use the QK-nano kernel (as opposed to the default "vanilla" kernel) simply by including the QK-nano interface `qkn.h` at the end of the `qpn_port.h` header file. Listing 12.17 shows the `qkn.h` header file.

Listing 12.17 QK-nano interface (file `<qp>\qpn\include\qkn.h`)

```
      #ifndef qkn_h
      #define qkn_h

(1)   #define QK_PREEMPTIVE   1

(2)   void QK_init(void);
(3)   void QK_schedule_(void) Q_REENTRANT;
(4)   void QK_onIdle(void);

(5)   extern uint8_t volatile QK_currPrio_;                    /* current QK priority */

      #ifndef QF_ISR_NEST
(6)       #define QK_SCHEDULE_() \
(7)           if (QF_readySet_ != (uint8_t)0) { \
                  QK_schedule_(); \
              } else ((void)0)
      #else
(8)       extern uint8_t volatile QK_intNest_;             /* interrupt nesting level */

          #define QK_SCHEDULE_() \
(9)           if ((QF_readySet_ != (uint8_t)0) && (QK_intNest_ == (uint8_t)0)) { \
                  QK_schedule_(); \
              } else ((void)0)
      #endif
```

```
          #ifdef QK_MUTEX
(10)          typedef uint8_t QMutex;
(11)          QMutex QK_mutexLock    (uint8_t prioCeiling);
(12)          void   QK_mutexUnlock(QMutex mutex);
          #endif                                                    /* QK_MUTEX */

          #endif                                                    /* qkn_h */
```

(1)   The qkn.h header file defines the macro QK_PREEMPTIVE, which configures the
      QF-nano real-time framework to use the QK-nano preemptive kernel rather than
      the cooperative "vanilla" kernel.

(2)   The QK_init() function performs CPU-specific initialization, if such
      initialization is necessary. This function is optional and not all QK-nano ports
      need to implement this function. However, if the function is provided, your
      application must call it, typically during BSP initialization.

(3)   The QK_schedule_() function implements the QK-nano scheduler. The QK
      scheduler is always invoked from a critical section but might unlock interrupts
      internally to launch a task.

NOTE

The macro Q_REENTRANT tells the compiler to generate ANSI-C compliant reentrant function code. See also Listing 12.11(8).

---

(4)   The QK_onIdle() callback function gives the application a chance to customize
      the idle processing (see also Listing 12.18(11)).

(5)   The global variable QK_currPrio_ represents the global systemwide priority of
      the currently running task. QK_currPrio_ is declared as volatile because it
      can change asynchronously in ISRs.

(6)   The QK_SCHEDULE_() encapsulates the invocation of the QK-nano scheduler at
      the exit from an interrupt to handle asynchronous preemptions.

(7)   When interrupt nesting is *not* allowed, the QK_SCHEDULE_() macro calls the
      scheduler only when the ready-set is not empty. That way, you avoid a function
      call overhead when all event queues are empty.

(8) The global variable `QK_intNest_` represents the global systemwide interrupt-nesting level. `QK_intNest_` is declared as `volatile` because it can change asynchronously in ISRs.

(9) When interrupt nesting *is* allowed, the `QK_SCHEDULE_()` macro calls the scheduler only when the ready-set is not empty and additionally the interrupt that is ending is not nesting on another interrupt (see also Section 12.6.4).

(10-12) When you define the macro `QK_MUTEX`, `qkn.h` defines the priority-ceiling mutex interface.

12.6.2 Starting Active Objects and the QK-nano Idle Loop

As shown in Listing 12.17(1), the `qkn.h` header file defines internally the macro `QK_PREEMPTIVE`, which causes elimination of the "vanilla" kernel (see Listing 12.16 (1)), replacing it with the preemptive QK-nano kernel. In particular, the function `QF_run()` has a different implementation under the QK-nano kernel, as shown in Listing 12.18.

Listing 12.18 Starting active objects and the QK-nano idle loop (file `<qp>\qpn\source\qkn.c`)

```
    /* Global-scope objects -----------------------------------------------*/
(1) uint8_t volatile QK_currPrio_ = (uint8_t)(QF_MAX_ACTIVE + 1);
    #ifdef QF_ISR_NEST
(2)     uint8_t volatile QK_intNest_;              /* start with nesting level of 0 */
    #endif

(3) extern QActiveCB const Q_ROM * Q_ROM_VAR QF_pCB_; /* ptr to AO control block */

    /* local objects -----------------------------------------------------*/
(4) static QActive *l_act;                                      /* pointer to AO */

    /*............................................................*/
    void QF_run(void) {
            /* trigger initial transitions in all registered active objects... */
(5)     static uint8_t p;                  /* declared static to save stack space */
(6)     for (p = (uint8_t)1; p <= (uint8_t)QF_MAX_ACTIVE; ++p) {
            l_act = (QActive *)Q_ROM_PTR(QF_active[p].act);
            l_act->prio = p;

    #ifndef QF_FSM_ACTIVE
            QHsm_init((QHsm *)l_act);                        /* initial transition */
```

```
      #else
              QFsm_init((QFsm *)l_act);                /* initial transition */
      #endif
          }

          QF_INT_LOCK();
 (7)      QK_currPrio_ = (uint8_t)0;        /* set the priority for the QK idle loop */
 (8)      QK_SCHEDULE_();                      /* process all events produced so far */
          QF_INT_UNLOCK();

 (9)      QF_onStartup();                              /* invoke startup callback    */

 (10)     for (;;) {                                  /* enter the QK idle loop */
 (11)         QK_onIdle();                      /* invoke the on-idle callback */
          }
      }
```

(1) The global variable `QK_currPrio_` represents the global systemwide priority of the currently running task. The QK-nano priority is initialized to a level above any active object, which effectively locks the QK-nano scheduler during the whole initial transient.

(2) The global variable `QK_intNest_` represents the global systemwide interrupt-nesting level. `QK_intNest_` is only necessary when nesting of interrupts is allowed.

(3) The `qkn.c` module reuses the global variable `QF_pCB_`, which is also used in `QActive_post()` and `QActive_postISR()` defined in `qfn.c` (see Listings 12.14 and 12.15).

(4) The static variable `l_act` holds a pointer to an active object and is shared among QK-nano functions.

(5) The local variable 'p' (active object priority) is declared `static` to save the stack space.

(6) All active objects in the application are initialized, exactly the same way as in Listing 12.16(6-11).

(7) The QK-nano priority is lowered to the idle-loop level.

(8) The QK-nano scheduler is invoked to process all events that might have been posted to event queues during the initialization of the active objects.

(9)   The `QF_onStartup()` callback function configures and starts interrupts. This function is implemented at the application level (in the BSP).

(10)   This is the *idle loop* of the QK-nano kernel.

(11)   The idle loop continuously calls the `QK_onIdle()` callback function to give the application a chance to put the CPU to a low-power sleep mode. The `QK_onIdle()` function is typically implemented at the application level (in the BSP).

NOTE

As a preemptive kernel, QK-nano handles idle processing differently than does the non-preemptive vanilla kernel. Specifically, the `QK_onIdle()` callback function is always called with interrupts *unlocked* and does not need to unlock interrupts (as opposed to the `QF_onIdle()` callback). Furthermore, a transition to a low-power sleep mode inside `QK_onIdle()` does not need to occur with interrupts locked. Such a transition is safe and does not cause any race conditions, because a preemptive kernel never switches the context back to the idle loop as long as events are available for processing.

12.6.3 The QK-nano Scheduler

The scheduler is the most important part of the QK-nano kernel. As explained in Section 10.2.3 in Chapter 10, the QK scheduler is called at two junctures: (1) when an event is posted to an event queue of an active object (synchronous preemption), and (2) at the end of ISR processing (asynchronous preemption). In the `QActive_post()` function implementation (Listing 12.14(7)), you saw how the QK-nano scheduler gets invoked to handle the synchronous preemptions. In the previous section, you also saw the definition of the macro `QK_SCHEDULE_()`, which calls the scheduler from an interrupt context to handle the asynchronous preemptions. Here I describe the QK-nano scheduler itself.

The QK-nano scheduler is simply a regular C-function `QK_schedule_()` whose job is to efficiently find the highest-priority active object that is ready to run and to execute it as long as its priority is higher than the currently serviced QK-nano priority. To perform this job, the QK-nano scheduler relies on two data elements: the set of tasks that are ready to run `QF_readySet_` (Section 12.4.1) and the currently serviced task priority `QK_currPrio_` (Listing 12.17(5)). Listing 12.19 shows the complete implementation of the `QK_schedule_()` function. You will certainly recognize in this function many elements that I already discussed for the nonpreemptive "vanilla" kernel.

Listing 12.19 The preemptive QK-nano scheduler
(file `<qp>\qpn\source\qkn.c`)

```
(1)     void QK_schedule_(void) Q_REENTRANT {
            static uint8_t const Q_ROM Q_ROM_VAR log2Lkup[] = {
                0, 1, 2, 2, 3, 3, 3, 3, 4, 4, 4, 4, 4, 4, 4, 4
            };
            static uint8_t const Q_ROM Q_ROM_VAR invPow2Lkup[] = {
                0xFF, 0xFE, 0xFD, 0xFB, 0xF7, 0xEF, 0xDF, 0xBF, 0x7F
            };
(2)         uint8_t p;            /* the new highest-priority active object ready to run */

                        /* determine the priority of the highest-priority AO ready to run */
        #if (QF_MAX_ACTIVE > 4)
            if ((QF_readySet_ & 0xF0) != 0) {                    /* upper nibble used? */
(3)             p = (uint8_t)(Q_ROM_BYTE(log2Lkup[QF_readySet_ >> 4]) + 4);
            }
            else                          /* upper nibble of QF_readySet_ is zero */
        #endif
            {
(4)             p = Q_ROM_BYTE(log2Lkup[QF_readySet_]);
            }

(5)         if (p > QK_currPrio_) {   /* is the new priority higher than the current? */
(6)             uint8_t pin = QK_currPrio_;                  /* save the initial priority */
(7)             do {
(8)                 QK_currPrio_ = p;   /* new priority becomes the current priority */
(9)                 QF_INT_UNLOCK();      /* unlock interrupts to launch the new task */

                                    /* dispatch to HSM (execute the RTC step) */
        #ifndef QF_FSM_ACTIVE
(10)                QHsm_dispatch((QHsm *)Q_ROM_PTR(QF_active[p].act));
        #else
(11)                QFsm_dispatch((QFsm *)Q_ROM_PTR(QF_active[p].act));
        #endif

(12)                QF_INT_LOCK();
                            /* set cb and a again, in case they change over the RTC step */
(13)                QF_pCB_ = &QF_active[p];
(14)                l_act = (QActive *)Q_ROM_PTR(QF_pCB_->act);

(15)                if ((--l_act->nUsed) == (uint8_t)0) { /* is queue becoming empty? */
                                                        /* clear the ready bit */
(16)                    QF_readySet_ &= Q_ROM_BYTE(invPow2Lkup[p]);
                    }
                    else {
(17)                    Q_SIG(l_act) =
                                ((QEvent *)Q_ROM_PTR(QF_pCB_->queue))[l_act->tail].sig;
        #if (Q_PARAM_SIZE != 0)
(18)                    Q_PAR(l_act) =
```

Continued onto next page

```
                            ((QEvent *)Q_ROM_PTR(QF_pCB_->queue))[l_act->tail].par;
        #endif
(19)                if (l_act->tail == (uint8_t)0) {              /* wrap around? */
(20)                    l_act->tail = Q_ROM_BYTE(QF_pCB_->end); /* wrap the tail */
                    }
(21)                --l_act->tail;                      /* always decrement the tail */
            }
                            /* determine the highest-priority AO ready to run */
(22)            if (QF_readySet_ != (uint8_t)0) {
        #if (QF_MAX_ACTIVE > 4)
(23)                if ((QF_readySet_ & 0xF0) != 0) {       /* upper nibble used? */
(24)                    p = (uint8_t)(Q_ROM_BYTE(log2Lkup[QF_readySet_ >> 4])+4);
                    }
                    else              /* upper nibble of QF_readySet_ is zero */
        #endif
                    {
(25)                    p = Q_ROM_BYTE(log2Lkup[QF_readySet_]);
                    }
                }
                else {
(26)                p = (uint8_t)0;                        /* break out of the loop */
                }
(27)        } while (p > pin);         /* is the new priority higher than initial? */
(28)        QK_currPrio_ = pin;                   /* restore the initial priority */
        }
(29)  }
```

(1)   The QK-nano scheduler must necessarily be a *reentrant* function. The scheduler is always invoked with *interrupts locked* but might need to unlock interrupts internally to launch a task.

(2)   The stack variable 'p' will hold the priority of the new highest-priority active object (task) ready to run.

NOTE

The 'p' variable *must* necessarily be allocated on the stack, because each level of preemption needs to compute a separate copy of the highest-priority active object ready to run. The compiler should never place the 'p' variable in a static memory. For that reason the QK_schedule_() function is declared as "reentrant" in line (1).

(3,4)   The highest-priority active object with a not-empty event queue is quickly found out by means of the binary-logarithm lookup table, exactly the same way as already explained in Listing 12.16(16-18).

(5)   The QK scheduler can launch a task only when the new priority is higher than the saved priority of the currently executing task.

> **NOTE**
>
> The QK-nano scheduler is an indirectly recursive function. The scheduler calls task functions, which might post events to other tasks, which calls the scheduler (synchronous preemption). The scheduler can also be preempted by ISRs, which also call the scheduler at the exit (asynchronous preemption). However, this recursion can continue only as long as the priority of the tasks keeps increasing. Posting an event to a lower- or equal-priority task (posting to self) stops the recursion because of the `if` statement in line (5).

(6)   To handle the preemption, the QK-nano scheduler will need to increase the current priority. However, before doing this, the current priority is saved into a stack variable `pin`.

> **NOTE**
>
> The 'pin' variable *must* necessarily be allocated on the stack because each level of preemption needs to save its own copy of the current priority. The compiler should never place the 'pin' variable in a static memory. For that reason the `QK_schedule_()` function is declared as "reentrant" in line (1).

(7)   The `do` loop continues as long as the scheduler finds ready-to-run tasks of higher priority than the initial priority `pin`.

(8)   The current QK-nano priority is raised to the level of the highest-priority task that is about to be started.

(9)   Interrupts are unlocked to launch the new RTC task.

(10,11)   The current event is dispatched to the active object state machine.

> **NOTE**
>
> Dispatching of the event into the state machine represents the *run-to-completion step* of the active object thread. Note that the RTC task is executed with interrupts unlocked.

(12)   Interrupts are locked again to update the status of the active object event queue after processing the current event.

(13)   The global pointer `QF_pCB_` is set to point to the control block of the active object `&QF_active[p]` located in ROM.

(14)   The static active object pointer `l_act` is set to point to the active object with priority 'p.'

> **NOTE**
>
> The active object pointed to by `l_act` is the same one that just finished the RTC step in lines 10-11. However, I did not set the `l_act` earlier because the local variable `l_act` could have changed when the interrupts were unlocked.

(15)   The number of events in the queue is decremented and tested against zero.

(16)   If the queue is becoming empty, the corresponding bit in the `QF_readySet_` bitmask is cleared by masking it off with the `invPow2Lkup[]` lookup table.

(17,18)   The next event is copied from the ring buffer at the `tail` index to the current event inside the state machine.

(19,20)   The `tail` index is checked for wraparound.

(21)   The `tail` index is always decremented.

(22)   If some active objects are still ready to run.

(23-25)   The scheduler finds out the next highest-priority active objects ready to run.

(26)   If the `QF_readySet_` turns out to be empty, the QK-nano kernel has nothing more to do. The variable 'p' is set to zero to terminate the `do-while` loop in the next step.

(27)   The `while` condition loops back to step 7 as long as the QK-nano scheduler still finds ready-to-run tasks of higher priority than the initial priority `pin`.

(28)   After the loop terminates, the current QK-nano priority must be restored back to the initial level.

(29)   The QK-nano scheduler always returns with interrupts locked.

12.6.4 Interrupt Processing in QK-nano

QK-nano can typically work with ISRs synthesized by the C compiler, which most embedded C cross-compilers support. However, unlike the nonpreemptive vanilla kernel, the preemptive QK-nano kernel must be notified about entering and exiting every ISR to handle potential asynchronous preemptions. The specific actions required at the entry and exit from ISRs depend on the interrupt-locking policy for ISRs.

When interrupt nesting is allowed (the macro QF_ISR_NEST *is* defined in qpn_port.h), the interrupt processing in QK-nano is exactly the same as described in Section 10.3.3 for the full-version QK. In particular, the interrupt-nesting counter QK_intNest_ must be incremented upon interrupt entry and decremented upon interrupt exit. The QK-nano scheduler can only be invoked when the interrupt-nesting counter is zero, meaning that the calling ISR is not nested on top of another ISR.

However, when interrupt nesting is *not* allowed (the macro QF_ISR_NEST is *not* defined in qpn_port.h), QK-nano allows for a simpler interrupt handling compared to the full-version QK. Specifically, when interrupts cannot nest you don't need to increment the interrupt-nesting counter (QK_intNest_) upon ISR entry, and you don't need to decrement it upon ISR exit. In fact, when the macro QF_ISR_NEST *not* defined, the QK_intNest_ counter is not even available. This simplification is possible because QP-nano uses a special ISR version of the event-posting function QActive_postISR(), which does *not* call the QK-nano scheduler. Consequently, there is no need to prevent the synchronous preemption within ISRs. Listing 12.20 shows the simplified interrupt handling when interrupt nesting is *not* allowed.

Listing 12.20 ISR structure when interrupt nesting is *not* allowed

```
void interrupt YourISR(void) {         /* typically entered with interrupts locked */
                                       /* QK_ISR_ENTRY() - empty */
    Clear the interrupt source, if necessary
    Execute ISR body, including calling QP-nano services, such as:
    QActive_postISR() or QF_tick()

    Send the EOI instruction to the interrupt controller, if necessary
    QK_SCHEDULE_();                                  /* QK_ISR_EXIT() */
}
```

12.6.5 Priority Ceiling Mutex in QK-nano

QK-nano supports the priority-ceiling mutex mechanism in the exact same way as the full-version QK (see Section 10.4.1 in Chapter 10). In QK-nano, the feature is disabled by default, but you can enable it by defining the macro `QK_MUTEX` in the `qpn_port.h` header file.

12.7 The PELICAN Crossing Example

The "Fly 'n' Shoot" game example described at the beginning of this chapter shows most features of QP-nano, but it fails to demonstrate QP-nano running in a really small MCU. In this section, I describe the pedestrian light-controlled (PELICAN) crossing example application, which demonstrates a nontrivial hierarchical state machine, event exchanges among active objects and ISRs, time events, and even the QK-nano preemptive kernel. I was able to squeeze this application inside the MSP430-F2013 ultra-low-power MCU with only 128 bytes of RAM and 2KB of flash ROM. Specifically, the accompanying code was tested on the eZ430-F2013 USB stick[3] from Texas Instruments. The eZ430-F2013 is a complete development tool that provides both the USB-based debugger and a MSP430-F2013 development board in a small USB stick package (see Figure 12.6).

The code accompanying this book contains several versions of the PELICAN crossing application for MSP430, Cortex-M3, and 80x86/DOS. For each target CPU I provide two versions: the nonpreemptive configuration with vanilla kernel and the preemptive QK-nano configuration. Refer to Listing 12.7 in Section 12.3.1 for the location of these examples.

Before I describe the PELICAN crossing controller state machine and the QP-nano implementation, I need to provide the problem specification. The PELICAN crossing (see Figure 12.7) starts with cars enabled (green light for cars) and pedestrians disabled ("Don't Walk" signal for pedestrians). To activate the traffic light change, a pedestrian must push the button at the crossing, which generates the `PEDS_WAITING` event. In response, oncoming cars get a yellow light, which after a few seconds changes to red light. Next, pedestrians get the "Walk" signal, which shortly thereafter changes to the flashing "Don't Walk" signal. When the "Don't Walk" signal stops flashing,

---

[3] At the time of this writing the eZ430-F2013 USB stick was available for $20 from the TI site (www.ti.com/eZ430).

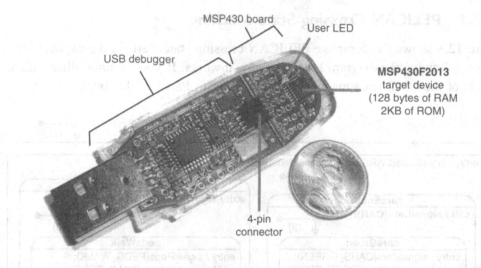

Figure 12.6: The eZ430-F2013 USB stick.

cars get the green light again. After this cycle, the traffic lights don't respond to the
PEDS_WAITING button press immediately, although the button "remembers" that it has
been pressed. The traffic light controller always gives the cars a minimum of several
seconds of green light before repeating the traffic light change cycle. One additional
feature is that at any time an operator can take the PELICAN crossing offline (by
providing the OFF event). In the "offline" mode, the cars get a flashing red light and
pedestrians a flashing "Don't Walk" signal. At any time the operator can turn the
crossing back online (by providing the ON event).

Figure 12.7: Pedestrian light-controlled (PELICAN) crossing.

12.7.1 PELICAN Crossing State Machine

Figure 12.8 shows the complete PELICAN crossing statechart. In the explanation section following the diagram I describe how it works. If you are unfamiliar with some aspects of the UML state machine notation, refer to Part I of this book.

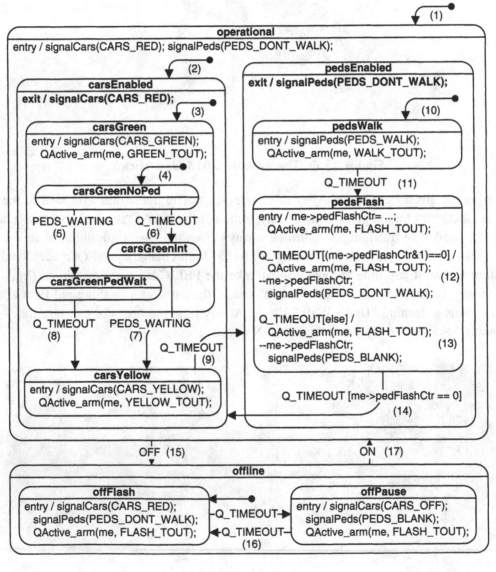

Figure 12.8: PELICAN crossing state machine.

(1)   Upon the initial transition, the PELICAN state machine enters the "operational" state and displays the red light for cars and the "Don't Walk" signal for pedestrians.

(2)   The "operational" state has a nested initial transition to the "carsEnabled" substate. Per the UML semantics, this transition must be taken after entering the superstate.

(3)   The "carsEnabled" state has a nested initial transition to the "carsGreen" substate. Per the UML semantics, this transition must be taken after entering the sperstate. Entry to "carsGreen" changes signals green light for cars and arms the time event to expire in the GREEN_TOUT clock ticks. The GREEN_TOUT timeout represents the minimum duration of green light for cars.

(4)   The "carsGreen" state has a nested initial transition to the "carsGreenNoPed" substate. Per the UML semantics, this transition must be taken after entering the superstate. The "carsGreenNoPed" state is a leaf state, meaning that it has no substates or initial transitions. The state machine stops and waits in this state.

(5)   When the PEDS_WAITING event arrives in the "carsGreenNoPed" state, the state machine transitions to another leaf state "carsGreenPedWait." Note that the state machine still remains in the "carsGreen" superstate because the minimum green light period for cars hasn't expired yet. However, by transitioning to the "carsGreenPedWait" substate, the state machine remembers that the pedestrian is waiting.

(6)   However, when the Q_TIMEOUT event arrives while the state machine is still in the "carsGreenNoPed" state, the state machine transitions to the "carsGreenInt" (interruptible green light for cars) state.

(7)   The "carsGreenInt" state handles the PEDS_WAITING event by immediately transitioning to the "carsYellow" state, because the minimum green light for cars has elapsed.

(8)   The "carsGreenPedWait" state, on the other hand, handles only the Q_TIMEOUT event, because the pedestrian is already waiting for the expiration of the minimum green light for cars.

(9)   The "carsYellow" state displays the yellow light for cars and arms the timer for the duration of the yellow light. The Q_TIMEOUT event causes the transition to the "pedsEnabled" state. The transition causes exit from the "carsEnabled" superstate, which displays the red light for cars.

The pair of states "carsEnabled" and "pedsEnabled" realizes the main function of the PELICAN crossing, which is to alternate between enabling cars and enabling pedestrians. The exit action from "carsEnabled" disables cars (by showing a red light for cars) while the exit action from "pedsEnabled" disables pedestrians (by showing them the "Don't Walk" signal). The UML semantics of state transitions guarantees that these exit actions will be executed whichever way the states happen to be exited, so I can be sure that the pedestrians will always get the "Don't Walk" signal outside the "pedsEnabled" state and cars will get the red light outside the "carsEnabled" state.

> **NOTE**
>
> Exit actions in the states "carsEnabled" and "pedsEnabled" guarantee mutually exclusive access to the crossing, which is the main safety concern in this application.

(10)   The "pedsEnabled" state has a nested initial transition to the "pedsWalk" substate. Per the UML semantics, this transition must be taken after entering the superstate. The entry action to "pedsWalk" shows the "Walk" signal to pedestrians and arms the timer for the duration of this signal.

(11)   The Q_TIMEOUT event triggers the transition to the "pedsFlash" state, in which the "Don't Walk" signal flashes on and off. I use the internal variable of the PELICAN state machine me->pedFlashCtr to count the number of flashes.

(12,13)   The Q_TIMEOUT event triggers two internal transitions with complementary guards. When the me->pedFlashCtr counter is even, the "Don't Walk" signal is turned on. When it's odd, the "Don't Walk" signal is turned off. Either way the counter is always decremented.

(14)   Finally, when the me->pedFlashCtr counter reaches zero, the Q_TIMEOUT event triggers the transition to the "carsEnabled" state. The transition causes execution of the exit action from the "pedsEnabled" state, which displays "Don't Walk" signal for pedestrians. The life cycle of the PELICAN crossing then repeats.

At this point, the main functionality of the PELICAN crossing is done. However, I still need to add the "offline" mode of operation, which is actually quite easy because of the state hierarchy.

(15)   The OFF event in the "operational" superstate triggers the transition to the "offline" state. The state hierarchy ensures that the transition OFF is inherited by all direct or transitive substates of the "operational" superstate, so regardless in which substate the state machine happens to be, the OFF event always triggers transition to "offline." Also note that the semantics of exit actions still apply, so the PELICAN crossing will be left in a consistent safe state (both cars and pedestrians disabled) upon exit from the "operational" state.

(16)   The Q_TIMEOUT events in the substates of the "offline" state cause flashing of the signals for cars and pedestrians, as described in the problem specification.

(17)   The ON event can interrupt the "offline" mode at any time by triggering the transition to the "operational" state.

The actual implementation of the PELICAN state machine in QP-nano is very straightforward and follows exactly the same simple rules as I described for the Ship state machine in Section 12.2.4. The source code for the PELICAN application is located in the directory <qp>\qpn\examples\msp430\iar\pelican-eZ430\.

12.7.2   The Pedestrian Active Object

The actual PELICAN crossing controller hardware will certainly provide a push-button for generating the PED_WAITING event as well as a switch to generate the ON/OFF events. But the eZ430 USB stick has no push-button or any other way to provide external inputs (see Figure 12.6). For the eZ430, I need to *simulate* the pedestrian/operator in a separate state machine. This is actually a good opportunity to demonstrate how to incorporate a second state machine (active object) into the application.

The Pedestrian active object is very simple. It periodically posts the PED_WAITING event to the PELICAN active object and from time to time it turns the crossing offline by posting the OFF event followed by the ON event. I leave it as an exercise for you to draw the state diagram of the Pedestrian state machine from the source code found in the file <qp>\qpn\examples\msp430\iar\pelican-eZ430\ped.c. Note that such "reverse engineering" of source code is very easy in QP applications because the code is always the precise *specification* of the state machine.

12.7.3   QP-nano Port to MSP430 with QK-nano Kernel

The source code for the PELICAN and Pedestrian active objects as well as the main.c module is actually identical for all target CPUs, but each target requires a specific QP-nano port. In this section I describe the QP-nano port to MSP430 with the preemptive QK-nano kernel. A QP-nano port consists of the qpn_port.h header file and the BSP implementation in the bsp.c source file.

Using the preemptive kernel in the PELICAN crossing example isn't really justified by the loose timing requirements of this application. I describe the preemptive QP-nano configuration mainly to demonstrate that the QK-nano kernel is very lightweight and can fit even in a very memory-constrained MCU, such as the MSP430-F2013. The code accompanying this book contains the nonpreemptive version of the PELICAN crossing example for the eZ430 target as well (see Listing 12.7).

The PELICAN crossing example for eZ430 with the preemptive QK-nano kernel is located in the directory <qp>\qpn\examples\msp430\iar\pelican-qk-eZ430\. Listings 12.21 and 12.22 show the qpn_port.h header file and the bsp.c source file, respectively. This port has been compiled with the free KickStart edition of the IAR Embedded Workbench for MSP430 v4.10A.

Listing 12.21 QP-nano Port to MSP430 with QK-nano
(file <qp>\qpn\examples\msp430\iar\pelican-qk-eZ430\qpn_port.h**)**

```
      #ifndef qpn_port_h
      #define qpn_port_h

      #define Q_NFSM
      #define Q_PARAM_SIZE        0
      #define QF_TIMEEVT_CTR_SIZE 1

      /* maximum # active objects--must match EXACTLY the QF_active[] definition  */
 (1)  #define QF_MAX_ACTIVE       2

                                        /* interrupt locking policy for IAR compiler */
 (2)  #define QF_INT_LOCK()       __disable_interrupt()
 (3)  #define QF_INT_UNLOCK()     __enable_interrupt()
 (4)  /*#define QF_ISR_NEST*/                    /* nesting of ISRs not allowed */

                                        /* interrupt entry and exit for QK */
 (5)  #define QK_ISR_ENTRY()      ((void)0)
 (6)  #define QK_ISR_EXIT()       QK_SCHEDULE_()

 (7)  #include <intrinsics.h>   /* contains prototypes for the intrinsic functions */
```

```
(8)   #include <stdint.h>    /* Exact-width integer types. WG14/N843 C99 Standard */
      #include "qepn.h"        /* QEP-nano platform-independent public interface */
      #include "qfn.h"          /* QF-nano platform-independent public interface */
(9)   #include "qkn.h"           /* QK-nano platform-independent public interface */

      #endif                                                           /* qpn_port_h */
```

(1)   The PELICAN crossing application uses two active objects (PELICAN and
        Pedestrian).

(2,3)   The IAR compiler provided very efficient intrinsic functions for locking and
        unlocking interrupts.

(4)   Nesting of ISRs is *not* allowed in this QP-port.

(5,6)   The interrupt entry and exit macro for QK-nano are defined consistently with the
        interrupt nesting policy (see Listing 12.20).

(7)   The IAR header file <intrinsics.h> provides declarations of intrinsic
        functions, such as __disable_interrupt() and __enable_interrupt().

(8)   The IAR compiler is C99-compliant and provides the standard header file
        <stdint.h>, which defines exact-width integer types.

(9)   The QK-nano is configured by including the qkn.h header file.

Listing 12.22 BSP for MSP430 with QK-nano
(file <qp>\qpn\examples\msp430\iar\pelican-qk-eZ430\bsp.c**)**

```
      #pragma vector = TIMERA0_VECTOR
(1)   __interrupt void timerA_ISR(void) {                         /* see NOTE01 */
(2)       QK_ISR_ENTRY();                         /* inform QK-nano about ISR entry */
(3)       QF_tick();
(4)       QK_ISR_EXIT();                            /* inform QK-nano about ISR exit */
      }
      /*..................................................................*/
      void BSP_init(void) {
          WDTCTL = (WDTPW | WDTHOLD);                                    /* Stop WDT */
          P1DIR |= 0x01;                                             /* P1.0 output */
(5)       CCR0   = ((BSP_SMCLK + BSP_TICKS_PER_SEC/2) / BSP_TICKS_PER_SEC);
          TACTL  = (TASSEL_2 | MC_1);                            /* SMCLK, upmode */
      }
      /*..................................................................*/
      void QF_onStartup(void) {
```

Continued onto next page

```
(6)      CCTL0 = CCIE;                                     /* CCR0 interrupt enabled */
    }
    /*.............................................................................*/
    void QK_onIdle(void) {                                            /* see NOTE02 */
(7)      __low_power_mode_1();    /* adjust the low-power mode to your application */
    }
    /*.............................................................................*/
    void BSP_signalPeds(enum BSP_PedsSignal sig) {
        if (sig == PEDS_DONT_WALK) {
            LED_on();
        }
        else {
            LED_off();
        }
    }
    /*.............................................................................*/
    void Q_onAssert(char const Q_ROM * const Q_ROM_VAR file, int line) {
        (void)file;                                     /* avoid compiler warning */
        (void)line;                                     /* avoid compiler warning */
        for (;;) {
        }
    }
```

(1)   QK-nano can use the ISRs synthesized by the C compiler. In MSP430, as in most CPUs, the ISRs are entered with interrupts locked.

(2)   The macro QK_ISR_ENTRY() informs QK-nano about entering the ISR.

> **NOTE**
>
> Even though this macro does not do anything in this particular port, I like to use it for symmetry with the QK_ISR_EXIT(). This also allows me to change the interrupt-locking policy without modifying the ISRs.

(3)   The ISR calls the QP-nano service designed for the ISR context.

(4)   The macro QK_ISR_EXIT() informs QK-nano about exiting the ISR.

(5)   The BSP initialization configures the system clock tick timer to tick at the predefined rate BSP_TICKS_PER_SEC (I set to 20Hz in bsp.h).

(6)   The startup callback enables the system clock-tick ISR. This is the only ISR in the PELICAN crossing application.

(7) The idle callback of the QK-nano kernel transitions to the LPM1 mode, which is just one of many low-power sleep modes available in the ultra-low-power MSP430 architecture. In your application, you should adjust the power mode to your particular requirements.

NOTE

When you apply low-power mode is MSP430, the QK_onIdle() function is actually called only once and the idle loop stops. This is because the MSP430 core keeps the power-control bits in the SR register of the CPU, which gets automatically restored upon interrupt return. So when a power-saving mode is selected, the CPU stops when returning to the idle loop. If you want to perform some processing in the QK-nano idle loop before going to sleep, you need to call __low_power_mode_off_on_exit() in every ISR to clear the power-control bits in the *stacked* SR register.

12.7.4 QP-nano Memory Usage

To give you an idea of the QP-nano memory usage, Tables 12.1 and 12.2 show the memory footprint of the QP-nano components for various settings of the configuration macros. The data for Table 12.1 has been obtained from the IAR compiler for MSP430 v4.10A (the KickStart edition), whereas data for Table 12.2 has been obtained from the IAR compiler for ARM Cortex-M3 v5.11 (also the KickStart edition). In both cases I have selected optimization level High/Size.

The first column of Tables 12.1 and 12.2 lists the configuration macros that are significant for the RAM or ROM usage in QP-nano. I have omitted the QF_ISR_NEST and QF_ISR_KEY_TYPE macros because they have virtually no impact on the code or data sizes shown in the tables (even though defining QF_ISR_KEY_TYPE somewhat increases the stack usage.)

Both MSP430 and Cortex-M3 offer good code density and the IAR compiler generates fantastic machine code for these CPU architectures. (I've seen much worse results for older CPU architectures, such as 8051 or the PIC). Therefore, you should treat the data in Tables 12.1 and 12.2 as minimum memory footprint of QP-nano rather than average results. The intent of Table 12.1 is primarily to give you a general idea for the *relative* cost of various options rather than to provide you absolutely accurate measurements.

> **NOTE**
>
> The Tables 12.1 and 12.2 show only the memory used directly by the QP-nano components but do *not* include the memory required by the application. In particular, you don't see the stack usage or the RAM required by active objects and their event queues.

Table 12.1: QP-nano memory usage in bytes for various settings of the configuration parameters (MSP430/IAR compiler/optimization-High/Size)

Configuration Number	Q_NFSM	Q_NHSM	Q_PARAM_SIZE	QF_TIMEEVT_CTR_SIZE	QF_MAX_ACTIVE	QK_PREEMPTIVE	qepn.c (RAM/ROM)	qfn.c (RAM/ROM)	qkn.c (RAM/ROM)	QP-nano Total (RAM/ROM)
1		√	0	0	4		0/110	6/420	N/A	6/530
2		√	2	0	4		0/110	6/474	N/A	6/584
3		√	0	0	8		0/110	6/504	N/A	6/614
4		√	2	2	8		0/110	9/578	N/A	9/688
5	√		2	2	8		0/634	9/578	N/A	9/1,212
6			2	2	8		0/722	9/578	N/A	9/1,300
7		√	0	0	4	√	0/110	3/207	4/246	7/563
8		√	2	0	4	√	0/110	3/238	4/268	7/616
9		√	0	0	8	√	0/110	3/207	4/292	7/609
10		√	2	2	8	√	0/110	6/313	4/314	10/737
11	√		2	2	8	√	0/634	6/313	4/314	10/1,261
12			2	2	8	√	0/722	6/313	4/314	10/1,349

The various QP-nano configurations are listed separately in Tables 12.1 and 12.2 for the nonpreemptive "vanilla" kernel (configurations 1-6) and the preemptive QK-nano kernel (configurations 7-12). Within each group, the simpler configurations come before the more expensive ones. For example, the absolutely minimal configuration number 1 eliminates the HSM code (so only basic FSM support is provided), uses no event parameters, no time events, and up to four active objects. This minimal

Table 12.2: QP-nano memory usage in bytes for various settings of the configuration parameters (ARM Cortex-M3/IAR compiler/optimization-High/Size)

Configuration Number	Q_NFSM	Q_NHSM	Q_PARAM_SIZE	QF_TIMEEVT_CTR_SIZE	QF_MAX_ACTIVE	QK_PREEMPTIVE	qepn.c (RAM/ROM)	qfn.c (RAM/ROM)	qkn.c + qkn.s (RAM/ROM)	QP-nano Total (RAM/ROM)
1		√	0	0	4		0/110	12/406	N/A	12/516
2		√	2	0	4		0/110	12/456	N/A	12/566
3		√	0	0	8		0/110	12/426	N/A	12/536
4		√	2	2	8		0/110	13/550	N/A	13/660
5	√		2	2	8		0/620	13/550	N/A	13/1,170
6			2	2	8		0/702	13/550	N/A	13/1,252
7		√	0	0	4	√	0/110	8/224	8/334	16/668
8		√	2	0	4	√	0/110	8/260	8/350	16/720
9		√	0	0	8	√	0/110	8/224	8/374	16/708
10		√	2	2	8	√	0/110	9/334	8/390	17/834
11	√		2	2	8	√	0/620	9/334	8/390	17/1,334
12			2	2	8	√	0/702	9/334	8/390	17/1,426

configuration is clearly very limited. However, the configuration number 4 is already quite reasonable. It still offers only nonhierarchical FSMs, but includes event parameter, time events, and up to eight active objects at a cost of less than 700 bytes of code space.

By far, the most expensive feature (in terms of ROM) is the HSM support, which costs about 650 bytes (e.g., compare configurations number 4 and 5 or 10 and 11). On the other hand, the QK-nano preemptive kernel increases the ROM footprint only by 50-100 bytes compared to the "vanilla" kernel. Obviously, the true cost of QK-nano lies in the increased stack requirements, which Tables 12.1 and 12.2 don't show.

In comparison, the full-version QP^4 compiled with the IAR compiler for Cortex-M3 requires 2,718 bytes of ROM (616 bytes for the QEP component and 2,102 bytes of ROM for the QF component) and 121 bytes of RAM with eight active objects configured. This corresponds roughly to the QP-nano configuration number 6 from Table 12.2.

12.8   Summary

Low-end MCUs are a very important market segment for embedded systems because many billions of units of these devices are sold each year. These small MCUs aren't well served by the traditional kernels or RTOSes, which simply require too much RAM. However, QP-nano demonstrates that an *event-driven* framework is scalable to very small systems, starting from about 2KB of ROM and some 100 bytes of RAM. Quite possibly, QP-nano is the smallest event-driven framework with support for UML-style hierarchical state machines and active objects in the industry.

---

[4] Both C and C++ versions have essentially identical footprint, the C++ version being bigger by insignificant 24 bytes of ROM, which represents less than 1% of the total code size.

Licensing Policy for QP and QP-nano

Any licensor of open source software should consider dual licensing options as a way of attracting new customers.
—Lawrence Rosen, general counsel and secretary of Open Source Initiative (OSI)

All software described in this book and available for download from the companion Website at http://www.quantum-leaps.com/psicc2 or any other sources is available under a *dual-licensing* model, in which both the open-source software distribution mechanism and traditional closed-source software licensing models are combined.

A.1 Open-Source Licensing

All software described in this book is available under the GNU General Public License Version 2 (GPL2), as published by the Free Software Foundation and reproduced in Section A.5 of this Appendix.

The GPL2 license is probably the best known and most established open-source license. It is fully compatible with the Open Source Definition, is endorsed by the Free Software Foundation, and has been approved by the Open Source Initiative [Rosen 05].

Note that GPL2 applies to software based not on how it is used but on how it is *distributed*. In this respect GPL2 can be restrictive because GPL2 Section 2(b) requires that if you *distribute* the original software or any derivative works based on the software under copyright law, you must release all such derivative works also under the terms of the GPL2 open-source license. GPL2 clearly specifies that distributing the original software or any derivative works based on it in *binary form* (e.g., embedded inside devices) also represents distribution of the software.

To read more about open-source licensing for QP or QP-nano or to contribute work to the open-source community, visit `www.quantum-leaps.com/licensing/open.htm` or contact Quantum Leaps, LLC via the following e-mail address: `dev@quantum-leaps.com`.

A.2   Closed-Source Licensing

If you are developing and distributing traditional closed-source applications, you might purchase one of the *commercial licenses*, which are specifically designed for users interested in retaining the *proprietary* status of their code. This alternative licensing is possible because Quantum Leaps LLC owns all intellectual property in the QP and QP-nano software and as the copyright owner can license the software any number of ways. The Quantum Leaps commercial licenses expressly supersede the GPL2 open-source license. This means that when you license the software under a commercial license, you specifically do not use the software under the open-source license and therefore you are not subject to any of its terms.

To read more about the commercial licensing options, pricing, and technical support and to request a commercial license, visit `www.quantum-leaps.com/licensing` or contact Quantum Leaps LLC via the following e-mail address: `info@quantum-leaps.com`.

A.3   Evaluating the Software

The open character of the QP and QP-nano software allows anybody to evaluate the software under the GPL2 open-source license. In this respect, evaluating dual-licensed software delivers a large advantage over highly supervised trial licensing practices still so common in the embedded systems marketplace.

Obviously, when you decide to *distribute* any portion of the original software or any derivative works based on it, you must either make your software available to the public, as required by GPL2 Section 2(b), or you can purchase one of the commercial licenses, as described in Section A.2.

A.4   Nonprofits, Academic Institutions, and Private Individuals

If you represent a nonprofit organization or an academic institution, you should consider publishing your application as an open-source software project using the GPL2 license. Thereby you'll be able to use the QP or QP-nano software free of charge under the GPL2 license. If you have strong reasons not to publish your application in accordance with GPL2, you should purchase one of the commercial licenses. Note that nonprofit organizations can apply for free commercial licenses, which will be liberally granted.

If you are a private individual, you are free to use QP or QP-nano software for your personal applications as long as you do not *distribute* them. If you distribute the software in any way, you must make a decision between the GPL2 and the commercial licenses.

Note that these rules apply even if you ship (distribute) a free demo version of your own applications.

A.5   GNU General Public License Version 2

GNU GENERAL PUBLIC LICENSE
Version 2, June 1991

Copyright (C) 1989, 1991 Free Software Foundation, Inc.,
51 Franklin Street, Fifth Floor, Boston, MA 02110-1301 USA
Everyone is permitted to copy and distribute verbatim copies of this license document, but changing it is not allowed.

Preamble

The licenses for most software are designed to take away your freedom to share and change it. By contrast, the GNU General Public License is intended to guarantee your freedom to share and change free software--to make sure the software is free for all its users. This General Public License applies to most of the Free Software Foundation's software and to any other program whose authors commit to using it. (Some other Free Software Foundation software is covered by the GNU Lesser General Public License instead.) You can apply it to your programs, too.

When we speak of free software, we are referring to freedom, not price. Our General Public Licenses are designed to make sure that you have the freedom to distribute copies of free software (and charge for this service if you wish), that you receive source code or can get it if you want it, that you can change the software or use pieces of it in new free programs; and that you know you can do these things.

To protect your rights, we need to make restrictions that forbid anyone to deny you these rights or to ask you to surrender the rights. These restrictions translate to certain responsibilities for you if you distribute copies of the software, or if you modify it.

For example, if you distribute copies of such a program, whether gratis or for a fee, you must give the recipients all the rights that you have. You must make sure that they, too, receive or can get the source code. And you must show them these terms so they know their rights.

We protect your rights with two steps: (1) copyright the software, and (2) offer you this license which gives you legal permission to copy, distribute and/or modify the software.

Also, for each author's protection and ours, we want to make certain that everyone understands that there is no warranty for this free software. If the software is modified by someone else and passed on, we want its recipients to know that what they have is not the original, so that any problems introduced by others will not reflect on the original authors' reputations.

Finally, any free program is threatened constantly by software patents. We wish to avoid the danger that redistributors of a free program will individually obtain patent licenses, in effect making the program proprietary. To prevent this, we have made it clear that any patent must be licensed for everyone's free use or not licensed at all.

The precise terms and conditions for copying, distribution and modification follow.

GNU GENERAL PUBLIC LICENSE
TERMS AND CONDITIONS FOR COPYING, DISTRIBUTION AND MODIFICATION

0.   This License applies to any program or other work which contains a notice placed by the copyright holder saying it may be distributed under the terms of this General Public License. The "Program", below, refers to any such program or work, and a "work

based on the Program" means either the Program or any derivative work under copyright law: that is to say, a work containing the Program or a portion of it, either verbatim or with modifications and/or translated into another language. (Hereinafter, translation is included without limitation in the term "modification".) Each licensee is addressed as "you".

Activities other than copying, distribution and modification are not covered by this License; they are outside its scope. The act of running the Program is not restricted, and the output from the Program is covered only if its contents constitute a work based on the Program (independent of having been made by running the Program).

Whether that is true depends on what the Program does.

1. You may copy and distribute verbatim copies of the Program's source code as you receive it, in any medium, provided that you conspicuously and appropriately publish on each copy an appropriate copyright notice and disclaimer of warranty; keep intact all the notices that refer to this License and to the absence of any warranty; and give any other recipients of the Program a copy of this License along with the Program.

   You may charge a fee for the physical act of transferring a copy, and you may at your option offer warranty protection in exchange for a fee.

2. You may modify your copy or copies of the Program or any portion of it, thus forming a work based on the Program, and copy and distribute such modifications or work under the terms of Section 1 above, provided that you also meet all of these conditions:

   a) You must cause the modified files to carry prominent notices stating that you changed the files and the date of any change.

   b) You must cause any work that you distribute or publish, that in whole or in part contains or is derived from the Program or any part thereof, to be licensed as a whole at no charge to all third parties under the terms of this License.

   c) If the modified program normally reads commands interactively when run, you must cause it, when started running for such interactive use in the most ordinary way, to print or display an announcement including an appropriate copyright notice and a notice that there is no warranty (or else, saying that you provide a warranty) and that users may redistribute the program under these conditions, and telling the user how to view a copy of this License. (Exception: if the Program itself is interactive but does not normally print such an announcement, your work based on the Program is not required to print an announcement.)

   These requirements apply to the modified work as a whole. If identifiable sections of that work are not derived from the Program, and can be reasonably considered independent and separate works in themselves, then this License, and its terms, do not apply to those sections when you distribute them as separate works. But when you distribute the same sections as part of a whole which is a work based on the Program, the distribution of the whole must be on the terms of this License, whose permissions for other licensees extend to the entire whole, and thus to each and every part regardless of who wrote it.

   Thus, it is not the intent of this section to claim rights or contest your rights to work written entirely by you; rather, the intent is to exercise the right to control the distribution of derivative or collective works based on the Program.

   In addition, mere aggregation of another work not based on the Program with the Program (or with a work based on the Program) on a volume of a storage or distribution medium does not bring the other work under the scope of this License.

3. You may copy and distribute the Program (or a work based on it, under Section 2) in object code or executable form under the terms of Sections 1 and 2 above provided that you also do one of the following:

   a) Accompany it with the complete corresponding machine-readable source code, which must be distributed under the terms of Sections 1 and 2 above on a medium customarily used for software interchange; or,

   b) Accompany it with a written offer, valid for at least three years, to give any third party, for a charge no more than your cost of physically performing source distribution, a complete machine-readable copy of the corresponding source code, to be distributed under the terms of Sections 1 and 2 above on a medium customarily used for software interchange; or,

   c) Accompany it with the information you received as to the offer to distribute corresponding source code. (This alternative is allowed only for noncommercial distribution and only if you received the program in object code or executable form with such an offer, in accord with Subsection b above.)

   The source code for a work means the preferred form of the work for making modifications to it. For an executable work, complete source code means all the source code for all modules it contains, plus any associated interface definition files, plus the scripts used to control compilation and installation of the executable. However, as a special exception, the source code distributed need not include anything that is normally distributed (in either source or binary form) with the major components (compiler, kernel, and so on) of the operating system on which the executable runs, unless that component itself accompanies the executable.

If distribution of executable or object code is made by offering access to copy from a designated place, then offering equivalent access to copy the source code from the same place counts as distribution of the source code, even though third parties are not compelled to copy the source along with the object code.

4. You may not copy, modify, sublicense, or distribute the Program except as expressly provided under this License. Any attempt otherwise to copy, modify, sublicense or distribute the Program is void, and will automatically terminate your rights under this License. However, parties who have received copies, or rights, from you under this License will not have their licenses terminated so long as such parties remain in full compliance.

5. You are not required to accept this License, since you have not signed it. However, nothing else grants you permission to modify or distribute the Program or its derivative works. These actions are prohibited by law if you do not accept this License. Therefore, by modifying or distributing the Program (or any work based on the Program), you indicate your acceptance of this License to do so, and all its terms and conditions for copying, distributing or modifying the Program or works based on it.

6. Each time you redistribute the Program (or any work based on the Program), the recipient automatically receives a license from the original licensor to copy, distribute or modify the Program subject to these terms and conditions. You may not impose any further restrictions on the recipients' exercise of the rights granted herein. You are not responsible for enforcing compliance by third parties to this License.

7. If, as a consequence of a court judgment or allegation of patent infringement or for any other reason (not limited to patent issues), conditions are imposed on you (whether by court order, agreement or otherwise) that contradict the conditions of this License, they do not excuse you from the conditions of this License. If you cannot distribute so as to satisfy simultaneously your obligations under this License and any other pertinent obligations, then as a consequence you may not distribute the Program at all. For example, if a patent license would not permit royalty-free redistribution of the Program by all those who receive copies directly or indirectly through you, then the only way you could satisfy both it and this License would be to refrain entirely from distribution of the Program.
If any portion of this section is held invalid or unenforceable under any particular circumstance, the balance of the section is intended to apply and the section as a whole is intended to apply in other circumstances.
It is not the purpose of this section to induce you to infringe any patents or other property right claims or to contest validity of any such claims; this section has the sole purpose of protecting the integrity of the free software distribution system, which is implemented by public license practices. Many people have made generous contributions to the wide range of software distributed through that system in reliance on consistent application of that system; it is up to the author/donor to decide if he or she is willing to distribute software through any other system and a licensee cannot impose that choice.
This section is intended to make thoroughly clear what is believed to be a consequence of the rest of this License.

8. If the distribution and/or use of the Program is restricted in certain countries either by patents or by copyrighted interfaces, the original copyright holder who places the Program under this License may add an explicit geographical distribution limitation excluding those countries, so that distribution is permitted only in or among countries not thus excluded. In such case, this License incorporates the limitation as if written in the body of this License.

9. The Free Software Foundation may publish revised and/or new versions of the General Public License from time to time. Such new versions will be similar in spirit to the present version, but may differ in detail to address new problems or concerns.
Each version is given a distinguishing version number. If the Program specifies a version number of this License which applies to it and "any later version", you have the option of following the terms and conditions either of that version or of any later version published by the Free Software Foundation. If the Program does not specify a version number of this License, you may choose any version ever published by the Free Software Foundation.

10. If you wish to incorporate parts of the Program into other free programs whose distribution conditions are different, write to the author to ask for permission. For software which is copyrighted by the Free Software Foundation, write to the Free Software Foundation; we sometimes make exceptions for this. Our decision will be guided by the two goals of preserving the free status of all derivatives of our free software and of promoting the sharing and reuse of software generally.

NO WARRANTY

11. BECAUSE THE PROGRAM IS LICENSED FREE OF CHARGE, THERE IS NO WARRANTY FOR THE PROGRAM, TO THE EXTENT PERMITTED BY APPLICABLE LAW. EXCEPT WHEN OTHERWISE STATED IN WRITING THE COPYRIGHT HOLDERS AND/OR OTHER PARTIES PROVIDE THE PROGRAM "AS IS" WITHOUT WARRANTY OF ANY KIND, EITHER EXPRESSED

OR IMPLIED, INCLUDING, BUT NOT LIMITED TO, THE IMPLIED WARRANTIES OF MERCHANTABILITY AND FITNESS FOR A PARTICULAR PURPOSE. THE ENTIRE RISK AS TO THE QUALITY AND PERFORMANCE OF THE PROGRAM IS WITH YOU. SHOULD THE PROGRAM PROVE DEFECTIVE, YOU ASSUME THE COST OF ALL NECESSARY SERVICING, REPAIR OR CORRECTION.

12. IN NO EVENT UNLESS REQUIRED BY APPLICABLE LAW OR AGREED TO IN WRITING WILL ANY COPYRIGHT HOLDER, OR ANY OTHER PARTY WHO MAY MODIFY AND/OR REDISTRIBUTE THE PROGRAM AS PERMITTED ABOVE, BE LIABLE TO YOU FOR DAMAGES, INCLUDING ANY GENERAL, SPECIAL, INCIDENTAL OR CONSEQUENTIAL DAMAGES ARISING OUT OF THE USE OR INABILITY TO USE THE PROGRAM (INCLUDING BUT NOT LIMITED TO LOSS OF DATA OR DATA BEING RENDERED INACCURATE OR LOSSES SUSTAINED BY YOU OR THIRD PARTIES OR A FAILURE OF THE PROGRAM TO OPERATE WITH ANY OTHER PROGRAMS), EVEN IF SUCH HOLDER OR OTHER PARTY HAS BEEN ADVISED OF THE POSSIBILITY OF SUCH DAMAGES.

END OF TERMS AND CONDITIONS

How to Apply These Terms to Your New Programs

If you develop a new program, and you want it to be of the greatest possible use to the public, the best way to achieve this is to make it free software which everyone can redistribute and change under these terms.

To do so, attach the following notices to the program. It is safest to attach them to the start of each source file to most effectively convey the exclusion of warranty; and each file should have at least the "copyright" line and a pointer to where the full notice is found.

<one line to give the program's name and a brief idea of what it does.>
Copyright (C) <year> <name of author>

This program is free software; you can redistribute it and/or modify it under the terms of the GNU General Public License as published by the Free Software Foundation; either version 2 of the License, or (at your option) any later version.

This program is distributed in the hope that it will be useful, but WITHOUT ANY WARRANTY; without even the implied warranty of MERCHANTABILITY or FITNESS FOR A PARTICULAR PURPOSE. See the GNU General Public License for more details.

You should have received a copy of the GNU General Public License along with this program; if not, write to the Free Software Foundation, Inc., 51 Franklin Street, Fifth Floor, Boston, MA 02110-1301 USA.

Also add information on how to contact you by electronic and paper mail.

If the program is interactive, make it output a short notice like this when it starts in an interactive mode:

Gnomovision version 69, Copyright (C) year name of author
Gnomovision comes with ABSOLUTELY NO WARRANTY; for details type 'show w'.

This is free software, and you are welcome to redistribute it under certain conditions; type 'show c' for details.

The hypothetical commands 'show w' and 'show c' should show the appropriate parts of the General Public License. Of course, the commands you use may be called something other than 'show w' and 'show c'; they could even be mouse-clicks or menu items-- whatever suits your program. You should also get your employer (if you work as a programmer) or your school, if any, to sign a "copyright disclaimer" for the program, if necessary. Here is a sample; alter the names:

Yoyodyne, Inc., hereby disclaims all copyright interest in the program
'Gnomovision' (which makes passes at compilers) written by James Hacker.

<signature of Ty Coon>, 1 April 1989
Ty Coon, President of Vice

This General Public License does not permit incorporating your program into proprietary programs. If your program is a subroutine library, you may consider it more useful to permit linking proprietary applications with the library. If this is what you want to do, use the GNU Lesser General Public License instead of this License.

Guide to Notation

The good thing about bubbles and arrows, as opposed to programs, is that they never crash.
—Bertrand Meyer

This appendix describes the graphical notation that I use throughout the book.[1] The notation should be compatible with the UML specification [OMG 07].

B.1   Class Diagrams

A class diagram shows classes, their internal structures, and the static (compile-time) relationships among them. Figure B.1 shows the various presentation options for classes.

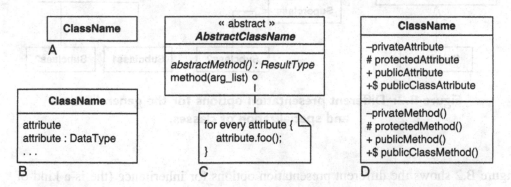

Figure B.1: Various levels of detail, visibility, and properties of classes.

---

[1] I prepared all diagrams with Microsoft Visio 2003. The code accompanying this book contains the Visio stencil that I used (directory `<qp>\visio\`).

- A class is always denoted by a box with the class name in bold type at the top. Optionally, just below the name, a class box can have an attribute compartment that is separated from the name by a horizontal line. Below the attributes, a class box can have an optional method compartment.

- The UML notation allows you to distinguish abstract classes, which are classes intended only for derivation and cannot have direct instances. Figure B.1C shows the notation for such classes. The abstract class name appears in italic font. Optionally you may use the «abstract» stereotype. If a class has abstract methods (pure virtual member functions in C++), they are shown in an italic font as well.

- Sometimes it is helpful to provide pseudocode of some methods by means of a note (Figure B.1(C)).

- Finally, a class box can also show the visibility of attributes and methods, as in Figure B.1(D).

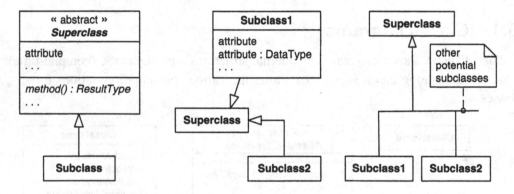

Figure B.2: Different presentation options for the generalization and specialization of classes.

Figure B.2 shows the different presentation options for inheritance (the is-a-kind-of relationship). The generalization arrow always points from the subclass to the superclass. The right-hand side of Figure B.2 shows an inheritance tree that indicates an open-ended number of subclasses.

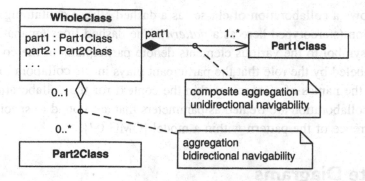

Figure B.3: Aggregation, navigability, and multiplicity.

Figure B.3 shows the aggregation of classes (the has-a-component relationship). An aggregation relationship implies that one object physically or conceptually contains another. The notation for aggregation consists of a line with a diamond at its base. The diamond is at the side of the owner class (whole class), and the line extends to the component class (part class). The full diamond represents physical containment; that is, the instance of the part class physically residing in the instance of the whole class (composite aggregation). A weaker form of aggregation, denoted with an empty diamond, indicates that the whole class has only a reference or pointer to the part instance but does not physically contain it. A name for the reference might appear at the base (e.g., part1 in Figure B.3). Aggregation also could indicate multiplicity and navigability between the whole and the parts.

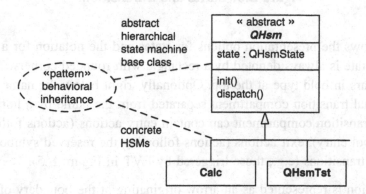

Figure B.4: Design pattern as a collaboration of classes.

Figure B.4 shows a collaboration of classes as a dashed ellipse containing the name of the collaboration (stereotyped here as a *pattern*). The dashed lines emanating from the collaboration symbol to the various elements denote participants in the collaboration. Each line is labeled by the role that the participant plays in the collaboration. The roles correspond to the names of elements within the context for the collaboration; such names in the collaboration are treated as parameters that are bound to specific elements on each occurrence of the pattern within a model [OMG 07].

B.2   State Diagrams

A state diagram shows the static state space of a given context class, the events that cause a transition from one state to another, and the actions that result.

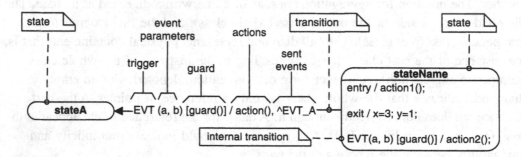

Figure B.5: States and a transition.

Figure B.5 shows the presentation options for states and the notation for a state transition. A state is always denoted by a rectangle with rounded corners. The name of the state appears in bold type at the top. Optionally, right below the name, a state can have an internal transition compartment separated from the name by a horizontal line. The internal transition compartment can contain entry actions (actions following the reserved symbol entry), exit actions (actions following the reserved symbol exit), and other internal transitions (e.g., those triggered by EVT in Figure B.5).

A state transition is represented as an arrow originating at the boundary of the source state and pointing to the boundary of the target state. At a minimum, a transition must be labeled with the triggering event. Optionally, the trigger can be followed by event parameters, a guard, a list of actions, and a list of events that have been sent.

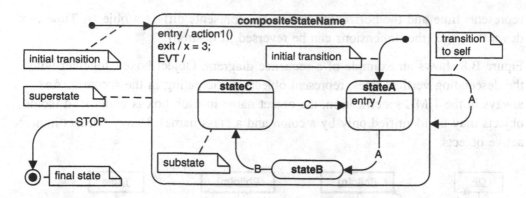

Figure B.6: Composite state, initial transitions, and the final state.

Figure B.6 shows a composite state (superstate) that contains other states (substates). Each composite state can have a separate initial transition to designate the initial substate. Although Figure B.6 shows only one level of nesting, the substates can be composite as well.

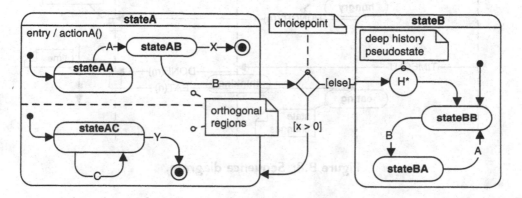

Figure B.7: Orthogonal regions and pseudostates.

Figure B.7 shows composite "stateA" with the orthogonal regions (AND-states) separated by a dashed line and two pseudostates: the choicepoint and the deep history.

B.3   Sequence Diagrams

A sequence diagram shows a particular sequence of event instances exchanged among objects at runtime. A sequence diagram has two dimensions; the vertical dimension represents time and the horizontal dimension represents different objects. Time flows down the page (the dimensions can be reversed, if desired).

Figure B.8 shows an example of a sequence diagram. Object boxes, together with the descending vertical lines, represent objects participating in the scenario. As always in the UML specification, the object name in each box is underlined (some objects may be identified only by a colon and a class name). Heavy borders indicate active objects.

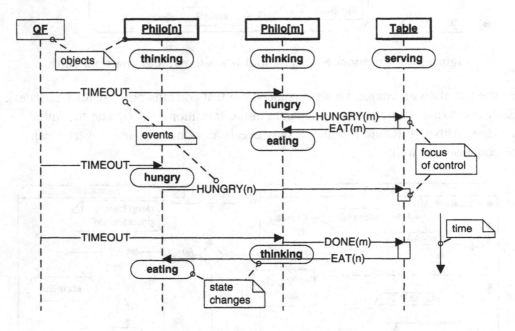

Figure B.8: Sequence diagram.

Events are represented as horizontal arrows originating from the sending object and terminating at the receiving object. Optionally, thin rectangles around instance lines can indicate focus of control. Sequence diagrams also can contain state marks to indicate explicit state changes resulting from the event exchange.

B.4 Timing Diagrams

A timing diagram shows the explicit changes of state in one or more objects along a single time axis. Figure B.9 shows an example of a timing diagram for multiple objects

(T1, T2, and T3). The timing diagram has two dimensions; time flows along the horizontal axis and the object state along the vertical axis. Each object is assigned a horizontal band across the diagram (a "swim lane") separated from other bands by dashed lines. Presentation options include deadlines, propagated events, and jitter.

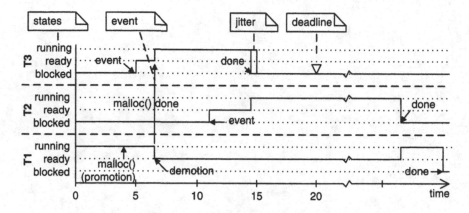

Figure B.9: Timing diagram.

Bibliography

[ARM 06a]   ARM, Ltd., *ARMv7-M Architecture Application Level Reference Manual*, 2006, www.arm.com/products/CPUs/ARM_Cortex-M3_v7.html.

[Bal Sathe 88]   Bal Sathe, Dhananjay, "Fast Algorithm Determines Priority," *EDN*, September 1988.

[Booch 94]   Booch, Grady, *Object-Oriented Analysis and Design with Applications*, Addison-Wesley, 1994.

[Brooks 95]   Brooks, Frederick, *The Mythical Man-Month, Anniversary Edition*, Addison-Wesley, 1995, ISBN 0-201-83595-9.

[Butenhof 97]   Butenhof, David R., *Programming with POSIX(R) Threads*, Addison-Wesley Professional, 1997, ISBN 0201633922.

[Butenhof 97]   Butenhof, David R., *Programming with POSIX(R) Threads*, Addison-Wesley Professional, 1997.

[Cargill 94]   Cargill, Tom, "Exception Handling: A False Sense of Security," C++ Report, November–December 1994, www.informit.com/content/images/020163371x/supplements/Exception_Handling_Article.html.

[Carryer 05]   Carryer, Edward J., "CMPE118/L: Introduction to Mechatronics: Event-Driven Programming," www.soe.ucsc.edu/classes/cmpe118/Spring05/.

[Clugston 07]   Clugston, Don, "Member Function Pointers and the Fastest Possible C++ Delegates," www.codeproject.com/cpp/FastDelegate.asp.

[Cormen+ 01]   Cormen, Thomas H., Charles E. Leiserson, Ronald L. Rivest, and Clifford Stein, *Introduction to Algorithms*, 2nd edition, MIT Press and McGraw-Hill, 2001, ISBN 0-262-03293-7; Section 6.5: Priority queues, pp. 138–142.

[Diaz-Herrera 93] Diaz-Herrera, Jorge L. *Software Engineering Education: 7th SEI CSEE Conference, San Antonio 1993*, Springer, 1994.

[Dijkstra 71]   Dijkstra, Edsger W., "Hierarchical Ordering of Sequential Processes," *Acta Informatica 1*, 1971, pp. 115–138.

[Douglass 01]   Douglass, Bruce Powel, "Class 505/525: State Machines and Statecharts," *Proceedings of Embedded Systems Conference*, San Francisco, Fall 2001.

[Douglass 02]   Douglass, Bruce Powel, *Real-Time Design Patterns: Robust Scalable Architecture for Real-Time Systems*, Addison-Wesley Professional, 2002.

[Douglass 06]   Douglass, Bruce Powel, *Real-Time UML Workshop for Embedded Systems*, Newnes, 2006.

[Douglass 99]   Douglass, Bruce Powel, *Doing Hard Time: Developing Real-Time Systems with UML, Objects, Frameworks, and Patterns*, Addison-Wesley, 1999.

[Douglass 99b]   Douglass, Bruce Powel, "UML Statecharts," *Embedded Systems Programming*, January 1999, pp. 22–42.

[Dunkels+ 06]   Dunkels, Adam; Oliver Schmidt, Thiemo Voigt, and Muneeb Ali, "Protothreads: Simplifying event-driven programming of memory-constrained embedded systems," in *Proceedings of the Fourth ACM Conference on Embedded Networked Sensor Systems* (SenSys 2006), Boulder, Colorado, November 2006.

[EC++ 01]   Embedded C++ Technical Committee, www.caravan.net/ec2plus.

[Gamma+ 95]   Gamma, Erich, Richard Helm, Ralph Johnson, and John Vlissides, *Design Patterns, Elements of Reusable Object-Oriented Software*, Addison-Wesley, 1995.

[Ganssle 98]   Ganssle, Jack. G., "The Challenges of Real-Time Programming," *Embedded Systems Programming*, July 1998, pp. 20–26.

[Gatliff 01]   Gatliff, Bill, "Embedding with GNU: Newlib," *Embedded Systems Design*, December 2001, www.embedded.com/15201696.

[GoF 95]   Gamma, Erich, Richard Helm, Ralph Johnson, and John Vlissides, *Design Patterns, Elements of Reusable Object-Oriented Software*, Addison-Wesley, 1995.

[Gomez 00]   Gomez, Martin, "Embedded State Machine Implementation," *Embedded Systems Programming*, December 2000, pp. 40–50.

[Harel 87]   Harel, David, "Statecharts: A Visual Formalism for Complex Systems," *Science of Computer Programming*, 8, 1987, pp. 231–274, www.wisdom. weizmann.ac.il/~dharel/SCANNED.PAPERS/Statecharts.pdf.

[Harel+ 98]   Harel, David, and Michal Politi, *Modeling Reactive Systems with Statecharts, The STATEMATE Approach*, McGraw-Hill, 1998.

[HDLC 07]   www.interfacebus.com/Design_HDLC.html.

[Hewitt 73]   Hewitt, Carl, P. Bishop, and R. Steiger, "A universal, modular actor formalism for artificial intelligence," Third International Joint Conference on Artificial Intelligence, pp. 235–245, 1973.

[Hoare 69]   Hoare, C. A. R., "An axiomatic basis for computer programming," *Communications of the ACM,* 12(10):576–585, October 1969.

[Horrocks 99]   Horrocks, Ian, "Constructing the User Interface with Statecharts," Addison-Wesley, 1999.

[Hunt+ 00]   Hunt, Andy, and Dave Thomas, *The Pragmatic Programmer*, Addison-Wesley, 2000.

[IAR 00]   IAR Systems, AB, IAR visualSTATE, www.iar.com/vs.

[ISO/IEC 9899:TC2]   WG14/N1124 Committee Draft—ISO/IEC 9899:TC2, May 6, 2005, www.open-std.org/jtc1/sc22/wg14/www/docs/n1124.pdf.

[Johnson+ 88]   Johnson, Ralph E., and Brian Foote, "Designing Reusable Classes," *Journal of Object-Oriented Programming*, June/July 1988, Volume 1, Number 2, pages 22–35.

[Kalinsky 05]   Kalinsky, David, "Mutexes Battle Priority Inversions," white paper, www.kalinskyassociates.com/Wpaper2.html.

[Kerninghan 88]   Kerninghan, Brian, W. and Dennis M. Ritche, *The C Programming Language*, 2nd edition, Prentice Hall, 1988.

[Labrosse 02]   Labrosse, Jean J., *MicroC/OS-II, The Real-Time Kernel*, 2nd edition, CMP Books, 2002.

[Lafreniere 98]   Lafreniere, David, "An Efficient Dynamic Storage Allocator," *Embedded Systems Programming*, September 1998, pp. 72–80.

[Li+ 03]   Li, Qing, and Caroline Yao, *Real-Time Concepts for Embedded Systems*, CMP Books, 2003.

[Luminary 06]   www.luminarymicro.com/products/kits.html.

[Maguire 93]   Maguire, Steve, *Writing Solid Code*, Microsoft Press, 1993.

[Mayer 97b]   North, R. D., T. DeMarco, J. Stern, and D. Morley, "When Software Is Treated Much Too Lightly," *Computer*, Volume 30, Issue 2, February 1997.

[Mellor 00]   Mellor, Steve, "UML Point/Counterpoint: Modeling Complex Behavior Simply," *Embedded Systems Programming*, March 2000.

[Meyer 97]   Bertrand Meyer, *Object-Oriented Software Construction*, 2nd edition, Prentice Hall, 1997.

[MISRA 98]   Motor Industry Software Reliability Association (MISRA), MISRA Limited, *MISRA-C: 1998 Guidelines for the Use of the C Language in Vehicle-Based Software*, April 1998, ISBN 0-9524156-9-0. See also www.misra.org.uk.

[Montgomery 06]   Montgomery, Paul Y., private communication.

[Murphy 01a]   Murphy, Niall, "Assertiveness Training for Programmers," *Embedded Systems Programming*, March 2001.

[Murphy 01b]   Murphy, Niall, "Assert Yourself," *Embedded Systems Programming*, April 2001.

[OMG 07]   Object Management Group, "Unified Modeling Language: Superstructure version 2.1.1," formal/2007-02-05, February 2007.

[Petzold 96]   Petzold, Charles, *Programming Windows: The Definitive Developer's Guide to Programming Windows*, Microsoft Press, 1996.

[Preiss 99]   Preiss, Bruno R., "Reference Counting Garbage Collection," excerpt from *Data Structures and Algorithms with Object-Oriented Design Patterns in Java*, www.brpreiss.com/books/opus5/html/page421.html.

[Queens 07]   www.cs.queensu.ca/Software-Engineering/tools.html.

[Rosen 05]   Rosen, Lawrence, *Open Source Licensing: Software Freedom and Intellectual Property Law*, Prentice Hall, 2005.

[Rumbaugh+ 91]   Rambaugh, James, Michael Blaha, William Premerlani, Frederick Eddy, and William Lorensen, *Object-Oriented Modeling and Design*, Prentice-Hall, 1991.

[Samek 02]   Samek, Miro, *Practical Statecharts in C/C++: Quantum Programming for Embedded Systems*, CMP Books, 2002.

[Samek 03a]   Samek, Miro, "The Embedded Mindset," *C/C++ Users Journal*, February 2003, www.ddj.com/cpp/184401620.

[Samek 03b]   Samek, Miro, "Who Moved My State?" *C/C++ Users Journal*, April 2003, www.ddj.com/cpp/184401643.

[Samek 03c]   Samek, Miro, "Dèjà Vu," *C/C++ Users Journal*, June, 2003, www.ddj.com/cpp/184401665.

[Samek 03d]   Samek, Miro, "An Exception or a Bug?" *C/C++ Users Journal*, August 2003, www.ddj.com/cpp/184401686.

[Samek 03e]   Samek, Miro, "Patterns of Thinking," *C/C++ Users Journal*, October 2003, www.ddj.com/cpp/184401713.

[Samek 03f]   Samek, Miro, "Back to Basics," *C/C++ Users Journal*, December, 2003, www.ddj.com/cpp/184401737.

[Samek 07a]   Samek, Miro, "Building Bare-Metal ARM Systems with GNU," Embedded.com Design Articles, published in 10 installments in July/August 2007, www.embedded.com/design/opensource/201802580.

[Samek 07b]   Samek, Miro, "Use an MCU's low-power modes in foreground/background systems," *Embedded Systems Design*, October 2007, pp 31–45, www.embedded.com/design/202103425.

[Samek+ 06]   Samek, Miro and Robert Ward, "Build a Super Simple Tasker," cover story, *Embedded Systems Design*, July 2006, www.embedded.com/showArticle.jhtml?articleID=190302110.

[Selic+ 94]   Selic, Bran, Garth Gulleckson, an d Paul T. Ward, *Real-Time Object-Oriented Modeling*, John Wiley & Sons, 1994.

[Serlin 72]   Serlin, O., "Scheduling of Time Critical Processes," *Proceedings of the Spring Joint Computer Conference*, Atlantic City, NJ, May 16–18, 1972, 925–932, Montvale, NJ: American Federation of Information Processing Societies, 1972.

[Stroustrup 00] Stroustrup, Bjarne, *The C++ Programming Language: Special Edition*, 3rd Edition, Addison-Wesley Professional, 2000.

[Sutter 07] Sutter, Herb, "Recursive Declarations," GotW.ca Website, www.gotw.ca/gotw/057.htm.

[Telelogic 07] http://modeling.telelogic.com/products/rhapsody.

[TI 07] Texas Instruments, Inc., "MSP430x20x1, MSP430x20x2, MSP430x20x3 Mixed Signal Microcontroller (Rev. D)," TI Datasheet, 2007.

[Turley 02] Turley, Jim, "Embedded Processors, Part 1," *ExtremeTech*, 2002, www.extremetech.com/article2/0,3973,18917,00.asp.

[VxWorks 01] www-sgc.colorado.edu/~dixonc/vxworks/docs/vxworks/ref/intArchLib.html.

Index

A

Abstract class, 631–632
Active objects-based application
  internal state machines, 446
  loose coupling, 445
  resource sharing, 445
  rules, 444
  sequence diagrams, 446
Active object state machine, 626,
    630, 653–654, 663
ANSI-C, 309, 336, 388
Arguments, 622
ARM7/ARM9 architecture, 637
ARM Cortex-M3, 310,
    321–322, 330
ARM Procedure Call Standard
    (APCS), 496
ARMv7 architecture, 276
Array, 15

B

16-Bit DOS platform, 4
8-Bit PIC MCUs, 656
Blocking tasks, 260
Block Started by Symbol
    (BSS), 625
Board support package (BSP), 5,
    307, 382–383, 412, 457,
    461, 613, 618, 628
  for "Fly 'n' Shoot"
      application, 628–630
  header file, 13
  Linux, 473–475

mC/OS-II, 471
"vanilla" kernel
    on Cortex-M3, 465–468
    on DOS, 461–464
Borland Turbo C++ 1.01
    compiler
  INSTALL.EXE program, 8
bsp.h header file, 13

C

Calculator
  cancel and off transition in, 94
  negative numbers,
      handling, 95
  statechart, 93
  submachines of, 94–95
Callback functions
  QF_OnIdle(), 630, 654–655,
      660, 675
  QF_onStartup(), 630, 653, 660
  QF_run(), 658
  QF_stop(), 630
  Q_onAssert(), 630
caps_locked state, 60
C compiler, 12–13
Central processing unit (CPU)
  architectures, 613, 616
  cycles, 260, 276, 312,
      354, 378
Cold Fire Altera Nios II, 310
Comma expression, 640
Constructors, 618, 622, 624
Context switching, 259–260

Controls
  capacitive touch sensing, 611
  lighting, 611
  motor, 611
  remote access, 611
Cooperative multitasking, 261
Cooperative "vanilla" kernel,
    379–380
  ports
    idle processing QF_onIdle(),
        418–420
    qep_port.h header file,
        414–415
    qf_port.h header file,
        415–417
    system clock tick QF_tick(),
        417–418
    qvanilla.c source file, 380–384
    qvanilla.h header file, 384–386
Cortex-M3 system tick timer, 38

D

Decimal Flag, 65–66
Defensive programming
    strategy, 295
Deferred event state pattern
  built-in UML mechanism,
      event deferral using, 220
  consequences, 229–230
  difficulties in implementation,
      228
  intent and problems, 219–220
  known uses, 230

Deferred event state pattern
 (Continued)
 real-time object-oriented
  modeling (ROOM)
  method, 230
 sample code
  DEFER.EXE, output
   generation, 222
  file defer.c implementation,
   223–229
  Q_NEW() macro file defer.
   c, request events
   generation, 228–229
 solutions, 220–221
 variations of, 227
Design by Contract (DbC)
 philosophy, 169–170
Dining philosophers problem
 (DPP), 545
 application execution
  Linux, 472–476
  mC/OS-II, 469–472
  "vanilla" kernel on Cortex-
   M3, 465–469
  "vanilla" kernel on DOS,
   461–465
 steps
  application termination,
   460–461
  requirements, 447
  sequence diagrams,
   447–449
  signals, events, and active
   objects, 449–451
  state machines, 451–457
  system initialization and
   application startup,
   457–460
__disable_interrupt(), 673
DOS screen, 7–8
Doxyfile, 316, 387

E
End-of-interrupt (EOI), 464
errno facility, 294
Error, 296

Event-action paradigm,
 56–59, 265
Event delivery mechanism
 direct event posting, 626
 event delivery guarantee
  policy, 648
Event-driven framework
 active object computing
  model, 266
 asynchronous
  communication, 269
 encapsulation, 269–271
 preemptive RTC kernel,
  276–279
 run-to-completion (RTC)
  fashion, 269
 sequential pseudocode, 271
 state machines, support for,
  271–272
 system structure, 267–269
 traditional preemptive
  kernel, 273
 vanilla kernel, 274–276
application design
 assigning responsibilities
  and resources, 17
 object-oriented
  technique, 16
CPU management and
 background loop,
  control flow, 257–258,
  260
 multiple tasks in, 259
 multitasking, 259–263
 traditional event-driven
  system, 263–265
 traditional sequential
  systems, 257–259
error and exception handling
 C and C++, assertions,
  297–300
 design by contract, 294–296
 errors and exceptional
  conditions, 296–297
 event delivery, 302–303
 shipping, 301

state-based handling,
  300–301
event delivery mechanisms,
 279
 direct event posting,
  280–281
 publish–subscribe model,
  281–282
event memory management
 brute-force approach, 284
 copying, 282–283
 event ownership, 288
 garbage collection, 284, 287
 memory pools, 289–291
 reference-counting
  algorithm, 286–287
 static and dynamic, 286
 zero-copy, 284–285
 inversion of control, 256–257
 software tracing, 303
 QS (Q-SPY) component
  in, 304
 time management
  system clock tick, 293–294
  time events, 291–292
Event-driven systems
 adding QS software
  tracing, 596
 application-specific trace
  records, 607–608
 initializing QS and setting
  up filters, 596–598
 platform-specific QS
  callbacks, 598–601
 QS dictionary records,
  604–606
 QSPY reference
  manual, 608
 QS target component,
  550–581
 QS timestamps, 601–604
MATLAB, 587
 and analyzing trace data,
  587–589
 matrices generated by
  qspy.m, 593–595

output file, 589–590
script qspy.m, 591–593
QSPY host applications,
    581–582
  installing, 582–583
  invoking, 585–587
  for Linux, 584
  with MinGW, 584
  with Visual C++,
      2005, 584
QS target component
  application-specific, trace
      records, 578–580
  dictionary, trace records,
      574–578
  general structure of,
      561–562
  global on/off filter, 562–564
  local filter, 564–566
  porting and configuring,
      580–581
  QS critical section, 560–561
  QS data protocol, 566–569
  QS platform-independent
      header files, 554–560
  QS source code
      organization, 552–553
  trace buffer, 569–574
quantum spy software-tracing
    system, 544–550
  examples for, 545–547
  human-readable trace
      output, 547–550
  software tracing concepts for,
      542–544
Event-handler functions, 263
  onA(), onB(), and onC(), 272
Event loop, 263
Event pool, 285
Events
  GREEN_TOUT, 666
  PEDS_WAITING, 666–669,
      671
  posting, types, 218
  Q_TIMEOUT, 669
  Q_TIMEOUT_SIG, 644

EV-LM3S811 board, 5–7
  LMI FTDI debugger, 9
  OLED display of, 6
Extended context switch
  asynchronous preemptions,
      520
  coprocessor registers, 521
  QK extended scheduler
      implementation, 522–524
Extended state machines, 67
  if or else, 65
  key_count, 63–64
  key_count ==, 0, 64
  Shift, Ctrl, and Alt keys, 66

F
Fine granularity library, 318
Finite state machine (FSM),
    59–60, 101
  graphical representation in, 61
First-in, first-out (FIFO) policy,
    218, 612
Flash programming process,
    Debug button, 10
Floating point coprocessor
    (FPU), 524
"Fly 'n' Shoot-type" game, 286,
    315, 328, 330
  active objects
    mine components, 29–31
    missile active object, 21–23
    ship active object, 24–26
    tunnel active object, 27–28
  BSP_init(), 14
  code, 8
  coding hierarchical state
      machines
    ship data structure, 39–41
    state-handler function,
        43–48
    state machine initialization,
        42–43
  comparison with, 50–52
  constant Q_USER_SIG, 35
  demo mode, 5–6
  design of, 16–20

DESTROYED_MINE(score)
    event, 20
  in DOS version for, 4, 6, 8–9,
      15–16
Esc key, 7–8
event-driven application
    design, 17
events, 32–36, 620–622
  generating, posting, and
      publishing events,
      36–38
  structure ObjectPosEvt, 35
execution model
  operating system and
      real-time operating
      system, 50
  plain vanilla, 48–49
  QK preemptive kernel,
      49–50
fire missile in, 5
game.h file, 33–34
game over screen, 5–6
HIT_MINE(score) event,
    20, 23
HIT_WALL event, 20, 23
ISR_ADC(), 38
ISR_SysTick(), 38
loading into, 10
Luminary Micro Quickstart
    application, 3
main.c file, 11–12
main() function, 11
MAX_PUB_SIG constant, 14
me-score data member, 26
in Microsoft Windows, 7
Mine
  bitmap, 19
  object, 30
  types of, 4, 6
MINE_DISABLED(mine_id)
    event, 28, 30
MINE_IMG(x, y, bmp)
    event, 19
MINE_PLANT(x, y) event, 30
Mine2 state machine
    diagram, 31

"Fly 'n' Shoot-type" game
    (*Continued*)
  MISSILE_FIRE(x, y) event,
    19, 22
  object-oriented technique
    of, 16
  PLAYER_QUIT event, 28
  PLAYER_SHIP_MOVE_SIG,
    46
  PLAYER_SHIP_MOVE
    signal, 32
  PLAYER_SHIP_MOVE(x, y)
    event, 20, 24
  PLAYER_TRIGGER button-
    press event, 286
  PLAYER_TRIGGER
    event, 19
  press button, 5
  publish-subscribe event
    delivery mechanism, 14
  QActive_postFIFO(), 29, 47
  QActive_postFIFO() and
    QActive_postLIFO(), 35
  QActive_start(), 15
  QActive_subscribe(), 35
  Q_ENTRY_SIG, 47
  QEvent structure, 32
  QF_init(), 14
  QF macro Q_NEW, 38
  QF_MAX_ACTIVE, 14
  QF_poolInit(), 14–15
  QF_publish(), 35, 38
  QF_run(), 15
  QF_tick(), 38
  QHsm_dispatch(), 29
  QHsm_top(), 46
  Q_INIT_SIG, 45
  Q_NEW() macro, 47
  QP (MAX_PUB_SIG), 35
  QP-nano application, building,
    630–631
  QTimeEvt_postEvery(), 28
  QTimeEvt_postIn(), 28
  Q_TRAN() macro, 45, 47
  Quickstart application, 51–52
  return Q_Handled(), 47

ScoreEvt and QEvent
    structure, 33
screen_saver, 28
screen time event (timer), 28
SCREEN_TIMEOUT
    transition, 28
&Ship_active, 48
Ship_flying(), 48
SHIP_IMG(x, y, bmp) event, 18
Ship structure, 39–41
signals and active objects
    for, 620
single inheritance in C, 32–33
sizeof(a)/sizeof(a[0]), 15
space bar, 7
state-handler functions, 43–48
state machine, 42–43
  in Stellaris EV-LM3S811
    evaluation board, 7
TIME_TICK event, 18–19,
    23, 286
tunnel state machine
    diagram, 27
UML sequence diagram of,
    17–18
Up-arrow and Down-arrow
    keys, 5
utility macro Q_DIM(a), 15
version of, 614, 620
Foreground/background
    architecture, 257
Framework, 255
  services, 282
Free Software Foundation, 679,
    681, 683–684

G
GAME-DBG.PRJ, 8
game-ev-lm3s811.eww, 9
game-ev-lm3s811 project, 9
game.h header file, 13
Garbage collector in QF, 330,
    340, 349–350, 369
Generic QEP event processor,
    comma expression,
    definition of, 136

Generic state machine interface
    concurrency model, 105–106
    events, 106–108
Gimpel Software, 309
GNU General Public License
    Version 2 (GPL2), 679, 681,
    683–684
Graphical notation guide
  class diagrams, 685–688
  sequence diagrams
    horizontal dimension, 690
    vertical dimension, 689–690
  state diagrams, 689
    static state space, 688
  timing diagrams, 690
    horizontal axis, 691
    vertical axis, 691
Graphical user interface
    (GUI), 56
  frameworks, 265
Guard conditions, 64
*Guidelines for the Use of the C
    Language in Vehicle-Based
    Software*, 309

H
Hardware abstraction layer
    (HAL), 581
Header files
  qepn.h, 616
  qepn_port.h, 616
  qpn_port.h, 616, 619, 627, 655
Head index, 645, 648
Help formats
  CHM, 633
  HTML, 633
Hierarchical event processor.
    See QEP event processor
Hierarchical nested states, 69–71,
    149. *See also* Hierarchical
    state machines (HSMs)
Hierarchical state-handler
    functions
  C version, 158–160
  C++ version, 160–161
  superstate designating, 158

Hierarchical state machines
(HSMs), 101
  calculator state machine
    enumerating signals,
      167–168
    implementing steps,
      183–191
    initial pseudostate
      handler, 165
    topmost initial transition,
      170–171
  implementation steps
    coding initial transitions,
      189–190
    entry and exit actions
      coding, 189
    enumerating signals, 185
    events, 185–186
    guard conditions coding,
      190–191
    initial pseudostate, 188
    internal transitions
      coding, 190
    regular transitions
      coding, 190
    specific state machine
      derivation, 186–187
    state-handler functions,
      188–189
  QHsm class
    C version, 162–163
    C++ version, 163–164
    entry/exit actions, 166–168
    events dispatch, 174–177
    generic state transition
      execution, 177–183
    nested initial transitions,
      166–168
    reserved events and helper
      macros, 168–170
    topmost initial transition,
      170–174
    top state and initial
      pseudostate, 164–165
High Level Data Link Control
  (HDLC), 552

Housekeeping code, 85, 255
Hypothetical state machine, 88

I
IAR 8051 compiler, 673,
  675–676
IAR C-Spy Debugger
  Go button, 10
  Reset button, 10
IAR EWARM toolset, 9
  Flash programming
    process, 10
  IAR C-Spy debugger, 10
Inheritance in C, 624,
  631, 640
Interrupt handling, 665
Interrupt locking levels
  ISR-Level, 635–637
  task-Level, 635–636
Interrupt-locking policies, 501,
  503–504, 509, 524,
  526–527, 529, 630, 634,
  637, 646, 665, 672, 674
Interrupt-nesting policy,
  counter, 655
Interrupt processing under
  "vanilla" kernel, 655
Interrupt service routines (ISRs),
  36–38, 105, 257, 489,
  628–630, 657, 665
  generating, posting, and
    publishing events in, 37
Interrupt types
  FIQ, 637
  IRQ, 637
  locking, 635, 672
  nesting, 635–636, 665
  unlocking, 635, 650,
    662, 672

K
Kernel, 259
  blocking, 310, 313, 330
  execution profiles of, 261
  nonpreemptive, 260–261,
    274, 276

  preemptive, 260–262, 269,
    273, 276, 279
  single stack, 276–278
Key board
  default and caps_locked state,
    60, 62, 67
  UML state diagram of, 78
Keystroke event, 66–67

L
Last-in, first-out (LIFO) policy,
  218–219, 227, 230
Least common ancestor (LCA),
  80, 177–180
Licensing
  policy for QP and QP-nano
    closed-source, 680
    GNU general public license
      version 2, 681–684
    nonprofits, academic
      institutions, and private
      individuals, 680–681
    open-source, 679–680
    software evaluation, 680
  types of
    closed-source, 680
    commercial licenses, 680
    copyright law, 679
    open-source, 679–680
Liskov Substitution Principle
  (LSP), 73–74
LM3S811 MCU, 11
Low-end 8-and 16-bit single-chip
  microcontrollers, 611
Luminary Quickstart
  application, 6

M
Machine interface functions
  dispatch(), 106–107, 112,
    115, 129, 131–133,
    135, 137, 142, 145,
    162, 631
  init(), 106, 111, 115–116,
    118, 125, 129, 131, 133,
    135, 631

Machine interface functions
    (*Continued*)
  QHsm_dispatch(), 154, 162,
    171, 174–178, 196
  QHsm_init(), 162, 165, 170,
    172–174, 183
  QHsm_isIn(), 154, 162–163
  QHsm_top(), 163, 165
  StateTable_empty(), 116, 135
Macro
  Alarm_dispatch(), 236
  Alarm_init(), 236
  LEFT(), 456
  OS_ENTER_CRITICAL(),
    424
  OS_EXIT_CRITICAL(), 425
  QACTIVE_EQUEUE_
    ONEMPTY_(), 406,
    434–435, 501–502
  QACTIVE_EQUEUE_
    SIGNAL_(), 406,
    434–435, 501
  QACTIVE_EQUEUE_WAIT_
    (), 406, 501
  Q_ALLEGE(), 173
  Q_ASSERT(), 170, 182
  Q_DEFINE_THIS_MODULE
    (), 409
  QEP_EXIT_(), 182
  QF_ACTIVE_DISPATCH_(),
    534
  QF_EPOOL_EVENT_SIZE_
    (), 406
  QF_EPOOL_GET_(), 407
  QF_EPOOL_INIT_(), 406
  QF_EPOOL_PUT_(), 407
  QF_EPOOL_TYPE_, 406
  QF_INT_KEY_TYPE, 402,
    412, 424, 434
  QF_INT_LOCK(), 404, 515,
    529
  QF_INT_LOCK()/
    QF_INT_UNLOCK(),
    561, 619
  QF_INT_UNLOCK(), 404,
    425, 515, 529

QF_ISR_KEY_TYPE, 637,
  650, 675
QF_ISR_LOCK()/
  QF_ISR_UNLOCK(),
  637
QF_ISR_NEST, 636, 655, 665
QF_MAX_ACTIVE, 617,
  633, 643
QF_TIMEEVT_CTR_SIZE,
  619, 627, 641–642, 672
Q_HANDLED(), 136,
  141–142, 160, 188–191
Q_IGNORED(), 136, 142
QK_EXT_RESTORE(),
  524, 530
QK_EXT_SAVE(), 523–524,
  530
QK_ISR_ENTRY(), 504,
  526–527, 529, 674
QK_ISR_EXIT(), 674
QK_MUTEX, qkn.h, 658, 666
QK_PREEMPTIVE, 657–658
QK_SCHEDULE_(), 501,
  503, 506–507, 509, 529
QK_TLS(), 510–511, 530
Q_NFSM, 619
Q_PAR(), 622
Q_PARAM_SIZE, 617, 620,
  626, 639
Q_PAR(me), 625
Q_ROM, 398, 412, 616
Q_ROM_BYTE(), 399
Q_ROM_VAR, 200, 616–617
QS_BEGIN(), 557, 562–563,
  566, 578–579, 607
QS_END(), 557, 562,
  578–579, 607
QS_FLUSH(), 559
QS_FUN(), 558
QS_FUN_DICTIONARY(),
  577
QS_FUN_PTR_SIZE,
  558, 581
Q_SIG(), 622
Q_SIGNAL_SIZE, 155, 200,
  399–400, 558

QS_INIT(), 600
QS_OBJ_DICTIONARY(),
  575, 605–606
QS_OBJ_PTR_SIZE, 558
Q_SPY, 553, 556
QS_QS_OBJ(), 557
QS_SIG(), 558
QS_SIG_DICTIONARY(),
  577, 605
QS_STR(), 607
QS_TIME_SIZE, 581
QS_U8(), 607
Q_SUPER(), 158,
  160, 173, 176–177,
  188, 201, 625–626,
  640
Q_TRAN(), 136–137,
  141–143, 160, 163, 166,
  173, 176, 188–191, 193,
  201, 625–626
return Q_HANDLED(), 142
RIGHT(), 456
TRAN(), 111–112, 116, 122,
  125, 129–130
Memory partition in QF event
  pool, 335
Memory pools, 289–291,
  369–377
  free(), 289
  malloc(), 289–290
  native, 313
Message queues, 282
  in active objects, QF,
    329, 359
  msgQReceive(), 283
  msgQSend(), 283
  safe mechanism, 283
Microcontrollers (MCUs)
  8-bit, 311
  16-bit, 311
Missile_ctor(), 620
Motor Industry Software
  Reliability Association
  (MISRA), 309
MSP430F2013 MCU [TI 07],
  612, 672

N

Native QF active object
  queue, 406
Nested switch statement, time
  bomb state machine
  disadvantages, 112–113
  implementation techniques,
    108–112
  variations, 113
Nested Vectored Interrupt
  Controller (NVIC), 321
Nonpreemptive kernel, 614
Numeric keypad, 74

O

Object-oriented programming
  (OOP), 270
  bird objects
    flying ( ), 72
  door_open state, 71, 76–77
Object-oriented state design
  pattern
  advantages and disadvantages,
    130–131
  implementation techniques,
    125–130
  UML class, 124–125
  variations, 131–132
OMG 07, 266, 269, 292
One-shot time event, reusing
  in QF real-time
  framework, 359
Opaque pointer, 236
Orthogonal component state
  pattern
  advantages, 244
  AlarmClock class and its UML
    state machine, 231
  consequences, 243–244
  intent and problems, 230–231
  known uses, 244
  sample code
    AlarmClock state machine
      definition (file clock.c),
      239–243

Alarm component
  declaration (file alarm.
  h), 236
Alarm state machine
  definition (file alarm.
  c), 237–239
common signals and events
  (file clock.h), 235–236
COMP.EXE, output
  generation, 234
solutions, 231–234
Orthogonal regions (OR), 74–75
osObject data member in active
  object, QF, 326

P

Pedestrian active object, 671
Pedestrian light-controlled
  (PELICAN), 666
  crossing controller, 661, 671
  crossing state machine,
    668–671
Philosopher active object
  event queue sizing, 478
  sequence diagrams, 448
  state machines, 451–452, 457
  system initialization and DPP
    application startup, 458
Platform abstraction layer (PAL),
  50, 309–310, 323, 386,
  388–389
Porting QK kernel
  FPU saving and restoring, 531
  qep_port.h header file, 525
  qf_port.h header file, 525–526
  qk_port.h header file, 526–531
  requirements, 524–525
Preemptive run-to-completion
  kernel, 655–666
features
  advantages, 514–515
  extended context switch,
    520–524
  priority-ceiling mutex,
    515–518

thread-local storage (TLS),
  515–520
implementation
  interrupt processing,
    503–506
  qk.c source file, 511–514
  qk.h header file, 498–503
  qk_sched.c source file,
    506–511
  source code organization,
    496–498
porting
  FPU saving and
    restoring, 531
  qep_port.h header file, 525
  qf_port.h header file,
    525–526
  qk_port.h header file,
    526–531
  requirements, 524–525
port testing
  asynchronous preemption,
    531–535
  dining philosopher problem
    (DPP), 531
  extended context
    switch, 539
  priority-ceiling mutex,
    535–536
  TLS demonstration,
    536–539
RTC kernel
  asynchronous preemption,
    488–491
  nonblocking, 487
  stack multitasking, 486–487
  stack utilization, 491–494
  steps, 484–485
  synchronous preemption,
    487–489
  traditional preemptive
    kernels, 494–496
Priority-ceiling mutex
  interface, 658
  mechanism, 666

Priority numbering system in QF, 331, 333
Programmable interrupt controller (PIC), 321, 527, 529
Publish-subscribe mechanism, 14–15

Q

QActiveCB, 615–616, 632, 644–645
QActive class, 268, 315–316, 327, 341
QActive_post(), 646–648, 660
QActive_postISR(), 649–650, 665
QEP event processor, 13, 297, 312, 315–316, 326, 333.
    See also Run-to-completion (RTC)
    features, 150–151
    hierarchical state-handler functions
        C version, 158–160
        C++ version, 160–161
        superstate designating, 158
    HSM class
        C++ (Class QHsm), 163–164
        C (Structure QHsm), 162–163
        entry/exit actions, 166–168
        events dispatch, 174–177
        generic state transitions, 177–183
        nested initial transitions, 166–168
        reserved events and helper macros, 168–170
        topmost initial transition, 170–174
        top state and initial pseudostate, 164–165
    HSM implementation steps
        coding initial transitions, 189–190

entry and exit actions coding, 189
enumerating signals, 185
events, 185–186
guard conditions coding, 190–191
initial pseudostate, 188
internal transitions, 190
regular transitions, 190
specific state machine derivation, 186–187
state-handler functions, 188–189
porting and configuring, 199–201
QEvent structure in C, 155–156
QEvent structure in C++, 157
QSignal, 154–155
structure
    QEvent class, 155–157
    QHsm class, 152–153
    source code organization, 153–154
usage guidelines
    accessing event parameters, 194–195
    ill-formed state handlers, 193
    inadvertent corruption of current event, 198–199
    incomplete state handlers, 192–193
    incorrect casting of event pointers, 194
    nonsubstate targeting in initial transition, 195–196
    run-to-completion (RTC) semantics violation, 198
    state transition inside entry or exit action, 193
    suboptimal signal granularity, 197–198
    switch statement, 196–197

QEP FSM implementation
    advantages, 142–143
    application-specific code, 137–142
    event processor structure, 132–134
    implementation files, 134–137
    state-handler function, 132
    structure, 133
    variations' disadvantages, 143–144
QEP-nano event processor, 612–613, 619, 631–632, 637–640
QEP-nano signals, 621, 637
QEQueue queue class, 316
QEvent.dynamic_ data byte in QF framework, 334
QEvent structure, 291
QF_active[] array, 615–619, 621, 625, 632–633, 644–645, 652–653
QF memory pool class (QMPool), 316
QF_onStartup(), 653
qf_port.h header file
    active object event queue operations, 406
    derivation of QActive, base class, 402–403
    event pool operations, 406–407
    include files used, 404–405
    interface used only inside QF, not in applications, 405–406
    linux (conventional POSIX-compliant), 432–435
    µC/OS-II (conventional RTOS), 423–425
    platform-specific QActive data members, types, 402
    QF critical section mechanism, 404
    QF framework, various object sizes, 403–404

unconditional interrupt
    unlocking policy, 416
QF, porting and configuring
  building QF library, steps
    required, 392
  Linux (conventional POSIX-
    compliant OS)
    qep_port.h header file, 432
    qf_port.c source file,
      435–441
    qf_port.h header file,
      432–435
  μC/OS-II (conventional
    RTOS)
    build script, 430
    idle processing, 431
    qep_port.h header file,
      422–423
    qf_port.c source file,
      425–430
    qf_port.h header file,
      423–425
    system clock tick (QF_tick
      ()), 430
  platform abstraction layer, 389
  ports, 389
QF_publish() function, 349
QF_publish() publishing
  events, 316
QF_readySet_, 645–646,
  652, 660
QF real-time framework, 13–14,
  255
  active objects, 324–328
    event queue of, 328–330
    internal state machine of,
      328
    thread of execution and
      priority, 330–333
  asynchronous event exchange
    QActive_postFIFO () and
      QActive_postLIFO (),
      36
    QF_publish () and
      QActive_subscribe (),
      36

  critical sections in, 318
    internal QF macros,
      323–324
    locking and unlocking
      interrupts
      unconditional, 321–323
    lock_key variable, 319
    saving and restoring the
      interrupt status,
      319–321
  design by contract (DbC),
    314
  event delivery mechanisms in
    direct event posting, 312,
      343–344
    publish-subscribe event
      delivery, 344–350
  event management in
    automatic garbage
      collection, 339–341
    deferring and recalling
      events, 341–343
    dynamic events allocation,
      335–339
    structure event, 333–334
  features of, 308
    assertion-based error
      handling, 314
    built-in software tracing
      instrumentation, 314
    built-in "vanilla" scheduler,
      313
    direct event posting and
      publish-subscribe event
      delivery, 312
    low-power architecture,
      313–314
    native event queues, 313
    native memory pool, 313
    open-ended number of time
      events, 312
    portability, 309–310
    scalability, 310–311
    source code, 309
    support for modern state
      machines, 312

    tight integration with QK
      preemptive kernel, 313
    zero-copy event memory
      management, 312
  FIFO policy, 312, 367
  LIFO policy, 312, 342
  low-power, 313–314
  macros
    QF_INT_KEY_TYPE,
      320–321
    QF_INT_LOCK(), 318,
      320
    QF_INT_UNLOCK(), 318,
      320
    Q_NEW(), 284
  memory management, 312
  native QF event queue,
    359–360
    active object Queue,
      362–367
    QEQueue initialization, 362
    QEQueue structure,
      360–362
    "raw" thread-safe queue,
      367–369
  native QF memory pool,
    369–372
    initialization of, 372–375
    memory block of,
      375–377
  native QF priority set,
    377–379
  PLAYER_TRIGGER event,
    19
  portability, 309–310
  QK preemptive kernel,
    integration with, 313
  scalability, 310–311
  structure of
    dispatch(), 316
    init(), 316
    missile, 315
    QEQueue, 316
    QF_publish(), 316
    QMPool, 316
    Q_NEW(), 316

QF real-time framework
  (Continued)
    ship, 315
    source code organization,
      316–317
    tunnel, 315
  time events, 312
  time management
    arming and disarming,
      356–359
    system clock tick and
      QF_tick() function,
      354–356
    time event structure and
      interface, 351–354
  TIME_TICK event, 18
QHsm base class, 268
QHsm_dispatch(), 639
QHsm_init(), 639
QK, 278
QK_init(), 657
QK kernel
  advantages, 514–515
  extended context switch
    asynchronous preemptions,
      520
    coprocessor registers,
      521
    extended scheduler
      implementation,
      522–524
  port testing
    asynchronous preemption,
      531–535
    DPP application, 531
    extended context switch,
      539
    pseudorandom number
      (PRN), 535–536
    TLS contexts, 536–539
  priority-ceiling mutex,
    515–518
  scheduler implementation,
    522–524
  thread-local storage (TLS),
    515–520

QK-nano
  idle loop, 658–660
  interrupt processing, 665
  kernal, 655, 660, 665,
    672–675
  ports, 657, 672
  preemptive Run-to-
    Completion, 655–666
  priority ceiling mutex, 666
  scheduler of, 650, 657,
    660–664
QK-nano Interface qkn.h,
  655–658
QK preemptive kernel, 307, 310,
  313–314
QK_schedule_(), 657–658, 660,
  662–663
QMPool structure elements and
  memory buffer relationship,
  370
Q_NEW() new dynamic events,
  316
QP
  developing guidelines
    heuristics, 445–446
    rules, 444–445
  DPP
    application execution,
      461–476
    application termination,
      460–461
    Linux, 472–476
    µC/OS-II, 469–472
    requirements, 447
    sequence diagrams,
      447–449
    signals, events, and active
      objects, 449–451
    state machines, 451–457
    system initialization and
      application startup,
      457–460
    "vanilla" kernel on Cortex-
      M3, 465–469
    "vanilla" kernel on DOS,
      461–465

event-driven platform,
    307, 310
  RAM/ROM footprints of, 311
  reference manual, 386–387
  sizing
    event pools, 479
    event queues, 476–479
    system integration, 480
  source code, 309
Q_PARAM_SIZE, 620
QP/C or QP/C++ systems,
  310–311
QP/C reference manual in HTML
  and CHM, 387
QP event-driven platform, 49–50
QP-nano, 311
  active objects, 640–642
  compiler, 615
  components of, 613
  configurations, 676–678
  constructor
    QFsm_ctor(), 642
    QHsm_ctor(), 642
  cooperative "vanilla" kernel,
    650–655
  critical sections, 634–637
  data structures of, 615
  event-driven
    applications, 611
    infrastructure, 612
    paradigm of, 611
    platform of, 611
  event queues, 644–650
    elements, 644
    QActive structure, 644
    storages, 644
  "Fly 'n' Shoot" game,
    implementation
  applications, 611, 613, 616,
    630, 631, 656
  example of, 614
  header files of, 618–622
  main() functions in,
    615–618
  signal limitations, 620
functions

QActive_postISR(), 629, 634, 636, 642, 655
QF_tick(), 629, 634, 636, 642–644, 652, 655
key features of
  arbitrary state transition topology, 612
  nesting, 612
  thread-safe event queues, 612
macro
  QActive_arm(), 642
  QActive_disarm(), 642
  QF_FSM_ACTIVE, 64–642
  QF_MAX_ACTIVE, 64–642
memory usages, 675–678
pedestrian active object, 671
PELICAN crossing example for, 666–678
ports, 615, 619
  QK-nano kernel, 619
port to MSP430 with QK-nano kernel, 672–675
posting events, 646–648
  ISR level, 649–650
  task-level posting, 647
processor, 615
QActive, 640
ready-set, 645–646
  QF_readySet_, 646
reference manual, 633
  sections, 634–637
  source code, examples, and documentation, 633–634
state machines, 637–640
structure, 631–633
system clock tick in, 642–644
time events, 626–628
version, 634
qpn_port.h, 618–620
QP platform abstraction layer, 255, 278
  building applications, 390–391
  building libraries, 391–392, 413

directories and files, 392–397
platform-specific QF callback functions, 412–413
qep_port.h header file, 398–400
qf_port.c source file, 407–411
qf_port.h header file
  active object event queue operations, 406
  active objects in application, 403
  derivation of QActive, base class, 402–403
  event pool operations, 406–407
  include files used, 404–405
  interface used only inside QF, not in applications, 405–406
  platform-specific QActive data members, types, 402
  QF critical section mechanism, 404
  QF framework, various object sizes, 403–404
qp_port.h header file, 411
system clock tick (calling QF_tick()), 413
QP Port, header file, 12–13
qp_port.h header file, 13
QPset64 data structure in native QF priority set, 377
QS software-tracing system
  interrupts, 548–550
  serial ports, 544–547
QS target components
  callback functions
    QS_getTime(), 602
    QS_initBuf(), 558, 569, 600
    QS_onCleanup(), 558, 580, 600
    QS_onFlush(), 559, 580, 601
    QS_onGetTime(), 559, 580, 598, 601, 603

QS_onInit(), 580
QS_onStartup(), 558, 569–570, 600
data
  compression, 551
  transmission protocol, 552, 567
filter, levels, 551–552, 562–566
header files
  gs.h, 553
  qs_dummy.h, 553
QS trace records, types of
  application-specific, 579–580
  function dictionaries, 577
  object dictionaries, 575–577
  signal dictionaries, 577
QS trace buffer
  functions
    QF_onIdle(), 574
    QS_getBlock(), 573–574
    QS_getByte(), 571–572
    QS_initBuf(), 569–570
    QS_onStartup(), 570
  initializing, QS_initBuf(), 569–571
  interface
    block-oriented, 573–574
    byte-oriented interface, 571–572
Quantum Spy(QS), 542
Quickstart application, 257
Q_USER_SIG, 621

R
Real-time embedded (RTE) system, 255, 308, 540
Real-time object-oriented modeling (ROOM), 266
  language, 69
  method, 230
Real-time operating system (RTOS), 50, 420
Reminder state pattern consequences, 218–219

Reminder state pattern
(Continued)
intent and problems, 211–212
sample code
file reminder.c
implementation,
214–218
REMINDER.EXE, output
generation, 213
solutions, 212–213
Ring buffer, 645, 648–649,
654, 664
events, 644
queue storage, 644
Root Directory qp, 4
Run-to-completion (RTC), 48,
484, 612, 645, 655, 663–664
kernel, 308, 331, 612, 645
asynchronous preemption,
488–491
asynchronous preemption
scenario, 492–493
interrupt processing,
503–506
nonblocking, 487
preemptive kernel, 498–499
qk.c source file, 511–514
qk.h header file, 498–503
qk_sched.c source file,
506–511
source code organization,
496–498
stack multitasking, 486–487
synchronous preemption,
487–489
synchronous preemption
scenario, 491–492
and traditional preemptive
kernels, 494–496
model, 67–68

S

Scheduler, 263, 274–276, 284
built-in "vanilla," 313
Semaphores, 260, 270, 291, 305
Sequential pseudocode, 271

Ship active object
implementation in QP-nano,
622–626
Ship structure, state-handler
functions, 39–41,
43–48
Single event-loop, 258
Software
bus, 281
licenses
copyright notice, 682
terms and conditions,
681–684
warranty, 681–684
tracing
macro Q_SPY, 314
QS, 307–308, 314
tracing concepts
data transmission
mechanism, 544
host-resident, 543
last-is-best policy, 543
target-resident, 543–544
timestamping, 544
tracing techniques, 304
printf() statements, 303
Spaghetti code, 65, 265
Standard state machine
implementations
C++ exception handling, 145
entry and exit actions, 146
generic state machine interface
concurrency model,
105–106
events, 106–108
guards and choice
pseudostates, 145–146
nested switch statement
advantages and
disadvantages,
112–113
implementation techniques,
108–112
variations, 113
object-oriented state design
pattern

advantages and
disadvantages,
130–131
implementation techniques,
125–130
UML class, 124–125
variations, 131–132
pointers to functions, role of,
144–145
QEP FSM event processor
advantages, 142–143
application-specific code,
137–142
files implementing,
134–137
state-handler function,
132–134
structure, 133
variations disadvantages,
143–144
state table
advantages, 122–123
generic event processor,
114–118
one-dimensional, 124
techniques, 118–122
two-dimensional, 113–114
variations, 123
time-bomb state machine,
102–103
State-based solution, 65
State-handler functions
AO_Ship, 624
final(), 210
generic(), 210
QActive_defer(), 227–228
QActive_post(), 626
QActive_recall(), 226, 228
QHsm_top(), 626, 639
QState, 638
QStateHandler, 638
Ship_active(), 626
Ship_flying(), 626
Ship_initial(), 624
specific(), 210
state-handler, 624

TServer_idle(), 228
typedef, 638
UltimateHook_specific(), 210
State machines, 311, 326, 328
  functions, 639
    QHsm_dispatch(), 639
    QHsm_init(), 639
    QHsm_top(), 639
  initialization
    constructor, 42–43
    Quickstart application, 50–52
    top-most initial transition, 42–43
  macro
    Q_HANDLED(), 639
    Q_IGNORED(), 639
    Q_PAR(), 639
    Q_SIG(), 639
    Q_SUPER(), 640
    Q_TRAN(), 640
  QFsm nonhierarchical, 312
  QHsm hierarchical, 312, 316
  types
    hierarchical, 637
    nonhierarchical, 637
  UML-compliant, 312
States
  basic machine concepts, 59–68
  diagram comparison of, 61–62
  event-driven programming, 63
  keyboard
    behavior of, 60
    UML state diagram with, 78
  machines, 63–64
  patterns
    deferred event, 219–230
    definition, 205
    essential elements, 204
    orthogonal component, 230–244
    reminder, 211–219
    summary of, 205
    transition to history, 244–251
    ultimate hook, 205–211

qualitative aspects and quantitative aspects, 65
  roles in, 79
  transformational programming, 63
  transition
    guard g (), 79
State table
  generic event processor
    event structure, 116–122
    secure code, 115–117
    state machines, 114–115
  implementing state machines
    advantages, 122–123
    one-dimensional, 124
    state machines, 114–115
    techniques, 118–122
    two-dimensional, 113–114
    variations, 123
Stellaris EV-LM3S811
  evaluation kit, 3–4
  ARM Cortex-M3, 4
  sample code for, 9
Structure
  active object, 616, 621–622, 631
  QActiveCB, 616, 644
System clock tick, 293–294
  jitter in, 293
  QF_tick()function, 293

T
Table active object
  DPP application termination, 460
  event queues sizing, 478–479
  Linux, 476
  sequence diagrams, 448–449
  state machines, 452–457
  system initialization and DPP application startup, 458
  "vanilla" kernel on Cortex-M3, 468
  "vanilla" kernel on DOS, 465
Tail index in native QF active object queue, 364–365

Task control block (TCB), 259
Task-level interrupt locking, 635
Task-level response, 261–262, 273, 276
Tasks/threads, 259
Telelogic rhapsody design automation tool, 273
Thread-local storage (TLS)
  definition, 518
  QActive_start( ) function, 514
  QK port testing, 536–539
  support in QK, 519–520
  uses of, 515
thread_routine() static function, 439–440
Time-bomb state machine
  FSM technique, 138–144
  implementation techniques, 102–103
  nested switch statement, 108–113
  object-oriented state design pattern, 124–132
  source code, 104–105
  state table techniques, 114–124
Timers, 291
  disarm() operation, 292
  postEvery(), 292
  postIn(), 292
  QEvent, 292
  QTimeEvt, 292
  rearm() operation., 292
  signals
    BSP_TICKS_PER_SEC, 674
    Q_ENTRY, 626
    Q_EXIT(), 626
    Q_INIT(), 626
Tools
  borland C/C++ compiler, 4
  Borland Turbo C++ 1.01, 8
  IAR EWARM, 9
Traffic lights controller, 666–667

Transition to History state pattern
  intent and problems, 245
  sample code
    consequences, 250–251
    (file history.c)
      implementation,
        247–250
      HISTORY.EXE, output
        generation, 246
    known uses, 251
    solutions, 245
Turbo C++ 1.01 compiler, 614,
  629–630

U
Ultimate Hook state pattern
  consequences, 211
  intent and problems, 205
  sample code
    (file hook.c)
      implementation,
        208–211
      HOOK.EXE, output
        generation, 208
    solutions, 206–207
    template method OO
      design, 211
Unified Modeling Language
  (UML), 685–686, 690
  internal transitions,
    ANY_KEY event, 78
  semantics of, 165, 170,
    669–670
  sequence diagram, 17–18
  specification, 190
  statecharts, transition D, 90
  state machines, 68–69, 668
    actions and transitions, 67
    architectural decay, 65
    behavior in, 69, 95–96
    diagrams and flowcharts,
      61–63
    elaborating composite
      states, 94–95

    entry and exit actions,
      75–77
    event-action paradigm,
      56–59
    event deferral, 83
    event-driven programming,
      55–56
    events, 66–67, 82–83
    extended machines, 63–64
    finalization of, 96
    guard conditions, 64–66
    hierarchically nested states,
      69–71
    high-level design, 92–93
    housekeeping code, 85
    if-else constructs, 55
    inheritance, 71–73
    internal transitions, 77–78
    limitations of, 86–87
    Liskov substitution
      principle, 73–74
    local and external
      transitions, 81–82
    orthogonal regions, 74–75
    problem specification,
      91–92
    pseudostates, 83–85
    run-to-completion execution
      model, 67–68
    scavenging for, 93–94
    semantics of, 87–91
    state variable, value of, 60
    toaster oven state machine
      with, 76
    transition execution
      sequence, 78–81
state machines and event-
  driven programming
  DOS version (in C), 7–8
  installation directory, 4
  main() Function, 11–16
state machines semantics
  DIGIT_0 case, 95
  operand2 equals. . ., 92

  operand1 operator, 92
  printf (), 87
  qhsmtst.c., 91
  QHSMTST.EXE.,
    87–88, 90
  transition G, 89
UNIX-like platforms, 16

V
"Vanilla" kernel low-power sleep
  modes in QF, 384
Vanilla scheduler, 309–310
Variables, 65
Visual basic calculator
  CE (Cancel Entry) button, 57
  before crash and after
    crash, 57
  current context, 58–59
  DecimalFlag, 58, 65–66
  entering_the_fractional_
    part_of_a_number, 66
  qpresourcesvbcalc.exe, 56, 58
  event-action paradigm, 59
  event-handler procedure, 58
  global variables and flags, 58
  LastInput, 58
  NumOps, 58
  operator events, 58
  OpFlag, 58
  priori, 66
  problems with, 56–57
VisualSTATE engine, 266

W
Wait-for-interrupt (WFI),
  callback function, 419

Z
Zero-copy event queues in active
  object, 312, 329, 333, 341,
  347, 387

Printed in the United States
by Baker & Taylor Publisher Services

Printed in the United States
by Baker & Taylor Publisher Services